AF843452

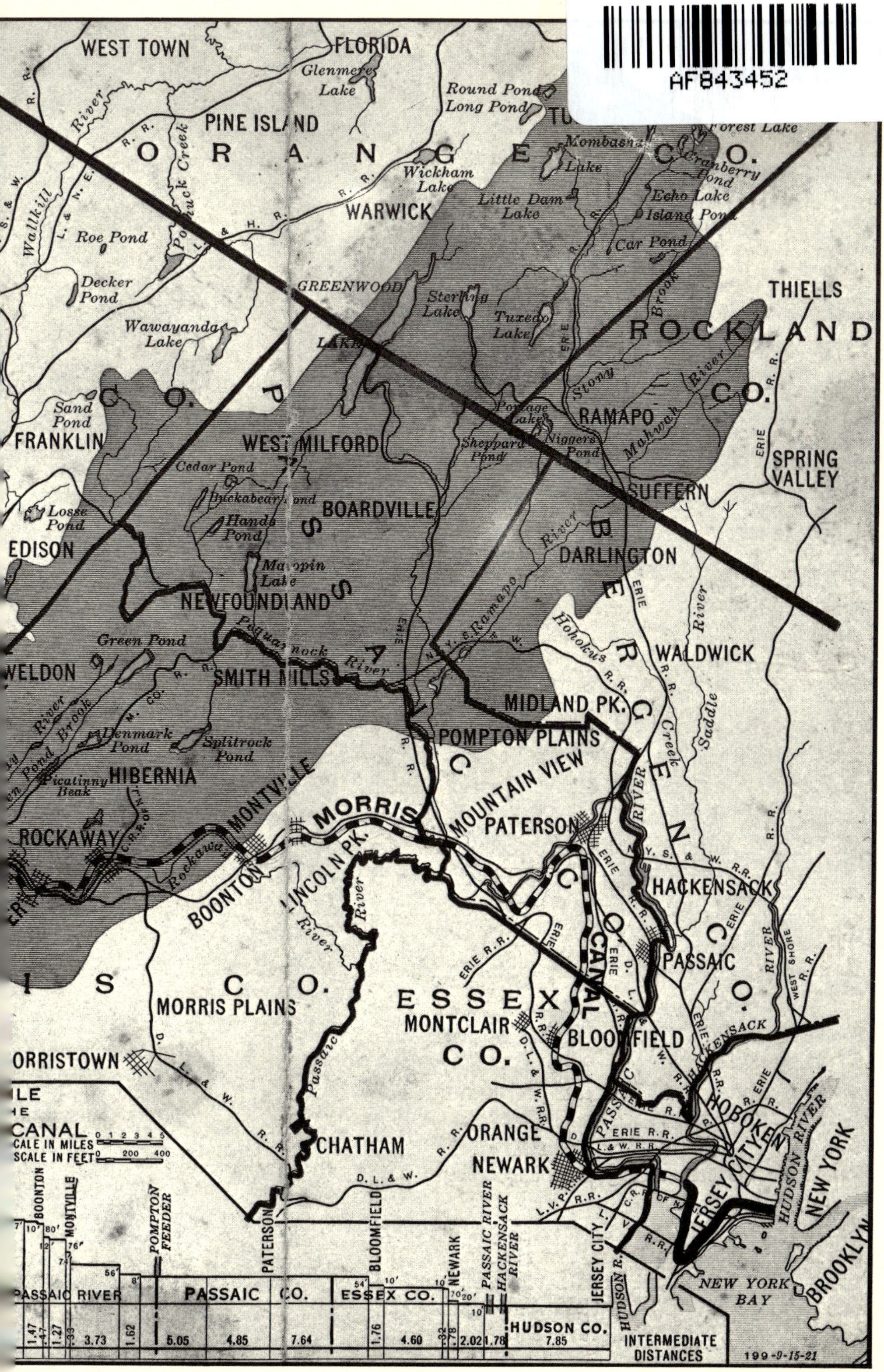
WEST TOWN
FLORIDA
Glenmere Lake
Round Pond
Long Pond
PINE ISLAND
O R A N G E CO.
Forest Lake
Mombasha Lake
Cranberry Pond
Wickham Lake
Little Dam Lake
Echo Lake
Island Pond
WARWICK
Wallkill River
Pochuck Creek
Roe Pond
Car Pond
Decker Pond
GREENWOOD LAKE
Sterling Lake
Tuxedo Lake
THIELLS
ROCKLAND CO.
Wawayanda Lake
Sand Pond
Stony Brook
Mahwah River
Potage Lake
RAMAPO
FRANKLIN
WEST MILFORD
Sheppard Pond
Niggers Pond
SPRING VALLEY
Cedar Pond
SUFFERN
Losee Pond
BOARDVILLE
Hands Pond
EDISON
DARLINGTON
Macopin Lake
NEWFOUNDLAND
P A S S A I C
B E R G E N CO.
Green Pond
Pequannock River
Ramapo River
Hohokus
WALDWICK
WELDON
SMITH MILLS
MIDLAND PK.
Denmark Pond
Splitrock Pond
POMPTON PLAINS
Saddle River
HIBERNIA
Picatinny Beak
MONTVILLE
MOUNTAIN VIEW
ROCKAWAY
MORRIS
PATERSON
Rockaway River
BOONTON
LINCOLN PK.
HACKENSACK
Passaic River
CANAL
PASSAIC
M O R R I S CO.
E S S E X CO.
MORRIS PLAINS
MONTCLAIR
BLOOMFIELD
MORRISTOWN
HOBOKEN
CHATHAM
ORANGE
NEWARK
JERSEY CITY
HUDSON RIVER
NEW YORK
BROOKLYN
NEW YORK BAY
ERIE R.R.
D. L. & W. R.R.
N. Y. S. & W. R.R.
L. & H. R. R.
WEST SHORE R.R.
L. V. R.R.
C. R. R. OF N. J.
CANAL
SCALE IN MILES
SCALE IN FEET
BOONTON
MONTVILLE
POMPTON FEEDER
PATERSON
BLOOMFIELD
NEWARK
PASSAIC RIVER
HACKENSACK RIVER
JERSEY CITY
HUDSON R.
PASSAIC RIVER
PASSAIC CO.
ESSEX CO.
HUDSON CO.
1.47
1.27
3.73
1.62
5.05
4.85
7.64
1.76
4.60
2.02
1.78
7.85
INTERMEDIATE DISTANCES
199-9-15-21

A HUNDRED YEARS
A HUNDRED MILES

New Jersey's Morris Canal

Barbara N. Kalata

Published by
Morris County Historical Society
68 Morris Avenue, Morristown, N.J. 07960

Copyright © 1983 by Barbara N. Kalata and the
Morris County Historical Society. All rights reserved.

ISBN 0-910301-07-7

Printed by Compton Press, Inc.

TO THE MEMORY OF

my uncle, George W. Keppler (1896-1947),
whose life-long interest in the Morris Canal
survived him to inspire this history

THIS BOOK IS AFFECTIONATELY DEDICATED

Preface

The purpose of this work is to tell the story of the remarkable Morris Canal and Banking Company and, coincidentally, its equally remarkable waterway. It is the story of the beginnings of our sprawling, brawling industrial economy, more or less confined in a triangle, roughly drawn from New York to Philadelphia and thence to Scranton. Within that triangle were Pennsylvania anthracite, New Jersey iron ore, and New York capital. Putting all of these elements together was the alchemical achievement of a handful of extraordinary men of vision and ability. Before they were finished, they had indeed turned base elements into gold and had helped establish the Northeast as the driving force behind America's rise to world prominence.

Ends usually determine the means we employ to achieve them. For me that has meant years of research, beginning with the extensive collection of Morris Canal materials and memorabilia left by my late uncle, George W. Keppler, and ending with a mind-numbing number and variety of publications and nonpublications. That source material is strewn across a three-state area in libraries, private collections, and Canal Company files, as well as in books in my own library. My intent, by citing accurately and at length from those sources, has been to compile a resource document that will save future researchers the time and trouble of gaining access to the originals. That is one reason for what *might* appear to be "over-documentation."

Another, and more important, reason for over-documentation is that some of my conclusions are unorthodox and must therefore be justified. For just as the ends help dictate the means, so the means often reshape the ends themselves. Thus, my means — the research — led me to take a new look at the old canal. Far from seeing the canal as a victim of railroad "villains," I came to see the Canal Company as a guiding force in the growth of those very railroads.

That the more efficient railroads obsolesced the canal itself is part of the story, to be sure. But another, more subtle and equally significant, part is that the bituminous-based steel industry moved westward beyond the Appalachians — close to the source of more abundant soft coal for coking so necessary for steel-making, and closer to the source of more abundant iron made cheap by open-pit mining in Minnesota's Mesabi, Menominee, Cuyuna, and Vermillion ranges.

History, itself, however, dictated the final chapter of this story. The growth of the New York-New Jersey metropolitan area by the turn of the century had made the canal's easternmost terminus the most valuable piece of waterfront real estate on the New Jersey side of the Hudson River, especially for the railroads. By then, the canal, *as a canal,* mattered little to the Canal Company. Under the leadership of the Lehigh Valley Railroad Company, its lessees, it had already gone into the water-supply business and already owned or had developed virtually every large reservoir in northern New Jersey.

For ease of reading, the cited documentation, on which my conclusions are based, has been placed at the ends of the chapters, where they will be less distracting but still convenient for reference. The appendices contain information about tonnage and revenues, as well as the names of the known stockholders, contractors, and suppliers who contributed to the construction of the canal. The bibliography is extensive, if not exhaustive, and will direct the future researcher to the works consulted as well as cited and, in many cases, to the places where they can be found.

There have been — literally — dozens of persons who have helped me in the preparation of this history. To them all I extend a heartfelt blanket "Thank you." But I must give special thanks to fellow historian, Edward T. Francis; to my editor, Virginia Hoff; to my proofreaders: Terry Getz, Meredith Jackowsky, and Virginia Manrodt; and to my supportive and tolerant husband, Steve.

Barbara N. Kalata

Lincoln Park, N.J.
December, 1983

Contents

Abbreviation Key For Illustrations

Credits for some of the illustrations that appear in this book are given individually; the rest are keyed to the abbreviations given below. The end papers show the route of the Morris Canal, as published by the Lehigh Valley Company in 1921.

BNK	Barbara N. Kalata
CLA	C. L. Andrews
CUL	Columbia University Library, School of Engineering
ECM	Easton Canal Museum
STF	Edward T. Francis
GWK	George W. Keppler Collection
JCPL	Jersey City Public Library
MPL	Morristown Public Library
PCHS	Passaic County Historical Society
RUL	Rutgers University Library

1

Readiness Is All

IN 1836, an aerial photograph of northern New Jersey — had one been possible — would have shown a sight few of us can comprehend or visualize: a canal that climbed mountains. From sea level at New York harbor, it panted its westward way up 914 feet, paused for breath near Lake Hopatcong, then laboriously descended 760 feet to the low-water mark at Phillipsburg on the Delaware River.

That was the Morris Canal. Men conceived it and built it; mule-muscles helped power it, with mountains hindering every step of its tortuous way. Men and mules, boats and cargo — they not only had to climb up those mountains; they also had to climb down them. That was a task! And it was accomplished. Spanning its hundred-odd miles, standing for about as many years, the Morris Canal was a monument to American know-how, to human ingenuity. It was the engineering feat of its ephemeral heyday, and then it was gone.

Near the close of its century, the Morris Canal lay stagnating in disrepute and disrepair, its fate the topic of political debates and, eventually, of legislative action. That fate, finally announced in 1924, was euphemistically called "reconstruction." In fact, however, the Morris Canal was abandoned, and its century came to an inglorious end.

The hundred years from 1760 to 1860 was the "century of the canal" in America, climaxing thousands of years of world-wide canal-building precedents. At least 3000 years ago, for example, the great Egyptian Pharaoh, Seti I, had trenched his way from the Mediterranean to the Red Sea. Remains of an ancient lock were still visible near Suez in 1935.[1] The ingenious Chinese had built their Grand Canal by 133 B.C., using inclined planes as well as locks in its construction.[2] Roman engineers left a legacy of canal-building skills wherever they stayed, and Charlemagne planned, but never completed, canal interconnections for the sprawling parts of his land-locked empire.[3]

Canals in Venice and the Netherlands were, and still are, both legion and renowned. And an early fifteenth-century unknown, unsung Italian engineering genius devised the double-gated lock, thereby opening an era of canal building on the Continent. The most famous and most exemplary was France's Languedoc Canal, completed during the reign of Louis XIV to join the Mediterranean Sea with the Atlantic Ocean while managing to bypass Gibraltar and the increasingly aggressive Spanish Armada. Its 119 locks helped raise the canal to its 620-foot maximum elevation.

Later, in Elizabethan England, the double-gated locks came to be called "pound locks."[4] But not until the reign of George III were Britain's early canals superseded by an improved, more efficient network of inland waterways, the system known today as the Grand Union Canal.[5] Still later, the Duke of Bridgewater had a railway constructed to carry the coal from his mines to the canal — a fact which could not well have escaped the notice of industrial planners in the new and growing United States at the beginning of the nineteenth century.[6]

When Robert Fulton was studying art and painting in London during the 1790s, he met Bridgewater, the canal-building duke. He doubtless met Brindley, a canal engineer, and he certainly met James Watt of steam engine fame. Perhaps through his association with such men, Fulton turned from art to engineering. His steamboat experiments are legendary, but he also devised and patented a way to supersede canal locks by means of inclined planes and, in 1796, published his *Treatise on the Improvement of Canal Navigation.*[7]

The century of the canal had reached maturity. Fulton returned to America to perfect his plans for a steamboat, and later to become a member of New York State's Canal Commission.[8] Through his recommendation of inclined planes instead of locks to overcome elevations on canal routes, he was to exert a profound influence over the construction of the Morris Canal.

Many canals had been proposed and some were even well under way in post-Revolutionary America, but they were essentially local in character — three-quarters of a mile to 30 miles in length. Not one had been planned as a major transportation route for the massive migrations which would one day penetrate westward into the American heartland, or for the abundant agricultural and commercial cargoes that would pour eastward from the Trans-Appalachian regions to the Atlantic seacoast. Only the Erie Canal fit that description and completed those tasks.

Governor Clinton's "big ditch" they called it. Under consideration before the War of 1812, under construction by 1817 and completed by 1825, the Erie Canal was an immediate financial success. The cost of approximately $6,000,000 was financed by taxes and state-run lotteries. That a population of little more than a million should take on the expected burden of $5 to 6 million to build such a canal through a sparsely settled wilderness without the aid of a single experienced engineer was a remarkable fact.[9] Nevertheless, the canal was built and within 12 years had yielded enough returns to repay its entire original cost.[10] That success triggered a reaction among other states that became a continental craze for canals. Frenzied canal building took place in several states with little thought of or regard for the costs involved, or for the possibilities of financial returns on initial investments.[11]

Prior to this, Washington himself had crusaded for canals as a means of drawing the new nation together despite its geographical, cultural, and political diversities. Albert Gallatin, Secretary of the Treasury under Jefferson, took up the cause and proposed that surplus federal monies be used to develop roads and canals. His 1808 report to the Congress recommended that a system of interconnecting waterways be built to bind the nation together, thus facilitating commerce and improving the financial status of the new country. Four years later, there was no surplus money, and canal proposals were necessarily dropped while the War of 1812 ran its course.

That "second war of independence" seriously shook complacent Americans' theories about both prosperity and military security. The significance of canals to facilitate troop and supply movements to, from, and throughout the back country was undeniable,[12] but a new surge of interest in canals as agents of development and exploitation was irresistible.

Out of that surge of interest came suggestions for a canal to be built across the Isthmus of Darien, or Panama.[13] By 1825, political maneuvering was begun to obtain an American right-of-way to build such a canal. France's dynamic Ferdinand de Lessups, who politically and physically engineered the Suez Canal in 1859, headed a French commission's abortive attempt to build the Panama Canal in 1879. However, it was American General George Goethals who later completed the task.

But compared with the Morris Canal, the Panama is — in a way — rudimentary: bold in size, but less ingenious and more pedestrian than New Jersey's mountain-climber. Next to the Morris, the Suez is nothing but a big flat ditch and straight as an arrow, at that. By comparison, even the Erie, with 77 locks raising it to its 700-foot summit, seems but an exercise in elementary engineering. And Britain's Grand Union leisurely climbs a staircase of 56 locks to its 393-foot summit at Cow's Roast a little more than 30 miles from London — pure laziness compared to the ambitious Morris.

Yet Britain's Grand Union remains, both as a major transportation system and as a way of life for hundreds of canaling families who choose to work and to live out their lives on it. The Suez is still operable, although the 1967 Arab-Israeli War brought it to a temporary halt. The Languedoc, under its new name, *Canal du Midi,* still functions much as it did 300 years ago. The Panama, at least for a while, will continue to accommodate a lion's share of the world's commercial products and most of its largest ships before it outlives its usefulness. Even the *old* Erie, long gone, survived in spirit and concept as part of the newer, larger, more efficient New York State Barge Canal System. Yet the Morris Canal is gone, and canal buffs

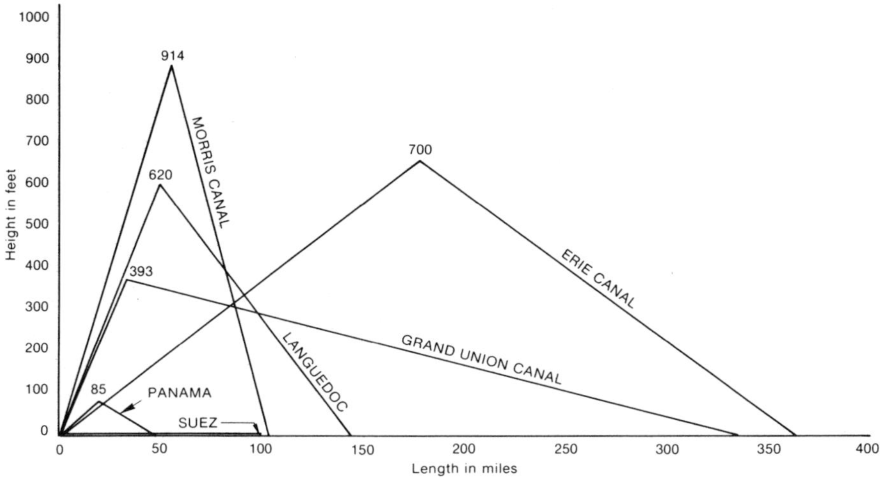

Chart showing comparative heights and lengths of various canals, prepared by the author.

ask "Why"? At least as interesting as that question is the one that asks: "How did the Morris Canal come to be built in the first place?"

The answer to the last question is not very complicated. It was readiness. Readiness on multiple levels.

By 1800, the American Revolution was history. A new nation of sovereign states had been born. The security of statehood had encouraged the ex-Colonies to solve their several intense internal post-war problems. Most of those problems were financial. Politically free, the new nation yet remained tied to Europe's economic apron strings. Only economic self-sufficiency could maintain political independence and guarantee economic opportunities and growth. The War of 1812 provided the impetus toward self-sustenance, and brought about the trend toward exporting more than was imported to achieve a balance of trade. Prosperity, still reluctant, was becoming a more frequent and most welcome guest, and the optimism of rising expectations characterized the commercially oriented portion of the population. Industrialization, said the planners, would be the means to fulfill those expectations.

New England had long led in American manufacturing, with adequate water power derived from her many rivers. She also had access to her own seacoast and harbors — an advantage denied to New Jersey by New York City, which controlled New York Harbor, claiming *all* of the Hudson River as its own up to the high water mark of the Jersey shore line. (The exact boundary between the two states would not be settled until 1834.) Even with three of its sides largely open to ocean commerce, New Jersey benefitted very little economically from that uniqueness. Politically, culturally, and geographically divided into East and West Jersey early in its history, the state was — and in some respects still is — New York and/or Philadelphia oriented. New Jersey served then — as it still does — as a depot for the transshipment of goods and persons from all over the world to the interior sections, and from the interior to the rest of the world.

After a brief post-Revolutionary attempt to compete commercially with their rival neighbors, New Jersey's merchants had settled into their roles as "middlemen." They simply lacked the funds necessary to enter the capitalists' competition for foreign markets. They had to be content with local trade, resigning themselves to remaining merchants instead of becoming magnates.[14] Since New York controlled the harbor, New Jersey shipping had to proceed from the freeports of Perth Amboy and Elizabethport to avoid paying heavy import and export tariffs to New York. Newark was not to become a port of entry until 1834, after the legal dispute about control of the Hudson River and New York Harbor was settled.[15] That dispute and its causes were to influence later decisions as to the eastern terminus of the Morris Canal.

A secondary role was not long to be played by New Jersey, however. Farsighted post-Revolutionary planners — Alexander Hamilton foremost among them — had seen in northern New Jersey an ideal spot for industrial development. The very geographical location that had kept the state economically subservient to New York and Philadelphia also made New Jersey ideally situated to transform subservience into competitive equality, if not supremacy.

Proximity to markets was an important factor, but paramount was the northern sector's ample water power — water power to turn the wheels of industry. In 1791 a private company, the Society for the Establishment of Useful Manufactures (SUM), received a charter for the purpose of founding an industrial city somewhere in New Jersey to utilize some of that ample water power.

Aquackanonck Landing (now Passaic) a thriving river port already well established, had seemed a likely location for the new industrial city. But the town lacked the water power that the Great Falls of the Passaic River, further upstream, could provide. The economic planners had considered transporting water from above the Falls to Passaic by means of a canal cut through the Great Notch, several miles above the city, employing L'Enfant to conduct a survey for that plan. The idea was rejected as too costly, however, and L'Enfant moved on to design our new nation's capital, Washington, D.C. At the foot of the Great Falls of the Passaic, the city of Paterson, "cradle of American industry," was born instead.

The SUM established its headquarters at Paterson and manufactories appeared one after the other in rapid succession. One of the city's foremost attractions in the eye's of New Jersey's planners was its potential for using water power to convert the products of the state's iron mines, furnaces, and forges into the much-needed tools of economic development.

Water power, of course, was essential to virtually all industries, however well-planned and wherever located within the infant economy. Iron industries were constrained to locate close to the source of their power — streams and forests. The water was needed to operate the bellows of furnace, forge, and bloomery; the forests provided the charcoal to feed their moloch-mouthed fires.

Charcoal was used simply because no other known fuel in North America could produce temperatures high enough to produce finished iron. A single forge is said to have required a thousand acres of forest annually in order to sustain itself.[16] Every ton of iron produced consumed about 400 bushels of charcoal,[17] yet no conservation was practiced, and forests surrounding these industries were rapidly depleted. One foreign visitor reported that the Union Furnace in the Musconetcong Valley had had to be abandoned after exhausting about 30,000 acres of woodland in less than 15 years.[18]

Although the natural constraints — motive power, transportation, fuel — acting upon all early New Jersey industries were similar, the growth potential of these industries differed significantly. While the textile and related textile machinery industries had begun to thrive in the post-Revolutionary period, the iron industries had declined as peace-time demands for their products fell dramatically below war-time levels.

Many New Jersey counties had discovered iron ore during the last quarter of the eighteenth century in sufficient abundance to justify the erection of many forges, furnaces, and bloomeries. Eventually, New Jersey rivaled even first-place Pennsylvania in iron production.[19] But the forested areas in New Jersey rapidly disappeared to become charcoal, and by the close of the century the iron industries lay idle, dormant, but waiting in a state of readiness. The emphasis here is on the state of readiness: the textile, the iron-working industries, and other fuel-starved industries of northern New Jersey were ready and waiting for a new source of power. They did not wait long, nor did they wait in vain. The answer lay in the coal fields of Pennsylvania.

In 1749, the Proprietary Government of Pennsylvania bought from the Indians of the Six Nations a tract of land about 125 miles long and 35 miles wide. In that roughly 500-square-mile area was concentrated an enormous quantity of the world's richest fuel: anthracite, the sun's energy stored eons ago in the form of almost pure carbon and deposited underground for safekeeping to await future discovery. The Proprietors paid $2500 for their tract. In view of what they got for their money, the transaction was a far better bargain than the proverbial purchase of Manhattan Island for $24.[20]

Geographically, the tract lies across nine northeastern Pennsylvania counties. Geologically, it is divided into four rather distinct coal districts or fields: the Northern, Eastern Middle, Western Middle, and Southern. The quality of coal from each of these fields is also distinct, for Nature arranged the deposits according to their carbon content through the mechanics of mountain building.[21]

For trade purposes, coal from the Northern Field was designated "Wyoming;" from the Eastern Middle Field, "Lehigh;" from the

Western Middle and Southern Fields, "Schuylkill" coal. Competition developed among the various coal transporters, and because superior Schuylkill coal was the most sought after, it was at the center of the anthracite arena.[22]

Wyoming coal was tapped very early by the Messrs. Wurts, with their Delaware and Hudson Canal. The Schuylkill Canal, also very early, serviced the Schuylkill coal region. But anthracite had also been discovered in the Lehigh Coal basin at Mauch Chunk in 1791, and Josiah White organized his Lehigh Coal and his Lehigh Navigation companies in 1819. In that same year, he began to tame the unruly Lehigh River.

When the British blockade during the War of 1812 effectively interrupted the shipment of both Liverpool coal from England and sulphur or soft coal from mines in Virginia, anthracite came to be widely advertised, and was eventually used as a substitute fuel.[23] By mid-June of 1820, huge quantities of anthracite were being floated 110 miles down the Lehigh and Delaware Rivers from Mauch Chunk to Philadelphia. White began a publicity campaign in the *American Daily Express,* the *National Gazette,* and the *Democratic Press* to convince housewives as well as the industrialists of the superiority of this modern fuel, brought to them on a "modern" highway, his newly built Lehigh Navigation Canal.[24]

Within five years, the potential of coal power, particularly for the dormant iron industries, was recognized by New Jersey's men of vision, seeking their El Dorado in iron and coal instead of gold. The primary problem was how to get this potential power from the coal mines to the iron industries, and to other markets.

Since the industries were located along the fall line of the state's rivers, why not ship the coal by way of those rivers? A look at a map shows why not: for the most part, northern New Jersey's rivers either flow the wrong way, or are unnavigable to heavy craft, or both. Why not then navigate the Delaware River? Take the coal downstream to Philadelphia for transshipment around Cape May and up to New Jersey ports by sloop. That was done, but on a small scale because it was such an arduous and dangerous task.

Clearly, a more expeditious means of transportation must be found. The building of the Lehigh Navigation Canal and the obvious

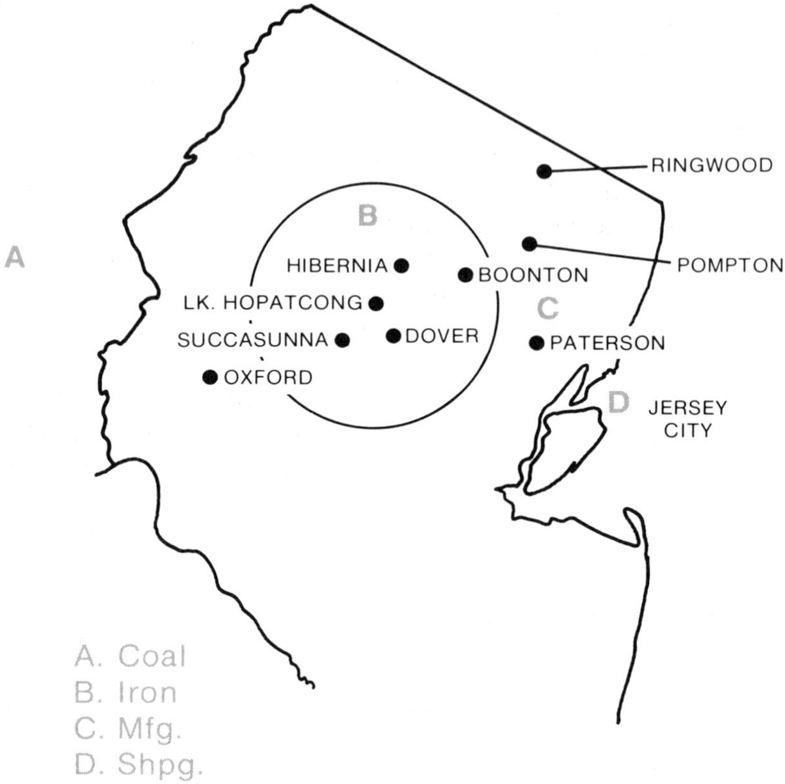

Map showing the principal coal, iron, manufacturing, and market areas served by the Morris Canal, prepared by the author.

success of the Erie Canal in the second decade of the nineteenth century had proved to New Jersey's planners and promoters that they were faced with possible economic exclusion from the commercial markets of the new country and the old. Once again New York and Pennsylvania were threatening to eclipse their smaller neighbor commercially and industrially. Shrewd minds turned to the task of finding an expeditious and economical means of transporting heavy freight — in particular, coal and ore for New Jersey's iron industries.

Ore had long been carried by horseback or pack mule in leather saddlebag arrangements, but never in the quantities needed and demanded by the anticipated general revival of the iron industries. The evolution of horse-drawn freight vehicles from the mid-1700s indicates that the need was fulfilled, however imperfectly.

The humble farm wagon was the model for most of the evolutionary changes. The farm wagon gave way to the oxcart. The oxcart, in turn, gave way to the "Jersey wagon" and then the stage wagon.* All of these general purpose freighters were shallow rectangular box-like carriers. The Jersey and stage wagons were lighter than their ancestors, however, and required less draught. Four- to six-horse teams pulled these wagons at respectable speed even when loaded. Speed, however, was not yet the prime consideration. Capacity was paramount.

Special wagons evolved to serve the special needs of the iron and coal industries. The farm wagon still served as prototype, but the "coal wagon" was much larger. Its interior was sheeted with iron for durability, and its express purpose was the transportation of charcoal to furnaces and forges. The finished product of forge and furnace, the bar or pig iron, was shipped from the Musconetcong Valley to markets as far away as Philadelphia by means of "body wagons." Without an iron lining, the body wagon obviously was lighter than its charcoal-carrying sister, often being used to carry charcoal during the winter months when road conditions made it poor going for the heavier coal wagons.[25] All of the vehicles mentioned were as brave and as sturdy as the beasts which hauled them. Varying in size rather than in appearance, the largest of them held but a 2-ton load.

Briefly — because a whole volume could be written about just the Conestoga wagon and its inventors — the huge wagon was built by early German (Pennsylvania Dutch) settlers to match their newly bred, massive draught horses. The Conestoga horse was too large for horseback riding, even too large for the wagons then in use.[26] The huge Conestoga wagon was the practical, ingenious outgrowth of

**Both the Jersey wagon and the stage wagon were indigenous to New Jersey according to Wheaton J. Lane, op cit, p. 48.*

that circumstance. But coal was too heavy; horses were too slow; and, in any event, a look at a map of early roads in the state shows that when they were there at all, northern New Jersey's roads, like her rivers, went the wrong way. Furthermore, depending on the previous day's weather, they were often as unnavigable as the rivers. Given a dry spell, the roads resembled the Sahara; given any amount of precipitation or a thaw, they became rivers of treacherous red Jersey mud.[27]

A dilemma, with no immediately apparent or forthcoming solution: how to move the new nation's heavy freight more efficiently. Before 1820, the practical application of steam to machinery was in the experimental stage. The word "railroad" was mentioned only in connection with a British invention for getting coal from the mines to the canal boats which would then transport the fuel to the waiting industries, or with John Stevens' hare-brained scheme for a "railroad." While Britain's problem was similar — transporting coal to the iron and manufacturing centers — her geography was not. Moreover, her population had long been settled and her proposed canal routes were already cleared. She also had trained engineers, experienced laborers and, above all, abundant capital.

Here New Jersey looked to her sister state, New York. In the Erie Canal lay the answers to her dilemma. It remained for knowledgeable and farsighted individuals to translate New York State's experiences into a workable New Jersey solution. The Erie Canal, ambitious in size, audacious in scope, and the training ground for all the specialized engineering and construction skills needed to civilize and canalize a virtual wilderness, was in truth the prototype of all American waterway systems.

The word "system" is crucial here, since a system of waterways presupposes certain economic and political conditions. Foremost is a central government covering an area large enough to require planned transportation facilities. That government must be able to tap a large and skilled supply of labor to execute a workable plan for that system. And, of course, it must have the power to command that labor force, as well as the funds with which to finance the project. Post-Revolutionary New Jersey had the unquestioned constitutional authority to execute a plan for transportation, and the Erie canal would provide the labor pool to tap. What was needed were a plan and the funds.

The anthracite-iron canals of Pennsylvania and New Jersey were thus the first attempt to provide a "system" of waterways in America, and that system grew out of the need to move anthracite coal from the mines to its consumers. That the "system" included feeder railways that would one day become full-fledged railroads is the tragic flaw in this drama, for the marriage of anthracite and industry spawned a family of canals and railroads in the northeastern United States. As in any family, this one was to become fraught with sibling rivalry.

In 1822, when the Morris Canal was allegedly conceived, "anthracite" was but a word in the dictionary to all but a handful of enlightened men. A "railroad" was simply a road on which rails had been placed to reduce friction and increase the speed of horses pulling heavy loads over them. A canal was more-or-less an expedient adopted where falls or rapids hampered travel on rivers.

The fortuitous *menage a trois* which was the anthracite tidewater canal and railroad system is what this book is all about. There *is* no history of the anthracite tidewater canals. Nor is there a history of the anthracite railroads. Indeed, there is no history of the anthracite trade in America unless it includes both the canals and railroads which serviced the coal fields. And to tell the story at all, sooner or later the narrator must touch on the Morris Canal and on the men whose names are most closely associated with it. Of those names, none is more famous than that of George P. Macculloch.

REFERENCES

1. Maynard Owen Williams, "The Suez Canal: Short Cut to Empires," *National Geographic Magazine*, November, 1935, p. 613.
2. "Canals and Canalized Rivers," *Encyclopedia Britannica* (1960), Vol. 4, p. 719.
3. Robert Payne, *The Canal Builders* (N.Y.: 1959), p. 51.
4. *Ibid.*, p. 103.
5. *Ibid.*, p. 138.
6. Eleanor Morton, *Josiah White: Prince of Pioneers* (N.Y.: 1946), p. 50.
7. "Robert Fulton," *Encyclopedia Britannica* (1960), Vol. 9, p. 913.
8. David Hosack, *A Memoir of DeWitt Clinton (N.Y.: 1829), p. 101.*

9. Julius Rubin, "An Innovating Public Improvement: The Erie Canal," in *Canals and American Economic Development*, Carter Goodrich, ed. (N.Y.: 1961), p. 65.
10. Payne, *op. cit.*, p. 148.
11. John D. Hicks, *The Federal Union* (N.Y.: 1937), p. 389.
12. Seymour Dunbar, *A History of Travel in America* (N.Y.: 1968), pp. 825-6.
13. Hosack, *op. cit.*, p. 101.
14. Richard P. McCormick, *New Jersey From Colony to State, 1609-1789* (New Brunswick, N.J.: 1964), p. 162.
15. Lucius Q. C., Elmer, *Reminiscences of the* [*N.J.*] *Bench and Bar During More Than Half a Century* (Newark, N.J.: 1872), p. 459.
16. Barbara Carper Lang, "New Jersey's Morris Canal," *The Royle Forum*, No. 104, Paterson, N.J., September 15, 1963, p. 7.
17. John Cunningham, *This Is New Jersey* (New Brunswick, N.J.: 1953), Rev. ed., 1968, p. 22.
18. Peter O. Wacker, *The Musconetcong Valley of New Jersey: A Historical Geography* (New Brunswick, N.J.: 1968), p. 114.
19. McCormick, *op. cit.*, p. 89.
20. *The Story of Anthracite* (N.Y.: 1932), p. 22.
21. *Ibid.*, p. 10.
22. *Ibid.*, pp. 13-14.
23. Wheaton J. Lane, *From Indian Trail to Iron Horse: Travel and Transportation in New Jersey, 1620-1860* (Princeton, N.J.: 1939), p. 223.
24. Morton, *op. cit.*, p. 96.
25. Wacker, *op. cit.*, p. 140.
26. Arthur Vernon, *The History and Romance of the Horse* (Boston: 1939), pp. 256-7.
27. Dunbar, *op. cit.*, pp. 192-3.

2

Macculloch's Dream

LEGEND has it that George P. Macculloch of Morristown was fishing at Lake Hopatcong one day in June, 1822, when he suddenly envisioned a plan for a canal to connect the Delaware River with the Passaic, using the waters of Lake Hopatcong to fill the artificial channel. Thus, allegedly, a dream was born. How it was conceived is quite another story. Like all legends, however, this one has elements of historical truth — and a logical explanation. For this dream of Macculloch's was no "Eureka" experience, no miraculous conception. Nor was Macculloch merely an idle dreamer.

George Perrott Macculloch arrived in Morristown, New Jersey, from England in the Spring of 1806.[1] Educated at Edinburg University, he was imbued with a knowledge of the then-popular political economy of Adam Smith's *Weath of Nations*.[2] He had also gained a wealth of experience while working for a London firm of international financiers. Most important, he transferred a sense of loyalty and patriotism to his adopted country, and set about establishing himself as one of Morris County's most dedicated citizens. Before 1820 was out, he had built his home, Macculloch Hall, and in it administered a boys academy. Meanwhile, he had been instrumental in organizing an agricultural society dedicated to the development of Morris County's largely rural economy, and to the improvement, thereby, of its collective intellect.

Morris County had been richly endowed by nature with iron ore deposits. The mines had been extensively worked during the Revolutionary period, as had the county's numerous forges and furnaces. With the general economic slump following the War of 1812, the mines, forges, and furnaces had shut down because foreign competition was too cheap and too keen, and because fuel was too scarce and too expensive. Moreover, transportation was too costly and too difficult. It was to this economic condition — potentially productive and therefore lucrative — that Macculloch addressed himself. It was at this time, 1823, that Josiah White was finishing his Lehigh Canal. And it was just about the time that Macculloch was conceiving his dream canal.

Canals in Pennsylvania were for the most part state projects, White's Lehigh Canal being a notable exception. When White applied to the Pennsylvania lawmakers for permission to "canalize" the Delaware River from Easton to Philadelphia, he was refused. Instead, the legislators opted to make a state canal paralleling the Delaware. Adding insult to injury, the new canal was apparently to be built with no outlet lock at Easton, so that White's desired anthracite through-route from Mauch Chunk to Philadelphia was not yet to be. Temporarily denied a practicable river avenue to the outlet for his coal, White turned eager and impatient eyes toward New Jersey, where the idea of a cross-state canal was under discussion. If he could not ship his anthracite to Philadelphia, White could — and would — ship it to New York City, if and when the Morris Canal were ever built!

The idea for a canal across northern New Jersey did not originate with George Macculloch. William Penn had begun the discussion of such a canal, and countless others repeated the theme throughout the years. By April of 1807, a series of articles about internal improvements was appearing in New York State's *Genesee Messenger.* Writing under the name "Hercules," Jesse Hawley pointed out that New Jersey had been "circumstanced mostly for domestic trade," and that the state was destined thereby "to be tributary to the first two commercial ports in the United States,

George P. Macculloch (MPL, photograph by Irving Fitzig).

situated on her right and left." He referred, of course, to New York and Philadelphia. A "water passage opened between the Delaware and the Hudson," he argued, would surely "be of essential service to the internal continental commerce during maritime wars." Thus, New Jersey, by being bisected by a canal would be serving the "good of the country," while "contributing to its own trade," not to mention being able to boast of such a "grand intersection." Hawley therefore suggested construction of a canal to terminate "somewhere about Easton," and to be maintained either by tapping the Delaware River or "by the use of the Mosconecunk [sic]."[3]

Macculloch was already in America when Hawley's articles appeared. It is not known, of course, that he was aware of Hawley's ideas, but it is a fact that canals were a paramount topic of discussion among the literate, interested citizenry of the day, and it is at least likely that Macculloch absorbed the essence of Hawley's suggestions simply by exposure to the climate of ideas. Readiness is all, and timing is a crucial part of readiness. However the idea for a canal may have occurred, having conceived it, Macculloch labored to deliver it to the public. His significant contribution lies in the fact that he was effectively on the spot at the right time with the right proposal.

Throughout the summer of 1822, Macculloch waged a publicity campaign for his canal.[4] Hawley had done the same on behalf of the Erie Canal, and in Pennsylvania, Josiah White had sought public support for his Lehigh Canal through the newspapers. By his own description, Macculloch engaged in "an endless catalogue of literary tasks," and later claimed that, as author "Agrestis," he had "single handed [sic] achieved the problem of rendering popular, and accomplishing, a scheme demanding vast resources and stigmatized as the dream of a crazed imagination."[5] In one of his articles, he wrote:[6]

> *Suppose, that by some miracle, the channel of the Muskonnetconk [sic] were rendered boatable from the Delaware up to Stanhope, that a canal were dug from thence to meet Rockaway River at some point between Dover and the Valley Forge, that the channel were rendered passable from thence down to Paterson, the falls obviated by a chain of locks, and thus a water communication opened between the Delaware and the*

Hudson, through the centre of Morris and Sussex Counties. What would be the effect of this miracle?

The mouth of the Muskonnetconk is close to inexhaustible mines of coal, which could be afforded, in New York, so cheap as to exclude not only Liverpool coal, but also wood which grows daily dearer; and for this reason also, it would be used instead of any other fuel, in distilleries, manufacture, steam boats, etc. The same reason would bring coal into domestic use throughout the districts of Newark and Elizabethtown, and for a certain distance along each bank of the canal . . . . The facility with which coal, iron ore and provisions could be transported, in a country full of water powers, would gradually introduce new forges, furnaces and manufacturers of every sort; Paterson might thus become the Birmingham, as well as the Manchester of America . . .

No man can say what the result will be, and were I to state my opinion, it is probable that at the present moment, it would be regarded rather as the offspring of warm imagination, than of cool reason.

By August, Macculloch's warm imagination had begun to smolder. Reasoning at first that the Great Pond, Lake Hopatcong, would be "a sufficiently capacious and bountiful reservoir for the summit level or cut between the Rockaway and Muskonnetconk [sic],"[7] he came to see that his canal need not follow any "particular stream," but would be free to follow the "easiest, shortest, and cheapest route" to its eastern and western termini — that is, *if* Green Pond were to be incorporated into the canal's reservoir system.[8] The only doubt remaining as to the practicability of this initial canal proposal lay in the terrain of the part of Sussex County to be traversed by the canal.

Originally, the western section of the canal was to have been artificial from the summit level to Stanhope, where the Musconetcong River would be made suitable for navigation to its mouth. However, the Pequest River presented another possibility for gaining access to the Delaware. By means of the first route, the Morris Canal would terminate below Easton; by the second, above it.[9] As the summer of 1822 wore on, Macculloch became convinced that, whatever the route, the canal was feasible, that it had a multitude of supporters, and that it would be well worth executing. The initial publicity campaign culminated in a public meeting, announced in the Morristown *Palladium of Liberty*:[10]

> *To the Intelligent and Public Spirited Citizens of Hunterdon, Sussex, Morris, Bergen and Essex Counties:*
>
> *Inland navigation is the parent of internal improvement. We accordingly see that the formation of Canals is in every instance, a primary object among those states which, with honorable emulation, are pressing forward to cultivate their natural resources.*
>
> *A commodious water conveyance can, in no part of America, produce effects more eminently beneficial than in the north of New Jersey; not only by invigorating agricultural enterprize, and enabling us to extract the metallic treasures of our mountains, but by developing the various manufacturing advantages for which our district stands unrivalled.*
>
> *The subscribers do not desire to expatiate on results so obvious, but state that a canal uniting the Hudson to the Delaware, and probably intersecting the counties of Essex, Bergen, Morris, Hunterdon, and Sussex, has lately become a subject of serious and general contemplation . . .*
>
> *We therefore most respectfully invite all patriotic persons of these counties to meet here, as a central point, at Drake's Tavern, on Wednesday, the 21st August, at 2 o'clock, for the purpose of conferring upon the measures necessary to form an accurate judgment and to take further measures in an undertaking, concerning the vast utility of which, every opinion must be in unison.*
>
> *Geo. P. M'Culloch*
> *Lewis Condict*
> *John Doughty*
> *Dan'l Phoenix*
> *Mahlon Dickerson*

Macculloch we have already met, but who were these other men whose names are interlinked with his? What did they stand to gain from a canal? Lewis Condict was a prominent Morristown citizen, a doctor, an Assemblyman, a Congressman and, from 1821-33, Speaker of the House. His interest in internal improvements is documented by his participation in the Morris Aqueduct, a corporation which supplied water to Morristown. While still a Morris County

Assemblyman (1807), he had been appointed to a Commission by the New Jersey Legislature to try to settle the dispute as to the location of the Hudson River boundary line between New York and New Jersey. Years later he was to become an active backer of the Morris and Essex Railroad.[11]

John Doughty, General of the Morris County Militia, was a county judge, justice of the peace, and one of Morristown's most prominent citizens. He was one of the corporators of the Newark-Morris Turnpike. Described as one of Washington's friends during the General's stay in Morristown, Doughty would no doubt have shared his views on internal improvements — in particular, canals. No doubt, too, he would have had an exceptionally keen interest in the development or sale of the iron-laden lands he'd been appointed to sell for the estate of iron-master John Jacob Faesch.[12]

Daniel Phoenix, a Morristown bank president, was an all-round good citizen. Munsell described him as a "contractor" who built the First Presbyterian Church of Morristown in 1791. Confirming his interest in transportation is the fact that he was a subscriber to the Union, the Paterson and Hamburg, and the Milford and Owego turnpikes.[13]

Perhaps one of the best-known of the subscribers to Macculloch's August 21st meeting was Mahlon Dickerson, a former Philadelphia lawyer and twice the governor of New Jersey. As inheritor of his father's large estate in Morris County, he owned the land on which was located the celebrated Dickerson Iron Mine at Succasunna.[14]

Whatever else united these men, certainly their common interest in internal improvements, in transportation, and in the development of the local iron industries united them behind Macculloch's canal proposal.

The meeting was well attended. The State's Governor was there, as well as "a number of the most respectable individuals from various parts of the district." The "district" in this case included the parts of Hunterdon, Sussex, Morris, Essex, and Bergen counties which were to be intersected by the canal. Two citizens from each county were selected by the meeting to join Macculloch in preparing a memorial for the legislature on behalf of the canal proposal. Nathaniel Saxton and Henry Dusenberry of Hunterdon County, Morris Robeson and

Gamaliel Bartlett of Sussex, Lewis Condict and Mahlon Dickerson of Morris, Gerard Rutgers and Charles Kinsey of Essex, and John Rutherford and William Colfax of Bergen were chosen for that task.[15]

It is useful here to recall that in August, 1822, Hunterdon County included part of what would in 1824 become Warren County; Passaic County would not come into existence until 1837, when it broke away from Essex County; Bergen County included all of Hudson County and parts of Passaic County; and Essex County, in addition to Passaic County, included all of Union County.

Lewis Condict and Mahlon Dickerson have already been introduced. Another member of the committee was Nathaniel Saxton of Hunterdon County, who owned extensive acreage in that part of the county which became Warren County, and in Sussex County. In matters of real estate, he was said to have been "without a peer in the State."[16] Morris Canal buffs will recognize his name since it has been affixed to a canal site that has been designated a State Park on the Musconetcong River between Waterloo Village and Hackettstown. I refer, of course, to Saxton's Falls.

The other Hunterdon County member, Henry Dusenberry, was originally a Philadelphia merchant, who settled first in the Musconetcong Valley at New Hampton, then moved to Belvidere. Whether or not Henry was related to William C. Dusenberry, founder of Port Colden and canal freight-line operator, is not known.

Committeeman Morris Robeson was owner and operator of the famous Oxford Furnace,[17] which Boyer claims was a Philadelphia-inspired enterprise.[18] The town of Oxford, located on a branch of the Pequest River, might well have been interested in a canal which would, with a minimum amount of difficulty in construction, have served it well.

Another Sussex committee member, Gamaliel Bartlett, was, among other things, the postmaster at Stanhope when that office was housed in his tavern, the Stanhope House (circa 1825). He also owned the upper forge and the grist mill at Stanhope, the forge having formerly belonged to Simeon Dickerson (until 1807), Joseph Jackson (until 1816), and Abraham Hathaway and Josiah Munson (until 1820).[19] Much later, in 1841, the Sussex Iron Company would buy up the land on which the forge was located, and there erect two anthracite blast furnaces.

The committee's Gerard Rutgers was a citizen of Belleville in Essex with enough concern for the public welfare to run for State Assembly and Council (later renamed "Senate").[20] His wife, Margaret Sarah Bayard, was the fourth daughter of Nicholas Bayard, Esquire, of New York City.[21] That the Bayard family was actively involved in the canal's early history will soon become apparent.

In 1801, memorialist Charles Kinsey, together with Israel Crane and a Mr. Fairchild, had leased the Essex Mill in the then-infant manufacturing city of Paterson. That mill revolutionized the paper industry. Up until that time, all paper had been made in sheets, and the Kinsey group devised a way of making it in rolls.[22] Kinsey patented his roll-paper machine, and later built a paper mill on the Second River at Bloomfield.[23] By then, his interest included transportation as well, and he became one of the corporators of the Paterson and Hamburg Turnpike.[24] But it was in his capacity as a member of the State's Legislative Council (Senate) that Charles Kinsey was to be most helpful to the Morris Canal project, as we shall see.

Committee member John Rutherford was proprietor of large land holdings in Sussex County — among them, the Andover Mine. However, he had inherited and lived on an estate in Bergen County, on the east bank of the Passaic River, 7 miles north of Newark. The estate, called "Edgerton," was located on the site of what is currently known as Rutherford. Mr. Rutherford was chosen as one of the memorialists from Bergen County. Apparently the Morris Canal, as it was eventually built, did not serve Rutherford or his interests, and he turned to railroads. In 1832, he became one of the corporators, and then president, of the New Jersey Hudson and Delaware Railroad.[25]

The last memorialist was General William Colfax, one-time close associate of George Washington. Colfax was another civic-minded citizen who lent his support and influence to further Macculloch's canal cause. The Colfax family operated a forge on the Morris County side of the Pequannock River, between Riverdale and Bloomingdale, but it is by no means certain that the forge was in operation at this time, or that the Colfaxes still owned it. At one time, too, the family had owned both a grist mill and a tan bark mill at Pompton. William Colfax, for many terms a Pompton town official, also served in the State Assembly and Council (Senate).[26]

Such were the varied interests and talents among the memorialists. In combination and concert, those talents would help to convince the State Legislature of the need, the desirability, and the practicability of a canal through the mountains of northern New Jersey. As a point of interest, note that, although one of the avowed purposes of the Morris Canal was to provide cheap transportation and raw materials as well as fuel to the manufactories of Paterson, no resident of that city represented Essex County (which included Paterson at that time) in the memorial to the legislature. Those chosen from Essex County were residents of Belleville and Bloomfield.

Antagonism and competition among the individuals and communities that were hoping desperately to be serviced by the canal were no doubt inevitable. But Macculloch had to fight more than local disharmony. He had to fight the stubborn sectionalism among New Jersey's legislators. Attempts to create *any* canal through the northern part of the state were resisted by representatives of the southern counties. Such sectionalism Macculloch held to be unconscionable. He addressed the issue — and its perpetrators — squarely and bluntly.

Writing to the editor of the Rahway *Museum,* a "southern" New Jersey newspaper,[27] Macculloch took exception to "a variety of facetious comparisons" which had been "used to illustrate the situation of New Jersey, placed between two neighbors who absorb the very marrow out of her bones and hold her as a humble tributary; a mere Lazarus, subsisting upon the crumbs which fall from the table of Dives . . ." Macculloch had been provoked by an article appearing in the *Museum,* which treated the Morris Canal "with some asperity and proscribed chiefly because it will be useful only to a part of the state . . ."

"This anti-social mode of thinking," Macculloch claimed, had been "engendered by our situation relative to Pennsylvania and New York, whom we look upon as everything and New Jersey as nothing. That our state spirit is extinct," Macculloch pointed out, "requires no better proof than that, when a measure is proposed which must raise us to our natural level, instead of feeling alive to a common interest, we forthwith sit down to calculate whether some other county will gain more than we . . ."

So sure of his ground, so dedicated to his dream was Macculloch that he concluded his article with a challenge: "A canal is only a strip of ten or fifteen yards wide; it can therefore come to the very door of only a few persons, but its route should be such as to enrich a large district. On this ground I am willing to place the Morris Canal in comparison with any other artificial navigation which can be traced on the map."

In an all-out effort to enrich his arguments on behalf of a canal, Macculloch solicited help from an unimpeachable source. He brought in an expert.[28]

> *In order to furnish every requisite elucidation on the subject of the intended canal, the most important and difficult points of the route have been visited by Mr. Renwick, Professor of Mathematics in Columbia College, New York, in company with Judge Lemuel Cobb and myself, assisted by the principal inhabitants of the district through which we passed. Mr. Renwick's high professional attainments as an engineer, and the circumstance of his having viewed the chief canals in America and Europe, give his opinion a conclusive weight on this subject.*

Judge Lemuel Cobb had been a surveyor for the East Jersey Board of Proprietors. While acting in that capacity, the Judge had been able to pick up pieces of real estate here and there throughout northern New Jersey until he had become a landholder of considerable importance. As the owner of Troy Forge at Parsippany (Cobb's Corner is named for him) and the Split Rock Forge, the Judge was understandably interested in reviving the iron industry and in improving transportation. Since 1809, he had been one of the incorporators of the Parsippany and Rockaway Turnpike. As the owner of lands at the outlet of Green Pond, as well as thousands of other acres that a canal company might be interested in or of service to, Cobb was pleased to offer his expertise and his services along with those of Renwick to Macculloch's cause.[29]

James Renwick, born in Liverpool, England in 1792, settled in New York City with his family when he was but a few years old. Although English-born, he was American-educated, graduating at age 15 from New York's Columbia College. After traveling throughout Europe with his close friend, Washington Irving, he returned to Columbia to complete the requirements for a Master of Arts degree in 1810. Renwick served as a major with the topographical engineers during the War of 1812, and in 1817 was commissioned colonel of the New York Militia's engineers. In 1820, he joined the faculty of Columbia College, being installed as Professor of Natural Philosophy and Experimental Chemistry, and lectured on such subjects as geology, mineralogy, practical mechanics, and astronomy. Because he was a recognized and respected authority on virtually every branch of engineering at that time, Renwick was called upon as a consultant for innumerable public or private engineering projects that were undertaken during his professional lifetime, and until his death in 1863.[30]

Equally inventive and famous in their own fields were Renwick's three sons. Henry, the eldest, was a prominent engineer. Edward, the youngest, was the inventor of a self-binder, designed to improve the McCormick reaper, and some poultry-incubator devices. James, the middle son, was an eminently successful architect as well as an engineer. Among the many outstanding structures credited to him, and still standing today, are New York City's Grace Church, Saint Patrick's Cathedral, and the Croton Reservoir. He also built Vassar College, and a host of hotels, hospitals, and other public buildings. Recently, in fact, the Corcoran Gallery of Art in Washington, D.C., was renamed the "Renwick Gallery," in honor of its designer, James Renwick, Jr. The junior Renwick was also among the nation's first railroad specialists. In 1836, when he was beginning his career, perhaps by following the iron rails laid down after his father's model of an inclined railway for the Morris Canal, he became one of the Erie Railroad's earliest engineers. The architecture came later; first, James Junior was a railroad man.[31]

Those iron rails of James Renwick's plan were mentioned by Macculloch in the September 26, 1822 issue of Morristown's *Palladium of Liberty* after a tour of the canal route with Renwick

James Renwick (CUL, photograph by Irving Fitzig).

and Judge Cobb. In his report on that tour, Macculloch optimistically observed that "until the necessary observations be completed and calculations formed, the hight [sic] of the Great Pond may be reckoned at 460 feet above the Passaic and 356 feet above the Delaware, at the points where the canal will terminate." From the summit level, "to Stanhope, a distance of nearly 3 miles, the fall is only a few feet, which one cheap dam there can probably surmount. From Stanhope to Andover there is a rapid fall of about 276 feet, of which at least 140 feet can be passed in two inclined planes, the total cost of which will probably not exceed $3000 for both. From Andover to the Delaware, very little cutting or locking will be required, as the river with only a very few exceptions is said generally to present a gentle stream or natural canal on which little more is wanted than a towing path."

On the eastern side of the summit level, Macculloch reported, the Rockaway River level might be reached by any one of "several routes of gradual descent. The ground on both this and the Musconetconk [sic] course is very remarkably favorable, not a single rock presenting itself which requires blasting, or no long hollows for embankment or aqueduct, *neither will any mill seat be touched* [italics Macculloch's], the immense reservoir of the Pond furnishing an ample supply if prudently administered: a leading principle is to preserve and if possible to increase the water powers of the country. On this side we may say that 100 feet fall can be cheaply gained by inclined planes . . . The elevation after deducting the inclined planes gives a total of about 500 feet of lockage on both sides . . . ."

Since more sophisticated (and less-proven) engineering designs as inclined planes were to be employed, Macculloch sought more expert engineering opinions. Early in October, 1822, when Renwick again toured the canal route, he was accompanied by General Joseph G. Swift, former Chief of the U.S. Department of Engineers at West Point. Renwick's purpose in the tour was to more accurately calculate with a mountain barometer the elevations to be overcome; Swift's purpose was to form an opinion as to the feasibility of using inclined railways. Renwick "examined the whole line from Dover to the Delaware, going one route, visiting Easton, and returning another." Presumably this meant the Musconetcong and Pequest

River routes. Swift, on the other hand, inspected only those "points where difficulties were to be apprehended." Both men were to report their findings to the State Legislators, who could not fail to be impressed by the standing each held "at the head of his branch of Mathematical Science."[32]

Meanwhile, Macculloch's publicity items continued to bombard the population. So caught up was he in the prevailing fervor about internal improvements that he begged his readers to open their maps of the U.S. and on them, "connect the Passaic with the Delaware by the Morris Canal; then to connect the Lehigh with the Nescopec and Susquehanna. Following this last river in its various branches," he maintained, "and also, the head waters of the Delaware, we will perceive that the countries whose produce will find their way through the Morris Canal are in extent or fertility hardly inferior to those intersected by the grand western canal of New York." Moreover, the Morris Canal, being more southerly than the Erie, would be free of ice three weeks earlier in the spring, and remain ice-free three weeks longer in the fall — a boating season six weeks longer than that of the rival Erie! In fact, Macculloch was sure that his Morris Canal would "equal in importance and surpass in profit any canal great or little in America."[33]

Macculloch, as we've seen, wrote under the name "Agrestis." There was an "Aristides" too, who contributed his thoughts to the canal discussions then being conducted in the *Palladium of Liberty.* In his September 5, 1822 remarks, Aristides declared that the canal would be a boon to the apple-growing Jerseymen along the canal's path. What a boon it would be, he argued, to be able to send cider to New York, "instead of being obliged to distill it into a poison; a poison which is beginning to be made so copiously at New Orleans, that we must in a very few years be undersold and driven out of every market. In a short period spirits must fall in price, and if we cannot send our cider to market by water, we may cut down half our orchards, or what is worse, be obliged to drink all our whiskey ourselves . . . ." Obviously, for Aristides, it would be far better for Morris County if the canal were built so that the Jersey Lightning could strike or be shipped elsewhere.

Macculloch didn't need the arguments of an Aristides to convince him that his canal should be built. He was already sure of the benefits the canal would bestow upon the districts it would traverse. He was sure, too, that once they knew of those benefits, individuals would be inspired to give land for the canal out of their united patriotism and self-interest.

One reason that Macculloch's original estimate of the probable cost of his canal was so low was that he really *did* expect people to give lands for the canal's right of way. He firmly maintained that $800,000 would "not be far from accomplishing this grand enterprize which must render our northern district the most manufacturing and richest portion of America." After all, he argued, labor should be cheaper here than in Europe, since here "we employ scrapers whenever we can, for which reason we work cheaper than they."[34]

In fairness to Macculloch, we must be aware that when he made this estimate of $800,000 to build the Morris Canal, he had planned an artificial waterway from Rockaway to Waterloo, with rivers supplying the remaining miles of navigation. He had planned, too, on using two inclined planes west of the summit, "the total cost of which will probably not exceed $3000 for both," and since he expected the eastern portion of the artificial canal to have only 100 feet of fall, he calculated that the fall there could be cheaply gained by inclined planes — probably two of them, at $3000 for both. It appears that Macculloch was estimating about $21.43 per foot of lift per plane, or 140 feet of lift in the two planes on the western division at $3000 = $21.42 + per foot lift. When Renwick compiled his report, he calculated the expense of the planes at $150 per foot lift.

Regardless of cost, Macculloch wanted that canal built. The Delaware and Raritan canal project had been revived and was being discussed in Trenton. Perhaps the State would underwrite part of the expenses of both canals. "The state may do the whole, or advance an equal sum for the Raritan and Morris each, leaving individuals to fill up the remainder."[35] Macculloch was sure that the people who lived along the proposed route of the canal would subscribe to the canal's stock. If they didn't, he was even surer that New York capitalists would come up with an even larger sum.

The big danger, as Macculloch saw it, was that the Legislature might "shipwreck the project" by refusing to do anything at all. If that august body declined to finance the canal, they must be induced somehow to grant a liberal charter, not one fraught with "illiberal restrictions" that would keep individuals from investing in the Morris Canal. Such, he was sure, had been the fate of the Raritan Canal several years earlier.[36]

The Morris Canal was not without friends in Trenton, as we have seen. On November 9, 1822, Macculloch's petition on behalf of the canal was acted upon by the Legislature. The petition had urged the lawmakers to make an appropriation available "for the purpose of causing a survey of a route contemplated for the cutting of a Canal from the Delaware to the Hudson . . . ." A bill was introduced by the committee that had been studying Macculloch's petition: "An act for ascertaining the most eligible route for, and probable expense of forming a Canal to connect the waters of the Passaic river with the waters of the Delaware river at or near the mouth of the Musconetcong River." The bill went through three readings, and on November 14, 1822, it was passed by a vote of 27 yeas and 9 noes.[37]

Three commissioners were appointed to implement the act and compile the information required by the legislature. It had begun to look as if New Jersey, like her sister states, was ready to go into the canal-building business.

The commissioners named were Macculloch, Charles Kinsey of Essex County, and Thomas Capner of Hunterdon County, representing the western, central, and eastern divisions of the proposed canal. Kinsey and Capner were, at the time, State Legislators. We saw that Kinsey's interests would be served by the proposed canal, and the same was true of Capner — if the canal were made along the Musconetcong River. The Capners owned "Mine Farm," so-called because it contained a vein of copper, situated near Flemington. The farm was also noted for its Blakewell sheep, which were bred there and sold in virtually every state in the Union.[38] Carding, weaving, and fulling shops on the farm turned the wool into worsted. Gordon's *Gazetteer* (1834) listed the Capner farm as a dairy, famous for its cheeses.

The Commission was empowered to obtain whatever services were needed to complete its task, and a $2000 allotment was reserved to them toward that end. In April, 1823, Macculloch traveled to Albany to confer with New York State's Governor DeWitt Clinton, its Canal Commission, and its Legislature.[39] Having considered the New Jersey proposal to connect the Delaware and Hudson Rivers with an inland navigation, New York's lawmakers saw that such a canal could "not fail to produce the most beneficial effects upon the agriculture, manufactures, and general industry" of New York. On April 23, 1823, the New York Legislature accordingly resolved that its state canal commission "send to the line of the . . . [Morris Canal], one of the engineers in the employment of this state [New York] . . . for the purpose of making a survey and estimate thereof, under the direction of the Commissioners appointed by the state of New Jersey, to whom he shall report, as also to the Canal Commissioners of this state."[40]

As a result of that legislative resolution, Macculloch proceeded with Judge Benjamin Wright, Chief Engineer of the Erie Canal, to Little Falls, New York, for the purpose of engaging Mr. Ephraim Beach "to take the levels and survey the route, having previously conversed with him, and agreed with Professor Renwick to entrust him with that task." Presumably, Macculloch returned immediately thereafter to New Jersey, for he described his activities during the remainder of that spring and summer as having been spent in "collecting topographical and statistic [sic] information, and also in reconnoitering the various routes, in company with the inhabitants of their vicinity . . . . In July 1823 Mr. Beach appeared for the first time on the scene of action, guided by Mr. Renwick, to whom the deliberative department was confided."[41]

Captain Beach, Major Beach, Mr. Beach, Esquire, Mr. Beach "The well-known engineer" — Who was this historically elusive man who accomplished so much yet remains so little known? Details are disappointingly scanty, but it is possible to piece together enough of the fragments from a variety of sources to form a respectable and recognizable profile.[42]

Ephraim was a direct descendant of and five generations removed from Thomas Beach, who had migrated from England to Connecticut with early religious dissenters. From Connecticut,

Thomas' son Zopher removed to Newark, New Jersey, where his presence is noted as early as 1685. Of Zopher's many children, only a few are cited here: son Thomas married Martha Board of Orange County, New York, just across the border from Ringwood; daughter Phebe also married into the Board Family, settling down with Joseph Board and producing Charles Board, one of Macculloch's canal supporters;* son Josiah married Annas Day. It is this Beach, Josiah, who concerns us here. Ephraim, son of Josiah and Annas Day Beach, married Deborah Johnson and sired Jedediah Beach. Jedediah, in turn, married Mary Post, settled in Hanover, Morris County, New Jersey, and gave us the Ephraim Beach of engineering fame.

Ephraim was born in Hanover, on October 11, 1783. Within ten years, Jedediah had moved his family to Springfield, New York, at the north end of Lake Otsego. Next to nothing is known about Ephraim's early education. He attended school once a week in nearby Cherry Valley, where Eliphalet Nott conducted classes from 1795 to 1797. Later, Nott went on to become President of Union College; Beach went on to become the celebrated canal and railroad engineer.

As for Beach's engineering background, all we can do is imagine what it might have been. Jedediah's Revolutionary War Pension papers reveal his "activities in construction work in New York and Long Island." After the War for Independence, it is possible, even probable, that Jedediah's construction expertise was put to work on the early attempts to dig a canal through New York's Mohawk Valley — the original "Erie Canal." I further submit that Ephraim might have worked right along with his father and learned his engineering skills on the job, that job probably being the pre-War-of-1812 Erie Canal.

In 1812, the outbreak of hostilities with England postponed work on the Erie Canal. Ephraim Beach, *wherever* he had worked, rounded up enough volunteers to form a company which he

**The Boards' progenitor was Cornelius Board, reputed to have been an agent of the celebrated Lord Stirling, and one-time owner of all of the Ringwood properties. Boardville, a settlement named for the family of Cornelius Board, was still shown on maps of New Jersey as late as the 1920s. The town was located between West Milford and Wanaque, and is no doubt covered by the waters of the Wanaque Reservoir.*

captained through the battles of Queenstown (1812) and Lundy's Lane (1814). By the time of his discharge, Captain Beach was promoted to Brevet-Major — hence the discrepant titles of rank usually applied to him: sometimes "Captain" Beach, sometimes "Major" Beach.

Shortly after his discharge from the Army at the close of the war, Beach returned to his native New Jersey. He is listed among the congregation of Morristown's First Presbyterian Church, and it was in that church that he took to wife Ann Ogden Lindsley on May 13, 1816. The marriage announcement in Newark's *Sentinel of Freedom,* however, gave Beach's place of residence at that time as Constantia, New York.

Ephraim and Ann had a son, Zenas Lindsley, born on March 12, 1819, and a daughter, Mary Elizabeth, born December 10, 1820. With the birth of their third child, Ephraim Ogden, on December 11, 1822, Ann's health failed, and she died on December 29.

Ephraim must have been working during all those early years, but documentation is lacking, or at least very well hidden. At the time of Ann's death, however, he was probably already employed as assistant engineer on the Schuylkill Canal. This much is known: The chief engineer of the Schuylkill Navigation was Thomas Oakes, the same Thomas Oakes who had come from England in 1802 to settle in Bloomfield, New Jersey. Thomas was a "consulting engineer," but his son David, born in Bloomfield in 1809, turned to manufacturing and established the famous Oakes' textile mills in his hometown.

On August 13, 1823, Thomas Oakes succumbed to an "unknown fever," probably cholera, in Reading, Pennsylvania. Henry King succeeded Oakes as "principal engineer" of the Schuylkill Canal. However, King was also struck down by the fever, and Ephraim Beach was named "principal engineer" of the Schuylkill in October, 1823. By July 5, 1824, the canal was opened for navigation from Reading to Pottstown. Since it was nearing completion, the canal no longer needed a full-scale principal engineer, and Beach left his position with the Schuylkill Navigation Company on November 12, 1824. He was succeeded by Ira Dodd, who was given the title "Superintendant." Later, Ira Dodd would reappear in Beach's life, when he and Caleb Baldwin combined efforts in the construction of the magnificent brownstone aqueduct at Little Falls, New Jersey, finished in 1829.

Beach seems to have been employed simultaneously on the Erie, the Schuylkill, and the Morris canals. Such a practice was no more uncommon then than it is today. Benjamin Wright, usually associated with the Erie Canal, was simultaneously involved in the projected Delaware and Raritan, the Morris, and the Ohio canals, as well as the Harlem River Canal Company in New York City. Canvass White, too, of Lehigh Canal fame, was involved with Wright in all of those projects. And James Renwick, while engaged as "consultant" on the Morris Canal, was one of a group called the New York Harlem Spring Water Company created to supply the burgeoning city with pure water. So, too, Ephraim Beach, already assistant engineer on the Erie Canal in April, 1823, was at work on the Erie's Little Falls locks when George Macculloch traveled from Morristown to solicit his services for the Morris Canal surveys. Yet by October, 1823, we find that Beach was chief engineer of the Schuylkill Navigation. Like other of his professional colleagues, Beach divided his time among many clients.

Such a heavy professional schedule undoubtedly kept widower Beach well-occupied. However, he was thrown together with the Morris Canal's backers, and it was inevitable that Beach should meet Elizabeth, daughter of Israel Crane, a substantial stockholder in the Morris Company, and turnpike transportation tsar in his own right. Eliza and Ephraim were married on September 7, 1826, and took up residence in Newark. Ephraim is cited in various *Records of Newark* during the 1830s as a member of the town council, as Surveyor of the Highways, and as a member of Newark's Third Presbyterian Church. As an engineer, Beach was given the task of ridding Newark of the pools of stagnant water that constantly collected in the unpaved streets.

Ephraim was made Chief Engineer of the Morris Canal in 1825, but he continued the practice of serving more than one master. In December of that year, for example, a Paterson newspaper noted that Beach was then working with Canvass White and Benjamin Wright in an attempt to determine the ultimate route of the Delaware and Raritan Canal which had been chartered at the same time as the Morris. However, he seems to have concentrated the bulk of his attention on the problem of determining an ultimate route for the Morris, and on supervising attempts at making the first experimental inclined planes perform according to expectations of them.

Once the Rockaway plane experiment had been demonstrated to be operational in 1826, the main question to be settled was that of an eastern terminal, and Beach was ordered to conduct a survey for a route from Newark to Jersey City. That was in 1827. The Associates of the Jersey Company, a corporation that owned all of Paulus Hook — about which much more later — engaged Beach to recommend a plan for "regulating the streets and curbing the side-walks." By January, 1830, the *New York Enquirer* reported that Jersey City's inhabitants had adopted the survey of Mr. Beach (Chief Engineer of the Morris Canal) with pleasing results. The city was said to be "very healthy; contagious or epidemical diseases are very seldom known there. The streets are wide, and run at right angles; the centre of the city being elevated, gives considerable descent each way to the river." That same year, 1830, Eliza and Ephraim gave a parcel of their Newark property to the Canal Company for use as a canal basin.

The year 1831 was apparently a very busy one for Beach. In May, while yet Chief Engineer of the Morris Canal, still struggling to make the canal ready for traffic from Phillipsburg to the Passaic River, he was hired by a group of Pennsylvania businessmen to prepare a survey. The Susquehanna and Delaware Canal and Railroad Company proposed to connect Pittston, on the Susquehanna River, with the Water Gap, on the Delaware River. Beach's report on the survey, dated December 31, 1831, indicated that a railroad 70.58 miles long would enable the businessmen to ship coal from Pittston to the Delaware River. From the Water Gap, another railroad could be made to Easton, from which point the Morris Canal could transport the coal to tidewater. Or a railroad from the Water Gap could be built to connect with the Canal at the summit level. In the event that the Orange and Sussex Canal were ever built, a railroad from the Water Gap to that canal would provide an outlet for Pittston coal at Newburg, N.Y., the proposed terminus of the Orange and Sussex.

To this writer, the most interesting thing about Beach's report to the Commissioners of the Susquehanna and Delaware Canal and Railroad Company was his proposal to use water-powered inclined planes to overcome the elevations between Pittston and the Delaware Water Gap. He was consistent. The plans were ambitious, reflecting both the experience and expertise of the engineer and the vision and ambition of the projectors. Many years later, one historian reviewed the early surveys that had been made for the Delaware, Lackawanna & Western (DL&W) Railroad, and noted that the line of that road's southern division was "much the same as that run by him [Beach] in 1831."

Beach could be found at work at still another locale in 1831. Remember the famous painting "Whistler's Mother"? Beach worked with Whistler's *uncle*, Captain William Gibbs McNeill of the Army's Corps of Engineers. Captain McNeill and his relationship to Ephraim Beach require and deserve a brief disgression.

The General Survey Act of 1824 had made it possible and legal for military engineers to participate in civilian attempts at internal improvements. President Monroe established a Board of Internal Improvements, consisting of General Bernard and Colonel Totten of the Army Corps of Engineers, and John Langdon Sullivan, a civilian civil engineer. The Board Members oversaw three topographical engineers: Major Abert and Captain McNeill, military engineers, and Daniel Shriver, a civil engineer. Each of these three, in turn, was assigned a team of five men, surveyors and assistants: military-officer teams for Abert and McNeill; a civilian team for Shriver. The engineers and their crews were to accompany the Board in its travels throughout the country as it tried to determine possible routes and probable costs of various roads, canals, and (later) railroads. One such engineering squad, McNeill's, was assigned to conduct a survey for a railroad to connect Catskill on the Hudson River with Canajoharie on the Erie Canal. When the group arrived in Catskill, the roster included the names of assistants Root, Fessenden, Johnson, White, Schenck and Beach, according to one source.

Just "how," or "why," or even "if," Beach became a subordinate to McNeill is a "history mystery," as yet unresolved by this writer. He had, after all, already achieved the status of chief engineer on both the Schuylkill and Morris Canals. An educated guess would place Beach in Catskill as an independent agent, a "consulting engineer" conducting an independent survey for yet another railroad that would have linked the Catskill and Canajoharie Railroad with one proposed to be built along the Susquehanna River Valley north to Lake Erie. He issued a report on that subject and its feasibility, which was reinforced by the confirming opinion by Benjamin Wright appended to his statement. While they were about it, both Beach and Wright wrote letters which added support to the Catskill and Canajoharie Railroad plan, their statements being published along with McNeill's report on his survey.

If Beach was not an independent agent at Catskill, as is proposed here, then a most interesting discrepancy arises. Beach is nowhere listed among the graduates of West Point and was not, therefore, a "military" engineer. Yet McNeill, according to the General Survey Act of 1824, would have had only military personnel under his direction for the Catskill and Canajoharie (C&C) Railroad survey. It is more likely that the independent survey Beach conducted for a road using the Susquehanna Valley and intersecting with the C&C Railroad had to do with the Pittston, Pennsylvania, group's plan to transport coal. This supposition is not so far-fetched. What the Delaware and Hudson Canal and railroad combination could do, the Susquehanna and Delaware could do, too. Even at this early date, engineers were thinking in terms of interconnecting transportation facilities, of systems of railroads, canals, and roads.

Captain McNeill was simultaneously engaged as "supervising engineer"for a survey being conducted for the Paterson and Hudson River (P&HR) Railroad, although his brother-in-law and first assistant, Lt. George Whistler, actually did the leg work and headed the survey party. Throughout 1831, work on the P&HR Railroad proceeded, and before the year was out, the Paterson company was authorized by a supplement to its charter to alter the easternmost part of its route where the road had to negotiate a solid traprock ridge at Weehawken. Originally slated to terminate at Hoboken to connect with the ferry service at that point, the Paterson Company opted for Jersey City instead. In an interesting bit of collusion yet to be explored by railroad historians, the Paterson Company threw in its lot with that of the New Jersey Railroad and Transportation's (NJRR&T) road which was trying to reach New York from New Brunswick.

The Paterson Junction Railroad, chartered February 3, 1831, only a few days after the Paterson and Hudson River road, was created to build a rail connection between the Morris Canal and the P&HR road at Paterson. What actually happened was a joint effort between the Paterson group and the NJRR&T Co. in building a junction which allowed both companies to use the same facility from Bergen across the trap rock ridge to Jersey City. The Morris Canal Company whole-heartedly subscribed to the idea and, it appears, ultimately to the stock. The Minute Books reveal that the Morris Canal's stockholders were to be given the privilege of buying one-half the stock of the NJRR&T Co. in exchange for the Canal Company's support on behalf of its application for a charter, pending in the Legislature. The NJRR&T Co. was, of course, trying to connect the Amboys, New Brunswick, Rahway, Elizabeth, and Newark with New York by means of rail and ferry. Beach served briefly as chief engineer for the New Jersey Railroad in 1833, working with McNeill and Whistler to bring the two railroads to tidewater, while he continued in his position as the Canal Company's chief engineer.

A Newark Directory continued to list Beach for the year 1835 as "Civil Engineer" with "offices at the Passaic River opposite the Lehigh Coal Docks." He was not to be so listed again for Newark, however, although he had been hard at work pushing for the completion of the canal's extension from Newark to Jersey City all through the early 1830's. Then, in November, 1835, he sold his remaining Newark properties and apparently left the employ of the Canal Company. The promoters of the Morris and Essex (M&E) Railroad hired him to build their railroad from Morristown to Orange. He also explored the possibility of a railroad from Morristown to Carpenter's Point, N.Y., a town no longer marked on modern maps, but which is known today as Port Jervis.

In any event, as of March 30, 1836, Beach's name no longer appears in the Canal Company's Minute Books as Chief Engineer. By that date, he had been replaced by the Canal's one-time Resident Engineer and Beach's subordinate, Roswell B. Mason.

At just about that point in time, 1836, Beach accepted the job of constructing the Catskill and Canajoharie Railroad. The sale of his Newark properties along the Passaic River had netted him the enormous sum of $75,000, and he appears to have removed his family to Catskill, where he invested his fortune in land there through his membership in the Catskill Land Association. He and Eliza joined the Christ's Presbyterian Church at Catskill on January 9, 1839, and although Ephraim's contractual association with the C&C Railroad ended by the close of 1840 when the road was completed — at least to Cookville — the family remained in Catskill until Ephraim's death in 1857.

The construction of the M&E Railroad, however, kept Beach professionally tied to Morris County. He was again employed to engineer the M&E's extension from Morristown to Dover, beginning construction there in April, 1846.

What did Ephraim Beach do in all of those intervening, unaccounted-for years? Both the Canal Company's Minute Books and local newspapers are silent on that subject except for references on August 24 and October 7, 1841, when Beach requested payment for "certain services rendered to this Company as witness" between the years 1832 and 1841. A third reference is on a deed which notes that on June 30, 1854, Ephraim Beach appeared in Essex County Court to swear that the signatures on that deed dated January 13, 1830, were genuine and had been given voluntarily.

Of the personal life of Ephraim Beach, little can be said because little documentation exists. The Beach family had grown since Widower Beach had been left with three infants after the death of his first wife, Ann. Eliza bore him six children between 1827 and 1839: James C., Anna M., Samuel Henry, Charles Edward, Robert J., and Fannie Caroline. Ephraim died in Catskill, New York on September 7, 1857 at the age of 73 years, 11 months and is buried in the Catskill Village Cemetery. Next to his name in the church records is simply the word: "Died." After Eliza's name is the entry: "Dismissed, August 11, 1857, to the Presbyterian Church, Bloomfield, N.J." Son James is known to have married Mary C. Butler on October 2, 1850, and to have built his home in Bloomfield shortly thereafter on — appropriately enough — Beach Street.

Hindsight, of course, is always gained with 20-20 vision, and conjecture must be considered in the light of what it really *is:* to wit, an *educated* guess, but still a guess. However, from fragments of his correspondence, from entries about him in the Canal Company's Minute Books, from his published professional reports, and from his statements in various contemporary newspapers, it seems justifiable to form certain reasonable conclusions and impressions about Ephraim Beach, the man and the engineer. He seems to have been a proud man, with a pride born of competence and success in his life's work. That he was competent is nowhere better seen than in the fact that the Morris Canal was built, and that the M&E Railroad was not only built but was to be the vital link in the steel chain that would one day become the DL&W Railroad, then the Erie Lackawanna, and today's Conrail. That he was successful can at least be inferred from the sum derived from the sale of his Newark properties, and from the fact that at one time he held 70 shares of Morris Canal stock at $100 a share, and may have acquired as many as 35 more shares by 1830.

That he was a man of intelligence and integrity can be seen in the way that Beach handled criticism of his work and his professional judgement. Pride made him answer his critics, but he couched his answers in the terms and tones of a reasonable and gentle man. His integrity and industriousness are best seen through an examination of his own words. Once, on June 8, 1834, when called upon to account for the progress then being made on the Canal's extension to Jersey City, Beach wrote to Secretary-Cashier of the Canal Company, Robert Gilchrist:

> *From your long acquaintance with me, you will know, sir, that on the question of performance, if it comes within the range of possibilities, the word cannot never belonged in my vocabulary. You also know that I have not been in the habit of promising what I could not perform . . . .*

According to Macculloch, Beach arrived in New Jersey for the Morris Canal surveys during July, 1823. At about the same time, Macculloch petitioned the War Department to have the Board of Engineers conduct a similar survey and submit a report to the three-man commission appointed by the New Jersey Legislature. In turn, the commission would submit the Army's findings to the New

Jersey Legislature, appended to its own report. Receiving their orders from the War Department on August 15, 1823, the Army Engineers began to compile information and form opinions as to the economic and physical feasibility of the projected Morris Canal.

Enter: The military, specifically the military engineers of the United States Military Academy at West Point, New York. The presence of Joseph G. Swift, one time Chief Engineer of the Academy, has already been mentioned in connection with one of Macculloch's preliminary surveys of the canal route. Swift had resigned from the Army in 1818, apparently jealous over the influence that "foreigners" had over the War Department. That "foreign influence" was to create a first-rate technical school out of the Military Academy.

Top flight professors had been imported from France's École Polytechnique to give much-needed instruction to aspiring native American military engineers. This was the "foreign influence" objected to by Swift, and probably others. Nevertheless, the French succeeded in educating the earliest West Point Cadets, and under the efficient administration of Superintendent Sylvanus Thayer, appointed in 1817, the USMA became the country's earliest school of technology and engineering. Training at West Point, however, had proved inadequate from a total-engineering standpoint. Experience on public works provided an invaluable education in civil engineering — a subject not yet taught there in 1823-4. The Morris Canal was one such public work which provided that kind of practical experience.

General Simon Bernard, one of the compilers of the Report of the Army Engineers on the Morris Canal, had been recommended to his post as Assistant Engineer of the Corps by no less than the Marquis de Lafayette and Albert Gallatin, Secretary of the Treasury. Lt. Col. Totten was a member of the Corps of Engineers whose special interest seems to have been coastal fortifications. Fort Totten guarding New York Harbor perpetuates his name.

Having been ordered to comment upon the merits of the Morris Canal, Bernard and Totten issued a lengthy and interesting document dated October 5, 1823. In the simplest terms, they thought, the purpose and object of the proposed canal was "to open a communication between the great beds of coal (anthracite), on the Lehigh River, and the iron works of New Jersey, the manufactures of Paterson, and the city of New York." They commended Renwick for having surveyed and located the canal, but they took some exception to the route he had mapped, and suggested that a better path for the waterway might be found from Dover eastward to Paterson.[44]

Noting the fact that 39 of Morris County's 93 forges were shut down because they had no fuel, and that imported Swedish iron costs were so high ($90 per ton, with 37,000 tons imported annually into New York), Bernard and Totten contended that all that was needed to bring down the cost of iron was a good supply of inexpensive mineral coal for the forges and a canal for the transportation of iron to New York Harbor. By such a canal, Paterson, then only 31 years old, would be enabled to expand its manufacturing operations beyond wool, cotton, and hemp, to include "articles of wrought metal, for which so much is now paid to foreign, to the detriment of native skill and industry." Such expansion would come as a result of cheap transportation, deliveries of ore and coal, and of course, communication with "the rich emporiums of New York." Moreover, the canal would afford additional water power for the city.[45]

Looking beyond the immediate local benefits to be derived from the proposed Morris Canal, as Macculloch did, the Army Engineers saw that the Lehigh River, already navigable to within nine miles of the extensive coal beds, could be easily connected with the Susquehanna River by a cut of about 12 miles. Thus, New York City would have a "direct communication" with "all the upper basins of the Delaware and Susquehanna Rivers; that is to say, with an extent of country more than 36,000 square miles."[46] This would have been the Nescopec Canal that Macculloch alluded to, a project that was under consideration at the time, but which eventually materialized in two quite different forms: the Lehigh and Susquehanna Railroad and the Nescopec Railroad.

Like Macculloch, the Army Engineers did not include the cost of land in their estimate of costs, since they "supposed that the owners, sensible of the advantages of having a canal at their doors, will demand little or nothing for it."[47]

Bernard and Totten thoroughly examined Renwick's novel introduction of the inclined plane for use on the Morris Canal. The fault of fixed locks, for so long the only known means for surmounting elevations on canals, had been corrected, they said. The corrections or "substitutes" were of two kinds: "first, inclined planes with parallel railways, on which the boats ascend and descend alternately; and secondly, vertical lifts, working in a similar manner," that is, gravity planes, where the preponderating weight is in the descending movable lock, causing machinery to turn to pull the ascending boat up. Mr. Renwick's, they identified as being of the first kind: "an inclined plane, on which are fixed two parallel sets of iron rail-ways . . . each set serving for a lock to ascend and descend." These types of devices used "the least possible water, since they contain no more than is sufficient to float the boat, which . . . is far from being the case with fixed locks."[48]

Where elevations permitted, however, fixed locks were recommended, and "a system of inclined planes combined with fixed locks, as proposed by Mr. Renwick" would be both preferable and desirable. Army Engineers Bernard and Totten recommended that Renwick's plan be sanctioned by experience. "Whether we consider the economy, the utility, or the durability of these inclined planes," they maintained, "ALL IS CERTAIN; and we look confidently forward to the day when their introduction will be regarded as a most important era in the history of canal navigation, and especially in this country, to which they are so peculiarly adapted."[49]

With such expressions of confidence and optimism, Macculloch might well have felt sure that the New Jersey Legislature would underwrite his dream canal. Yet he sought still another battery of expert opinions: that of DeWitt Clinton, President of the Board of Canal Commissioners of the State of New York, and of Judge Benjamin Wright, Chief Engineer of the Erie Canal. Visiting the line of the proposed Morris Canal with Macculloch and Beach, Clinton and Wright formed opinions that were later expressed in a letter from Clinton to the three Canal Commissioners of New Jersey.[50]

A canal between the Delaware and Passaic Rivers would benefit not only New Jersey, Clinton wrote, but would also be of mutual interest to New York and Pennsylvania. In words that have a peculiarly modern "Port-of-Authority" tone, Clinton reasoned that "the interests of one state has [sic] a beneficial influence on the prosperity of all." New York would, after all, gain an incalculable commercial advantage with the completion of the Morris Canal. The Erie Canal was a pipe line for the agricultural wealth of the Finger Lakes region and for the unknown products of the Middle West which was just opening up. The Morris Canal would be a funnel through which would pour the fuel to spark New York's industries, as well as all of the natural and manufactured products of New Jersey, Pennsylvania, even beyond.

As for the extremes of elevation that the Morris Canal must overcome, Clinton felt sure that the inclined plane, used in England and elsewhere on the continent, but modified for use on the American canals, would be quite serviceable. In fact, the only problem Clinton foresaw for the builders of the Morris Canal was money. Raising enough money could be more of an insurmountable barrier to success than the mountains that stood in the path of the proposed canal.

Clinton's letter discussed four aspects of the Morris Canal proposal: the physical practicability, the financial practicability, the inducements or motives for adopting the canal as proposed, and the agent best suited to building the canal. As to the physical practicability of such a canal, Clinton saw no impediment. There was a plentiful supply of water for canal operations, and even the extremes of altitude could be overcome by several series of locks, if inclined planes could not be made to work properly. In fact, he felt sure that "the practicability of the work" was "as certain as any future event can possibly be, whose accomplishment is not yet realized." Financially, too, the canal could be practicable if the money were borrowed on the credit of the state. The anticipated income of the canal would be sufficient to pay the interest and principal on the loan. Lotteries and banks, along with other "auxiliary expedients," would surely obviate the need for taxation.

The State of New Jersey, Clinton suggested, should be motivated to undertake the construction of the canal if for no other reason than the enormous potential to be realized through increased industrial activity, based on an abundance of cheap, available fuel. It was, he thought, humiliating to tour New Jersey and see "foreign iron worked by foreign coal" when the state's own mountains were full of ore, and the coal supply lay but a day's journey away.

In examining the subject of "who shall build?" Clinton touched on three possible agents: an individual, an incorporated company, or the state itself. His unequivocal conclusion was that the state exclusively ought to undertake the project. Of course, state backing was precisely what Macculloch had in mind; but first he had to demonstrate the practicability of his canal for the New Jersey Legislasture.

When all the reports were in, Macculloch, Kinsey, and Capner prepared their statement for presentation to the New Jersey Legislature. The Army Engineers had estimated that the canal would cost $1,148,103.65 if it were to be built using locks to overcome the elevations. Locks, they calculated, would cost $400 per foot of lift. A canal using locks and Renwick's planes, on the other hand, was figured at a cost of $824,263.65. Obviously, there would be a savings of $323,840.00 by adopting Renwick's plan.[51]

Ephraim Beach reported that after he'd completed surveying the Morris canal route, he had inspected the Lehigh Coal and Navigation Company's progress with both its slack water canal and its coal mining operations. His express purpose, however, according to the instructions given him by the Commissioners, had been "to ascertain the practicability of constructing an ascending and descending boat navigation to the coal region; the extent of the coal mines, and the probably benefits which might be derived from the use of Lehigh coal, for manufacturing purposes."

There was little doubt as to the benefits of using anthracite for domestic purposes. Nor was there reason to doubt that with but few changes of furnace design, iron could be made very successfully with Lehigh coal. But perhaps Beach's most interesting — and revealing — observation concerned the possibility of making still another canal. He had been authorized by the managers of the Lehigh Coal and Navigation Company to assure the three-man commission that "the distance from the Lehigh to the Susquehanna river at Wilkesbarre does not exceed twelve miles," and that a canal could be "constructed with comparatively small expense, and without fear of deficiency of water, which, connected by the Lehigh with New Jersey canal," would effect a complete navigation from Wilkesbarre to New York.[52]

Bearing in mind Macculloch's own thoughts on the subject of a *system* of canals which would make the Morris the equal of the Erie Canal,[53] let us examine a useful discussion by Totten and Bernard, who had also subscribed to the idea of a system of waterways. The Army Engineers distinguished between two classes of canals:

1. Those made with a view to the general interest of the country; the revenue being a secondary object.
2. Those in which the revenue is the principal object.

For canals of the first class, only the public treasury should undertake construction. Canals of the second class might be undertaken by either the government or private associations. As far as Bernard and Totten were concerned, the Morris Canal was definitely a "class one" canal.

The concept of a canal for the general good of the country versus one for private gain is stressed here only to reinforce Macculloch's stand, and to indicate the breadth and scope of his thinking. His was no narrow, limited vision for Morris County and New Jersey alone. A catalogue of the State's minerological wealth, as reported by Renwick and Charles Kinsey, seemed endless, but the iron ores alone would be sufficient reason to build the canal. The magnetic ores that could be mined along the route of the canal were equal in quality to the Swedish iron which was then being imported at $90 per ton at New York. The availability of cheap fuel delivered by cheap transportation and the shipment of finished iron by the same means would drop the price of iron to $55 per ton for comparable, if not superior, quality.

There is no question that Macculloch intended the State of New Jersey to build the Morris Canal — but he urged that it be done at no expense to the taxpayers! He described himself as "an enemy of taxation of every sort," and felt that the canal should be part of a "system of internal improvements, on the sole ground of its practicability."[55] There is no question, either, that Bernard and Totten recommended state responsibility for the canal's construction. And DeWitt Clinton claimed that that was about the only way to keep New Jersey from being "bound hand and foot by the shackles of a non-resident company."[56]

Macculloch, Kinsey, and Capner reported to the state's lawmakers and, as one author put it: "received the thanks of the Legislature; but the latter could not be induced to make it a State affair, and left it to private enterprise." Macculloch called the defeat the result of "local interests, jealousy, and most laudable dread of public debt."[57] By local interests and jealousies, Macculloch meant those lawmakers who couldn't see why *their* counties and constituents should be made to pay for canal that would not yield them any returns, since, after all, the waterway was not planned to penetrate their boundaries. He also meant the faction of lobbyists who were simultaneously trying to get a charter for a canal across the waist of New Jersey, from the Delaware River to the Raritan, even and perhaps particularly to the exclusion of the Macculloch proposal.

The 48th New Jersey Legislative session had rejected both northern and southern canal proposals because it feared wild speculation and external interference in the state's financial workings. But by the time the 49th Legislature convened, the opinions of the lawmakers had been modified by the extensive publicity campaign waged in local newspapers by both the proponents and opponents of the Morris and the Delaware and Raritan, and by the pressures exerted both by lobbyists and by their constituents. Instead of urging the State to undertake the building of those canals as public works, however, the legislators pushed for private charters with liberal provisions for banking privileges designed to attract private capital. In short, neither the Morris nor the Delaware and Raritan Canal won government support initially. Whatever comprehensive reason accounted for the defeat of both bills, the dread of public debt must indeed have been one of the weighty factors.

REFERENCES

1. *Biographical material for George P. McCulloch was excerpted from a variety of sources:* John Brett Langstaff, *New Jersey Generations: Macculloch Hall, Morristown* (N.Y.: 1964); Francis Bazley Lee, *Genealogical and Memorial History of the State of New Jersey*, Vol. 3 (N.Y.: 1910), pp. 932-3; Munsell's *History of Morris County, New Jersey: 1739-1882* (N.Y.: 1882); *Biographical Encyclopedia of New Jersey of the Nineteenth Century* (Philadelphia: 1877), p. 285.
2. Langstaff, *op. cit.*, p. 43.

3. David Hosack, *A Memorial of DeWitt Clinton* (N.Y.: 1829), pp. 299-342.
4. *Biographical Encyclopedia of New Jersey*, p. 286.
5. Munsell's *History of Morris County, New Jersey*, p. 69 (hereinafter cited as Munsell's *Morris*).
6. Morristown (N.J.) *Palladium of Liberty*, June 27, 1822 (hereinafter cited as *Palladium of Liberty*).
7. *Ibid.*, July 4, 1822.
8. *Ibid.*, July 8, 1822.
9. *Ibid.*, August 1, 1822.
10. *Ibid.*, August 8, 1822.
11. Munsell's *Morris*, pp. 26, 69, 74, 76, 77, 78, and Lucius Q. C. Elmer, *Reminiscences of the New Jersey Bench and Bar During More Than Half a Century* (Newark, N.J.: 1872) p. 450.
12. Munsell's *Morris*, pp. 29, 45, 55, 67, 75, 76, 79.
13. *Ibid.*, pp. 133, 147, 149, 162, and Harry B. Weiss and Grace M. Weiss, *The Early Lotteries of New Jersey* (Trenton, N.J.: 1966), pp. 132-3.
14. *Biographical Encyclopedia of New Jersey*, p. 28.
15. *Palladium of Liberty*, August 29, 1822.
16. James P. Snell, *History of Hunterdon and Somerset Counties, New Jersey* (Philadelphia: 1881), p. 207 (hereinafter cited as Snell's *Hunterdon*).
17. James P. Snell, *History of Sussex and Warren Counties, New Jersey* (Philadelphia: 1881), pp. 78-9 (hereinafter cited as Snell's *Sussex*).
18. Charles Boyer, *Early Forges and Furnaces of New Jersey* (Philadelphia: 1931), p. 148.
19. *Paterson Intelligencer*, October 19, 1825; Boyer, *op. cit.*, pp. 220-1, 469.
20. *Paterson Intelligencer*, September 14, 1825.
21. Rutgers Bible and Prayer Book, Morristown Public Library.
22. Charles A. Shriner, *Four Chapters of Paterson History* (Paterson, N.J.: 1919), p. 78.
23. L.R. Trumbull, *A History of Industrial Paterson* (Paterson, N.J.: 1882), p. 42.
24. Munsell's *Morris*, p. 67.
25. *Biographical Encyclopedia of New Jersey*, pp. 451-3; Snell's *Sussex*, pp. 79, 166-7.
26. Munsell's *Morris*, pp. 31, 67, 75, 76, 160, 274, 279; *Biographical Encyclopedia of New Jersey*, p. 555.
27. *Palladium of Liberty*, September 19, 1822.
28. *Ibid.*, September 26, 1822.
29. Munsell's *Morris*, pp. 46, 56, 218, 219, 223, 335.
30. "James Renwick," *Encyclopedia Britannica* (1960), Vol. 19, p. 156.
31. National Trust, *Preservation News*, May 1972, p. 6.
32. *Palladium of Liberty*, October 17, 1822.
33. *Ibid.*, September 26, 1822.
34. *Ibid.*
35. *Ibid.*
36. *Ibid.*

37. *Journals of the Legislature of New Jersey,* November 9, 1822.
38. Snell's *Hunterdon,* p. 328.
39. Munsell's *Morris,* p. 68.
40. *Report of the Commissioners Appointed by the Legislature of the State of New Jersey, For the Purpose of Exploring the Route of a Canal to Unite the River Delaware, Near Easton, with the Passaic, Near Newark, with Accompanying Documents* (Morristown, N.J.: 1823), Appendix D, p. 1 (hereinafter cited as *Report of the Commissioners,* 1823).
41. Munsell's *Morris,* p. 68.
42. Beach biographical material was excerpted from a wide variety of sources: *The Beach Family Magazine,* Vols. 1 and 2, 1926 and 1927; Morristown's First Presbyterian Church Records; Newark's Third Presbyterian Church records, and *Records of the Town of Newark;* Catskill, N.Y., Christ's (Presbyterian) Church records; Morris Canal and Banking Company Minute Books, files, deeds, correspondence, etc.; several newspapers; and William Helmer's *Rip Van Winkle Railroad* (Berkeley, Cal.: 1970), various pages.
43. Historical background on the USMA at West Point was gleaned from Cullom's *Biographical Register of the Officers and Graduates at West Point, New York, from Its Establishment in 1802 to 1890, with the Early History of the United States Military Academy.* (Boston, 1892); R. Ernest Dupuy's *Men of West Point: The First 150 Years of the USMA (N.Y.: 1951);* Forest G. Hill's *Roads, Rails and Waterways: The Army Engineers and Early Transportation* (Norman, Okla.: 1957).
44. *Report of the Commissioners, 1823,* "Report of the Army Corps of Engineers," pp. 6-7.
45. *Ibid.* pp. 8-10.
46. *Ibid.,* p. 10.
47. *Ibid.,* p. 23.
48. *Ibid.,* pp. 30, 36.
49. *Ibid.,* p. 43.
50. Hosack, *op. cit.,* pp. 209-215.
51. *Report of the Commissioners, 1823,* "Report of the Army Corps of Engineers," p. 36.
52. *Ibid.,* p. 48.
53. *Palladium of Liberty,* September 26, 1822.
54. *Report of the Commissioners, 1823,* "Report of the Army Corps of Engineers," p. 10.
55. *Palladium of Liberty,* April 1, 1824.
56. Hosack, *op. cit.,* p. 215.
57. Munsell's *Morris,* p. 68.

3

Certificate of Birth

THE newly formed federal government assumed the Revolutionary War debts of the several states, leaving them more or less able to cope with their post-war financial ills. Dunbar estimates that after its other necessary expenditures, the federal government was left with revenues of less than $22 million per year to pay all its operating expenses, thus making national projects such as interstate canal systems out of the question.[1] States less-developed and less-populated than New York and Pennsylvania were unable to undertake large-scale public works, even though out of debt. That some states, notably New York and Pennsylvania, did execute such plans is perhaps less a commentary on their farsightedness than on their fiscal prowess, more an index of their private money-hunger than of their public-spiritedness. That New Jersey was reluctant to thus burden herself is probably as much an indication of her laissez-fairism as it is of the widespread effectiveness of special-interest groups of private citizens called "societies."

By the turn of the eighteenth century, societies had become proper vehicles for the concerted action of citizens concerned for both public welfare and private gain. They were to vie even with governments in encouraging manufacturing and in stimulating the economy.[2] In

1789, Pennsylvania had a society for "Promoting the Improvement of Roads and Inland Navigation,"[3] and in 1799, a society for the "Promotion of Agriculture, Manufactures, and the Useful Arts,"[4] not unlike the Morris County Agricultural Society organized by Macculloch in 1819.[5]

The War of 1812 had virtually destroyed American commerce and reduced the importation of goods to a minimum. The Embargo and Non-Intercourse acts as well as the British blockade of the eastern seaboard had stopped the flow of imported manufactured goods, and also the cargoes of the coastal trading ships. Thus, manufacturing and the interstate transportation of manufactured goods became the paramount problems to beset the political and private populations. During this period, states, counties, municipalties, societies, politicians, and statesmen urged increased domestic production as a patriotic duty, and attractive bounties were offered as encouragement.[6] Significantly, societies were regarded as official groups.

From the private society to the government-chartered private company authorized to do its own banking is not a big step. When governments prove to be unable financially to undertake public works, two alternatives present themselves. Either private groups finance the public works, or the private groups advance money to the governments for those purposes, the advances being made in the form of investments or loans at interest. In England, at the end of the seventeenth century, a financially weak government had provided an opportunity for the privately owned Bank of England to arise, and with it the joint stock principle.[7] A group of financiers offered to lend the Crown more than a million pounds sterling in exchange for the right to establish a private, joint-stock bank.[8]

The real progenitor of the joint-stock principle, however, had been the medieval *societas,* which was but one of many ways that had been devised to circumvent the prohibition of usury (or interest) imposed by the Roman Catholic Church. The *societas* provided for mutual profit to be derived for a mutual investment by a multiplicity of investors. Since the partners of a *societas* shared the possibility of loss, they shared rights in whatever gains might be made on the initial sum invested. When extended business ventures (long-term investments) required more capital, more partners were sought to provide the needed funds. Profits were distributed according to the risk (amount invested) involved. The most significant ramification of the joint-stock principle was the divorcing of capital from management, making for a sort of absentee landlordism, wherein management of the common funds is delegated to a person or group of persons whose risk is small compared to the total investment.[9] Societies such as Hamilton's Society for the Establishment of Useful Manufactures (SUM)* were mutual or joint-stock companies, requiring the investment of funds by many partners, all sharing in the risk of loss or hope of gain according to the amount of their individual investments.

Many prominent persons of the times seem to have favored action through such groups. Robert Fulton urged the establishment of a society in each state to consider the propriety of canals.[10] Copies of his *Treatise on the Improvement of Canal Navigation* were sent to both Washington and Gallatin, and the latter is said to have incorporated much of Fulton's thought in his *Report on Roads and Canals.* Washington, of course, was one of the founders of the Cincinnati in 1783, a society devoted to "patriotic, beneficent, and social purposes." In New Jersey, the lives of many locally prominent men who were members of The Society of the Cincinnati were also interwoven with the fortunes of the Morris Canal. Among them were:

General John Doughty	Aaron Ogden
Theodore Frelinghuysen	Elias B. D. Ogden
Robert Gilchrist	Samuel Southard
Abraham Kinney	Richard Varick

Thus, in the young United States, when the embarassment of the federal government was not unlike that of seventeenth-century England's, and similar circumstances provided an opportunity for the joint-stock principles to arise, private societies flourished, and many of *those* became chartered companies.

Corporations in New Jersey were at that time chartered exclusively by the Legislature. Groups seeking charters had first to petition the

**New Jersey's first business corporation charter (1791); other charters preceding it had been for religious and educational groups.*

legislative body. Where public lands or rights were to be involved or infringed upon, it was incumbent upon the petitioners to convince the legislature of the public good or benefit to be derived from their proposal. One approach used to convince lawmakers, perhaps one of the least desirable, was lobbying. Shortly before the Morris Canal and Banking Company (MC&BCo.) received its charter, one newspaper wit reported from Trenton that ". . . the Wall street bees, attracted hither by the savor of New Jersey honey, are buzzing around the State-House . . . Many very splendid offers have been made for our canals, and bank charters, by these gentlemen, who profess a great regard for the welfare of the state; but which some of our plain unlettered Jerseymen uncharitably interpret to mean, a warm wish for the welfare of their own purses, and to make their fortunes out of the monopoly of our public favors if they can."[11]

Not all support for the Morris Canal came from New York lobbyists, of course. Reviewing the economic success and progress of rival canals in rival states, one July Fourth celebrant in Paterson was moved to toast:[12] "The Canal System, in full operation in the state of New York — May it admonish us of our duty in regard to the Morris Canal." Indeed, the still-unfinished Erie Canal was progressing so well that a steam boat had already been built at Utica "to ply on the canal between Rochester and Schenectedy for the conveyance of passengers." In Pennsylvania, coal was moving in arks down the Lehigh and Delaware Rivers in ever-increasing quantities. The Lehigh Canal was also tapping the Susquehanna valley region, so that the editor of Morristown's *Palladium of Liberty* was moved to write:[13]

> *Among the many visionary views of which we have been accused, is an opinion announced 18 months ago, that the River Lehigh would conduct the produce of the Susquehanna, in to the Morris Canal.*

Citing a Philadelphia paper that had reported on the arrival of goods from the Susquehanna area, the Morristown editor felt vindicated:

> *Thus the produce of the Susquehanna, falls into the Delaware, precisely at the mouth of the Canal, as we predicted, forming an*

immense addition to its tolls, and thus a revenue regarded hitherto as conjecture, becomes fixed and ascertained.

Finally, after arduous "campaigning," a charter was issued to the Morris Canal and Banking Company on December 31, 1824. The terms, limitations, and licenses built into the charter were to cause serious disagreement between Macculloch and the other members of the newborn company. Originally, according to Macculloch, the charter was to have provided for two state-appointed canal-company directors from each of five counties: Bergen, Essex, Morris, Sussex, and Warren*, all of which were touched or traversed by the canal. As it was issued, the charter named *as incorporators:* Jacob S. Thompson of Sussex; Silas Cook of Morris; John Dow of Essex; and Charles Board of Bergen. As commissioners and directors of the company, the charter specified: George P. Macculloch and John Scott of Morris; Israel Crane of Essex; and Joseph G. Swift, Henry Eckford, and David B. Ogden — all of New York City. The altered charter allowed for the election, rather than appointment, of board members by stockholders. Thus control passed into the hands of those owning the most stock.

In fact, the charter underwent such drastic changes in intent and implication that Macculloch was enraged. As he pointed out in a letter to Cadwallader D. Colden, a later president of the MC&BCo:[14]

It may be well here to remark that, anticipating the danger of throwing the whole concern into the control of mere foreign capitalists, the draft of a charter provided that a certain number of directors should be chosen resident in each county penetrated by the canal . . . Several gentlemen from Wall Street had volunteered their good offices and very kindly took post in the Trenton lobby after my departure. Upon their suggestion the draft of the charter was transformed into its present shape, nor did I receive the most distant hint of any alteration until the bill was finally passed. A company was formed and myself included in its direction. The precarious position of a canal coupled to a bank and directed by men of operations

**Warren County had been formed during 1824.*

exclusively financial was obvious. The interests of the country and the development of the iron manufacture were merged in a reckless stock speculation. I did all in my power to arrest this perversion, but soon found myself a mere cipher, standing alone, and responsible in public opinion for acts of extravagant folly, which I alone had strenuously opposed at the board of directors . . . .

I clung to the sinking ship until every hope of safety had vanished, and then vacated my seat by selling out, thus saving myself from ruin, if not from loss. From the moment the charter, altered without my knowledge, was obtained, the whole affair became a stock-jobbing concern, the canal a mere pretext; my efforts to recall the institution to its duty were regarded as an intrusion, and every pains was taken [*sic*] *to force me to retire.*

With New Jersey capital so scarce, Macculloch might have expected New York's moneymen to become heavy investors in the Canal Company. He directed his wrath, however, against the banking aspects of the charter, and against the known stock manipulators who were involved in the lobbying during his absence. In all fairness, it should be pointed out that the small capital reserve in New Jersey was held by small businessmen and farmers who were not, by nature, speculators. No wealthy merchant class had yet arisen in the state, nor had big businessmen or factory owners become a dominant factor in the state's economy. Indeed, one of the main purposes of the Morris Canal was to *create* capital and capitalists. Nowhere is that objective more clearly stated than in the preamble to the charter:

Whereas, the construction of a Canal, to unite the river Delaware, near Easton, with the tide waters of the Passaic, will be of great public benefit and advantages to the people of New Jersey; now, therefore, in order to secure to the state the results of this public work, and also as an inducement to a company to engage in this important undertaking, and in consideration of the risks and expenditures which they may encounter in its execution and operations; therefore — be it enacted by the Council and General Assembly of this State, and it is hereby enacted, by the authority of the same.

Thus, the purpose of this act of incorporation was to confer certain privileges upon those willing to take the risks of investment to build a canal.

Of the 28 sections of the act, four provided for the establishment, organization, and governance of the company; nine specified the functions and powers of both the canaling and banking aspects of the company; eleven placed restraints or limitations upon all of the company's powers and operations; one both limited and enabled the company's actions; two provided for the ultimate dissolution of the company and the disposition of its rights and properties; and one was "neutral," simply identifying the act as a "public law" which needed no special pleading in a court of law. The brief examination of the provisions of this charter that follows may help the reader to better understand Macculloch's dream, and perhaps his frustration. It will certainly help in understanding the rather ruthless methods often employed by the Canal Company in building its waterway, and it will clarify the ultimate problems of dissolving the company and destroying the canal itself.

Interestingly enough, the first Charter Supplement was passed immediately — on the very same day as the original act — December 31, 1824. Its provision was that $100,000 had to be spent on the canal itself before $100,000 worth of banking monies could be called in. However, the company had to submit proof that $100,000 had actually been spent on the canal by the end of one year in order to exercise that privilege. Failure to submit proof would result in the revocation of the banking privileges which had been conferred by the charter.

There were many supplements and amendments to the MC&BCo's charter. Since each reveals a serious defect in the original law and each affords an insight into the historical conditions out of which it arose, they will be dealt with as they occur throughout the canal's story.

The charter, as passed, covered the two aspects of MC&BCo's operations. As a *canal* company, it was empowered to raise money to finance construction of the waterway — $1,000,000's worth of $100 shares, with the right to sell $500,000's worth additional shares to be sold if and when construction progress justified and necessitated. As a *banking* company, it was authorized to issue its own currency, as were other banks at the time; in the form of bank notes, printed one-side only, in various denominations from one dollar to 2,000 dollars. It was also entitled to receive subscriptions for an additional $1 million in bank stock, although the Company's banking privileges were intended to remain in effect only until the end of 1855.

However, the most unusual aspect of the charter, by far, was its provision for a trust company. Colden later called it one of the "most extraordinary privileges" that had ever "before or since been granted on one company," probably never to be conferred again. He was referring to Section XXIV of the charter, which he saw as a charter in its own right, creating a third company. While he regarded the banking powers of the Morris Company a "valuable privilege," Colden felt that in view of the canal's chronic indebtedness, the power to execute trusts would be even more valuable, "particularly as the trust property is by the charter entirely separated from the bank and the canal, and is not liable for the debts of the company."[16]

For the time being, however, as of January 1, 1825, the most immediately valuable privilege the Canal Company had was to sell shares in its waterways, and *that* it proceeded to do.

Summary Analysis of the Morris Canal Company 1824 Charter

ESTABLISHMENT, ORGANIZATION, AND GOVERNANCE
Sections I, II, III, VIII

Section I	Named the incorporators, Thompson, Cook, Dow, and Board, and stated that as a Company they would be enabled to "purchase, receive in donation, possess, enjoy and retain, demise, grant, alien, and sell, all such lands, tenaments, hereditaments, water streams, and water privileges, rights, goods, chattels and effects of every description whatsoever, as may be necessary for carrying into effect any of the provisions of this act and under the restrictions and limitations therein mentioned . . . *Provided,* the same be not contrary to the constitution or laws of the United States of the state of New Jersey."
Section II	Named the commissioners and empowered them to receive subscriptions for the sum of one million dollars "to constitute the capital stock for the canal aforesaid, in shares of one hundred dollars, each; and that the said commissioners, or a majority of them, shall and do open books for that purpose, at such time and in such place or places within this state, as they shall designate by public ad-

Section II

vertisements to be previously inserted for at least three weeks in a public newspaper printed in Morristown, Newark, and the city of New York, respectively; and shall continue the same open until the said capital stock shall be subscribed, or at their discretion close the same, after they shall have remained open two days, and again open the same at other time or times, place or places, giving public notice thereof as aforesaid. And the sum of ten percent upon each share so subscribed shall be paid in specie, or in bills of banks which redeem their bills with specie, by each subscriber, at the time of subscription, to the said commissioners, or a majority of them; and the amount so received at the time of subscription, shall be the said commissioners or a majority of them, be deposited in such bank or banks, as the said commissioners or a majority of them, may elect in their joint names, there to remain until the same shall be by them paid over to the directors of the said company, to be appointed as hereinafter directed; that the said directors, when so appointed, shall have power, and they are hereby authorized, to call upon said subscribers for the payment of further installments, in such sum or sums, at such time or times, and under such forfeiture or forfeitures, as they may deem expedient, until the whole amount of the said shares so subscribed shall have been fully paid."

Section III	Provided for the management of the Company by 15 directors to be selected from the stockholders. From their own members, the directors were to elect a president by a simple majority vote. Each stockholder was to be entitled to one vote for every share of stock he held. Elections were to be held yearly on the first Monday of April, beginning in 1826. The commissioners named by the charter were to serve as directors until that first election.
Section VIII	Provided for the election of the Canal Company's personnel by the president and directors of the Company, or a majority of them. The president and directors were empowered to "charge and collect tolls and rates" for all cargo and passengers, but the charter set a limit upon the rates to be charged. Coal, lime, gypsum, iron, iron ore, and stone should "at no time" be shipped at a rate higher than 3 cents per ton per mile. This section further stated that if the Capital stock of the Company were to be increased, those already holding stock should be given preference in subscribing. Such an increase in capital was not to exceed $500,000.

FUNCTIONS AND POWERS
Sections IV, V, VI, VII, X, XI, XII, XIV, XXIV

Section IV	Specified that the Canal Company was to be exempt from taxation on any "estate or property . . . possessed, occupied and used by the said company for the actual and necessary purposes of said canal navigation."
Section V	Gave the Canal Company the authority to build its canal, but clearly specified that the artificial waterway must "connect the waters of the Delaware River, near Easton, with the tidewaters of the Passaic River, and *passing through the county of Morris."* (Italics mine) It is entirely possible that this precise provision for the canal's passage through Morris County is the basis for calling the waterway the "Morris," rather than the "Easton and Newark," or "the Delaware and Passaic" canal — which, in fact, it *was* called in one *Palladium of Liberty* advertisement. However, the name "Morris" has other connotations: Gouverneur Morris had been the first man to suggest the use of an inclined plane (for the Erie Canal) in America; Robert Morris had been an avid advocate of canals as internal improvements. Then, too, as "Paterson," the city, commemorated a New Jersey

Section V

Governor's name, "Morris" would have commemorated Lewis Morris, first New Jersey Governor.)

This section also spelled out the authority of the Company's president, directors, agents, engineers, superintendants, contractors, and "any other person or persons employed in the service of said company" to enter *all* lands to explore and survey the canal route, and to locate the "locks, works, devices, wharves, toll-houses, and offices necessary for the use of said canal," so long as they did no "unnecessary damage." Once a survey of the route had been deposited with New Jersey's Secretary of State, it would then become lawful for the persons named above to "enter upon, take possession of, and use, all and singular such lands, water, and streams, subject to such compensation to be made" as specified by the charter.

Section VI

Arranged for the acquisition of lands, waters, and streams to be used for the canal in the event that owners should *not* make a free gift of them to the Canal Company. Quite simply, the procedure was called "condemnation." If, as an owner, you didn't give lands or water freely to the Canal Company, or sell them at a price you and the Company considered fair, a State Supreme Court Justice was required to appoint three "discreet and dis-

Section VI

interested freeholders, chosen from some county or counties not intersected by the canal," and whose function it would be "to make a just and equitable estimate and appraisal of the value of the lands, and damages to the several owners . . . ." The Company was then to pay the appraised amounts. As a safeguard, the Justice and the appraisers, of course, could not be stockholders in the Canal Company. Another safeguard was that a Canal Company engineer had to affirm under oath that the properties in question were necessary "for the purpose of carrying into effect the object set forth" by the charter: i.e., the construction of the canal. Moreover, the Justice was required to certify the Canal Company's survey of necessary properties and file that survey with the clerk of the county in which the properties lay.

Once the appraised fee was paid by the Company to the owners, "immediately, thereupon, the estate, right, property and interest, in and to the premises so appropriated, described and appraised, shall be vested in the company, to be by them held so long as they shall be used for the purposes of said canal."

Section VII

Stated that it was to be lawful for the president, directors, superintendants, engineers, workmen, and laborers of the Canal Company, together with

Morris Canal and Banking Company.

Jersey City, March 29th, 1828.

Sir,

You are hereby required to take notice; That the President and Directors of the Morris Canal and Banking Company, will exhibit before Charles Ewing, Esquire, Chief Justice of the State of New-Jersey, at his Chambers in Trenton, on Tuesday the Twenty-ninth day of April next, at Ten O'clock in the forenoon of that day, the Surveys, Maps, and Field-Books, of the Lands, Waters, and Streams, required to be occupied by the said Company for the purpose of constructing, making, and executing the Canal from the waters of the Delaware River near Easton, to the tide waters of the Passaic River; And will also, at the same time and place, apply to the said Justice to appoint Three discreet, and disinterested Freeholders, to make a just and equitable estimate, and appraisal, of the Value of the Lands and Damages, to the several Owners, Proprietors, or parties interested in the premises so required for the purposes aforesaid. Pursuant to the Act, entitled An Act to Incorporate a Company to form an artificial Navigation between the Passaic and Delaware Rivers; passed at Trenton, the 31st. day of December, 1824.

By Order of the Board of Directors.

R. Gilchrist Secretary.

To Jacob De Mott

Notice regarding right-of-way, issued by the Morris Canal and Banking Company (BNK).

Section VII	their "carts, wagons, and other carriages, and with their beasts of draught and burden, and all necessary materials, tools and implements, to enter upon all lands contiguous or near to the route or tract of the canal . . ." Not only could they enter those lands, but they could appropriate and use whatever material they needed for the purposes of construction, maintaining, or repairing the canal and its works.
	The personnel so named were required only to do as little damage as possible, and repair breaches in any enclosure that may have been made. The owners or occupants of the lands thus entered were to be compensated a fair amount for whatever materials were taken or damage done "if claimed within twelve months"! Again, if the amount claimed was not mutually acceptable to both owners and Canal Company, the three appraisers were to be called upon to settle the dispute. The appraisal was to be considered "conclusive evidence of the amount of damages in any suit to be brought therefor . . . ."
Section X	Named all capital stock as "proper personal property," transferable on the company's books so long as all debts to the Company were "satisfied previously out of said stock" before

Section X the transfer was made, and before any dividends were paid to the shareholder so transferring.

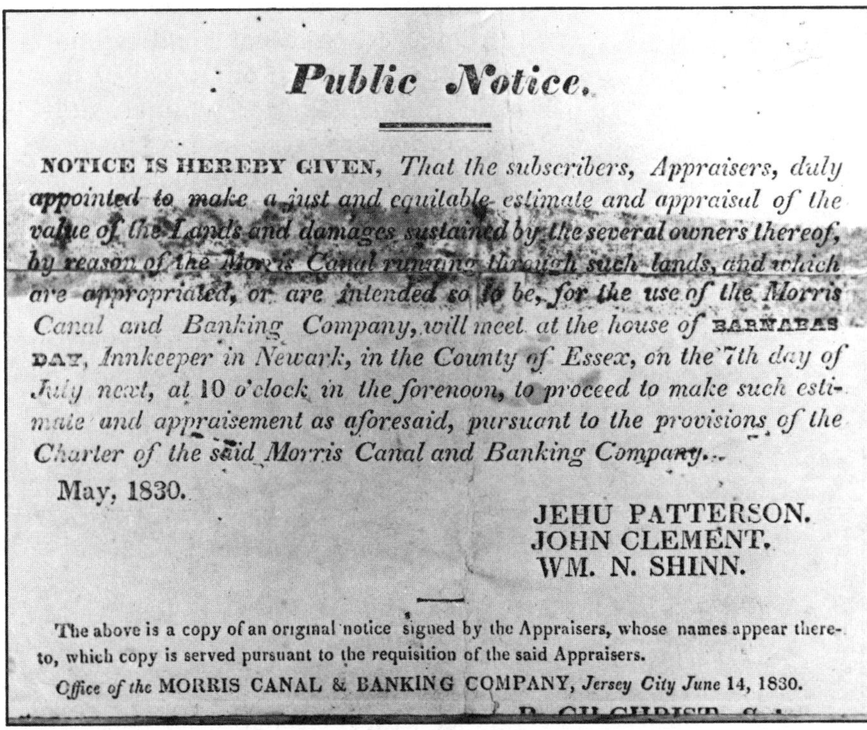

Public Notice.

NOTICE IS HEREBY GIVEN, *That the subscribers, Appraisers, duly appointed to make a just and equitable estimate and appraisal of the value of the Lands and damages sustained by the several owners thereof, by reason of the Morris Canal running through such lands, and which are appropriated, or are intended so to be, for the use of the Morris Canal and Banking Company, will meet at the house of* BARNABAS DAY, *Innkeeper in Newark, in the County of Essex, on the 7th day of July next, at 10 o'clock in the forenoon, to proceed to make such estimate and appraisement as aforesaid, pursuant to the provisions of the Charter of the said Morris Canal and Banking Company.*

May, 1830.

JEHU PATTERSON.
JOHN CLEMENT.
WM. N. SHINN.

The above is a copy of an original notice signed by the Appraisers, whose names appear thereto, which copy is served pursuant to the requisition of the said Appraisers.

Office of the MORRIS CANAL & BANKING COMPANY, *Jersey City June* 14, 1830.

Public Notice regarding estimates and appraisal of damages issued by the Morris Canal and Banking Company (Courtesy of Mr. and Mrs. Richmond Chipman).

Section XI	Stated that canal route, as submitted to the Legislature to get a charter for the Canal Company, was to be flexible. The Company was empowered to alter "at any time, any part or parts of the route of the canal," so long as compensation was made to land owners for damages. No canal could be built "within ten miles of any point of the main trunk of the Morris Canal, without the consent of the said company," but a canal less than 10 miles in length could be built so long as it did not come within two miles of the canal. This section provided for the damming of Green Pond as well as Lake Hopatcong to ensure a water supply for the canal's operation. Property damaged or flowed was to be compensated for by the Canal Company at an amount agreeable to the owner of the land, and in accordance with the provisions of the charter.
Section XIII	Specified that no person was to tamper with the canal or its "parts or works" on threat of a $25.00 fine, or two-year maximum jail term or both. Moreover the Canal Company still had the right to sue for damages in a civil suit in spite of the criminal prosecution called for by the charter.

Section XIV	Empowered the Canal Company to increase its capital stock, within certain limitations, for its banking operations. The sum of $200,000 had to have been spent *first* on the *canal,* its lands, waters, materials, and construction before the president and directors could call from the company's stockholders the sum of $200,000 (or less) to be used for banking purposes. In fact, for every $200,000 spent on the canal, $200,000 could be added to the banking capital of the Company, until the maximum sum of $1,000,000 had been reached.
Section XXIV	Specified that banking capital could be used in "buying and selling bills of exchange, or the stock of any body or bodies corporate and politic, and in loans upon bonds and mortgages . . ." The Company might act as trustee for both real and personal property left or conveyed to it "solely for the purpose of the protection of property so conveyed to them," of course, and "not for the purpose of aiding or promoting any traffic in any kind of stocks, or in any kind of goods, wares and merchandise . . . ."

RESTRAINTS AND LIMITATIONS
Sections IX, XII, XV, XVI, XVII, XVIII, XIX, XX, XXI XXV, XXIII, XXVII

(The Legislature was well aware of the power it was placing in the hands of a very few persons. The right to take lands by condemnation is, in legal theory, the prerogative of the state. By delegating that right and authority to a private corporation, the legislators had transferred the right of eminent domain to this newly created agent, duly authorizing it to act on behalf of the state. To ensure that the Company did not overstep the privileges and rights thus conferred, the Legislature placed equally heavy restrictions both on its activities and on the banking aspects of those activities.)

Financial Aspects

Section IX	Required the Canal Company to report once a year to the general meeting of the stockholders on the affairs of the Company. At that time, the dividends to be paid were to be declared.
Section XV	Specified a penalty for false oaths and/or statements made by the president or cashier of the company. If the banking capital were increased after the required amount of money had been spent on the canal, an annual financial statement had to be sworn to and sumitted to the Secretary of State.
Section XVI	Called for, in the event of failure to make an annual statement, the for-

Section XVI	feiture of the Company's banking privileges until the proper accounting had been made. This provided a necessary safeguard for Sections XV and XVII.
Section XVII	Specified that a willful false statement of the Company's financial status and operation would be considered perjury, subject to indictment, trial, and conviction.
Section XVIII	Covered reporting of banking operations conducted in Jersey City to the Secretary of State once every two years. Failure to make an accounting of the canal company's stock, specie, notes in circulation, and debts both due and owing would result in a suspension of the banking operations by legislative action.
Section XIX	Restricted the use of Company funds to those provided for by the charter, never to be employed in "any other operations, nor for any other purpose."
Section XXI	Provided that the State should derive some direct benefit from the canal company's banking privileges. Eleven years after the passage of the charter (December 31, 1835), the Canal Company was to pay to the State a bonus of one-half of one percent on its banking capital.

Banking house of the Morris Canal and Banking Company, Green Street, Jersey City (JCPL).

Section XXIII	Stated that if a subscription of $1,000,000 had not been received by November 1, 1835, or if the canal were not completed and operating within 15 years (by December 31, 1839), the charter itself would become null and void.
Section XXVII	Forbade the Canal Company to "take or appropriate to the use of said canal, or under color or pretence that the same are necessary therefor, any lands, waters, or streams of water, but such as are actually necessary for the erection and use of said canal, for the purposes of navigation only, and its necessary towing paths and works . . ." Under no circumstances was the Company authorized to use or appropriate water "for mills, or any other purpose than the mere navigation of said canal. . . ." Moreover, any waters needed for navigation had to be allowed to flow uninterrupted "into their ancient and natural courses and channels . . . . (This adamant statement seems to have expressed the fears of canal opponents, manufacturers, riparian holders, etc. In effect, it appears to have put an unnecessarily severe limitation on the right of the company to construct reservoirs. As we will see, the charter had to be amended to empower the company to build holding basins and to appropriate the waters with which to fill them.)

Canalling Aspects

Section XII	Stated that the Canal Company was to build and maintain "good and sufficient bridges" across the canal.
Section XX	Guaranteed the right of any person to take action against the Canal Company for damages resulting from the erection of the canal in cases of disputes between property owners and the Company over damages. This applied in the event that the damage was sustained by a person with whom the Canal Company had not already reached an agreement. (This section, along with XII, reflected the fears of at least some legislators, or perhaps lobbyists, that the rights of individual property owners were going to be completely submerged as a result of this unusually strong charter.)
Section XXV	Dedicated the canal as a public highway, open to the transportation of anyone's goods provided the users or shippers observed the Company's rules and regulations, and, of course, paid the tolls, the rates of which the Legislature reserved the right in establishing. It further provided for the State's appointment of three commissioners to work with Canal Company appointees to determine these tolls.

DISPOSITION OF COMPANY RIGHTS AND PROPERTIES
Sections XXII, XXVI

Section XXII	Covered the disposition of rights and properties, and stated that banking privileges might be exercised for a period of 31 years only, and would expire by law after December 31, 1855.
Section XXVI	Specified that at the end of 99 years the State might choose to take possession of the canal and all of its appurtenances, at a price to be determined mutually by agents of the Canal Company and the appointed agents of the state. The state, then, was to have one year, from January 1, 1924, through December 31, 1924, to decide whether or not it would buy back its rights. Should the state decide not to do so, the charter was to remain in effect for another 50 years, so that on December 31, 1974, with the expiration of the charter, the canal would cease to exist and all of its rights and properties would simply revert to the State.

LEGAL STATUS
Section XXVIII

Section XXVIII	The last section of the charter simply declared that the charter was a public act requiring no special pleading in a court of law.

REFERENCES

1. Seymour Dunbar, *A History of Travel in America* (N.Y.: 1968), pp. 714-15.
2. John D. Hicks, *The Federal Union* (N.Y.: 1937), p. 322.
3. Dunbar, *op. cit.,* p. 784.
4. Frederick B. Tolles, *George Logan of Philadelphia* (N.Y.: 1953), pp. 211-12.
5. John Brett Langstaff, *New Jersey Generations: Macculloch Hall, Morristown* (N.Y.: 1964), p. 42.
6. Hicks, *op. cit.,* p. 322.
7. "Banking," *Encyclopedia Britannica* (1960), Vol. 3, p. 42.
8. E. A. J. Johnson and Herman E. Krooss, *The Origins and Development of the American Economy* (N.Y.: 1953), p. 55.
9. *Ibid.,* pp. 42-3.
10. Cadwallader D. Colden, *The Life of Robert Fulton* (N.Y.: 1817), p. 334.
11. Trenton *True American,* December 18, 1824.
12. Morristown *Palladium of Liberty,* July 15, 1824.
13. *Ibid.*
14. Munsell's *History of Morris County, New Jersey 1739-1832* (N.Y.: 1882), p. 69.
15. *Laws of New Jersey,* 1824.
16. *Report to the Stockholders of the Morris Canal and Banking Company,* August 23, 1832, reprinted in the *American Railroad Journal,* September 29, 1832, p. 627.

4

Men To Match His Mountains

IN the Spring of 1825, the MC&BCo. stock was made available to investors and speculators alike. On March 16, newspaper reports painted a picture of phrenetic eagerness at Hayden's Tavern in Morristown, where potential buyers with bulging wallets must have been queued up for hours before the subscription books were opened. Newark's *Sentinel of Freedom* on March 22 observed that "multitudes" of such investors poured into Morristown to indulge in an orgy of $10-down, $10-installment purchases of stock. Paterson's *Intelligencer* (January 14, 1829) recalled that ". . . New Jersey subscribers were indignantly thrust aside," on that occasion, in favor of foreign investors from across the Hudson. Morristown's *Palladium of Liberty* (March 17, 1829) was doubtless a bit embarrassed to have to report an oversight on the part of the Canal Company. The charter (Section II) demanded that advertisements be run for "at least three weeks in a public newspaper printed in Morristown, Newark, and the city of New York" before stock subscriptions could be received. Since that requirement had not been met, the Canal Company was forced to close its books and rerun the advertisements before the sale of stock might recommence. When the subscription books

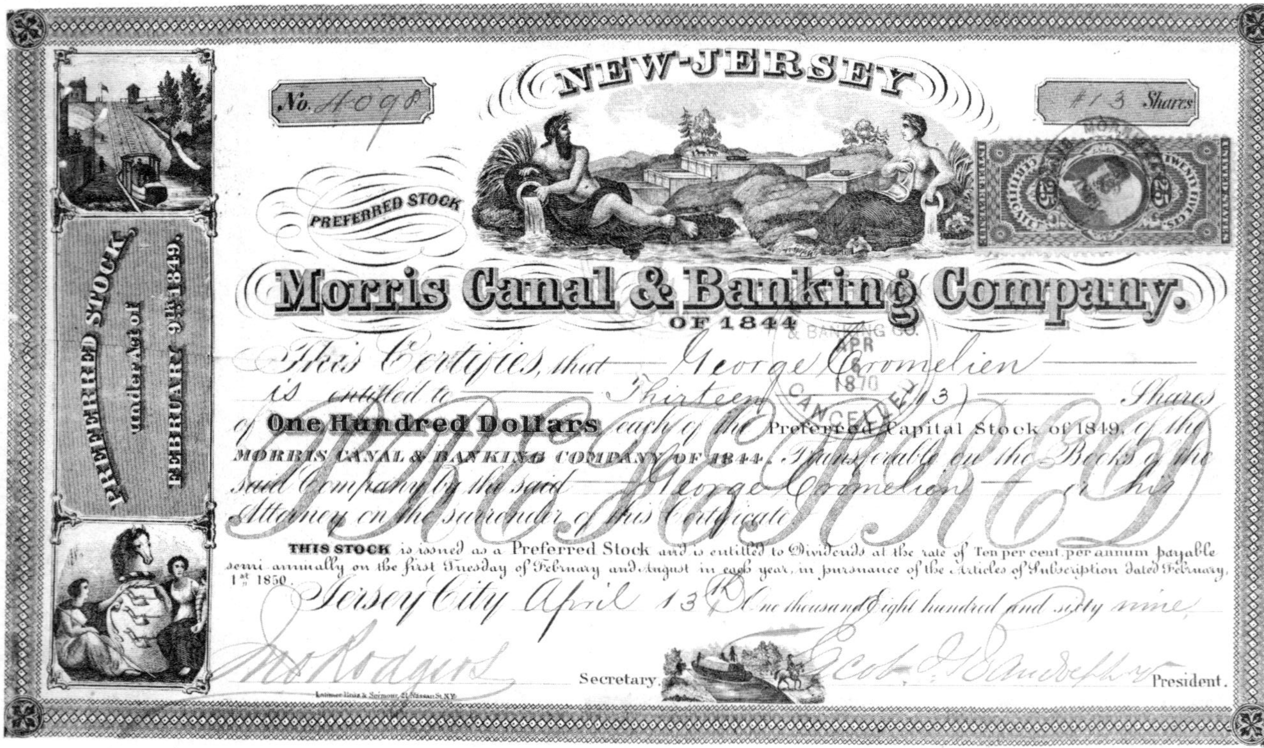

Stock certificate of the Morris Canal and Banking Company. Note the stock Transfer Tax stamp in the upper right corner, and an early version of an inclined plane in the left corner (BNK).

re-opened on April 26, however, subscriptions were received at the company's Jersey City banking house offices.

The oversubscription to Morris Canal stock is legendary. The charter authorized the issuing of stock worth one million dollars at $100 per share; seven million dollars' worth was subscribed during the frenzied buying at Jersey City. As we have seen, however, the buying was marginal — an initial $10 down payment with $10 installments to be paid in at intervals to be determined by the company.

Despite the frenzy and the confusion attending the initial sale of stocks, enough cash was in hand by July, 1825, to warrant beginning construction. By September 14, the *Paterson Intelligencer* was reporting that the men already at work on the canal were being paid with bills of the MC&BCo. By October 5, the same newspaper estimated that some 700 men were employed on the canal, and that excavation at Brooklyn Pond (Lake Hopatcong) would be "mostly done during the winter."[1]

Alvin Harlow remarked in his *Old Towpaths* that ever since the dedication of the Erie Canal had taken place on July 4, 1817, it had become "almost an ironclad custom to begin internal improvements and especially canal jobs on the nation's birthday. The canal project which did not turn its first earth on the Fourth was considered to have made a bad start."[2] Morris Canal construction was *not* officially started on July 4, 1825, and we, some 150 years later, with perfect 20-20 hindsight vision might be inclined to say: "Ah hah! *That* explains it!" Work did begin in mid-July, 1825.[3] However inauspicious the date, the official dedication ceremonies for the Morris Canal took place on October 15, 1825. A crowd of some 800 invited guests were present for the groundbreaking at the outlet of Lake Hopatcong into the Feeder canal. New York and Pennsylvania guests stood shoulder-to-shoulder with Jerseymen to hear Canal Company President William Bayard address them on behalf of the directors:[4]

> *The mines of New Jersey; the value of the coal of Pennsylvania to work these mines, and for fuel; the easy means of transporting the products of your farms to market; the importance of supplying our country with*

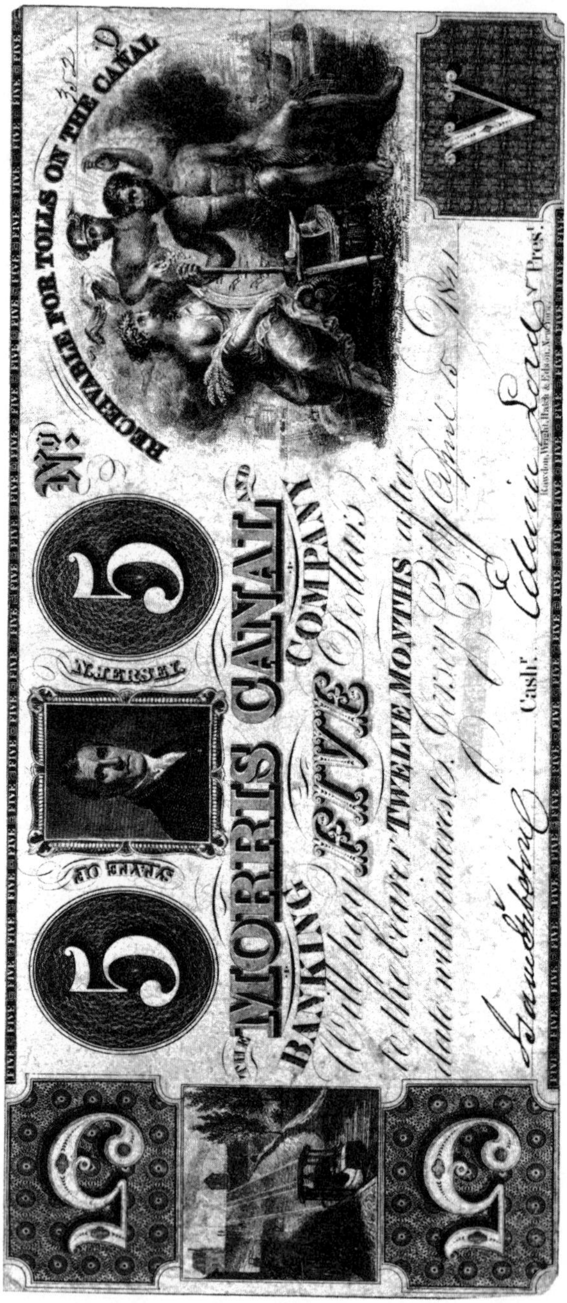

Five-dollar note of the Morris Canal and Banking Company (BNK).

staple articles from our own resources, rather than depend on Europe for iron and for coal, induced a number of our friends in this state to ask a charter of the New-Jersey legislature, to construct this Canal through the mining and farming regions of Sussex and Morris counties, uniting the Delaware river at the mouth of the Lehigh, with the waters of New-York harbor.

The legislature of this state has liberally granted us this Canal charter. This Lake which now lies before us, is found to be a principal and ample source to feed this Canal with water on its slope to the Delaware and to the Passaic; in order to facilitate the construction of the Canal, Bank privileges were granted. Some have supposed that Banking would be pursued and the Canal neglected. Not so, my fellow citizens — it has been determined since the day the stock was subscribed, to prosecute the Canal to its completion as our first and greatest object; and as an evidence of that fact, the excavation of thirty odd miles, nearly one half of its whole length, has been already contracted for and commenced — part of which is finished, and we have no doubt the whole will be completed, and that boats will pass from the Delaware river through this County to New-York, in less than three years.

We believe that a more proper occasion could not be found to express our mutual congratulations for these important privileges, than the commencement of cutting the feeder from this Lake, to supply the water to our Canal — for that ceremony we are here assembled; and fellow citizens, in the name of the Company, I congratulate you; I shall now proceed to commence this important work, which will be prosecuted by hands more able than mine to complete it.

With the digging of the first spadeful of earth, refreshments were served by Mr. Lyon of Jersey City, and the inevitable toasts were proposed:[5]

By the President — *The Legislature of the State of New Jersey.*
By Col. Ogden. — *The Alchymists of New-York, who extract gold from the iron mines of New Jersey.*

By Hon. M. Dickerson. — *The state of New-York; rich and munificent; first in public improvements; may her example be followed by other states, as fast as their means will permit.*

By Mr. Eckford. — *The President of the United States.*

By Gen. Smith of Sussex. — *The President and Directors of the Morris Canal — their spirited commencement assures us of a speedy completion of this important work, and it will ensure to them the gratitude of the state.*

By Mr. M'Farlan. — *Capt. Beach, the Engineer of the Morris Canal; a man of talents and industry, and a gentleman.*

By Mr. Russell of Morris. — *The memory of the Duke of Bridgewater, the father of Inland Navigation in the land of our ancestors.*

By David B. Ogden, Esq. — *The mines of New-Jersey, they will one day make her importance known to the Union.*

By Joseph Scott of Hanover. — *Health to the Liberator of the water power of North America, DeWitt Clinton; his essay will perpetuate our union and prosperity.*

By Mr. Cheesbrough. — *Lake Hopatcong now before us; its arms embrace the Delaware and the Passaic.*

By Mr. M'Culloch. — *The capitalists, present and absent, of New-York — May the success of the present enterprize induce them to embark in future plans for the prosperity of New-Jersey.*

Macculloch's enthusiastic toast to the New York capitalists is difficult to reconcile with the statements he made later about those "foreigners." Yet, there he was at Lake Hopatcong with many such foreigners from Pennsylvania and several New Yorkers, many of the latter having been elected directors of the Canal Company, and one of whom had been elected first president of the MC&BCo.

William Bayard, one-time president of the Bank of America and of New York City's Chamber of Commerce, was known as one of the nation's wealthiest men. Together with Herman Le Roy and James McEvers, his cousin, he headed the very successful shipping firm, Bayard & Le Roy, in New York. During the War of 1812, the firm had run a fleet of privateers that successfully eluded the British blockade — a very profitable enterprise. At the close of that war, Bayard was an ardent promoter of the Erie Canal project being revived by the New York State Legislature. He used his extensive influence to try to procure from Amsterdam a loan for the Erie Canal's construction.

Bayard's sons, also active in their father's business, gained a measure of notoriety for jerry-building two frigates for the Greeks during their war for independence from Turkey in 1825. Bayard Senior was also to fall under the shadow of suspicion when news of a Canal Company stock scandal broke shortly after ground was broken at the Lake Hopatcong ceremonies.

By the end of 1825, contracts had been let to construct the western portion of the canal from a point about 7 miles east of the Delaware River to the summit level at Hopatcong, and from there east to the town of Rockaway, where the first experimental plane was being built. Both eastern and western termini were left undetermined for the moment until the vigorous and vocal controversy over final locations could be settled.

It is important to keep in mind the many changes that had affected Macculloch's original route for the canal. The November 15, 1822 Act of the Legislature that had authorized a survey of the canal's route was based on the assumption that the canal would "connect the tide waters of the Passaic with the waters of the Delaware river, at or near the mouth of the Musconetcong river."[6] There seemed to be little doubt by 1825, however, that the canal would end farther north, at Phillipsburg, opposite the mouth of the Lehigh River, the projected source of all the Morris Canal's coal cargoes. But the final location of the terminal was contingent upon decisions yet to be made by the Pennsylvania Legislature regarding the Delaware River and the Delaware Division Canal. Meanwhile, the route of the Morris through its eastern division, and its tide water terminal, were far less certain, far more controversial.

Wheaton Lane makes a case for the glacial terminal moraine as a "natural" path across northern New Jersey. Just as the Erie Canal pushed westward along the natural gap through the Appalachians (the Mohawk Valley), so the Morris Canal would push along its "obvious course of transit," the terminal moraine, which had been "followed in part by Indian trails, highways . . . , and the Morris and Essex Railroad."[7] However, the actual choice of a canal route was far from being that simple or obvious for the engineers whose job it was not only to locate a practicable route for a canal, but also to satisfy the greatest number of potential future users of that canal.

As already noted, the earliest suggested routes for the Morris Canal were as varied as they were numerous. To settle the question of location for the eastern portion of the canal, the company invited engineers Benjamin Wright and Canvass White to lend the weight of their authority and experience to the task. Before 1825 was over, the *Paterson Intelligencer* expressed the hope that those noted, skilled and supposedly objective engineers would present their findings so as to "produce satisfaction to the public, and conduce to the best interests of the Company."[8] Unfortunately, Wright and White were not able to attend the task and unidentified "others" were assigned to complete it. The same editor again expressed the hope that those "others" would be able to perform their duties "faithfully and accurately," as well as satisfactorily, to all parties. Moreover, the editor inferred, these all-important surveys should be conducted by men of the "highest skill and strictest candor," so that the route ultimately decided upon would be chosen according to "clean" evidence, and not on the personal bias of the surveyors.[9] One such "other" surveyor was most likely Ephraim Beach.

During 1825, Ephraim Beach had been appointed Chief Engineer of the Morris Canal. During that year, too, the controversy over the easternmost terminus of the canal was nearing its peak. Without altogether revealing his "personal bias," Beach confined his attention to determining the path the canal was to follow. Requested by the board of directors of the Canal Company to give a professional opinion as to the best route, Beach summarized his views and findings in a report that is, unfortunately, undated.[10] An educated guess, however, makes the time of its writing toward the end of 1825, since a December 20, 1825, issue of the *Paterson Intelligencer* mentions that the route had already been decided upon.

Four possible paths for the canal were described by Beach in his report. All of them began at Rockaway, site of the first experimental inclined plane, then proceeded to Denville, where they diverged. The time has arrived, Beach wrote, "when it becomes necessary for the advantageous prosecution of the construction of the Morris Canal [that] a decision should be made of the route from where it is at present located to its eastern termination.

"A partial survey of several routes has been made from Rockaway to tidewater, the levels carefully taken, and the most difficult and objectionable points of each examined, so that the nearest practicable estimate of the expense of each might be made."

For the benefit of the Directors, Beach divided his report into four segments covering:

1. A description of each route
2. The probable expense of each route
3. The facilities of supplying each route with water
4. His opinion as to which route should be adopted.

The *Southern, or Morristown-Elizabeth, Route* could "be excavated from Rockaway through Denville on tolerably good ground to the Whippany River near Morristown," the fall being "overcome by three inclined planes and one or two intervening locks." The Whippany River would be crossed by culvert or aqueduct and an eighteen-foot high embankment "near the site of the old Powder Mill." Following that level past Charles Ford's, the canal would cross the turnpike west of Bottle Hill (Madison) and approach the Passaic River "upon the brow of a high bank near Jonathan Crane Bonnel's saw mill." That level could be kept upon an "embankment one hundred feet high, and an excavation of about one hundred feet depth, through Short Hills, near Bishop Hobart's."

Or, the Passaic River could be approached on the west side by means of a 100-foot-high inclined plane, the river itself being crossed by "a comparatively small embankment and a tunnel made through Short Hills about 180 feet below the summit." From that level, "the soil is generally favorable for a Canal, through Springfield, Connecticut farms (Union) and Elizabethtown to the point where there is a favorable site for a basin." The fall from the summit, he felt, "may be principally overcome by inclined planes," although "a few locks," would be "indispensable."

The *First Middle Route* would take the canal from Rockaway to Denville, then follow the same direction as the Southern Route until it passed Hinchman's mill, "from thence taking a more northerly direction through Little Town [Littleton] and Troy." Then, crossing the "peat meadows near Tobias Boudinots Esq [sic] and the Rockaway River near where the turnpike crosses it," the route would "pursue the level to the Passaic at Pine Brook," crossing the "lower lands and that river by the side of the turnpike road." Pursuing the line of the big ditch from that point, the canal would then proceed "nearly to Little Falls," passing that town "between Messrs Millers dwellings and mills." The "Peckmans brook" would be crossed above Jacobus' Tavern, and the canal would follow that level to a point "opposite Paterson." From that point, the country was "favorable for the location of a Canal terminating either at Aquaknok [sic], Bellville [sic], or Newark."

These two distinctly different routes thus far described by Beach diverge in their points of termination as well as in their paths. The Southern Route favored Elizabethtown; the First Middle Route, Passaic, Belleville, or Newark.

The *Second Middle Route* differed from the First only in its path. From Rockaway and Denville, the waterway would follow the Rockaway River valley "to Capt. Scott's Forge," near Boonton Falls, crossing the Rockaway at his pond, then passing down the river's north side "through Montville," intersecting the First Middle Route "about one mile west of Pine Brook bridge."

The *Northern, or Pompton Route* would diverge from the Second Middle Route "about a mile below Boonton falls [sic]," then run "northeasterly to Pompton plains [sic], then along the margin of the plains at the foot of a ridge" until it approached the Pompton River "near George Ryerson's Tavern." After crossing the Pompton River there, it was to pass through "Singack to Little Falls," cross the Passaic "below Esq. Millers factory," where it would intersect either of the two Middle Routes.

Beach's cost estimates for the routes follow:

Southern Route from Rockaway; 33 miles, $380,110

Middle Route through Troy to Newark: 37 miles, $223,810

Middle Route by Boonton and Pine Brook to Newark: 37 miles, $205,035

Northern Route by Boonton, Pompton Plains to Newark: 37 miles, $205,211

Canal extension from Newark using the Northern Route: additional $36,600, $241,811 total.

Before giving his opinion as to best route, Beach examined the problem of water supply for each of the proposals.

The Southern, or Morristown-Elizabeth, Route would have to take water from the Rockaway River above Denville. Moreover, the Whippany River might be taken in a feeder below the Speedwell works near Morristown, but Beach thought that since that river was already extensively used for manufacturing, suits for damages would be great. The same situation held for waters that might be taken at Springfield.

The Middle Routes would also take water from the Rockaway River above Denville; then, near the peat meadows at Troy, they could take that river itself through construction of a 3-mile-long feeder from Boonton Falls. Should the Rockaway be diverted from its course to supply a canal that followed either the Southern or Middle Route, however, the nature of the porous soil would depreciate the water in that stream at Boonton and at other works below, thus precipitating heavy property damages.

The Northern, or Boonton-Pompton, Route could take water from the head of Boonton Falls. A navigable feeder could be "constructed with comparitively [sic] small expense more than 20 miles up . . . [the Pompton] river, by means of which the waters of that stream and the business of the country may be brought into the Morris Canal." Moreover, by using the Pompton Route, a canal could also be "excavated at a moderate expense from Chatham" to intersect the Northern Route at Little Falls." If that were done, Beach felt, "a navigation would be made from that place [Chatham] to the Morris Canal and probably some thousand acres of land reclaimed," which he described as then "rendered useless by inundation."

In approaching the fourth segment of his report to the Directors, i.e., his professional opinion as to the "proper route" for the canal, Beach carefully explained his position. Each of the several proposed routes had its supporters, since the ultimate eastern termination of the canal was involved. Having witnessed "the excitement manifested in those interested on [sic] each route," Beach had wished "to avoid volunteering an opinion on the subject." However, he'd been called upon by his employers to so volunteer, and this he did "cheerfully":

Not being sufficiently acquainted with the products of the country through which the several routes pass, I am not in possession of data from which to form any consistent calculations of the amount of business which might be offered on either route. You are therefore refered [*sic*] *for the claims of Morris and Elizabeth routes to George P. Macculloch, Esq. report on that subject.*

The Middle Route, Beach felt, would partake "a very considerable share of the merits of both" the others. However, the Northern (Boonton and Pompton) Route, would enable the development of a water power of "about 300 feet in the space of 3 miles in the Boonton area. From that point to Little Falls, the country traversed by the canal was not only "susceptible of improvement," but also "but little inferior to any neighboring district" of comparable improvement (or development). As for the district from Little Falls to Newark, the district was "unrivaled in this country for its improvements in agriculture and manufactories."

Summing up his reasons for recommending the Northern Route rather than any of the others, Beach cited the "facilities" available at Boonton, Little Falls, Paterson, and Newark "for the erection of extensive manufactoring [sic] establishments." Further, the numerous establishments already in existence at Bloomfield would be afforded "invaluable accommodations" by the presence of the canal through that town. But while Beach was concerned about the future development of industry, he placed even more emphasis on the transportation potential of the Northern Route. He had already mentioned the possibility and desirability of an intersecting canal between Chatham and Little Falls, and a navigable feeder to connect the Pompton River with the canal. At the close of his report he noted again the "feasibility of constructing intersecting collateral canals through different districts of country," and in particular, *"up the branches of the Pompton River."* Here, if anywhere, the Morris Canal would best serve the iron industry, for iron mines were plentiful up those "branches of the Pompton River." And those iron mines, "those sources of wealth," were said to have been "the ostensible objects of this Canal." Clearly, "the interests of the Company and those of the public would be most combined in the

Boonton and Pompton route," and Ephraim Beach, Engineer, did not hesitate to opt "decidedly in favor of that route."

At the risk of oversimplification, it is easy to see that each of the routes would have satisfied a different faction. The Southern Route, of course, would have serviced Morristown, making the Macculloch faction happy indeed. The First Middle Route through Littleton and Troy would surely have delighted Lemuel B. Cobb, who operated a forge and probably a sawmill at Cobb's Corners, Parsippany. [Cobb didn't live long enough to be bitterly disappointed with Beach's choice; he died on April 1, 1831.] The Second Middle Route would have serviced the forge of Colonel John Scott at Powerville, and probably would have accommodated Silas Cook's distillery in Montville. The Northern Route would have made possible the re-emergence of the Ryerson Family in the iron industry of the Pompton area. And that it definitely did do, as we will see, while servicing Scott and Cook as well.

But Beach obviously had in mind far more than the construction of a navigable feeder to bring water into the canal from Greenwood Lake. A moment's reflection on those branches of the Pompton River, and a reference to Beach's mention of going "more than 20 miles up that river, brings the realization that Beach was thinking toward a canal *system.* Picture such a system: an intersecting canal from Chatham to Little Falls using the Passaic River with other canals using the Pequannock River to reach northwest almost as far as Hamburg; the Ramapo, intersecting the state's northern border at Suffern, New York; and the Wanaque, which would have serviced Boardville, home of some of Beach's iron-mining relatives, and Ringwood — even Sterling Furnace in New York State. What an internal improvement that would have been for virtually the entire northern part of the state.

Macculloch, too, had similar thoughts on the construction of collateral canals, if only to service the Morristown area. He pictured, too, a navigation or other communication that would extend northward from Lake Hopatcong or from Waterloo (Andover) by means of a "chain of ponds" that led to Newton and then the Delaware River, where the "Lackawaxen Canal"* was to end. Coal could thus

**Here he meant the Delaware and Hudson Canal.*

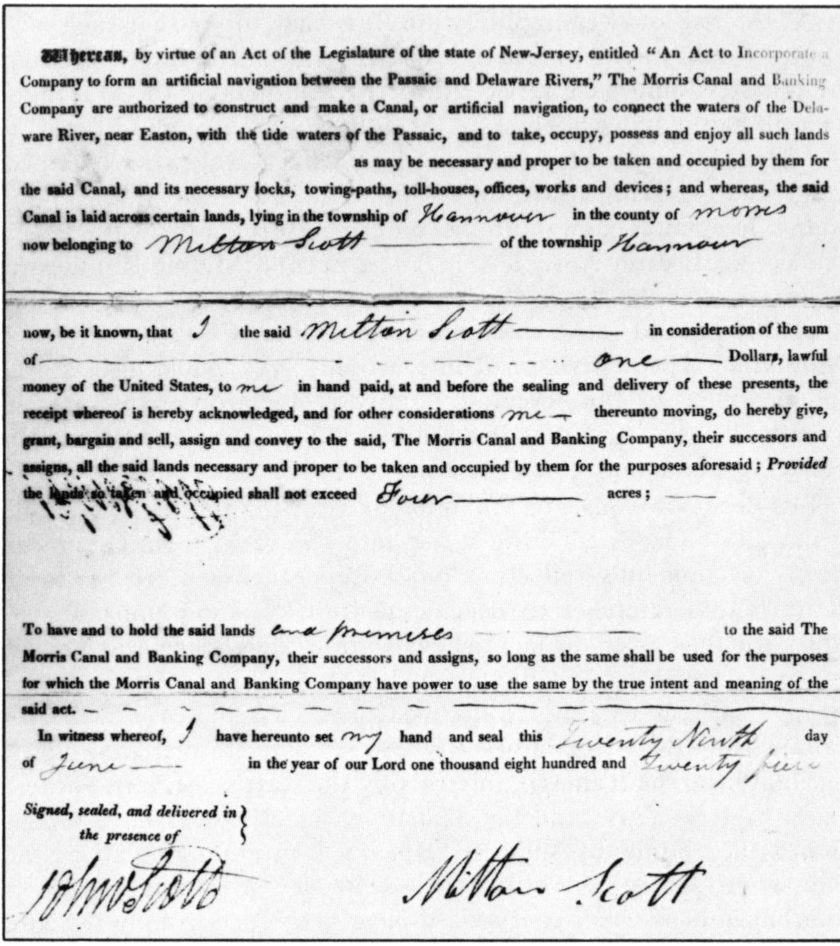

Whereas, by virtue of an Act of the Legislature of the state of New-Jersey, entitled "An Act to Incorporate a Company to form an artificial navigation between the Passaic and Delaware Rivers," The Morris Canal and Banking Company are authorized to construct and make a Canal, or artificial navigation, to connect the waters of the Delaware River, near Easton, with the tide waters of the Passaic, and to take, occupy, possess and enjoy all such lands as may be necessary and proper to be taken and occupied by them for the said Canal, and its necessary locks, towing-paths, toll-houses, offices, works and devices; and whereas, the said Canal is laid across certain lands, lying in the township of Hannover in the county of Morris now belonging to Milton Scott of the township Hannover.

now, be it known, that I the said Milton Scott in consideration of the sum of One Dollars, lawful money of the United States, to me in hand paid, at and before the sealing and delivery of these presents, the receipt whereof is hereby acknowledged, and for other considerations me thereunto moving, do hereby give, grant, bargain and sell, assign and convey to the said, The Morris Canal and Banking Company, their successors and assigns, all the said lands necessary and proper to be taken and occupied by them for the purposes aforesaid; *Provided* the lands so taken and occupied shall not exceed Four acres;

To have and to hold the said lands and premises to the said The Morris Canal and Banking Company, their successors and assigns, so long as the same shall be used for the purposes for which the Morris Canal and Banking Company have power to use the same by the true intent and meaning of the said act.

In witness whereof, I have hereunto set my hand and seal this Twenty Ninth day of June in the year of our Lord one thousand eight hundred and Twenty five

Signed, sealed, and delivered in the presence of

John Scott

Milton Scott

Deed of Lands to the Morris Canal and Banking Company, with reverter clause and signature of Canal Commissioner John Scott (Courtesy of Peter Wendt, Sr.).

be brought into the canal from yet another region of northeastern Pennsylvania. Obviously a champion of the Southern Route described by Beach, Macculloch later commented:[11]

> *The location of our line has not been exempted from the rivalries and dissensions which always embarrass similar operations. The route through Morristown presented many advantages, both in distance and revenue, which however, were counterbalanced in the opinion of a large majority of our Board, by the appalling elevation of Springfield mountain, and the possibility of a failure of water on that route.*

Somewhat prophetically, Macculloch continued: "The district, populous, manufacturing, and agricultural, which this line *would have traversed* [italics mine], will, in a few years, demand some access to markets more facile and cheap than its rugged turnpikes." Eventually, he hoped, the Canal Company would provide such access, "as part of a system of *railways or lateral canals,* of which the Morris Canal will act as a common center." He could foresee a time when the canal would be so burdened with such a variety of cargoes, that coal, if it were to be transported by the Canal Company at all, "must be upon a *Railway or additional canal* . . ." Morristown citizens were to be prominent in procuring a charter for the Morris and Essex Railroad, less than ten years after Macculloch made this statement.

Macculloch wasn't the only Elizabethtown-terminal supporter, by any means. His remarks cited above were published on July 1, 1826. By October 5, Morristown's *Palladium of Liberty* reported that Newark and Elizabethtown newspaper editors were still very much "disputing about the termination of the Morris Canal — the former insisting upon Newark, and the latter upon Elizabethtown." Indeed, the Morristown paper mused, "men are bad judges in their own case." Macculloch, "the father of the Morris Canal," had definitely indicated Eliabethtown as "the proper point of termination." But by means of the Southern Route, the canal would have pushed through Morristown, Macculloch's adopted home. How could the *Palladium* have missed the implication of that fact! It could very well have affected Macculloch's judgment in *his* own case.

It must be admitted, however, that Macculloch's reasoning on behalf of the Elizabethtown Route was persuasive. To this writer, the fact that Elizabeth was a free port on deep water, outside the authority of the New York Harbor authorities, and thus free of the duties charged on cargoes by that body, would have seemed reason enough. But Macculloch took a different tack.

> *Taking it for granted that the words 'tidewaters of the Passaic' will carry us to the dock of Elizabethtown Point, at which place this river may be said to mingle with the joint waters of the Raritan, the Hudson, and the Ocean. Such was the spirit of the act, both in the mind of the framer* [*himself*] *and of the Legislature.*
>
> *Be the canal terminated where it may, the trade must double the Cape of Bergen, and the line of transportation will be of equal length, whether the canal closes at Aquacnock* [*sic*], *Bellville* [*sic*], *Newark, or Elizabethtown Point. It being recognized that canal is less hazardous, more expeditious, certain and cheap than river navigation, it follows that the public interest points to the longest canal, or to a termination at this last station."*

Persuasive as Macculloch's argument was, months earlier, in January, Beach was already completing the surveys for the canal's path through Newark, presumably with the idea of ending it there. The *Newark Eagle* reported that Beach had actually made two surveys in order to determine where the canal would enter the Passaic River. The *Eagle* believed, however, that "it is the intention to bring it through Market Street (the principal cross street in the town) and continue it on nearly in a straight line, til it meets the river." Excavating contracts, the paper said, would be given out for the section of the canal between Rockaway and Little Falls by the 31st of January, 1826.[12] By June, the Newark *Sentinel of Freedom* placed the easternmost limit of the canal's location only "as far as the Newark Quarries, at the north end of town."[13] Beyond the quarries, the route of the canal was still in doubt. Other places, however, were progressing: "The contractors have commenced operating in the vicinity of Paterson, and others are expected to break ground between this and Bloomfield in a few days. From the summit level to

Rockaway, a very considerable part of the excavation was effected last year [1825] . . . ."

In his *Annual Report upon the Morris Canal for the Year Ending July 1, 1826,* Macculloch summed up the construction progress from July, 1825 to July, 1826:[14]

> *The excavation begins about 6 miles from the western termination near Easton, and continues, with very short undug intervals, to Rockaway, a distance of 42 miles. We had the pleasure to sail along 4 miles of this line of canal, and two other levels of 14 miles will be navigable this fall. Between Rockaway and Newvillage* [sic], *one inclined plane, two guard and five lift locks, ten culverts, and the masonry of two aqueducts are completed; two locks, two dams, one inclined plane are in a state of great forwardness. The materials for the great dam and guard lock upon Hopatkung* [sic], *as also for two lift locks, are collected. From Rockaway to Little Falls, a distance of 18 miles, the excavation is in great forwardness, and from thence to the diverging point of the Newark and Elizabethtown routes, the digging is under contract, and in many places opened. Between Rockaway and the vicinity of Bloomfield, four culverts and the masonry of a large aqueduct are finished; two locks, five aqueducts, and three culverts have been, or will be, immediately commenced. Thus a line of 76 miles from Newvillage to near Newark is progressing as rapidly as 1100 men can urge it . . . .*

As we can see, the controversy was far from being settled, at least for Macculloch. Nor was the route controversy the only problem being faced by the Canal Company during its first year of existence. By the end of 1826, the Company was left holding more than 5800 shares of its stock that had reverted to the Company when their holders failed to make installment payments on time, or that had been rejected by the buyers and sold back to the Company as the reputation of the Morris Canal's banking aspect became more and more tarnished by scandal.

As if that weren't enough, a labor shortage kept step with the money shortage. It takes more than noble ideals, patriotic goals, good intentions, and ready cash to build a canal: It takes labor, and labor was a scarce commodity in the infant United States. This was an era of manual labor, an era not yet industrialized, much less automated; an era of brawn, without bulldozers and pushbuttons. And in the America of the early 1800s, laborers, skilled and experienced, were not easily come by. Some locals undoubtedly left the farms and probably the mines of northern New Jersey to work on the Morris Canal, but for the most part, New Jersey had to import labor.

The immigrants who earned their fortunes and in some cases their citizenship on many of the nation's first canals are said to have been Irish, almost to a man. Not the Scotch-Irish, who were already here, having fled the religious persecutions in Scotland to live in Northern Ireland before embarking for the New World, but the Gaelic-Irish. Moreover, they were the Irish Roman Catholics in a WASPish civilization — a situation that caused sporadic but vicious trouble in this country, not unlike the current riots in Northern Ireland. In 1824, Orangemen (Irish Protestants) made an attack upon Roman Catholic Irishmen in Paterson. New York City also had a similar riot in July, 1833. Newspapers labeled the fracas "disgraceful," because even the women took part![15]

Aside from their religious persuasion, these immigrants who turned their "spades and blades" and brawn and brains to dig our canals filled a special niche. Dickens saw a pair of them in New York City during his American travels. With his typical flair, he limned them, and commented: "It would be hard to keep your model republics going without the countrymen and countrywomen of those two laborers. For who else would dig, and delve, and drudge, and do domestic work, and make canals and roads, and execute great lines of internal improvement."[16]

What precisely had brought them here, these Irish laborers, is subject enough for a separate treatise, but a few observations are possible and pertinent. Many Irish "bog-trotters" had been brought from Old World peat bogs, where they cut fuel, to work on the Erie Canal. For an excellent and lively treatment of this immigrant labor, see Walter Edmunds' *Erie Water*.[17]

Many of those nameless heroes had been brought from the Old World on the strength of promises. One Irishman testified in 1837 as to the recruiting tactics used by agents employed by many New

York shipping companies, such as the Black Ball Line.[18] The Black Ball Line was owned by Jonathan Goodhue, soon to become a director of the Canal Company.[19] Such agents, according to the testimony, placed handbills, placarded on every corner, tree, and pump and public place in the city of Dublin, and for 40 or 50 miles in the surrounding country, stating, in substance, that the people were "fools not to leave the country, where there was nothing but poverty staring them in the face." Since laborers were so much in demand in America, wages were high and even food was supplied. At least one or more such agents were operating in "every principal town in Ireland, receiving a commission for collecting and forwarding emigrants to Liverpool, where they take ship for America." The Irishman who testified, ironically, had given up a good job as a canal engineer in Ireland.

Many Irish who arrived here uninvited by company agents were doubtless met at the New York City docks and hired on the spot. They were offered "'roast beef guaranteed twice a day, regular whiskey rations, and wages eighty cents'." Some workers are said to have been paid a dollar a day — good pay for unskilled labor at that time.[20]

Serious competition existed among employers for this unskilled and semiskilled labor. Although the Erie Canal was officially opened just after work had begun on the Morris Canal, hundreds of men remained at work completing Clinton's Ditch. Moreover, during 1826, about 3500 men were at work on the two Ohio canals under construction. And the Delaware and Hudson Canal, also being built, was employing about 1200 men at the time, and advertising in New Jersey newspapers for 3000 *more* such men to work during that spring of 1826![21]

Many were no doubt "imported" from New York State, bringing with them whatever skills and experience they'd gained in digging and blasting Clinton's ditch all the way from Troy, on the Hudson River, to Lake Erie. They'd cut their teeth on that canal, blazing a trail that would be followed in fact and in concept through solid rock and across swampy wastelands in other states. They connected the mouth of the Hudson with the heartland of the nation by way of the Great Lakes and paved the way for progress and growth.

It is more than likely that whole work crews owing allegiance to one gang boss or foreman migrated from the Erie, Schuylkill, or other canals, to work on the Morris. Such a practice was not uncommon. Thomas Telford, the great British engineer and canal builder, took his own labor force with him as he traveled from one engineering job to another in Britain and on the Continent.[22] DeWitt Clinton remarked to the Ohio Canal Commissioners that once his canal was completed, "Ohio could avail herself of the aid of able engineers and skillful contractors" in the building of her canals.[23]

In any event, and from whatever source, canal laborers were hired for the Morris Canal's construction. Most sources agree that they were housed by the Canal Company, usually being paid half in cash, half in whiskey, and were worked in gangs of 40 to 50 under section chiefs or foremen. Living in close quarters in hastily built, barrack-like shanties along the canal route, their numbers were periodically decimated by cholera epidemics. Such shanties had been built on Nathaniel Saxton's lands on the Musconetcong River near Waterloo. The shacks remained for two years while the workers used Saxton's sawmill to cut lumber for canal construction.

They must have been a boisterous, brawny lot, these sons of Erin, spending most of their wages for "Jersey Lightning" (apple jack) and, when it was available, Irish whiskey. To the communities along the canal, they were also a troublesome lot, especially on paydays, when they converged on the local taverns and wreaked their own peculiar havoc. Many a farmer's daughter was locked up while these weekend gladiators roamed the countryside. Miners, mostly Welshmen and Poles, also went into towns like Dover to spend their pay, and a meeting with the canal-building Irish usually resulted in hand-to-hand, even foot-to-head, combat and casualties without number.

With unflagging energy, the brawny, lusty Irish toiled away from sunup to sundown, six days a week, putting all their muscle and sweat and blood into the canal. It has been estimated that less than a dozen wheelbarrows were in use from Phillipsburg to Newark and, except for a pitifully few primitive tools, most work was done with hand labor or with a little help from livestock.

Robert Fulton had clamored for the invention and use of mechanical devices and sophisticated tools in New York's canal construction, but with a few notable exceptions, work remained a matter of hand labor and elbow grease. One such exception was a hand-worked lifting device designed by Orange Dibble to lift rock blasted from the Erie Canal locks at Lockport, New York. Because of the way the device dipped and bobbed its "head," workers dubbed it a "crane."[24] It is not known whether or not the crane was used in digging the Morris Canal, but, according to Alvin Harlow, a stump puller was used with a team of horses to remove stumps wherever they interfered with the canal's path. The device consisted of "an axle twenty inches in diameter and thirty feet long, supported on wheels sixteen feet in diameter. Midway on the axle was mounted a drum fourteen feet in diameter. When the outer wheels were chocked, a chain wound around the axle and fastened to the stump, and a team of horses or oxen hitched to a rope which encircled the central drum several times, the stump was easily pulled. The outer wheels were then released and the same machine hauled the stump away. One of these machines operated by seven men and a team could pull and move thirty to forty stumps per day."[25] A "stone sled" pulled by horses was also used to drag boulders away from or to the construction sites. The "scoop" was a curved wooden device dragged by horses to level or to remove earth from the canal, the towpath, or the berme bank.[26]

The invention of sophisticated construction equipment had to await a more technologically sophisticated age, the development of steam power, and a more demanding socioeconomic structure. Meanwhile, it was scrapers instead of bulldozers, elbow grease instead of bucket loaders, black powder instead of dynamite. The Irish seemed ideally suited for such physically demanding and dangerous work. One of the Edmunds' characters mused: "An Englishman's all right. The only trouble with him is he hasn't ever considered an Irishman. You put an Irishman against a stone mountain and give him plenty of blasting powder and he'll go through. An Englishman would do it scientific, but an Irishman would just bust loose."[27]

Rock blasting was a particularly dangerous occupation for the early construction crews. Dynamite had not yet been invented. Ordinary black powder, far less dependable or predictable, was put into a hole drilled in rock. A fuse was tamped into the powder and held in place with a bit of clay. The fuse was lighted, and the blaster ran for his life to a spot safe from flying rock missiles. Many boaters on the Delaware and Hudson Canal refused to take their boats near Port Jervis, where the Erie Railroad blasters were trying to gouge a foothold out of a sheer rock face. The blasters were lowered over the edges of cliffs by baskets and ropes. They set their charges and were pulled quickly — all too often, not quickly enough — out of harm's way. Rocks from the ensuing blasts rained down on the Delaware and Hudson Canal below the rock ridge, causing no little damage.[28]

The *Easton Centinel* reported that a rock 40 feet high and 25 feet wide was blasted out of the Delaware and Hudson Canal's path at Marbletown. The feat was accomplished in a single blast by an Irishman using 70 pounds of powder poured into a hole in the rock, then setting it off. Parts of the rock, the paper said, were picked up "several miles distant," while one chunk, "three feet square fell within two feet of the spectators," sinking so deep in the frozen ground that it was not visible.[29]

At Kingston, New York, the incident was reported a little bit differently. The April 20, 1826 *Palladium of Liberty* cited the blast as "one of the greatest ever made in a rock in the United States." The "Irishman", here, is identified as "Patten," and the rock seems to have grown somewhat in the month that elapsed between the blast and the reporting — now being estimated at 40 feet high and 20 to 30 feet wide. Patten, it appears, was a passing visitor who happened to be on the spot when the canal contractor was offering $5 to blast the rock. By both accounts, however, 70 pounds of powder effected the blast.

Another hazard in canal construction, cave-ins, seems to have taken the lifes of several laborers:[30]

> *A young man, son of Joseph Lord of Hunterdon County, was killed near Hackettstown in Warren County, while working on the Morris Canal. The father was so much hurt at the same time that he is not expected to*

recover. They were undermining a large bank of earth, with a view of afterwards prying it down, when the bank gave way and it is judged 2000 weight of earth fell upon them.

In Paterson, at the end of July, 1826, a similar accident befell "Christopher Flood, a native of Ireland, from which place he had lately arrived." He was "crushed to death beneath a large rock which loosened from the bank above where he was at work, and rolled upon him."[31]

On August 16 of that same year, an unidentified workman was killed by a cave-in near J.W. Hornbaker's in Mansfield Township (Washington, New Jersey), according to a report by the *Paterson Intelligencer* of August 23, 1826. And on January 23, 1828, the same newspaper reported that three more men had been killed in a cave-in while working on the canal below "Succasunny [sic] Plains." Two others were seriously injured in that accident.

Injuries to livestock, too, were far from uncommon. Lured by the prospect of money, a relatively scarce commodity in their lives, farmers released their livestock to work on the canal. Teams were unhitched from farm wagons and plows to be hitched to scrapers along the canal prism or tow-path, and to stump-pullers which removed tree stumps. But as the farmers saw their favorite horses spavined by the strenuous work, or watched them being beaten to make them work still more strenuously, they became more and more reluctant to hire out their livestock. They probably took back their animals and returned them to the fields and furrows of the farm. Eventually, the Company was forced to hire teams from contractors.

Manpower, fortunately, was a little more plentiful than livestock, and work proceeded at Lake Hopatcong, which almost everyone agreed would be more than sufficient to supply the canal with water. Early estimates by Renwick indicated 17 million cubic yards of water demanded by the canal; Bernard and Totten in their "Report of the Army Engineers" had estimated that 10 million cubic yards would suffice. Macculloch reported to the stockholders that the Canal Company had adopted "the average of those two authorities," and had calculated that the canal would demand 13.5 millions. Lake Hopatcong was thought to discharge 55 million cubic yards of water, or four times what was needed to operate the canal from the Delaware to the Passaic. When the new dam was completed, the lake was expected to be raised an additional four feet, thus guaranteeing an even more capacious canal reservoir. Moreover, the enlarged lake thus created would afford navigation from one end of Lake Hopatcong to the other and "convince the most incredulous" doubters as to the capacity of that reservoir.[32]

The Canal Company had bought a parcel of land at Brooklyn, at the outlet of Lake Hopatcong. Included in the parcel were the "forges, mills, farm and woodland," which Macculloch was sure would give the Company the "entire control over all the operations on the summit level, besides affording a manufacturing seat of the highest value." With that control over "an unlimited command of surplus water," and the water power thus created at its inclined planes, Macculloch felt that the Company could expect the income from manufacturing to "equal the revenue of the canal itself" within twenty years.[33] For the moment, he appears to have forgotten that the canal's charter expressly forbade using canal-appropriated water for manufacturing or mills.

Beach, too, was certain that if the Company were willing to expend "a trifle more money" to create such water powers at the inclined planes, they would "probably be worth twenty times more" than they would cost.[34]

Water supply clearly would not be a problem, Macculloch assured the stockholders. Not only did Lake Hopatcong promise four times the amount needed for the canal, but there was Green Pond, which "could easily be made to reserve a large supply," since it already served as a feeder for the canal. Obviously, with Great Pond, Green Pond, and the Rockaway River acting as feeders, "all fears" that there might be a scarcity of water "must originate in want of information [ignorance]."[35]

As we have seen, many paths had been considered for the canal to follow in its eastward trek to tidewater. The original plan of 1823 had "contemplated ending the canal at tidewater, at [Passaic] Aquackanonk."[36] Another proposal would have led the canal from Lake Hopatcong to Dover, then turned southward to Morristown and continued to Elizabethport. Another was to have had the canal "fall into the Passaic near the Great Falls — then to render the river navigable by locks and dams, and to make a towing path on its border." Still another was to go by way of Paterson and Belleville, then Newark.

It was probably the early idea of ending the canal at Aquackanonk (Passaic) or of going to Elizabeth without touching Newark or Paterson that provoked spirited citizens from those areas to clamor for still another canal. Thus on October 26, 1825, Mr. Pennington submitted a petition to the House of Assembly in Trenton from Essex, Morris, and Bergen County residents, asking "for a law to authorize a canal from Acquackanonk, by Belville [sic] and Newark, to unite the waters of the Passaic and Hackensack with the Hudson, in order to effect a more easy communication with New-York." The petition was committed.[37]

By December 20, 1825, the *Paterson Intelligencer* picked up a Morris Canal news item from the *Palladium of Liberty:* "We have to announce that the Directors of the Canal Company have at length adopted a course through Boonton and Pompton, to the Little Falls, and ordered the same forthwith proceeded upon. This decision is bottomed upon a report of their Engineers, stating that the Morristown route would cost about $180,000 more than the Boonton, beside incurring great danger of want of water. If these points be correct, the determination of the Board is unquestionably correct; it is only a matter of serious regret that by much the largest portion of this County, in extent, population, and trade, is thus excluded from all benefit in this public improvement."

The route decided upon, as noted later by the *Paterson Intelligencer* (January 25, 1826) did indeed bypass Morristown. It was to proceed from Boonton to Paterson by way of "Pompton" (that is, the Pompton River at Mountain View), and Little Falls. After passing "near" Paterson, the canal was to "proceed through Stone House Plains, Bloomfield, etc., to Newark," where the *still* undecided route would "probably intersect the Passaic near the foot of Market-street." This route placed the eastern division of the canal within the watershed area of the Passaic River. At Dover, and again at Boonton, the canal actually joined with the Rockaway River, one of the principal feeders of the Passaic River, and tapped that water supply. Another important tributary was the Pompton River, which fed waters from the Ramapo, Wanaque, and Pequannock Rivers into the Passaic, and *that* river the Canal Company planned to tap by means of a feeder canal. Moreover, while they were "certain" that Lake Hopatcong would provide ample water for the canal, the Company also had an eye on Green Pond as a potential reservoir, since Green Pond ultimately discharged into the Rockaway River at Wharton. And so, the Canal Company was thrown into direct and open conflict with Paterson's SUM, whose own charter, granted much earlier than the Canal Company's, had given it the exclusive water rights over *all* the Passaic River and its tributaries.

In an effort to forestall the impending "irrepressible conflict," the Canal Company pushed for the completion of a substantial dam at the outlet of Lake Hopatcong to impound even more water. Mr. Fred Eckhart, formerly of the State's Bureau of Parks, conducted a line survey which places the original elevation (circa 1750) of Lake Hopatcong at 912.09 feet above sea level. A dam subsequently built about 1800 raised the water to a level of 917.09 feet above sea level. The Canal Company dam, built in 1826, brought the level still higher, to 922.09 feet.

The writer has been unsuccessful in uncovering any "eyewitness" accounts of construction difficulties, but the following, from an unpublished novel, can be taken as a fair description of what it must have been like at the Summit Level during 1825 and 1826.[38]

> *The digging there was the hardest on the canal because of the nature of the terrain. The right of way extended through an area of swamp land covered with trees whose roots ran deep into the ground, winding around each other in such a maze and tangled mess that it was impossible to clear but a small space each day. Then there was an undergrowth of vines, bramble bushes and poison ivy growing over huge boulders and clinging to the trees, all buried in soft marshland. Whole strata of trap rock had to be blasted out in some places, and the muck was up to the men's armpits.*

While the digging proceeded, the men at work on the lock at the outlet of the lake uncovered "the skeleton of an Indian, the arm bone of which from the wrist to the elbow was eighteen inches, and from the elbow to the shoulder the same. The remaining parts of the skeleton were of the same proportions. These bones would imply a height of nearly eight feet to this man of the forest."[39]

Not long afterward, in mid-July, 1827, the skeleton of a mammoth was found while excavating near Schooley's Mountain, Hackettstown. The remains were found about three feet below the surface, and were considered to be in a "remarkable state of preservation." One of the tusks was said to have weighed 150 pounds, and the width of the grinding surface on one tooth was measured at 3½ inches, with a length of 7 inches. The tooth weighed four pounds![40]

Construction from Paterson was underway as far as Israel Crane's quarries outside of Newark, by June, 1826,[41] and it was at this time that Paterson decided to object to the very existence of the canal. If Lake Hopatcong were really sufficient to supply the canal with water to operate, the SUM demanded, why didn't the Canal Company build its canal *over* the Rockaway River; cross it on aqueducts, rather than enter the river at Dover and Rockaway to use that water to help supply the canal? What if, during the dry summer months, the manufacturers at Paterson didn't have enough water power to keep them in operation because the canal had diverted too much of the Passaic River to its own use? The use of the Passaic River and all its tributaries had, after all, been guaranteed to the SUM by the State Legislature. What if the Canal Company were trampling on the rights of the prior charter of the SUM? On the basis of those objections, the SUM sought an injunction against the Canal Company, to cause the Morris Canal not to be built. It wasn't until 1829 that Chancellor Isaac H. Williamson handed down an opinion, saying, in effect, that until the loss of water was actually experienced by the manufacturers, and unless such loss could be directly attributed to the Canal Company, he would have to deny the injunction. The issue was settled, but only temporarily.

One item that settled the issue for Chancellor Williamson was a letter from Roswell L. Colt, Governor of the SUM, to Cadwallader D. Colden, President of the MC&BCo., in December, 1826, when the route of the canal at Paterson had just been determined. In that letter, Colt "objected to the present location of the Morris Canal, only on

Rogers' Locomotive Works in Paterson, circa 1839. The SUM's lower and middle raceways are at the right; the Morris Canal is in the background against Garret Mountain (from Shriner's Paterson Illustrated).

the ground that this route would not afford the expected accommodation to the Town of Paterson." In brief, Paterson's manufacturers had expected the canal to enter the city itself; the route determined by the Canal Company's engineers placed the canal up on the side of Garret Mountain, "thirty feet above the tops of the highest buildings in the town," as that indefatigable traveler, Mrs. Trollope, observed.[42] Suspended up there, the waterway would be far less serviceable to those Paterson manufacturers who had products to ship to metropolitan markets. With the route decided upon, they would first have to wagon those products part way up the mountain before putting them on canal boats.

Out of this very early hostility between Paterson (the SUM) and the Canal Company came an 1828 survey for a railroad which was to parallel the canal across the state, then cross the Delaware River, and connect with the Susquehanna River at Pittston. The road was expressly for the transportation of anthracite coal. Although that railroad was not built, the initial surveys provided a basis for the Paterson and Hudson River Railroad, which *was* built. But more about that later.

Although railroad building was not yet to become a serious threat to the Morris Canal, there was an ample supply of other and different problems that had to be solved before construction could be completed. First and foremost was the development of a truly efficient inclined plane, for it was on the success of the inclined planes that the success of the entire canal rested. And the first of the experimental planes had been ordered at Rockaway, to be "built upon a plan which, by a new machinery, adopts the principle of Fulton's medium inclined plane." The enterprise was expected to effect "a most beneficial revolution in inland navigation."[43]

REFERENCES

1. *Paterson Intelligencer,* October 5, 1825.
2. Alvin Harlow, *Old Towpaths* (N.Y.: 1926), p. 51.
3. *Report to the Directors of the Morris Canal and Banking Company; made by the President, Cadwallader D. Colden, May 1st, 1827: With a Statement of the Work Then Done Upon the Canal, A Report from the Chief Engineer, and a Communication from Governor Clinton* (N.Y.: 1827), p. 11.

4. *Paterson Intelligencer,* October 26, 1825.
5. *Ibid.*
6. *Laws of New Jersey,* 1822, p. 79.
7. Wheaton J. Lane, *From Indian Trial to Iron Horse; Travel and Transportation in New Jersey, 1620-1860* (Princeton, N.J.: 1939), p. 10.
8. *Paterson Intelligencer,* December 7, 1825.
9. *Ibid.,* December 14, 1825.
10. Morris Canal and Banking Company Docket Files, Trenton, N.J.
11. *Annual Report Upon the Morris Canal for the Year Ending July 1, 1825* (N.Y.: 1826), pp. 15-20 (hereinafter cited as *Annual Report, July 1, 1826*).
12. *Paterson Intelligencer,* January 26, 1826.
13. *Palladium of Liberty,* June 22, 1826.
14. *Annual Report, July 1, 1826,* pp. 3-4.
15. *Paterson Intelligencer,* July 31, 1833.
16. Charles Dickens, *American Notes* (N.Y.: c. 1868), p. 78.
17. Walter D. Edmunds, *Erie Water* (N.Y.: 1933), p. 433.
18. Robert G. Albion, *The Rise of New York Port* (N.Y.: 1939), pp. 340-1.
19. *Ibid.,* p. 247.
20. Bernard A. Weisberger and the editors of American Heritage Publishing Company, *American Heritage History of the American People* (N.Y.: 1971), p. 152.
21. *Palladium of Liberty,* March 16, 1826.
22. Robert Payne, *The Canal Builders* (N.Y.: 1959), p. 131.
23. David Hosack, *A Memorial of DeWitt Clinton* (N.Y.: 1829), p. 218.
24. Edmunds, *op. cit.* p. 433.
25. Harlow, *op. cit.,* p. 53; Edmunds, *op. cit.,* p. 227.
26. Cadwallader D. Colden, *The Life of Robert Fulton* (N.Y.: 1817), pp. 13-4.
27. Edmunds, *op. cit.,* p. 321.
28. For an excellent treatment of the rock blasting by Erie Railroad workmen see Stewart H. Holbrook's *Story of American Railroads* (N.Y.: 1947), pp. 86-7.
29. *Palladium of Liberty,* April 20, 1826.
30. *Paterson Intelligencer,* January 4, 1826.
31. *Palladium of Liberty,* August 9, 1827.
32. *Annual Report, July 1, 1826,* p. 5.
33. *Ibid.,* p. 17.
34. *Ibid.,* p. 20.
35. *Ibid.,* p. 5.
36. *Paterson Intelligencer,* January 31, 1826.
37. *Ibid.,* November 2, 1825.
38. G. W. Keppler Collection, unpublished manuscript, "Goodman Novel," p. 48.
39. Munsell's *History of Morris County: 1739-1882* (N.Y.: 1882), p. 233.
40. *Paterson Intelligencer,* August 1, 1827.
41. *Ibid.,* June 14, 1826.
42. Frances Trollope, *Domestic Manners of the Americans* (London, 1927), p. 308.
43. *Annual Report, July 1, 1826,* pp. 6-7.

5

If At First You Don't Succeed

MODERN historians have generally attributed the invention of the Morris Canal's inclined planes to Professor James Renwick. George Macculloch credited Renwick with the engineering ingenuity that initially made the dream canal a reality, but he was precise in identifying the inventor of the planes as Robert Fulton. Renwick, then, adapted Fulton's plane design for use on the Morris Canal.

Critics of the Morris Canal, such as Rutherford, claimed that Fulton's design had been devised for only descending navigation; that is to say, for loaded boats traveling down an incline. The illustrations in Fulton's *Treatise on the Improvement of Canal Navigation* (1796), however, clearly show both ascending and descending boats. Still, Fulton's plane system transported boats that were but "two to five feet wide, containing from two to five tons burthen," and it was considered more economical for Morris Canal boats to carry 25-ton cargoes. Presumably, Renwick's task was to adapt the principles of the inclined plane to that end.

Renwick was convinced that an engineer builds a reputation not by searching after "novelties, but rather from an adaptation, successfully, to useful purposes, of known machinery, and received philosophical principles." He readily admitted his debt to the genius of his predecessors: Leach, Reynolds, Ducart, Anderson, Rowland, Pickering, and the Duke of Bridgewater, in England; De Solages, Bossert, and Forey, in France. Curiously, the name "Fulton" does not appear in Renwick's list of the inventors of prototype planes. Neither does the name "Fussell."[1]

Renwick submitted a working model of his inclined plane for the third annual exhibition (and competition) of American Manufactures, sponsored by Philadelphia's Franklin Institute in 1826. A committee, examining the model, noticed an obvious similarity between Renwick's design and that of an Englishman named Fussell, who had patented his mechanism in 1798. One of the Institute's committeemen, G.W. Smith, called upon Renwick to defend the originality of his plane design, whereupon Renwick disclaimed any knowledge of Fussell's invention.[2]

Apparently satisfied with Renwick's answer, the committee awarded him the Institute's silver medal as a premium for his "improvements in the inclined plane." According to Renwick's own description, the plane that won the silver medal consisted of two sets of parallel strap-iron rails, secured to rows of 200-pound stones laid atop the smoothly graded surface of an incline. On each pair of the rails were triangular frames set on four pairs of graduated wheels which were fitted to the rails so as to make the tops of the frames level with the horizon while the bottoms of the frames were parallel to the rails.[3]

Atop the frames were wooden troughs, or wet basins, 60 feet long, 9 feet wide, and 3 feet deep. Actually the troughs were movable, or floating locks, gated both front and rear to admit and release boats and water. Filled, the basins held 45 tons of water, or a boat and the amount of water left after the boat had displaced its liquid equal.

One frame and its lock were positioned at the lower level of the canal, at the foot of the incline; the second lock was positioned at the top, or upper level. Both frames were connected to horizontal shafts by means of 1-inch iron chains, which were wound around drums attached to the shafts. The shafts themselves were interconnected by means of spur wheels, which turned each other in opposite directions. A clutch lever engaged or disengaged these gears. As Renwick explained it: "If the chains, suspending these [frames], be so connected, that one shall be drawn up, by the descent of the other, the discharge of a small quantity of water from either [movable lock], will set the apparatus in motion.[4]

In short, Mr. Renwick's inclined plane was a kind of gravity plane, the weight of the descending lock-basin being sufficient to make the horizontal shafts, spur wheels, and drums revolve, thus winding or unwinding the chains, which moved the basins and their cargoes.

A boat ready to ascend the plane was admitted to the lower lock-basin, with only enough water to float it. The upper lock, filled to capacity, weighed 45 tons. With boats at that time carrying cargoes not exceeding 25 tons, the boat in the lower lock-basin, with its complement of sufficient water to float the boat, weighed considerably less than the upper lock-basin, which consequently had a tendency to start downward. A movement of the clutch lever allowed it to do so, according to the laws of gravity. As it moved down the plane, its chain wound around the drum on its horizontal shaft, while the chain attached to the ascending boat-filled lock-basin unwound from *its* drum. In the upper level of the canal at the top of the incline, the front gate of the lock-basin was opened, and the boat was allowed to float into the canal prism. Both locks, upper and lower, were then in position to handle the next boat, ascending or descending. One thing Renwick was especially proud of was that his mechanism, unlike those of his predecessors, required no extrinsic motive power.[5]

General Joseph G. Swift had examined Renwick's design and reported that he was "entirely satisfied that the planes can be easily constructed, and that they would be safe, and sure." Macculloch described the device as "combining cheapness with the most beautiful simplicity," one that was sure to be of the "highest importance to the science of inland navigation." Henceforth, he concluded, "the most mountainous regions, and the most appalling elevations will be traversed with ease and economy, both of money, time and water."[6]

Macculloch remarked that without Renwick's "zeal, talent and science," the Morris Canal "would not within our day and generation have emerged beyond a scheme transmitted to a more liberal and enlightened posterity."[7] But, *was* Renwick's plane actually used on the Morris Canal? The answer must be a *qualified* "yes."

The 1823 Report of the Three-Man Commission contains a map

Inclined plane design, circa 1823 (a vignette from a map in the Report of the Three-Man Commission).

of the canal route as it was planned in 1823. In the upper left-hand corner of the map is a vignette showing an inclined plane that certainly fits the verbal description Renwick gave of his device. Granted that the picture is an artist's rendition, the triangular frames and movable troughs atop those frames seem to fit Renwick's word-picture.

When Renwick answered the Franklin Institute's questions about the originality of his design, he was writing in October, 1826, several months *after* the first experimental inclined plane had been tested at Rockaway. He mentioned that the chains *being used* on the plane were *being made* and tested at Dover. Even without further documentation — and the newspapers were considerably negligent in covering this vital event — that statement alone, in his report to the Franklin Institute, indicated that it was, indeed, his plane design that was first implemented at Rockaway.

However, such local newspaper coverage as there was of the Rockaway experiment identified Ezekiel Kitchell, a Whippany mill-wright, as the designer and builder of the inclined plane. Other sources explain that Kitchell had the aid of his son-in-law, Calvin Howell.[8]

Renwick's plan was tried, and then rejected in favor of Kitchell's design. Renwick was, after all, the *Consulting* Engineer for the Canal Company, a fact that does not necessarily mean that he was a *working* engineer. Moreover, Renwick was involved in other engineering projects at the same time he was affiliated with the Morris Canal. On April 17, 1826, while the first experimental plane was being built, Renwick and Anson G. Phelps, and their associates, incorporated the New York-Harlem Spring Water Company, to bring fresh, pure water to the City of New York.[9]

The New York *Evening Post* reported that, on July 13, 1826, when the Rockaway plane was tested on a fall of 52 feet, the plane was 630 feet long, and *"notwithstanding the imperfection of the water wheel* at the first starting, the expectations of the friends of the plan were realized in the most satisfactory manner." The car "with and without the boat, ascended and descended several times, containing a large number of persons who had assembled to witness the interesting spectacle. The time required in ascending and descending was from eight to nine minutes, *with less than half the head of water that could have been put on the wheel."* (Italics mine.)

Water would not have been needed "on the wheel" of Renwick's plane, since his design did not use a water wheel. A "car" would not have been used on Renwick's plane, since his design called for a trough full of water. The open-work framework car was a later device, developed "after some experimentation."[10] It appears that that "experimentation" took place at Rockaway, following, in part, Renwick's original plan and that, after that plan proved unworkable, it was discarded in favor of Kitchell's.

If there were still any question as to whether or not Renwick's plane was ever built, it was negatived by John L. Sullivan's statement that "Kitchell's method" was "preferred" to that of Renwick's.[11]

Just why Rockaway was chosen for the grand experiment was a question whose answer was elusive for some time. The moderate lift of 52 feet could not have been the prime consideration, for there were several other sites along the canal route where the same or less elevation would have had to be overcome. Nor could available water power have been the reason, since the Rockaway River flowed in at the foot, not the head, of the plane. Although Renwick's plane did not require water for a water wheel, it would have needed water to fill its movable locks. Kitchell's plane with its water wheel obviously needed water at the head of the plane, too.

The answer seems to lie at both ends of the 4-mile stretch of canal that was first completed from Dover to Rockaway, for at Dover there *was* water. The feeder lock and dam at Hopatcong were not completed until late in 1826, so that water could not be brought down from the summit level. But the Rockaway River flowed right through Dover. A short stretch of canal could be — and was — built first of all to conduct water from the river at Dover to the head of the inclined plane at Rockaway. It was this 4-mile stretch of canal that was navigable before July 10, 1826, when Macculloch wrote his report on the canal. He mentioned "sailing" along that section at the very outset of his report.

Yet water wasn't the principal reason for choosing Rockaway for that plane; "Iron" was. At the Dover end of that 4-mile canal was the firm of Blackwell and McFarlan, iron manufacturers; at the Rockaway end were the rolling mill and forge of Colonel Joseph Jackson. And Henry McFarlan was a director of the Morris Canal and Banking Company.

Colonel Joseph Jackson and his brother, William, had an iron-bound ancestry. Their father, Stephen, had operated the Upper Rockaway Forge until 1812, when he turned it over to two of his sons, John D., and William. Joseph Jackson, a third son, who already owned the Lower Rockaway Forge, bought the Upper Forge from his brothers shortly after 1812. Colonel Joe also owned the Allen, Teabo, and Swedes mines. Together with brother William, he owned an iron works in Paterson, where, in 1820, they were the first in the United States to roll round iron. So successful were they that they built the rolling mill at Rockaway, owning it jointly until 1826. At that point, brother William invested heavily in the Clinton Furnace. Ransom claims that it was poor planning on William's part to build (or rebuild) a furnace where there was no fuel — the reason for its failure.[12] But it seems a certainty that both the Jacksons fully expected the Morris Canal to supply them with ample fuel — anthracite coal. Colonel Joe's expectations were fulfilled at Rockaway; apparently William's at Clinton Furnace were not.

Joseph Jackson was virtually the sole owner of the town of Rockaway, its grist and saw mills, its store, and its two forges, not to mention its rolling mill. His forges were among the handful of iron works that had continued to operate in New Jersey during the 1820s, despite the post-War-of-1812 economic depression and the fuel shortage. With the coming of the Morris Canal, he was in an excellent position to benefit from its location at Rockaway.

Renwick had mentioned in his letter to the Franklin Institute that the chains being used on the incline plane at Rockaway were being made at Dover, at the other end of that first stretch of canal. The iron works at Dover were at that time already owned and operated by the New York partners, Blackwell and McFarlan. The forge and furnace at Dover had been owned by Israel Canfield and Jacob Losey. That firm had failed in the widespread depression following the War of 1812, and Blackwell and McFarlan bought them out about 1816.

Messrs. Blackwell and McFarlan were New York iron importers. With the firms of Boorman, Johnson & Co., and Phelps & Peck, Blackwell & McFarlan received about half of all the iron imported into the United States from Sweden, Russia, and England. Headquarters for the firm were at 8 Coenties Slip, New York, but their interests included an iron works at Monroe, New York (then known as Southfield), and Valley Forge in Rockaway township — one of Canfield & Losey's properties they acquired in about 1816. Like the Jacksons in Rockaway, Blackwell & McFarlan were the virtual owners of Dover. Both towns and their owners anticipated a rebirth of the iron industry promised by the advent of the Morris Canal.

On July 4, 1827, the *Paterson Intelligencer* published William Jackson's advertisement to sell 20 to 40 building lots in Rockaway, all at the foot of the inclined plane, where two basins had already been built, and "another large one" was to be "formed for the accommodation of boats." Only 30 miles from New York, there were "very few situations on the line of the Morris Canal, possessing greater advantages than Rockaway, having a very extensive Manufacturing and Agricultural district laying [sic] north and west of this place, and a great proportion of the trade of the back counties will be directed to this spot." The ad "particularly recommended" Rockaway "to the notice of Mechanics and Merchants."

A Blackwell and McFarlan notice appeared in the *Paterson Intelligencer* on September 12, 1827:

> *Dover has long been the seat of an extensive iron manufacturing concern, it is here that Messrs. Blackwell and M'Farland have their large rolling and slitting mills. The place is handsomely situated in the valley of the Rockaway, as it expands below the gorge of the mountains, into a broad and beautiful plain. The house named above [Blackwell and McFarlan] owns the site of the village which is now springing into existence, in consequence of the Canal. Besides the hospitable mansion of Mr. Losey, the agent of the works, there is a large stone hotel (built by the proprietors), several stores, the manufactories, and the houses for the families of the workmen. The canal leads directly through the village, and has been handsomely walled with stone, upon both sides, at the individual expense of Messrs. Blackwell and M'Farland. This cannot prove otherwise than valuable property, as it is situated at the very entrance of the hilly mineral regions of New-Jersey, and the water power being abundant, it will grow into an important manufacturing town. About eighty workmen are constantly employed.*

The inclined plane at Rockaway was operating when Macculloch visited the line of the canal, and had "sailed" along that 4-mile stretch that was completed on July 10, 1826. The plane had been tested for effectiveness at least as early as April 6, 1826, when the *Palladium of Liberty* reported that "an inclined plane or railway of Mr. Ketchell's [sic] projection, situated at Rockaway, we understand, was last week so far tested as to afford satisfactory evidence of its complete success."

The "official" trial of the Rockaway Plane was set for July 4, 1826, but "the breaking away of a few yards of the bank of the canal let the water out of that level." The breach was repaired "in a couple of days, at an expense of perhaps $70," after which "the level from Dover to Rockaway, of 4 miles, was passed in boats with four feet water," and the canal had "every appearance of standing firmly."[13]

Another report on the incident placed the breach in the "dam" at Rockaway, and stated that the damage had been "considerable."[14]

Macculloch didn't mention the breach at all, perhaps mindful of the inauspciousness of missing still another fourth of July.

Some newspapers adopted a tone of such scepticism that Beach was moved to protest and defend the workmanship of the canal and its appurtenances. He took any slur against the canal as a slur against his abilities as a supervisor of construction. In a July 17, 1826 retort to the *Palladium of Liberty,* Beach took exception to that paper's insinuations, which he felt were "calculated to form improper and unjust prejudices in the minds of the public against the inclined plane constructed at Rockaway." He considered "those insinuations (or aspersions) premature, futile, and unjust," since a subsequent trial of the plane, after the *Palladium's* article had appeared, "was attended with the most complete success."

Furthermore, Beach argued, "a heavy boat, with considerable burthen, was elevated from the bottom to the top — a perpendicular height of 50 feet, and delivered into the upper line of the Canal, in seven minutes." He did not deny that "improvements would be made upon the system of inclined planes," but insisted that "the inclined plane projected by Mr. Kitchell, and constructed at Rockaway, will answer the purpose of elevating or lowering boats of any necessary burthen from one level to the other." He was confident that "inclined planes will supercede locks in overcoming large elevations on canals of ordinary size."[15]

Shortly thereafter, the *Palladium of Liberty* noted — not without a touch of smugness — that the Kitchell plane "machinery gave way while operating upon a boat partially loaded." The smugness was born of a prediction by a member of the *Palladium's* staff that "the principle of the machine was faulty at the precise point where it subsequently broke."[16]

The problems of plane construction were, indeed, a long way from being solved, and the planes were still the subject of criticism. Even a year later, the Easton *Centinel* published an apology for a statement made in that paper that had apparently been misinterpreted as a condemnation of the Rockaway Plane. After all, Governor Clinton had been to see the plane in operation, and his testimony as to its efficiency was "sufficient to remove all doubts on the subject."[17]

Meanwhile, in an effort to bring forth some budding mechanical genius to solve the problems of plane construction, the Canal Company offered a $200 premium "for a working model of the best machine or invention, for passing boats from one level of a canal to another." The invention "must be so new and so far superior to any in use, either in Europe or America, or described in books of science, that the inventor would be entitled to a patent for it under the laws of the United States."[18] That was in December, 1826.

Two months later, in February, the Company announced that the $200 premium had been awarded to Ezekial Kitchell, Esq., of Whippany, Morris County, apparently for a substantial revision of his own design.[19] Kitchell, helped by his son-in-law, Calvin Howell, in building the Rockaway Plane, was later identified by William H. Talcott, Chief Engineer of the Canal, as a "millwright." Howell was elsewhere identified as a carpenter. Yet, in 1826, it was said that the Rockaway Plane "was built under the superintendance of the inventor, who is not a professed mechanic. The contractor is a house carpenter, and it may be easily imagined that the machine is not finished in the style that it might have been, if it had been made by a millwright or professional mechanic."[20]

For all the accolades given Kitchell for the performance of his device, and the premium awarded him by the Canal Company, the Rockaway Plane was, at best, a qualified success. In 1860, Talcott described it as a "summit plane with the machinery placed on a framework over the track at the apex of the plane." A glance at Fulton's *Treatise,* at the illustration of his plane machinery, shows virtually the identical configuration. Talcott labeled Kitchell's plane "an entire failure," and the problems remained.[21]

In all, 19 competitors had submitted models of inclined planes by the end of 1826 hoping to be awarded the $200 premium offered by the Canal Company.[22] But in retrospect it mattered little who won the contest or if it was won at all, for by the end of 1826 the MC&BCo. was in serious trouble of quite another kind.

Many other "money" institutions were in serious trouble during 1826. Too much speculation, too much wheeling and dealing, too many get-rich-quick schemes! One of the companies that failed and could not meet payments on its notes was the Life and Fire Insurance Company of New York. Its president, Henry Eckford, also owned a very profitable boat-building concern and was a director of both the Fulton Bank *and* the MC&BCo.

To save the Life and Fire Insurance Company from financial ruin, Mr. Eckford "temporarily borrowed" 2500 shares of Morris Canal stock worth $250,000.[23] President William Bayard was out of town at the time, and his son, William, Junior, was serving as president *pro tem.* At the request of the Canal Company's finance committee, the junior Bayard signed blank stock certificates before he, too, went out of town. On May 10, 1826, the Canal Company's cashier, J. Talman, also at the request of the finance committee, issued the certificates to the Fulton Bank, neglecting, however, to enter the transaction on the stock ledger of the Canal Company.

The Fulton Bank was to give the Canal Company 2000 of its shares of stock in exchange. The Life and Fire Insurance Company agreed to give the Canal Company $250,000 worth of its bonds as security for the return of the Morris Canal stock. In this rob-Peter-to-pay-Paul scheme, the Canal Company was to "receive a facility of fifty thousand dollars in money in consequence of said transaction," in other words, a commission, or interest on a loan. Morris Canal and Banking Company stock certificate numbers 356 and 357, each for 1250 shares, were thus issued to the Fulton Bank on May 10, 1826. The only notation in the Canal Company's books was that the cashier, Talman, was authorized to make loans of Canal Company stock to the Life and Fire Insurance Company, at a rate of 1/4 of 1% per month commission, and bonds in the appropriately corresponding amount as security for the loan's repayment.[24]

In the center of all this financial wizardry was the figure of Jacob Barker, a wealthy New York shipper, merchant, and financier.* Barker had been asked by several financially shaky firms to supply funds to cover their embarrassments. He was no altruist, just an honest businessman, as he called himself. Of course he expected to receive a "fair remuneration" for his efforts — a broker's percentage

**Barker's interest in transportation led him to import the first steam engine used by Fulton in a steamboat (New York Times, December 27, 1871).*

— on behalf of the Fulton Bank, the Tradesmen's Bank, the Dutchess County Insurance Company, the Hudson Insurance Company, the Life and Fire Insurance Company, *and* the MC&BCo.[25]

When the hypothecation of Morris Canal stock to the Fulton Bank became known, the roof fell in on virtually all of the then-embryonic Wall Street. Three Canal Company directors, Thomas Vermilye, William P. Rathbone, and Joseph G. Swift, plus Henry Eckford, the Canal Company's Assistant President, resigned their posts when they were indicted, along with several others, "by a New York grand jury for conspiracy to defraud. . . ."[26]

The judge presiding over the trial forbade the newspapers to publish the testimony given in court, presumably because it was too inflammatory and would have had adverse economic effects in the commercial world.[27] Fortunately for posterity, Barker, claiming a miscarriage of justice, carried his case to the public in an effort to reestablish his good reputation. He issued a series of pamphlets containing the unpublished testimony of several of the witnesses and several of the indicted co-conspirators, hoping that *their* words would clear his name.*

Barker had been badly used once before — by the United States Government. He had put his personal fortune at the disposal of the government to help finance the War of 1812. He was in the Capitol when the British set fire to the still-unfinished White House, and helped Dolly Madison to rescue Stuart's portrait of George Washington from the flames. Despite his patriotism and generosity, the government never repaid the monies borrowed from Barker, and he had had to begin all over again to rebuild his fortune.[28]

Now, in 1826, he claimed he was being badly used again, being made the scapegoat in the financial free-for-all staged by others. Although the grand jury was dismissed by October 25, 1826, unable to reach a unanimous decision as to who was or was not guilty, a new trial opened in November. The new indictment did not name Rathbone, Swift, or Eckford. Barker, already ruined in both reputation and fortune, left New York for New Orleans and yet another new start.

**See Reference 23.*

Meanwhile, the Canal Company's directors had ordered an inspection of the Company's books, and appointed a new finance committee for that purpose. The Newark *Sentinel of Freedom* carried this statement by Messrs. Bayard, McFarlan, and Abraham Ogden, members of the committee:[29] "Affairs for the moment are under a passing cloud, but we look forward with firm hope to the final completion of the enterprise." That Committee of Finance subsequently published its findings on September 21, 1826:[30]

> *Certainly enough has been accomplished to demonstrate the practicability of completing a direct inland communication between the Passaic and Delaware in the course of the next year, at an expense not exceeding the means contemplated by the charter. Although these means have been greatly abridged by the large amount of stock, recently thrown back on the company, the undersigned have great satisfaction in stating to the stockholders, that the losses which have been otherwise sustained may be safely estimated at less than seven thousand dollars; and that a full equivalent has been received in efficient work for every dollar which has been expended on the canal.*

In a response to this statement from the committee, the stockholder's own committee decided to express, by resolution, their confidence in the report, in the "skill and fidelity of the Engineer," and in the progress made to date on the canal. They then resolved "that in consequence of the defalcation of the former subscribers to the stock of the company, it is expedient to offer a new subscription in such manner as not to be obligatory, unless 3000 shares be subscribed."[31]

After all, the stocks that had been hypothecated to the Fulton Bank had all been returned to the Morris Company. All was now well, and it was time to get on with the work at hand.

Another stockholder's committee of investigation had reported earlier, on September 6, that even the conduct of J.T. Talman, the cashier, had been blameless, and that his part in the "transfer of 2500 shares of Morris Canal stock to the Fulton Bank, for account of the Life and Fire Insurance Company was made in virtue of a general authority from the Finance Committee," and since that entire committee had already resigned, things were right again.[32] Blameless or not, the very next day, September 7, 1826, the *Palladium of Liberty* announced that Talman, too, had resigned, and was replaced by Robert Gilchrist, Esq., as cashier.

To further convince the public that all was well with the MC&BCo. and with the canal itself, Cadwallader D. Colden, prominent New York attorney, was invited to review the canal and publish his thoughts on the waterway. The distinguished background of Mr. Colden, it was hoped, would help quell public doubts. He had been mayor of New York City from 1818 to 1820, then served in the New York State Legislature. He had also been New York City's District Attorney. In fact, Colden had even served as counsel for the defendant, Henry Eckford, during his trial for conspiracy to defraud.[33]

Colden's grandfather, Cadwallader Colden, had been New York's Surveyor General, and one of the most outspoken proponents of the Erie Canal. Following family tradition, Mr. C.D. Colden now took up the gauntlet on behalf of the Morris Canal. He reported on November 2, 1826, that he had closely examined the affairs of the company, and liking what he saw, felt "induced" to take some stock in the company and to visit "the whole canal line from the Passaic to the Delaware."[34]

His *Report to the President and Directors of the Morris Canal and Banking Company* is an eyewitness account of the progress made on the canal, as well as an expression of confidence in the company. Enthusiastically, he reported that there would be 22 inclined planes "by which an ascent and descent, together amounting to 1500 feet, will be passed." One, at Rockaway, was already completed. "Besides those, locks for 150 feet lift will be required. Of these, eight are completed," as well as "two important guard locks and several dams." In fact, "two-thirds of all the excavation is done, and forty miles, or nearly one half of the whole canal is entirely completed, and is ready to receive water." Of course the 40 miles were not contiguous.

The Rockaway Plane, Colden thought, was "perfectly satisfactory as to the evidence it affords of what may be done by this means, yet it is obvious that it is not so perfect a machine of this kind that might be made." In any event, "the superintendents have considered this an experiment. It was necessary that it should be made with all possible expedition, because, upon its success depended the means to be adopted of overcoming the elevations in different parts of the canal. The work, therefore, bears marks of the haste with which it has been executed."

Within five days of his report, on November 7, 1826, stockholder Cadwallader D. Colden was elected a director of the MC&BCo., along with Frances Baretto, Jr., Billopp B. Seaman, and Lambert Suydam, as replacements for the directors who had been forced to resign. Later that same day, the directors unanimously elected Colden president of the Company.[35]

The Canal Company that Colden now headed was quite different from the kind of company Macculloch had envisioned. Many of the directors for 1826 represented interests that seem far removed from his efforts to serve New Jersey's economy. Macculloch complained that the powers granted to the Canal Company had been intended for New Jersey citizens, but had now "become vested on the opposite shore of the Hudson."[36] Indeed, reviewing the addresses and occupations of the 1826 directors, the reader can see that New Jersey was represented by only Macculloch and Israel Crane. Moreover, only three directors, D.B. Ogden, Samuel Gouverneur, and Henry McFarlan, ostensibly had anything to do with New Jersey iron. And while the Morris Canal was said to be a blow to the coasting vessel trade in coal from Philadelphia bound for New York, the presence of so many directors with shipping interests suggests that the shipping trade made a quick adjustment to the idea of the coal-carrying Morris Canal.

William Bayard, Sr., N.Y.	Shipbuilder, ship owner
Thomas H. Smith, N.Y.	Canton trader, tea shipper, fleet owner
Joseph G. Swift, N.Y.	Revenue Inspector, Port of N.Y.
Israel Crane, N.J.	Freight forwarding; turnpike owner
Thomas Vermilye, N.Y.	Officer of several money institutions
Henry Astor, N.Y.	Brother of John Jacob; meat dealer
Duncan P. Campbell	Unknown stockholder
Abraham Ogden, N.Y.	Nephew of Gouverneur Morris; first U.S. Attorney for N.J.
Henry Eckford, N.Y.	Shipbuilder; financier

William Bayard, Jr., N.Y.	Shipbuilder; lawyer
D.B. Ogden, N.Y.	Lawyer; part owner of Ringwood Iron with Samuel L. Gouverneur
George P. Macculloch, N.J.	Canal promoter; stockholder
William P. Rathbone, N.Y.	Eckford's partner in shipbuilding
Samuel L. Gouverneur, N.Y.	Part owner of Ringwood Iron with D.B. Ogden; son-in-law of President Monroe
Henry McFarlan, N.Y.	Iron magnate; owner of Dover Iron

This group of directors and its new president, Colden, set about extricating itself from the embarrassments of the stock scandal. There had been a reduction of public confidence in the canal enterprise, and a great reduction of working capital for canal construction since the Canal Company had to take back 6000 of the 10,000 shares of stock originally sold. Half of those 6000 shares were resold on a new subscription for 10- and 20-percent part-payments. The remaining 3000 shares were left with the Company. As Colden later speculated: "Had the stock, instead of so large a portion of it being disposed of as an object of speculation, been distributed among those who subscribed to the amount of $7,000,000, when the books were opened, the company would have been able to complete the canal this season," i.e., during the 1827-28 season.[37] But then, the Morris Canal would have remained a New Jersey, not a New York concern, for the original subscribers of 1825 were said to have been Jerseymen.

This "derangement of the affairs of the company" had made it necessary to discharge a "great part of the workmen" in September, 1826, just a little more than a year after construction had begun. Beach estimated that, originally, 1100 men had been employed on the canal construction. By the time of Colden's May 1, 1827, report to the directors, only 530 men were at work.

Accompanying Colden on his May 13, 1827, inspection tour of the canal was DeWitt Clinton. The men were "welcomed at different places, by the inhabitants, by the discharge of artillery, and such other demonstrations of respect as were justly due to the distinguished patron and promoter of internal improvements, by canals."[38] Writing to Colden after their tour, Clinton claimed that since he personally had "participated in a passage up and down" the Rockaway Plane, he could "speak with confidence on the subject" of its effectiveness. The plane could doubtless "be greatly improved," but "in its present state, it affords unequivocal testimony in favor of the utility, the practicability and the economy of the erection, and completely silences all cavils and objections."[39]

Colden reported satisfaction "that a well-constructed machine, made sufficiently strong and powerful, upon the same principles with this one [at Rockaway], would answer the purpose." He seemed particularly impressed by the fact that the plane worked "by the mere force of the water power."[40]

We read of other designs for inclined planes in the newspapers of the period. A Mr. Robert Graves, of New York City, had patented an invention which would "transport a boat of any weight from the lower to the upper level" of a canal, "at the same time bringing one, if required, from the upper to the lower level, 90 feet, and launching each into the adjoining canal in the space of 12 minutes," and operating entirely upon water power.[41]

Morristown millwright, William Knight, also presented "a perfect model of a new Inclined Plane of his invention. The rail-ways are similar to others, but the cars are so constructed that a boat floats from the canal into it from the upper or lower level, and displaces as much in weight of water from the car as the boat and cargo, so that one car, whether it contains a boat or not, completely counter-balances the other." The machinery was described as "very simple and not liable to get out of order."[42]

All of the planes described so far, Renwick's, Kitchell's, Graves', and Knight's were double-tracked planes. It is by no means implied here that either Graves' or Knight's plane was ever used on the Morris Canal, merely that their designs were at least similar to the early Morris Canal devices — and that there *were* other designs.

The next experimental plane to be built on the Morris Canal was designed and built by Canal Commissioner Colonel John Scott at Boonton Falls in 1827. It, too, was a double-tracked plane, but it apparently differed substantially from the Renwick and/or Kitchell designs, being "constructed upon a different plan from that at Rockaway."[43]

Beach reported in May, 1827, that the ground was being prepared for the masonry work to begin at Boonton. Originally, the elevation was to have been 90 feet, "the highest on the canal," but when built, it was actually an 80-foot lift.[44]

Talcott described Colonel Scott's plane as having "the machinery placed underneath the track at the apex of the plane" and working with an endless chain.[45] The *Paterson Intelligencer* reported that the Boonton Plane had been tested on August 2, passing a boat "from the lower to the upper level in 20 minutes." Its operation was described, then, as "highly satisfactory."[46]

The plane, itself, was "an improved" plan, but "owing to the very bad construction of the walls, they gave way after the boats had passed over the planes." Still, Colden reported, the passages had been viewed as "entirely satisfactory to those who observed them, among whom were several scientific men."[47] The walls were subsequently rebuilt, and on December 13, 1828, a Canal Company inspection committee passed over the plane in a boat 60 feet in length, and 8-1/2 feet in width. In the boat with the committee were "about 100 passengers," and "18 tons of stone." The passage took 14 minutes. While the entourage ascended the plane, "an empty car descended. Had the descending car contained a loaded boat, the ascent would no doubt have been accomplished in a shorter time."[48] From this description, we learn that not only was the Boonton Plane double-tracked, but also that the descending car and boat acted as a counterweight and boost for ascending traffic.

In mid-1827, in the *Report to the Directors of the Morris Canal and Banking Company,* President Colden and Chief Engineer Beach described the progress made to date on the canal's construction.[49] All of the "principal embankments" were completed, as were the "great dam and locks at the outlet of Hopatcung lake [sic]." However, the three-quarter-mile feeder canal that would bring water from that lake to the main canal was but half done. The half-mile-long Beaver Brook feeder was nearly completed, providing the basis for some later, complicated ligitation.

Two-thirds of the 19 other locks were done, as were four culverts on the western, and five on the eastern division, the one at Bloomfield being described as "a very heavy work." The eastern division already had 22 completed road and farm bridges, and there were 31 completed on the western division; while timber and materials were already provided for several others.

In addition to the Rockaway Plane, which was then considered finished, the Boonton plane site was ready for masonry work, and the "walls for another [plane] of thirty-eight feet lift" were already laid at Drakesville. Some materials were provided for other planes, but none of them had yet been started.

Of the aqueducts, all the mason work was completed at Brass Castle Creek, Hurd's Brook, Beaver Brook, Rockaway River [Denville], and Peckman's River at Little Falls. Also at Little Falls was the site of an aqueduct to cross the Passaic River, but no work had yet been done; nor had the Pompton River structure near Mead's Basin (Mt. View, Wayne) been started. It was expected that both of those large aqueducts could "be completed without any difficulty by the Fall."

The report on this substantial progress was dated May 1, 1827. By July 4, the *Paterson Intelligencer* remarked that much further progress had been made. In fact, "the aqueduct over the Passaic has been commenced, and is rapidly going forward to its completion." The Canal Company had advertised for bids from contractors for the aqueduct on February 15, announcing that specifications for that work were available "at the Canal Office of the Company, at the City Bank, New York, at their Banking House, Jersey City, or on application to the Engineer at Mr. Israel Crane's, Bloomfield, or to the Canal Commissioner at Dover, N.J." Proposals from contractors were to be "directed to John Scott, Canal Commissioner."[50] The firm of Ira Dodd and Caleb Dodd Baldwin undertook the building of the aqueduct, and it was completed late in 1829.[51]

In 1827, it was expected that the canal, when completed, would transport 232,502 tons of coal, at an estimated cost of $3.15 per ton or less. The demand for coal was expected to be "greatly increased" because of its cheapness, and it was suggested that "a sufficient number of boats could not be passed on the Morris Canal to supply one-third of the demand." Moreover, "the products of agriculture and manufactures" would also be increasing, the transportation of which would "be even more profitable than that of coal."[52]

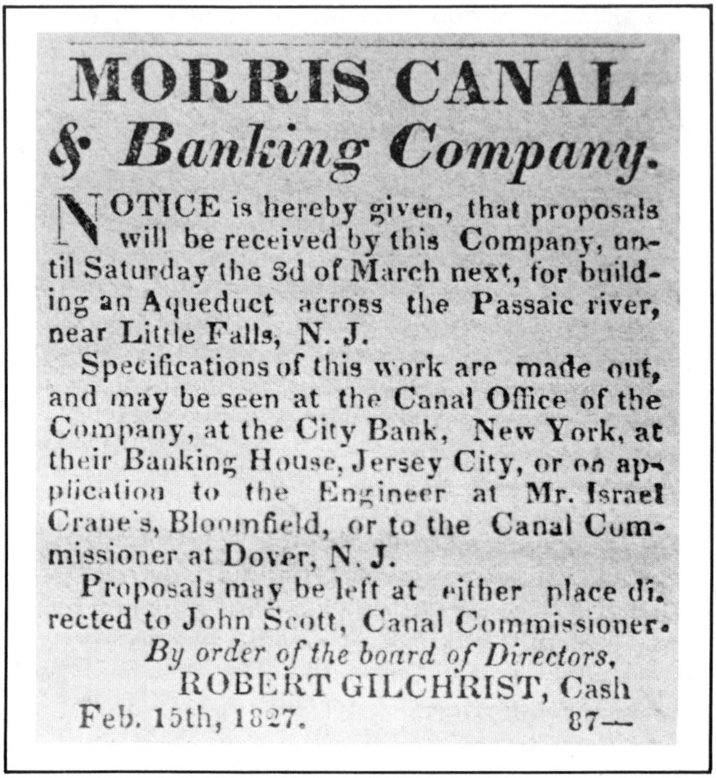

MORRIS CANAL
& *Banking Company.*

NOTICE is hereby given, that proposals will be received by this Company, until Saturday the 3d of March next, for building an Aqueduct across the Passaic river, near Little Falls, N. J.

Specifications of this work are made out, and may be seen at the Canal Office of the Company, at the City Bank, New York, at their Banking House, Jersey City, or on application to the Engineer at Mr. Israel Crane's, Bloomfield, or to the Canal Commissioner at Dover, N. J.

Proposals may be left at either place directed to John Scott, Canal Commissioner.

By order of the board of Directors,
ROBERT GILCHRIST, Cash

Feb. 15th, 1827. 87—

Advertisement for bids to build the Little Falls Aqueduct (Morristown Palladium of Liberty, February 22, 1827, photograph by Irving Fitzig).

A "disinterested and impartial observer" from New York had prophesied that the Morris Canal would "monopolize the transportation of Lehigh coal to New-York, as well as for the manufacturing villages on and near its line, where it must be extensively used, as the anthracite is now getting into use in preference to other coal in most branches of iron work, as well as for heating in other manufactories, although it has not yet been found to answer for smelting ores."* Furthermore, he predicted that the Morris would be "fair competition with the Hudson and Delaware [Canal] in the coal trade."[53]

Another New York observer was sure that "the immense quantities of coal which will be furnished to New-York by the Lehigh will always prevent the necessity of returning empty; on the contrary, our goods going back to Easton will be considered as backloading to boats in the regular coal trade which this route must secure to us."[54]

An interested, collective, Canal Company eye was watching the progress of the Lehigh Navigation Canal, too. The Morris' engineers had visited Mauch Chunk, Pennsylvania, in 1826, in order to see for themselves the "mysterious" switchback gravity railroad that Josiah White was experimenting with. And they wanted to see for themselves, too, exactly "how their watercourse would, and could, tie with the 'Lehigh Canal'."[55] The Lehigh's managers were likewise keeping an anxious eye on the progress of the Morris Canal, since Josiah White was counting heavily on the tidewater markets the Morris would provide for Lehigh coal. The earliest annual reports of the Lehigh Coal and Navigation Comany contain many references to this desirability of the northern New Jersey route to New York.

Just about everyone seems to have been anxious for the canal's completion so that coal would become plentiful and therefore cheap. Colden had long been involved in a project called the Fuel Saving Fund, which he sponsored for New York's Pauperism Society to provide fuel for the city's poor. Coal had also been used for lighting the city's gas lamps since 1823. It took so much less space to store large quantities of coal in coal yards, as contrasted with the amount of space need to store cord wood.

All that was needed to perfect the coal-burning process in the home was the development of superior cast iron stoves or furnaces. By 1832, President Eliphalet Nott of Union College, Schenectedy, New York, had invented such a stove.† It was then introduced into most of New York's churches and public offices, and was described

**The use of anthracite in smelting ore was initiated by David Thomas of the Crane Iron Works in the Lehigh Valley, but not until 1841.*

†*The reader will recall that Professor Nott had been Ephraim Beach's teacher.*

by one man as "secure against accident from fire," economical in consumption of coal, and yet keeping temperatures indoors at an equal level, day and night.[56]

Clearly, the markets were ready and waiting to receive coal, and the suppliers were anxious to ship it. There were truly great expectations for the Morris Canal. But the canal, two and a half years after construction was begun, was a long way from completion, much less from successful, profitable operation. Those devilish inclined planes!

REFERENCES

1. *The Franklin Journal, and American Mechanics Magazine,* Vol. II, No. 5, December, 1826, pp. 329-32.
2. *Ibid.,* p. 332.
3. *Ibid.,* November, 1826, pp. 260-3, -5.
4. *Ibid.,* p. 263.
5. *Ibid.,* December, 1826, p. 330.
6. *Report of the Commissioners Appointed by the Legislature of the State of New Jersey, For the Purpose of Exploring the Route of a Canal to Unite the River Delaware, Near Easton, With the Passaic, Near Newark, With Accompanying Documents* (Morristown, 1823), p. 7.
7. Munsell's *History of Morris County, New Jersey 1739-1832* (N.Y.: 1882), p. 68.
8. *Biographical and Genealogical History of Morris County* (N.Y.: 1899), pp. 43-4.
9. I.N. Phelps Stokes, *Iconography of Manhattan Island, 1498-1909, Compiled From Original Sources* (N.Y.: 1918), p. 1657.
10. Cornelius C. Vermeule, Jr., *Final Report of Consulting and Directing Engineer, June 29, 1929* (E. Orange, N.J.: 1929), p. 68.
11. John Langdon Sullivan, *Report on the Origin and Increase of the Paterson Manufactories and the Intended Diversion of Their Waters by the Morris Canal Company, Also on Post Rail Roads, as the Means of Cheap Conveyance Throughout New-Jersey, of Bringing to Susquehanna Coal to the Iron Mines and Forges, and to Supply Patterson and New-York: Also on a Method of Supplying the City of New-York with Water from the Great Falls of the Passaic.* (Paterson, N.J.: 1828), p. 15.
12. James M. Ransom, *Vanishing Ironworks of the Ramapos* (New Brunswick, N.J.: 1966), p. 102.
13. *Palladium of Liberty,* July 13, 1826.
14. *Paterson Intelligencer,* July 12, 1826.
15. *Ibid.,* August 2, 1826.
16. *Palladium of Liberty,* August 3, 1826.
17. Easton *Centinel,* June 27, 1827.
18. *Paterson Intelligencer,* December 13, 1826.
19. *Ibid.,* February 21, 1827.

20. *Ibid.*, November 15, 1826.
21. *Report to the Stockholders of the Morris Canal and Banking Company* (N.Y.: 1861), p. 16 (hereinafter cited as *Report to the Stockholders, 1861*).
22. *Paterson Intelligencer*, June 6, 1827.
23. This account of the Canal Company "stock scandal" has been gleaned from a variety of sources, chief among them Jacob Barker's *Letters Developing the Conspiracy Formed in 1826 for His Ruin* (N.Y.: 1827).
24. *Ibid.*, p. 5.
25. *Ibid.*, various pages.
26. New York *American*, August 14, 1826.
27. *Paterson Intelligencer*, October 18, 1826.
28. Robert G. Albion, *The Rise of New York Port* (N.Y.: 1939), p. 436.
29. Newark *Sentinel of Freedom*, August 22, 1826.
30. *Paterson Intelligencer*, September 21, 1826.
31. *Ibid.*
32. *Ibid.*, September 6, 1826.
33. *Palladium of Liberty*, August 17, 1826.
34. *Paterson Intelligencer*, November 15, 1826.
35. *Ibid.*
36. *Ibid.*, December 24, 1828.
37. *Report to the Directors of the Morris Canal and Banking Company . . ., May 1, 1827 . . .* (N.Y.: 1827), p. 11 (hereinafter cited as *Report to the Directors, 1827*).
38. *Palladium of Liberty*, May 17, 1827.
39. David Hosack, *A Memorial of DeWitt Clinton* (N.Y.: 1829), pp. 221-2.
40. *Report to the Directors, 1827*, p. 5.
41. *Paterson Intelligencer*, July 4, 1827.
42. *Ibid.*, August 27, 1828.
43. *Ibid.*, August 13, 1828.
44. *Report to the Directors, 1827*, p. 18.
45. *Report to the Stockholders, 1861*, p. 16.
46. *Paterson Intelligencer*, August 13, 1828.
47. Cadwallader D. Colden, *An Answer to Mr. John Sullivan's Report to the Manufacturing Society in New Jersey* (N.Y.: 1828).
48. *Report of the Joint Committee of Council and Assembly Appointed to View the Morris Canal and Inclined Planes* (N.Y.: 1828), p. 7.
49. *Report to the Directors, 1827*, pp. 13-18.
50. *Palladium of Liberty*, February 22, 1827.
51. *Paterson Intelligencer*, November 25, 1829.
52. *Ibid.*, July 4, 1827.
53. *Ibid.*, September 20, 1826.
54. *Ibid.*, February 21, 1827.
55. Eleanor Morton, *Josiah White; Prince of Pioneers* (NY.: 1946), p. 158.
56. *Letters from John Pintard to His Daughter Eliza Noel Pintard, 1816-1833* (N.Y.: 1940), Vol. IV, pp. 12-13.

6

Try, Try Again

THE Upper Montville Plane, No. 8 East, was placed under construction during 1827-28. It was to be built "upon the plan and under the direction of Robert P. Bell."[1] Bell, a New York City dry goods merchant at 11 Front Street, had been made a Canal Company director on November 14, 1826.[2] His career with the Canal Company, as Superintendent of the Western Division, as director, as member of the land-buying Canal Committee, will occupy us later. Central to this chapter is the fact that his plane mechanism was considered a "successful experiment," giving "much satisfaction at the time."[3]

A "Mr. Miller" who was the "mechanic" at the Montville inclined plane undertook the construction of Bell's design.[4] It was described as "a summit plane with the machinery placed under the track like the Boonton plane, and worked by two water wheels on one shaft in place of a clutch to change the motion."[5]

Nathaniel Ogden Condit, a Montville carpenter, is said to have "built the first water wheel on the inclined plane of the Morris Canal at Montville."[6] Timothy Condit, brother of Nathaniel, and Aaron Baldwin were suppliers of materials for Mr. Miller at the Montville and Beavertown planes, No. 8 East and No. 10 East.[7] The Condits later established a tannery at the foot of Plane No. 8 East.

The same "Mr. Miller," now identified as Eliphalet Miller, was the builder of the Beavertown Plane, No. 10 East. That plane stood at the border between Towaco and Lincoln Park, known then as Whitehall and Beavertown. Most often the plane was called simply the "Pompton Plane," because it connected with the level of the Pompton River at its foot. Eliphalet Miller's mechanism was altogether different from the planes built at Rockaway, Boonton, even at Montville. They were all "summit" planes, with a "brow" or "hump" of earth at the top of the plane to prevent water in the canal's upper level from running out and down the plane. Miller's design was a "lock" plane, but no description of it is known to exist.[8]

Aaron Baldwin and Timothy Condit supplied Miller with materials, as we have seen, and Charles Ford supplied iron plates "for the ways of the inclined plane" at Beavertown.[9] It is not impossible, or even unlikely, that Nathaniel Condit supplied the waterwheel for this plane, too.

When the Canal Company offered its $200 premium for new inclined plane designs, Ephraim Morris was said to have been one of the 19 competitors who submitted designs for plane mechanisms, and to have won the premium.[10] Morris' suppliers were Caleb Dodd Baldwin and Ira Dodd, who operated the ancient and famous Morris Mill in Bloomfield, in which they produced iron parts and the machinery for the Bloomfield Plane, No. 11 East. The reader will recall that Dodd and Baldwin were the builders of the Little Falls Aqueduct over the Passaic River, and that Ira Dodd had been one of Ephraim Beach's assistant engineers on the Schuylkill Canal.

Morris' inclined plane was a "lock" plane, as was Miller's Beavertown Plane, but "differing materially" from it. Morris is credited with having devised a "new plan called a lock plane because the boat was received into a lock at the head of the plane" instead of going up over the brow and descending into the upper level of the canal "on the car as was done at the summit planes."[11]

Like the Rockaway plane site, Bloomfield's was a 52-foot elevation. It had a length of 624 feet, and a lift-ratio of one foot in twelve. It was described in the August 6, 1829, *Paterson Intelligencer* as being

> *constructed on the balance principle — that is, a chain passes round a pulley wheel at the head of the plane, and a*

car, on which a boat and cargo is to be transported, is attached to the end of a chain. The pulley-wheel is connected with a water-wheel by a common gearing so that the motion may be instantly and easily reversed by throwing spur-wheels in and out of gear. The descending assists the ascending body, just as the descending bucket in a well assists that which is ascending. And the water-wheel at the head of the plane (to continue the familiar comparison) supplies the place of the man at the crank of the wheel; and as the man at the crank would have to hold back, if the descending bucket were the heaviest, so at the plane, if the descending boat is the most weighty, the water-wheel will be used to retard the motion, instead of propelling it. The trade from the summit level to Newark will be chiefly a descending trade — that is, the descending freight of coal, iron, ore, lumber, plaster, agricultural products, &c. ninety-nine times out of a hundred will be the heaviest. And in such cases no water will be used on the water-wheel.

Note that Morris' plane also operated in part as a gravity plane. Note, too, that the water wheel was located at the head of the plane.

The greatest difficulty on previously tried inclined planes on canals had been "to pass the boat from the apex, or highest part of the plane, into the water of the upper level, or vice versa, from the water to the plane." Double planes had been tried, whereby "the plane from the lower level was carried up to some height above the water in the lower level, and from the highest point a plane was made to descend into the water of the upper level upon which the boat was to be launched, and to descend by gravity into the water. When a boat was to descend, she was to be drawn up by machinery to the apex, and thence to descend to the lower level by her preponderance alone or with the assistance of a water wheel. Another mode which has been tried, is, to carry the boat up in a movable lock."[12]

Whatever difficulties the double plane and movable lock presented, Morris' "very simple machinery" had overcome them. Morris' device, the *Intelligencer* reported, consisted of "two lock's (or rather two half locks) upon the inclined plane at its head. Into one of these locks the ascending boat is drawn by the pulley-wheel, while the descending boat is leaving the other lock and descending to the lower level. When the ascending boat reaches the top of the plane, she is completely within the lock, into which the water is let from the upper level. And during the operation, by a most ingenious contrivance, as the water rushes in, it raises a gate behind the boat, closing the lock, which then fills with water till it rises to a level with that of the upper level. The upper gates of the lock then open, and the boat proceeds on her voyage."

For the New York *Commercial Advertiser's* reporter, the most "beautiful part of the machinery of this plane" deserving mention was the governor, by means of which "the possibility of any great injury in case the chain should break, or any other part of the machinery give way, is prevented." The governor "is so applied to the wheels that the moment their velocity is unduly accelerated, large wedges drop upon them instantly, and stop their motion." To Ephraim Morris, the contractor and inventor, belong the credit for the governor and the "self-acting water gate," both of which were "proof that great ingenuity and ability" had "been employed in the construction of this plane" at Bloomfield.[13]

A few months later, Niles' *Register* gave a more detailed picture of Morris' invention:[14]

> *The boats to be passed along this plane, are introduced into a cradle of the proper dimensions, suspended by vertical iron rods from the framework of a large car. The car is supported by 4 cast iron wheels of about 8 feet diameter, the rims or circumferences of which are made concave, to correspond with the convex surface of the rails which sustain and direct the car in its motion. There are two rail tracks, or in other words there are 4 distinct and parallel lines of rails which extend from the summit to the foot of the plane. A car is provided for each track, and they are connected, the one with the other, by means of a strong iron chain passing around an iron drum or wheel at the head of the plane. The length of this chain is such, that when one car is at the foot of the plane, the other is at the summit. The drum or wheel around which the chain passes, is placed horizontally, and has a diameter equal to the*

perpendicular distance between the centre lines of each track. Directly under the drum is a transverse horizontal shaft, extending out on one side of the plane, and supporting, at its other extremity, a water wheel of 18 or 20 feet diameter. This shaft is connected with the drum through the intervention of a double set of cog-wheels, so proportioned as to enable it to act upon the drum at a great mechanical advantage. The water wheel, which is the first mover, is propelled by the water drawn from the trunk of the canal on the upper level, and motion is communicated to the cars in either direction, by merely shifting the gearing agreeably to the means provided in the double set of wheels above mentioned.

[*Descending boats and their cars*] *approaching the foot of the plane, are made to plunge into the water of the canal at the extremity, hence the boats which they contain are easily discharged, and those which are to convey upwards, are as easily introduced into them. At the head of the plane, they are made to enter lock chambers of a suitable size, which being filled from the canal above, the boats are enabled to enter upon, or pass off from them with the utmost facility. The water which is discharged from the lock-chambers, as well as that which passes on from the water wheel, is conveyed along a raceway, formed on the side of and parallel with the plane, into the lower level.*

[*The rails were described as*] *bars of cast iron, about 2-1/2 feet in length, 3-1/2 or 4 inches in breadth, and about 1-1/2 inches mean thickness, and . . . connected with each other by ball top joints, and onto the timbers on which they rest by iron spikes.*

As for Mr. Morris' lock chambers, the upper ends were "closed in the usual manner," by double hinged gates, "but the lower ends have but one gate, each running square across, with a horizontal axis, and when open, are made to be flat down in a recess formed in the bottom of the chamber."[15]

President Colden and a joint committee of Canal Company directors and State Councilmen and Assemblymen visited the canal to inspect the work in general, the inclined planes in particular. At Bloomfield, the group saw "a very considerable number of workmen" employed on the plane, and it was expected that the device would be completed by April 11, 1829. At Paterson, the Committee viewed the "deep cut" at Garret Mountain, "averaging about 16 feet, through a hard rock of blue trap, extending 16 rods [264 feet]." This spot was "far above the highest trees at the base of the ridge, and much higher than the houses and factories in the town of Paterson."[16]

The Little Falls Aqueduct was not yet finished when the committee passed through Little Falls, but the less imposing, wooden aqueduct at Mountain View was ready for first-hand inspection that day, December 13, 1828. Two boats awaited the committee at Mead's Basin,* where they boarded "to navigate 20 miles of canal completed to Dover."[17] They crossed the Pompton River on the 236-foot-long aqueduct and inspected the stone lock at Lincoln Park (then Beavertown), about a mile beyond and west of the river.

Colden was careful to note that "from Mead's Basin to Dover, the canal was supplied with water from the great reservoir at the summit, the Hopatcong Lake." When the waters of Hopatcong were seen to mingle with those of the Rockaway River at Dover on August 18, 1828, there had been quite a celebration in the town.[18] However, as Colden's report makes clear, the canal was not completed above Dover at the end of 1828.

"The water of that lake," he explained, "for the supply of the canal, is brought into the pond at Dover, and passes from the pond into the canal. It is, in like manner, conducted into, and taken out of, a pond at Powerville, and at Booneton [sic] Falls there is a dam and feeder, by which water from the Rockaway river, may be taken into the canal. These are the only connexions that the canal has (except that it must, of necessity, take in some inconsiderable runs) with the Passaic, or any of its tributaries."[19]

In the two small boats they had boarded at Mead's Basin, the joint committee members were then transported from Lock No. 13 East, as the Lincoln Park lock was then numbered, "to the foot of the

**A post office was established at Mead's Basin in January, 1829, with Mr. H. Mead appointed postmaster.*

Montville Planes, always on the canal, except that the boats were carried, and the committee walked around the unfinished Pompton Plane." At the Montville Planes, the committee "walked a distance of about 800 yards, to the head of those planes," where they boarded "one of the boats, which the company have built to be used upon the canal," and were then "carried up the inclined planes [at Boonton?], and along the canal to the unfinished plane at Rockaway, and from the head of that plane, in another canal boat, to Dover."[20]

From Colden's report we learn that at the end of 1828, the Pompton, or Towaco, Plane was unfinished, and that the Montville Planes were also unfinished. Only the Boonton Plane, then, seems to have been ready to receive boats, for the committee were carried up the incline and then to the foot of the Rockaway Plane. From the head of that plane, they were then transported in yet another canal boat to Dover.

At Lake Hopatcong, the committee met Mr. Randolph, who was charged with the "care and superintendance of this part of the work: the dam, the waste gate, and the lift lock into the lake." Since one of the gates of that "lift lock" had been damaged during "the last winter," the waters of the lake had been lowered so that repairs could be made.[21] Other locks, however, were apparently in good working order, and the committee had the distinction of passing through their water-filled chambers for the first time.

On its way westward, the committee stopped overnight at Hackettstown on December 3, 1828, and then resumed its journey to Phillipsburg. At Phillipsburg, at a "spot directly opposite to the mouth of the Lehigh, and opposite to the Lehigh Canal, where the Morris Canal will fall into the Delaware," they saw where the Commonwealth of Pennsylvania planned to erect "a dam across that river." Such a dam would form a basin which would serve the Lehigh Canal, the Delaware Division Canal, and, of course, the Morris Canal — all of which had their "termination" there.[22]

In February, 1828, a group of interested citizens had met at Easton to express their views on the subject of using the waters of the Delaware River to feed canals in both New Jersey and Pennsylvania. The "interested citizens" were in favor of such a dam so long as the up-river trade could be accommodated by locks or planes.[23]

But in December, 1828, the joint committee's chief concern — indeed, its reason for being — was the inclined plane intended for use on the Morris Canal. By its own admission, only the Boonton Plane was operating successfully when it made its inspection tour. The Rockaway Plane, while operating, was not considered "completed," but, rather, very much in need of improvement. The Bloomfield Plane, of Ephraim Morris' design, was, perhaps, the one to watch, when a few months hence it would be tested.

Colden had said that the Bloomfield Plane would be completed by April 1829. Its efficiency was "partially tested" towards the end of that month, when 25 tons of boat, car, and presumably cargo, had passed the plane "in the space of 27 minutes, in the presence of the commissioners of the canal, and others, who had assembled to witness the experiment." The testing of the Bloomfield Plane before there was any water in the canal below Rockaway was made possible by using water taken from adjacent millstreams.[24]

For one reporter, "considerable doubt" remained as to the adaptability of the inclined plane "to all the purposes of navigation." In brief, he was puzzled as to why the Company hadn't built a railway "throughout the whole route."[25]

The Canal Company directors authorized the building of two more scows in order to test the Bloomfield and Towaco Planes. There were no millstreams adjacent to the Towaco Plane, so William Scott, who owned the water rights in the Rockaway River at Boonton, was paid $100 to bring water to the head of the plane.[26]

Ephraim Morris continued to have difficulty with his plane in Bloomfield, and made several requests that the Company give him more money to complete construction of Plane No. 11 East.[27] The Bloomfield Plane No. 11 East was one of four experimental planes that had been scheduled for completion in early spring 1829. They were "experimental" because each of them differed "in some measure, in its plan, from the other." Once they were built, the Boonton, Montville, Towaco, and Bloomfield planes would serve as models, "so that the company may select that which, upon trial, shall prove to be the best," making that one the standard "for building their other planes."[28]

Colden, anxious for the successful trial of the planes, was anticipating a visit from a delegation of Lehigh Coal and Navigation Company principals, including Josiah White. The reason for the visit was to promote interest in the canal, so that the Lehigh men would agree to a loan of $500,000 towards the completion of the canal. Colden returned from his personal inspection at Towaco and Bloomfield, "very apprehensive neither of said planes would be finished in time" for the delegation of potential investors. He decided to invite the Philadelphia group anyway, hoping that White would agree to join them, bringing with him his "scientific and practical knowledge."[29]

The Lehigh delegation *did* view the canal construction and the planes. Although it is not known whether or not White was among the reviewers, the group appears to have been well satisfied. The Morris Canal host-committee traveled back with them as far as Mauch Chunk, to see for themselves the coal mining and shipping operations there.[30] A subscription of $26,000 was made at that time, and by August 3, the whole $500,000 had been subscribed.[31] "Foreign," Philadelphia capitalists now had a stake in the Morris Canal, but at least the planes could be built.

Thanks to the efforts of Renwick, Kitchell, Howell, Miller, Bell, Scott, and Morris, the inclined planes of the Morris Canal *were* made to work, however haltingly. Still, on August 13, 1829, the Canal Company's directors authorized Colden to write to Major Douglass, a professor at the United States Military Academy at West Point, "requesting his opinion and advice respecting the inclined planes of the company."[32] On September 8, a committee was appointed to "make an arrangement with Major Douglass so as to secure to the company the benefit of his services in deciding upon the plan of inclined planes to be adopted, and in superintending the building of any that shall hereafter be erected."[33]

Within a week, Douglass offered to work for the Company as chief engineer for the construction of inclined planes. His terms were:[34]

1. $4000 a year, to include expenses and all demands on the company for previous services.
2. He was to have the entire responsibility for the design and execution of the planes.

3. He was to devise and mature all the plans and mechanical arrangements for same.
4. He was to designate the species, quality, and quantity of the materials and workmanship required by every contract, being the judge of the fulfillment of all contracts.
5. All workers on the planes were to receive orders only from him; he was to do all the necessary hiring of those workers and would set their wages.
6. He would make every effort to complete the planes in the best and quickest way, reporting periodically to the board of directors.

The board approved of Douglass' terms and decided to hire him — provided he was able to get a leave of absence from West Point.

Born in Pompton, Bergen County (now Passaic County), in 1790, son of Nathaniel Douglass, David Bates Douglass was graduated from Yale in 1813. He served as head of the Department of Mathematics at West Point from August 29, 1820, to May 1, 1823, when he was made head of the Department of Civil and Military Engineering.[35]

Douglass applied for a leave of absence from his West Point duties to devote all his energies to the construction of the Morris Canal's inclined planes. His most pressing task was to determine which of the experimental planes was to be used as a model for the construction of the remaining planes. It was imperative that the canal be finished as quickly as possible so that the Morris would be in a position to compete for the Lehigh coal trade. And competition there already was, with more promised for the immediate future.

The Delaware and Hudson Canal, organized on March 8, 1825, only a few months after the Morris, had been delivering coal to New York City since December 10, 1828, when its first shipment had reached the city.[36]

The Delaware Division Canal, although badly constructed when it was first built in 1827, was being repaired and enlarged to accommodate Lehigh coal boats bound for Philadelphia by midsummer of 1830.[37]

During 1827, the Erie Canal, then only two years old, collected $859,058.48 worth of tolls, an increase of almost $100,000 over 1826.

On May 9, 1829, the Erie collectors received $3,507 — one day's receipts![38] New Jersey's Morris Canal promoters were anxious to begin collecting tolls on such a scale.

During 1826, the Lehigh Company had shipped 30,000 tons of anthracite to Philadelphia, while New York City and New Jersey's iron industries idled for want of fuel.

The Lehigh Company expected the Morris Canal to be completed by mid-1830, at about the same time the Delaware Division Canal would re-open. If the Morris Canal were not ready to receive Lehigh coal, it was feared, and rightly so, that the Lehigh Company would seek another outlet for its anthracite.

The Delaware and Raritan Canal, chartered on February 4, 1830, was originally intended to carry Schuylkill coal, but with the rebirth of the Delaware Division Canal, and the probability that the Lehigh Company would ship over both the Delaware Division and the Delaware and Raritan (D&R) canals, the latter canal also posed a very real threat to the Morris. In May, 1834, when the D&R did finally open, "the first cargoes transported over it were lumber and stone for the new State Penetentiary, and coal brought down the Delaware Division Canal from the Lehigh fields."[39] That was but the beginning.

The Morris Canal and Banking Company's *Minutes* during this period, 1829-1832, give the impression of almost frenetic activity. Only occasional notes of caution creep in, in such entries as: "In the exercise of the power hereby given [by the board of directors to the canal committee], be not governed solely by the lowest offer [or bid, on a contract], but that they shall take into consideration the character of the party offering and his ability to perform any work he may propose to undertake."[40]

The planes that had already been built as experiments still weren't working well. In January, 1830, William Scott, Canal Commissioner John Scott's brother, submitted a proposal to the Canal Company, offering "to make the Inclined planes at Pompton and Montville (known as Miller's and Bell's planes) working planes."[41] In March, Major Douglass submitted a plan for inclined planes that the Canal Company's directors accepted. They ordered him to "proceed at once in completion of the plane" (No. 12 East), and appointed a committee to "make arrangements" with Douglass as to "hiring assistants and subordinates" he thought necessary.[42]

Whereas the original agreement with Major Douglass had been on *his* terms, the Canal Company now proposed a $2500 annual salary, to be paid quarterly, unless he resigned his West Point position. If and when he resigned, his annual income was to be $4000, as previously agreed upon. Douglass was to be allowed to hire two or three assistant engineers and twelve superintendants of planes. All his other duties were to conform to the specifications set forth in the Company's *Minutes* for September 15, 1829.[43]

Once Douglass' plan for an inclined plane was accepted by the board of directors, he made a major decision — no pun intended. His application for a leave from West Point had been denied. But he seems to have become so caught up in the challenge of the inclined planes of the Morris Canal, particularly Plane No. 12 East at Newark, which he was building, that he resigned his position at the Military Academy on May 1, 1831, just as he had said he would if it became necessary.[44]

Near the middle of November, 1829, the New York *Morning Herald* announced that about 60 miles of canal, from the summit level to Newark, would be opened to navigation on November 18, and that a number of boats were waiting at Newark to depart.[45] That announcement was premature, and perhaps dangerous to the Canal Company's cause at the moment, since they were in litigation with Paterson's SUM about the use of Rockaway River water. Cashier Robert Gilchrist quickly corrected the impression that the canal was ready for boats.[46] Neither November 18 nor any other date, Gilchrist declared, had been "fixed upon for opening the gates of the Morris Canal." Gilchrist was "not aware that any boats are in readiness to depart from Newark." And "although the water is now coming into the level between Montville and Newark and comes from the summit level, it by no means follows," he argued, "that the canal will be navigable from that point to Newark. The canal will be open and navigable without interruption, in the course of eight or ten days, from Newark to Montville, a distance of about 30 miles, and it is not expected to have any greater distance navigable this fall; from thence to Dover it will be open very early in the spring, as nothing remains to be done but test inclined planes which will be in a great state of forwardness this winter. All the unfinished work from Dover to Easton is now under contract, and it is confidently expected that the whole line will be finished in all next August."

As Gilchrist predicted, the canal was opened for navigation from Newark to Montville within eight to ten days. On November 25, the *Paterson Intelligencer* reported that "the works . . . are found to be well constructed," and that navigation on the 30-mile section would continue "uninterrupted until closed by ice." Although Gilchrist had disclaimed any knowledge of boats ready to depart from Newark, "several loads of wood" had "already been brought down" to Paterson on the canal.[47]

The New York *Courier* reported that the canal was[48]

> *filled with water from its summit level to Bloomfield,* [*and that*] *the beautiful barge Maria Colden, owned and commanded by Mr. Morris, is plying between Bloomfield and Mead's Basin, passing by Paterson and crossing the magnificent aqueduct at Little Falls, a truly sublime and picturesque scene. The novelty and beauty of this route induces a great number of persons to take the trip. Last Thursday* [*November 26*] *this boat was met returning to Bloomfield, having a large party on board, in fine style, with a band of music and colours flying; and last Saturday* [*November 28*] *a large party of gentlemen and ladies went to Mead's Hotel on an excursion of pleasure.*
>
> *We understand there is no doubt but that the water is now down at Newark. If so, we understand by the proprietor that he intends making the Maria Colden a regular daily packet between Newark and Paterson for the remainder of the season.*

The *Maria Colden* had been named for President Colden's wife. The "Mr. Morris," owner and commander of the packet is said by some sources to have been Ephraim while others identify Stephen as the operator of the boat, with Ephraim the builder.[49] Ephraim is cited as the master of the *Emeline,* another packet boat that ran between Newark and Paterson during the 1831 boating season.[50] In any event, Ephraim Morris was to have a very long, very dis-

The Maria Colden ascending an inclined plane. This barge made daily trips between Newark and Paterson (Dunbar, A History of Travel in America, p. 662).

tinguished, and very creative and productive career. He was later made Superintendant of the Planes of the Morris Canal on May 1, 1833, after Douglass left the Canal Company.[51] The position paid $1100 a year in addition to the salary he earned as tender of his plane, No. 11 East, at Bloomfield. He would remain a Morris Canal superintendant until 1843, when he left to open the firm of "Morris and Cummings," makers of dredging equipment.[52] Where that firm was located in 1843 is not presently known, but in 1856 he established his works at the Little Basin at Jersey City. Morris' dredging invention was described as "a clam-shell bucket, for digging and dredging channels and removing submarine obstructions."[53]

Among Morris' other achievements was a railroad braking mechanism, for which he received a patent on September 19, 1838, the first in America to receive a patent for such a device.[54] And, although his obituary claimed he was "the inventor of the inclined planes used on the Morris Canal in place of locks,"[55] the facts, as shown later, do not fully substantiate that claim.

Still other ingenious Morris devices were "a machine for yarding coal," used for years by the Pennsylvania Railroad and the Delaware and Hudson Canal Company, and "an automatic coal-weighing machine for weighing coal in vessels."[56] Whether or not any or all of his inventions were used on the Morris Canal is not reported in Canal Company records, but it is likely that they were. He died at his South Bergen home on June 9, 1865.[57]

Throughout 1829 and 1830, the Canal Company's board of directors pressured its engineering staff to push for completion of the canal. Beach, Scott, and Douglass were asked to make the "most persevering and energetic exertions" to get the canal ready for navigation between the Delaware and the Passaic rivers" during the 1830 boating season. Beach, as Chief Engineer, was told to see that every contract in his department was faithfully executed, and that there be no delays in their execution. If the contractors needed more men in order to finish on time, they should be instructed to get those men, or their contracts were to be voided, and new ones issued to other contractors.[58]

On the Western Division of the canal, west of Lake Hopatcong, construction was well underway, but far from completed. Drake and Bartlett, contractors for several of the western inclined planes, were urged by Beach, on instructions from the directors, to get moving so that they would be finished by the fall of 1830.[59] At Port Colden, or Anderson, as it was then called, Thomas and Isaac Shields had been hard at work building Plane No. 6 West since 1828, but they were not to finish construction until 1831. Isaac Shields later became the Canal Company's toll collector at Phillipsburg, according to Snell.[60] However, in 1830 and for many more years, the collector's office was in South Easton.

However hard Beach or Douglass pushed the contractors and laborers, the canal *couldn't* be completed to the Delaware River during the 1830 boating season. A spot for the western terminus had not yet been located. Beach had been instructed to buy a 7- or 8-acre strip of land between the proposed canal route and the Delaware "at a price not to exceed $50 an acre."[61] The land belonged to Michael Roseberry, a Phillipsburg farmer, which land was described as lying "near the proposed terminus."[62] However, Roseberry had held out for "$100 or more" per acre, and the directors felt that their agreeing to that price would encourage other Phillipsburg land owners to think that their properties were at least that valuable.[63]

To pay Roseberry $100 or more per acre for his lands would also have been premature since in mid-1830 the Pennsylvania Legislature considered a law that would permit the construction of a dam across the Delaware River. Such a dam would very much affect the Canal Company's choice of a spot to locate its westernmost inclined plane, No. 11 West, which was to convey boats into and out of the Delaware. The purchase of land necessary to the plane's construction was thus deferred in mid-1830, pending the Pennsylvania Legislature's decision.

The Lehigh Coal and Navigation Company had already built a dam for slack-water in the Lehigh River, and an outlet lock into the Delaware River. But the Delaware Division Canal, now under construction, would have to be fed somehow with water. The public was in favor of a dam across the Delaware River so that the resulting pool would provide good landing on all sides of the borough of Easton.[64] However, the Pennsylvania Canal Commission decided that a new dam should be placed across the Lehigh, just below the Lehigh Company's dam. A new outlet lock into the Delaware River would thus be necessary.

According to the Lehigh Canal and Navigation Company's annual report, dated January 11, 1830, a contract had been signed on June 18, 1829, for the Lehigh Company to build the new dam and outlet lock to supply feedwater to the Delaware Division Canal and access for canal boats. Construction was scheduled for completion by January 1, 1830. A Pennsylvania Canal Commissioner's report dated December 21, 1830, stating that the guard lock from the Lehigh River into the Delaware Canal and the outlet lock into the Delaware River were both 22 feet wide by 100 feet long, indicates that the new dam and outlet lock were indeed built. However, there is evidence that a dam across the Delaware was built, too, for a Morris Canal boat was swept over the Phillipsburg Falls early in May, 1832.[65] The boat had been caught in the Delaware River current, and being swept downstream, narrowly missed being dashed to pieces on the New Jersey side of the river. A letter cited in Josiah White's biography, *Prince of Pioneers,* states that the "Morris Dam" had been completed, and that the Delaware Division Dam was to be completed very shortly. No date is given for the letter, but the chapter containing it covers generally the period of 1831-32.[66]

The Easton (Pennsylvania) *Whig* reported in 1839 that the Pennsylvania Canal Commissioners had decided to build "a new dam across the Lehigh at Easton, for the purpose of feeding the Delaware Division of the Pennsylvania Canal. The State Engineer, who has for some time past been examining the streams for a good location, has settled upon a place immediately below the present dam. When the new dam therefore is completed, the interval between it and the old one is to be filled up with stone."[67]

Looking south, circa 1830: dam at Easton, outlet lock from the Lehigh River into the Delaware, and, left of center, the Morris Canal (BNK).

In any event, even though they did not yet know *where* the Morris Canal would enter the Delaware River, the board of directors decided *how* it would do so. Despite the fact that Major Douglass was the Chief Engineer in Charge of Inclined Planes, the directors decided to have the canal "terminate at the Delaware by a plane and a guard lock agreeably to the recommendations of Major Beach as contained in his letter of the 22d May last [1830]."[68]

Meanwhile, at the eastern end of the Morris Canal, there was even more indecision about a point of termination for the canal, and for just as good reasons as on the western end. Beach's original surveys had included the canal path between Newark and the Hudson River, at or near what is now Jersey City. That path he found to be unsuitable for a canal because of the high ridge of trap rock, a continuation of the Palisades, which ran south through Bergen township to Bayonne, and which would have made any kind of canal prohibitively expensive. A new survey he'd been ordered to make in 1827 showed that there was a ravine "three and a half miles south of the Bergen church," which would accommodate a canal "without more than a deep cutting of 20 feet for a quarter of a mile; so that it [the canal] may be made from the City of Jersey to the upland near Newark, on the level of tidewater — the distance not exceeding 12 miles, and at an expense per mile *less* than the average cost of the other parts of the route."[69]

Beach reported that the distance from Newark's commercial dock to the tidewaters of the Hudson at Jersey City, along a route suggested to him by Colden, was fixed at 11-1/2 miles,[70] and the Canal Company made ready to petition the Legislature for an amendment to its charter, enabling them to extend the Morris Canal to Jersey City. The law was passed on January 26, 1828, and its passage was to have a profound effect on plans for the canal's path through the town of Newark.

Meanwhile, Major Douglass had been at work completing the Newark inclined plane. "The first boat to go up the inclined plane at Newark was the barge *Dover of Dover* under the direction of Major Douglas [sic] the engineer, on December 10, 1830." Six and one-half minutes were required to complete the passage, traveling 1040 feet and overcoming a 70-foot elevation. The *Dover of Dover* returned that afternoon and "descended, meeting and passing the boat *Maria Colden* midway on the plane."[71] Douglass' device was a summit plane, and, like the others built at this early stage, double-tracked, hence the meeting of the *Dover* and the *Maria Colden* as they passed one another on the plane.[72]

As to *how* the *Dover* got down the Newark Plane in order to be the first boat up, the *Jerseyman* reported that on November 18, 1830:[73]

> [*a*] *full and successful trial was made on the locks and inclined planes on the Morris Canal, between Dover and Newark. Five boats left Dover for Newark . . . , laden with iron and ore, and passed the inclined planes with so much ease and facility that the concourse of people assembled on the occasion were fully satisfied of their superiority over locks for a great ascent in a canal. The first boat that passed the plane at Rockaway, was The Easton of Dover with a cargo of 12-1/2 tons, and 42 passengers. It passed in 15 minutes, a descent of 52 feet. The machinery, being new, a full force of power was not applied. The Dover of Dover with a cargo of 18 tons and several passengers, next passed in seven minutes, and the remaining boats in the same time. Two boats from Newark, with merchandize for Losey & Wood, and stone coal, etc., for the Dover Company, ascended the plane in the average time of 7-1/2 minutes. At Booneton Falls* [*sic*], *where the extensive iron works of Messrs. Green and Wetmore are established, the second boat from Newark ascended the plane in eight minutes, a lift of eighty feet. The superiority of inclined planes over lockage, for heavy lifts, has now been made manifest.'*[73]

While the *Jerseyman's* article simply noted that the voyage had taken place on "Thursday last," the article appeared in Paterson's paper on November 24, 1830, and since the *Intelligencer* was printed only on Wednesday, we could infer that "Thursday last" meant November 18. However, an advertisement in Newark's *Sentinel of Freedom* suggests a different date and lists only four boats:[74]

Arrived, at Newark from Dover, the following boats, with Iron and Iron Ore, consigned to Jonathan Cory.

Packet boat	*Dover*	Capt. Bregnelle*
	Newark	Wilmott
	Eagle	Morris
	Easton	Casterline

The above boats belong to Messrs. M'Farlane & Ayers, and are in charge of Mr. Elias Van Winkle, with 45 tons Iron Ore for Mr. Charles Stuart, Monmouth County, N.J., from the mine of Hon. Mahlon Dickerson, Suckasunny.

Newark, [Monday] Nov. 15, 1830

Still more confusion is provided by Alvin Harlow, in his *Old Towpaths,* when he places the trial of the eastern slope of the canal, between Dover and Newark, on November 1, 1830.[75]

Confusion is further compounded by another report that has the *Dover of Dover* going to Newark for the first time in 1831.[76] Freshly painted and decorated, the *Dover* was captained by Byram Pruden, veteran of the War of 1812. Aboard, as supercargo, was Judge Freeman Wood, making the trip to bring back provisions for the store which he owned jointly with Israel Losey in Dover. According to Munsell, Wood was nicknamed "Supercargo Wood" everafter.[77]

Freeman "Supercargo" Wood was a director of Dover's Union Bank in 1832, as well as a store manager with Losey. The Union Bank, it is claimed, was almost entirely owned by Anson G. Phelps, wealthy iron merchant of New York City.[78] He was the "Phelps" of Phelps & Peck Company, colleagues of Blackwell & McFarlan. Phelps' son later associated himself with the Scrantons, and with William E. Dodge, one-time owner of Morris County's Weldon Forge, with whom he formed the world famous firm of Phelps-Dodge.

Long before Douglass directed the *Dover* up the Newark Plane, long before Beach's plan for the Delaware River Plane had been adopted, the seeds had been planted for several controversies

**This might be William Pragnall, whose boatyard at Dover is said to have built the Dover of Dover.*

affecting the Canal Company, not the least of which was the full-blown emnity that blossomed between the Morris Canal and Paterson's SUM. Then, there was the publicly pitched professional battle between Beach and John L. Sullivan, early engineering spokesman for the SUM and for the railroad that Paterson was threatening to build in competition with the canal. A more private and subtle animosity was developing, too, between Ephraim Beach, Chief Engineer of the Morris Canal, and David Bates Douglass, Chief Engineer of the Inclined Planes of the Morris Canal.

As if the Canal Company didn't have enough problems, threats of litigation increased as more and more property owners submitted claims for damages to their lands by the construction of the canal, by the taking of building materials from their lands, or by the back-flooding of farms and pastures due to dams.

Then, too, myriad construction difficulties continued to plague the plane builders. Ephraim Morris submitted a bill for $92 in excess of the figure Beach had estimated would need to be spent on Plane No. 11 East. The cashier was instructed to pay the amount Beach had estimated and not one penny more.[79] Apparently the Montville Plane was not satisfactorily completed, either, for on April 27, 1830, the Canal Committee was authorized by the directors to employ as many masons as needed, at wages the Committee thought reasonable, to work there.[80]

The Montville and Bloomfield planes were not alone in their struggle for completion. The Pompton Plane had also been a problem for a year or more. One of Douglass' first "official" visits along the line of the canal in 1829 had been to the Pompton Plane to find out why Eliphalet Miller, the builder, had submitted a bill for $932, requesting the Company to reimburse him. The directors had already informed Miller that he wouldn't be paid until the plane was completed "according to contract."[81]

Douglass inspected Plane No. 10 East to get an estimate of how much work needed to be done and at what cost, before Miller qualified to be paid. The Major's opinion was that $787 had yet to be spent on the plane.[82] Miller's request for payment of the $932 already owing to him included the sum of $332 that *he* owed to Charles Ford for the iron materials Ford had supplied for the plane.

Rather than have Ford sue the Company for his money, the directors decided to pay him. However, there was a condition to the payment. *If* Mr. Ford would give to the Company a guarantee of the title to the house at the plane, and *if* Miller and Ford would give satisfactory receipts in full of all demands they were making, the cashier would be instructed to settle with Mr. Ford.[83]

In July, 1830, Beach was told that the dry walls being constructed in Newark were "considered inferior," and he was instructed "to take immediate and effectual measures to cause it to be done in a better manner."[84]

Other pesty problems of construction arose, too. For example, in Little Falls, six panels of iron railings manufactured by Messrs. Jackson had been "lost" on their way to the aqueduct.[85] When the Jacksons did not "find" the railings after six days, Cashier Gilchrist ordered new ones to be made by "Mr. Freeman, the contractor."[86] Also in Little Falls, a Mr. Everett H. Van Ness had filed a claim against the Company for stone taken from his quarry for the aqueduct. A team of Essex County Appraisers — freeholders, according to the Canal Company's charter — had set the value of the stone at $595.25, and the cashier was authorized to pay Mr. Van Ness "on demand."[87]

In Boonton, the firm of Green and Wetmore was building its New Jersey Iron Works at the site of Plane No. 7 East. There had been an agreement formed among New Jersey Iron, the Canal Company, and William Scott, owner of the water power at Boonton Falls, which, somehow, New Jersey Iron had "departed from." Both Beach and Douglass were dispatched to Boonton to "examine the works now in progress by Green and Wetmore and see that the same correspond in all respects with the agreement, and suffer no encroachment on the privileges guaranteed to the Company by the said William Scott."[88]

Douglass reported "a most unjustifiable and impudent interference" with company works at Boonton by Mr. Ashton, the agent of the New Jersey Iron Company at that place. The Canal Committee was ordered to visit the site with Douglass to "examine and decide whether the Morris Canal Company can with safety allow any departure to be made from the original agreement between William Scott and this company, and what or whether it is not prudent to insist on a rigid compliance with the terms of said agreement on the part of the said Iron Company."[89]

Green and Wetmore had bought 200 acres of property from William Scott for their enterprise. Originally a New York firm of iron and hardware dealers, the company had incorporated in New Jersey on November 17, 1829, with William Green, Jr., Apollos R. Wetmore, and David W. Wetmore at its head. When the Wetmores left the firm, Green was joined by Lyman Dennison.[90] As early as September 15, 1829, before their works were built, Green and Wetmore had requested a commutation of tolls for cargo shipped to them in New York. The Canal Company declined to reduce tolls, however, because the directors didn't care for the "present shape" of Green and Wetmore's proposition.[91]

At Lake Hopatcong, litigation was being threatened by land owners whose lands had been flowed adjacent to the lake and who now claimed damages. Colden agreed that the appraisers appointed by law should appraise the said damages, "not withstanding any objections which there may be for want of a preliminary survey and certificates required by the charter."[92]

In Rockaway, Colonel Jackson had requested the right to construct a basin and bridge. Ephraim Beach and John Scott reported that they could see no objection to Jackson's constructing a "basin on the south side of the towing path with a cut through the path for boats," provided he paid for it, and provided that he also "build and maintain a drawbridge for the tow path." Neither Scott nor Beach could see any reason why the Canal Company should build that bridge at its own expense. The Company directors authorized Colden to settle with Jackson on his request if Jackson would release the Company from claims of "all damages for lands occupied by the canal."[93]

In Denville, the Canal Company had built a half-mile-long, unnavigable feeder canal, a work "of considerably expense and labour," to bring water from Beaver Brook to the canal.[94] The brook joins the Rockaway River from the north, about a mile below the town of Rockaway, fed by three lesser streams: White Meadow, Beach Glen and Meriden Brooks.

In November, 1829, the Canal Company had received "a long and angry letter" from Matthias Kitchell, owner of property along Beaver Brook, where the Company had built a dam.[95] Beach was asked to examine the site where Kitchell's lands had been flowed, and give his opinion on "altering or removing the dam erected by the company at Beaver Brook," and "on the present construction of which said Kitchell makes so much complaint."[96] This complaint was one of the longest-termed against the Canal Company, not being settled until April 27, 1838. The Canal Company then agreed to pay Kitchell $1500.00 "for all damages sustained or to be sustained by reason of the company's dam across Beaver Brook." Upon Kitchell's release, a weir gate was to be put into the feeder to supply Kitchell with water, but to be maintained by the Company. The Company further agreed that no other dam was to be built, nor other obstruction placed "in or across the brook of any greater height than the one there" at that time. When the navigation season was over, the Company "agreed in future to remove or keep open the gates in the dam so that the waters might flow in their natural and accustomed course and channel."[97]

In April, 1830, Beach was ordered by the directors not to take any water from the Rockaway River for the canal until water had been brought from Lake Hopatcong, enough to fill the canal from Dover eastward. He was also ordered *not* to take the Hopatcong water, either, until the board of directors authorized him to do so. When he did receive authorization, he was *still* not to let water into the canal "if any part of the canal in progress would be harmed, or if it cost too much, or if to do so he would require the labor of persons who ought to be otherwise employed."[98]

By mid-June, the directors "heard" that Beach was going to fill the canal on the Eastern Division, and he was notified to stop his assistants and himself from "letting water into any part east of Dover until further ordered."[99] That order was close to being given, but first Beach had to report what sections of the canal had become ready for permanent navigation, and to which sections water could be brought down without interfering with the works in progress. Meanwhile, Douglass was to make similar reports about the planes, and to make sure that the tail races were ready between Dover and Newark.[100]

Note that, in April, Beach was constrained not to use laborers who should be employed elsewhere on the canal, and that, in June, he was ordered not to let water into the canal, as the directors had "heard" he was ready to do. By this time, Beach and Douglass had clashed and the board of directors called both men in for a discussion about their problems. From the meeting emerged a strict definition of the duties of each engineer. One can imagine that the basis for the clash was originally Civil Engineer *versus* Military Engineer, or the age-old controversy between book-learning and on-the-job-training. Whatever the case, Douglass was given exclusive authority to direct the plan and execution of planes and to connect planes with levels of canal. All workmen on the planes were answerable to him. With Beach's consent, Douglass might take men from his work force, those men then being considered responsible only to Douglass. All estimates, plans, and minutes regarding the planes were to be turned over to Douglass. To Beach were reserved all other rights and duties. John Scott, Canal Commissioner, was no longer to be responsible for paying Beach's men; Beach would be their paymaster.[101]

On July 5, 1830, Beach reported that the canal was ready for water from Powerville to the basin above Newark, but Douglass contradicted, saying that water should be let in only from Montville down. Accordingly, Beach let the water from streams and brooks into the canal below Montville.[102]

Later in July, Beach was ordered to gauge the quantity of water in the Passaic River from time to time so long as the discharge from Lake Hopatcong was stopped while the lock there was being repaired. Douglass was ordered to gauge the Rockaway River, and the waters in the lake itself. Lock repairs were not finished by August 3, 1830, and the board ordered Beach to complete the lock quickly, so as to be ready for navigation on the eastern division by Fall.[103]

Water was let into the canal from the Rockaway River at Dover to the plane below, at Rockaway. But the directors had resolved that no water "be taken from any of the natural streams till as much water as shall be required for the canal be brought from the Lake at the Summit or from Green Pond."[104]

In one day, September 7, 1830, the board of directors heard complaints from Douglass, that Beach had been interfering with his corps of men, *and* from McFarlan and Ayres Iron Company, that Beach had ordered 120 cast iron paddle gates for canal locks. President Colden was given the task of seeing both Beach and Douglass, to try to reconcile their differences. He was also directed to ask Beach why in the world he had ordered so many paddle gates.[105]

All was not wrong with the canal, however, and a subscriber to the *Paterson Intelligencer,* initials "J.G.O.," took the time to write to the editor about a pleasant experience on the canal. He had gone on[106]

> *an excursion on the Morris Canal, from the Little Falls to Mead's Basin. By the indefatigable exertions of Messrs. Bone and Speer, of the former place, their very handsome Canal Boat was launched in the Canal a few days since, and on Friday last* [*September 10, 1830*] *not only proceeded to Paterson and brought the Canal Commissioners to the Falls, but also with a large party and Band of Music, went as far as Mead's Basin. The inhabitants along the Canal were much pleased with the appearance of the boat, and a greater confidence of its success is now entertained.*
>
> *The boat is intended as a regular packet from Newark to Dover, and so through to Easton, and commanded by Captain C. who is exactly suited to please.*

A committee of directors of the Company visited the inclined plane at Montville, and "witnessed a loaded boat pass up and down the plane and into the levels at its head and foot." They were "entirely satisfied with the operation of the machinery and the solidity and durability of the whole work."[107]

As luck would have it, 1830 was a year of "great drought," making the waters of the Passaic River at Paterson very low. Paterson, however, was sure that it wasn't the drought alone that was the cause of the low water. Nine Paterson mills had had to close for want of water to turn their wheels, and many men had been thrown out of work as a consequence. "The greatest scarcity that was experienced last year [1829, when the SUM filled a bill for injunction] was at the time the Morris Canal was filling; but neither then nor at any other preceding season have the mills been as short of water as they were last week . . . . The whole of the Rockaway had, for several days, been taken into the Canal at Boonton, and retained in it . . . ."[108]

That was the very situation that Paterson and the SUM had feared and anticipated. The existence of the SUM, the city of Paterson and all of its industries were totally dependent upon the Passaic River and its tributaries for manufacturing. The existence of the Morris Canal, totally dependent upon water for navigation, was finding it impossible to avoid using at least some of the tributaries of the Passaic. The problem was inherent in the charters of both the SUM and the Canal Company, since the Legislature had conferred virtually identical riparian rights on each. The scene was set and the time was right for a bitter confrontation.

REFERENCES

1. *Report to the Stockholders of the Morris Canal and Banking Company* (N.Y.: 1861), p. 16 (hereinafter cited as *Report to the Stockholders, 1861*).
2. *Paterson Intelligencer*, November 22, 1826.
3. *Report to the Stockholders, 1861*, p. 16.
4. Morris Canal and Banking Company *Minutes*, April 14, 1829 (hereinafter cited as MC&BCo. *Minutes*).
5. *Report to the Stockholders, 1861*, p. 16.
6. *Biographical and Genealogical History of Morris County* (N.Y.: 1899), p. 183.
7. MC&BCo. *Minutes*, August 1, 1829.
8. *Report to the Stockholders, 1861*, p. 16.
9. MC&BCo. *Minutes*, April 28, 1829.
10. Jos. Fulford Folsom, ed., *Bloomfield Old and New* (Bloomfield, N.J.: 1912), p. 105.
11. *Report to the Stockholders, 1861*, p. 16.
12. *Paterson Intelligencer*, August 6, 1829.
13. *Ibid.*
14. Niles' *Weekly Register*, October 17, 1829, p. 124.
15. *Ibid.*
16. *Report of the Joint Committee of Council and Assembly Appointed to View the Morris Canal and Inclined Planes* (N.Y.: 1828), p. 5 (hereinafter cited as *Report of the Joint Committee*).
17. *Ibid.*, p. 6.
18. *Jerseyman*, August 20, 1828.
19. *Report of the Joint Committee*, p. 6.
20. *Ibid.*
21. *Ibid.*, p. 8.
22. *Ibid.*, pp. 10-11.
23. Easton *Centinel*, February 29, 1828.

24. MC&BCo. *Minutes,* April 28, 1829.
25. Niles' *Weekly Register, loc. cit.*
26. MC&BCo. *Minutes,* May 12, 1829.
27. *Ibid.,* May 19 and June 30, 1829.
28. *Report of the Joint Committee,* p. 7.
29. MC&BCo. *Minutes,* July 3, 1829.
30. *Ibid.,* July 21, 1829.
31. *Ibid.,* August 3, 1829.
32. *Ibid.,* August 17, 1829.
33. *Ibid.,* September 8, 1829.
34. *Ibid.,* September 15, 1829.
35. George W. Cullum, *Biographical Register of the Officers and Graduates of the U.S. Military Academy, from Its Establishment in 1802 to 1890, with the Early History of the United States Military Academy (Boston: 1892).*
36. *The Story of Anthracite* (N.Y.: 1932), pp. 64-6.
37. Lehigh Coal and Navigation Company Annual Report to the Stockholders, 1830, p. 15.
38. *Paterson Intelligencer,* January 16, 1828 and May 27, 1829.
39. Wheaton J. Lane, *From Indian Trial to Iron Horse; Travel and Transportation in New Jersey, 1620-1860* (Princeton, N.J.: 1939), p. 262.
40. MC&BCo. *Minutes,* October 20, 1829.
41. *Ibid.,* January 26, 1830.
42. *Ibid.,* March 12, 1830.
43. *Ibid.,* March 16, 1830.
44. *Ibid.,* November 18, 1829.
45. New York *Morning Herald,* November 14, 1829.
46. *Ibid.,* November 17, 1829.
47. *Paterson Intelligencer,* November 25, 1829.
48. *Ibid.,* December 2, 1829.
49. Folsom, *op. cit.,* p. 107.
50. Newark *Sentinel of Freedom,* June 7, 1831.
51. MC&BCo. *Minutes,* May 1, 1833.
52. Folsom, *op. cit.,* pp. 23, 105.
53. William H. Shaw, *History of Essex and Hudson Counties, New Jersey* (Philadelphia: 1884), p. 877.
54. "New Jersey Days," N.J. State Department of Education Bulletin, Trenton, N.J., September, 1972.
55. Paterson *Daily Press,* June 13, 1865.
56. Shaw, *op. cit.,* p. 877.
57. Paterson *Daily Press,* June 13, 1865.
58. MC&BCo. *Minutes,* May 18, 1830.
59. *Ibid.,* August 3, 1830.
60. James P. Snell, *History of Sussex and Warren Counties* (Philadelphia: 1881), p. 730.
61. MC&BCo. *Minutes,* November 3, 1829.

62. MC&BCo. *Minutes,* November 3, 1829.
63. *Ibid.,* May 5, 1830.
64. Easton *Centinel,* February 29, 1828.
65. Newark *Daily Advertiser,* May 31, 1832.
66. Eleanor Morton, *Josiah White: Prince of Pioneers* (N.Y.: 1946), p. 183.
67. Easton *Whig,* August 7, 1839.
68. MC&BCo. *Minutes,* June 2, 1830.
69. *Paterson Intelligencer,* May 30, 1827.
70. *Ibid.*
71. Newark *Sentinel of Freedom,* December 14, 1880.
72. *Report to the Stockholders, 1861,* p. 17.
73. *Paterson Intelligencer,* November 24, 1830.
74. Newark *Sentinel of Freedom,* November 15, 1830.
75. Munsell's *History of Morris County, New Jersey 1739-1882* (N.Y.: 1882), p. 316 (hereinafter cited as Munsell's *Morris*).
76. *Ibid.,* p. 306.
77. Alvin Harlow, *Old Towpaths* (N.Y.: 1926), p. 197.
78. Munsell's *Morris,* p. 315.
79. MC&BCo. *Minutes,* April 6, 1830.
80. *Ibid.,* April 27, 1830.
81. *Ibid.,* August 13, 1829.
82. *Ibid.,* November 3, 1829.
83. *Ibid.,* December 19, 1829.
84. *Ibid.,* July 6, 1830.
85. *Ibid.,* July 20, 1830.
86. *Ibid.,* July 27, 1830.
87. *Ibid.,* September 7, 1830.
88. *Ibid.,* July 13, 1830.
89. *Ibid.,* September 25, 1830.
90. Munsell's *Morris,* p. 59.
91. MC&BCo. *Minutes,* September 15, 1829.
92. *Ibid.,* July 27, 1830.
93. *Ibid.,* September 14, 1830.
94. *Report to the Directors of the Morris Canal and Banking Company: Made by the President, Cadwallader D. Colden, May 1st, 1827: With a Statement of the Work Then Done Upon the Canal, a Report from the Chief Engineer, and a Communication from Governor Clinton.* (N.Y.: 1827), pp. 16-17.
95. MC&BCo. *Minutes,* November 18, 1829.
96. *Ibid.,* March 8, 1830.
97. Morris Canal and Banking Company *Docket Files.* Dep
98. MC&BCo. *Minutes,* April 20, 1830.
99. *Ibid.,* June 17, 1830.
100. *Ibid.,* June 19, 1830.
101. *Ibid.,* June 12, 1830.
102. *Ibid.,* July 6, 1830.

103. *Ibid.*, August 3, 1830.
104. *Ibid.*, August 17, 1830.
105. *Ibid.*, September 7, 1830.
106. *Paterson Intelligencer,* September 15, 1830.
107. MC&BCo. *Minutes,* October 27, 1830.
108. *Paterson Intelligencer,* October 13, 1830.

7

Riparian Wrongs

THE final location of the Morris Canal at Paterson was not decided until the end of 1826. A very early possibility, called the "lower" route, had been to end the canal at the Passaic River near the Great Falls, and make the river below that point navigable by means of locks and dams, with a towing path along its bank. A "middle" route, surveyed by Beach in 1825, would have raised the SUM's dam above the Falls by 20 feet, and would have meant an excavation of the rocks adjoining the Falls to form a canal bed, with an embankment 26 feet high forming a basin there. An alternative path, or "upper" route, would be a continuation of a level which began about 6 miles above the *Little* Falls, where the canal would cross the Passaic River on an aqueduct. It would then pass Paterson above the *Great* Falls, continuing along the face of Garret Mountain, on the same level, to Bloomfield, and then to Newark, where it would fall into tidewater.

Paterson, or at least the SUM, apparently had been quite satisfied with the "middle" route. But when the ultimate location of the canal was adopted according to the "upper" route surveyed by Beach back in 1825, the SUM's president, Roswell Colt, was moved to write to the Canal Company's new president, Cadwallader D. Colden, in

December, 1826, protesting the new location of the canal. He claimed that it would not accommodate the city of Paterson as had been expected, and that in bypassing the Falls, and the river, the canal would use still more water. Instead of replenishing the flow of the Passaic above the Falls where canal and river would have joined, the canal would further deplete the river since it would be several miles longer than originally planned, thus requiring more water.[1]

Colt took no legal action at the time, but the Canal Company, anxious to confirm the wisdom of adopting the "upper" route, enlisted the engineering expertise of Benjamin Wright, former chief engineer of the Erie Canal. Wright was not told of the "lower" route as a possibility, however, "it not having been thought necessary by the Board [of Morris Canal Company directors] to direct his attention to the route by the river from Paterson to Newark." As Colden expressed it, there was so little to recommend the lower route "that it does not appear to have been seriously thought of by any person." The distance along the discarded "lower" route between Paterson and Newark would have been nearly 20 miles. The "middle" route, by way of Aquackanonk and Belleville, would have reduced that distance to 16, while the "upper route further reduced the mileage to 12 miles. Actually, Colden claimed, "the strongest argument in favor of the middle route" was that "it would accommodate the inhabitants of Paterson."[2]

Ever mindful of public opinion since the recent stock scandal, the Canal Company "took much pains to ascertain the opinion of the inhabitants of Paterson on this subject." After all, a number of Morris' stockholders lived in the city. The results of the opinion poll indicated that "nine-tenths of the inhabitants" favored the upper route, believing that it would "accommodate them quite as well as any other would do." Still, it would be well to get a confirming opinion from an unimpeachable expert, Judge Wright. Accompanied by Canal Company Director Robert P. Bell, Canal Commissioner Colonel John Scott, and Chief Engineer Ephraim Beach, Wright examined the middle and upper routes near Paterson, and the course each would follow on the way to Newark. In a report dated

January 8, 1827, Wright reduced the problem to a two-part question*:[3]

> *1st. A matter of economy, as regards the cost of construction.*
>
> *2d. As to the advantages and disadvantages to the growing and important town of Paterson, as securing their trade, and thereby the interests of the Canal Company.*
>
> *The original plan of 1823, contemplated ending the Canal at tidewater, at Aquackanonk, and now some persons say it ought to end at Belleville, and thereby avoid some shoals between the latter and former places. The upper route, which I designate as the location of 1825, being on a much higher level than that of 1823, passes from Paterson direct to Newark. The distance from Paterson to Belleville, or Paterson to Newark, is computed the same, viz., 12 miles.*
>
> *1st. As to economy in the execution of the work, no person possessing the eye of an experienced engineer, and a knowledge of all the levels, can doubt for a moment that the direct route to Newark is much the cheapest, and that the canal will be less liable to breaches on the banks of this route . . . .*
>
> *It has been suggested to pass the river at Aquackanonk on an aqueduct. This probably would not be permitted, as the tide navigation extends above that point some distance.*
>
> *The plan of passing the great falls at Paterson, on the lower level, is by raising the present dam 4 to 6 feet, and continuing that level through the gap in the mountain, where the present water to feed the factories passes through, and at an elevation of 15 to 20 feet above the surface of the present water at that point. As it must, after passing this gap, be kept at the same elevation along the face of the Trap Rocks . . . and after passing the road to Little Falls, follow round the hill, according to the plan of Capt. Beach in 1823; or by a heavy embankment . . .*

**Keep in mind Wright's ignorance of the "lower" route, as Colden presented it. Wright knew of only the middle and upper, which he called lower and upper.*

form a large and spacious basin; and . . . follow along the declivity of the ground, take its course to Aquackanonk.

There are strong objections in my mind against this plan:

1st. It is intended to change the whole feeding water of the present manufacturing establishments, which are of immense importance to the town, and the whole of these now erected, and those hereafter to be erected, are to receive their supply of water from this level of the canal; and the risk of breaks and interruptions to these very important works will be more than quadrupled (I may say ten times as great), because the present canal has settled into permanancy, and little danger of difficulty on that head [*sic*].

2d. It is difficult to conceive or devise a plan for effecting this work, which will not be at great risque on its first trial; and it is difficult to devise a plan to execute the work and change the whole feeding water of these factories without stopping the whole of these establishments for many days, and the liability to have breaches in these high embankments is certainly great, and when they do take place, it will require many days to repair, and in all these times the Canal Company will be liable to heavy damages.

Large manufacturing establishments can never be worked successfully without their movements being almost as regular and uninterrupted as the sun. They give bread to thousands in the present state of this village; and the time is probably not distant when this village will contain a population of 10,000 souls, and 9,000 of these will be dependent for their daily living from their daily labour in these factories.

And since nature has formed the country so as to avoid collision as to water, it cannot be doubted for a moment, that this is the least of the evils in passing these falls.

So strongly do I view this question as to changing the plan of feeding all these establishments, and so great do I consider the risque and injury I should sustain if I owned one of them, that I would protest in a formal manner against it being done by the Canal Company in this way.

Again — I have seen the trial on the Erie Canal of a kind of indivisible, undefinable partnership in water to be used partly for hydraulic purposes, and partly to feed canal, that I would avoid it, at almost any reasonable expense, when I could do it. Suppose Paterson to contain 9,000 manufacturing souls (at least that number directly and indirectly gaining their living by it), and suppose a long continued drought which should cut short a supply for all, and the canal also, can it be believed that these people would not take every drop of water in the canal rather than go without bread?

The advantages of the upper level in separating the canal entirely from the factories, are great, and should not be abandoned.

Much has been said as to the higher level, making it necessary to transport further to and from the factories: this is true.

Starting from the middle of Congress Street, opposite Dunlap's Factory, and taking the road to the Little Falls, we found the distance to the basin on the upper level, to be 704 yards; and starting from the same point, and going as direct as the ground will admit to an accessible point . . . on the lower level, where a basin may be used, was found to be 671 yards.

It may be said with truth, that the basin on the road to the Little Falls is only approachable by rising a bad hill. This would better accommodate the trade of the country as a depot, than the business of the town.

A cursory examination will show, that a basin may be easily formed a little west of the turnpike road, in a point of wood, and this basin I should estimate at 12 or 1300 yards from the town. This extra distance for the manufacturers to cart their goods and receive their supplies, can never be brought as an argument that they will not use the canal. If they do not find it for their interest to use it, they will not do it, whether it is 600 or 1300 yards carriage.

In brief, Wright endorsed the "upper route," and ended his report of January 8, 1827, with some complimentary words on behalf of the "skill, prudence, and judgment" shown by Ephraim Beach in locating the canal's path at Paterson on the "upper" level.

Reporting to the Canal Company's directors in May, 1827, Colden remarked that it had been "predicted by some; who, probably, have not of late heartily wished success to the Canal, that the water would not rise" in Lake Hopatcong, behind the Canal Company's new dam, "or be retained in this reservoir." He was pleased to report, however, that the water had "already risen to twelve feet above the bottom of the discharge gates, although seven water wheels to move machinery are constantly supplied from the Dam." In his opinion, Colden said, the lake would "afford a supply five times greater than would be necessary for all the locks that would be requisite on the whole line" of the canal. In fact, the lake provided "a super abundance of water."[4]

By November 7, 1827, the SUM's Roswell Colt presented a petition to the State Legislature, "complaining of the actual and contemplated encroachments of the Morris Canal and Banking Company on the vested rights of the said Society."[5] Judge Wright had argued that the canal route, as finally adopted, was to Paterson's advantage, since the canal and manufacturing waters would be kept separate, and the final route, while "inconvenient" for Paterson, was certainly not an encroachment. Why then did Colt seek redress for the SUM? On the surface it appears to be competition for water supplies. Beneath the surface, the competition threatened to spill over into the area of manufacturing and to the very control of the sources of water power.

The first solid complaint emerges from the SUM faction over the use of the Rockaway River at Dover and Powerville, points where the canal took in water from that important tributary to the Passaic River.

A description of the Morris Canal that appeared in New York's *Commercial Advertiser* in September, 1827, may have alerted Colt to the intentions of the Canal Company.[6]

> *From Montville, the canal leads to Boonton falls, upon the Rockaway River. It will here ascend by another in-*

clined plane of 80 feet. Here the scenery assumes a still wilder and bolder aspect. The stream of the Rockaway dashes down a dark and rocky ravine, while rocks, and hills, and forests, and mountains, in all their broken and grotesque appearances, rise on every side. Above the falls a dam of the most durable structure, has been built, so that every drop of the river can be used as a feeder, if necessary, and the company [*the Morris Canal and Banking Company*] *will here possess water power to any extent that may be required.*

The Beaver Brook feeder canal also took into the canal water that was ultimately destined for the Rockaway, and thus the Passaic. But it was the Canal Company's avowed plan to tap Green Pond, still another Passaic tributary that seems to have brought the matter to a head, for the waters in question were to be used by the Canal Company not to enhance navigation, but to establish mill seats at the inclined planes. The canal had already been lengthened 8 miles beyond Aquackanonk to Newark, and perhaps might even reach Elizabethtown. The longer the canal, the more water would obviously be needed for its operation. Worse, there were plans afoot for "an improvement of Pompton River to Ramapo, and of the Passaic to Chatham," as Macculloch had put it in his July 1, 1826, report to the Canal Company's president and directors. But that the Canal Company should, in effect, announce plans to enter into manufacturing competition with the SUM, *and* do it with the SUM's own water supply — that was unthinkable. Confrontation was inevitable.

When Macculloch wrote his first annual report on the canal's progress (July 1, 1826), William Bayard was still president of the Canal Company, remaining in that position until his death in September of that year. When Cadwallader D. Colden assumed the presidency that November things began to change rapidly for the canal. In *his* first report to the Company's directors (May 1, 1827), Colden announced that no longer would Newark be the ultimate eastern terminus for the canal; Jersey City was to receive that honor. Beach had been ordered, he said, to survey the route of that extension from Newark to Jersey City and to report on the best site available for a basin there, opposite New York City.

Although Colden did not announce it, still another "enlargement" was being considered for the Morris, as the Company's engineers were assigned to begin a survey for a branch canal on the Western Division, up through Belvidere to the Delaware.[7] There, presumably, the branch canal was to intersect a proposed canal that was to begin at the Delaware, at or near the Blue Mountain in Sussex County, and connect with the Hudson River near Newburg, Orange County, New York.[8] Governor Clinton was reported to be considering a branch for the Erie Canal to push southward to intersect the Orange and Sussex Canal.

Admittedly, what the Canal Company did on its Western Division with the waters of the Musconetcong River was of less concern to the SUM. But the SUM seems suddenly to have realized that the Canal Company was a potential rival, as well as a threat to its survival in Paterson. Very well, then, if the Canal Company wanted to engage in manufacturing, the SUM would enter into the field of transportation! John Langdon Sullivan was hired to investigate the possibilities of that field. Sullivan gave his findings in a pamphlet dated July 21, 1828, and entitled: *Report, on the Origin and Increase of the Paterson Manufactories, and the Intended Diversion of Their Waters by the Morris Canal Company: Also on Post Rail Roads, As the Means of Cheap Conveyance Throughout New-Jersey, of Bringing Susquehanna Coal to the Iron Mines and Forges, and to Supply Paterson and New-York: Also on a Method of Supplying the City of New-York with Water From the Great Falls of the Passaic.* One wonders how the SUM thought it could continue to supply the mills in Paterson if it piped water to New York City, when it claimed it could not operate because the Morris Canal was depleting its water. One wonders, too, how the SUM thought it could both operate navigable canals using the same water as the Morris was to use, and maintain the water power needed for its mills.

Sullivan's diatribe is even more wordy than his title. (We met him earlier as the civil engineer appointed to the Board of Engineers for Internal Improvements, which had been created by the General Survey Act of 1824.) In his capacity as "Civil Engineer to the Society for the Establishment of Useful Manufacturers in New-Jersey," he viewed "the natural obstructions to navigation, both above and below the Great Falls of the Passaic, and in the branches of this river." Moreover, he examined the "ground favorable to the formation of the requisite communications with the interior, and the great markets, from and to Paterson, the principal seat of the Manufactories planted here under the protection of the charter." In fact, he had "visited the whole line of the Morris Canal, and seen the way in which its complicated constructions indicate the manner in which the Company intends to supply the canal with water from the streams which flow to supply the Manufactories at Paterson."

Sullivan was convinced that it was "of immediate importance to the Society," for him to report "the obvious facts" which led him "to believe that the Morris Canal *cannot,* as now constructed in the parts and places I shall name, go into operation without doing great injury to the Society." Knowing that the Society had every desire "that the Canal should prosper in its original plan," he would attempt to answer the Society's inquiry: "How the Canal might be altered . . . to avoid the trespass evidently meditated" by the Canal Company.[9]

The very safety and existence of the SUM, Sullivan felt, depended upon certain essential alterations to the canal. In effect, that meant no more than restoring the Canal to its original plan. According to his understanding of the "original" plan, that meant that the Canal should come down from Lake Hopatcong to the Little Falls and into the Passaic River, *and* should do so by crossing the Rockaway River by aqueduct at Dover and Powerville and passing "Beaver Brook under the canal by a culvert." Furthermore, the Canal Company ought "to desist from making the feeders from Chatham to Pompton."[10]

Actually, Sullivan claimed, what with the construction of the Delaware Division Canal from Easton to Philadelphia, and the proposed Delaware and Raritan Canal which he felt would soon be building, there was almost no need to even complete the Morris Canal, at least its western half, since the canal would no longer serve the purpose intended. If, however, it *were* to be completed as *he* recommended, from Lake Hopatcong to Little Falls, then the canal that the SUM was going to make from Paterson to tidewater would fill the same need. Moreover, the preliminary surveys he had made on behalf of the SUM for a railroad from Hoboken on the Hudson to Paterson revealed "no considerable obstacle" to that plan.[11]

Still another reason for not completing the canal's western division, according to Sullivan, was the railroad projected by a Pennsylvania group, headed by Henry W. Drinker, which would bring coal from the Scranton-Wilkes Barre-Pittston area to the Delaware Water Gap. A continuation of that road on the Jersey side of the Delaware, reaching from the Gap to Lake Hopatcong, would make it possible to convey coal right to the canal's summit level, where canal boats could receive transshipments bound for Paterson.* From Paterson, the coal could be once more transshipped to be sent to Aquackanonk by way of the SUM's projected canal from Paterson to that place, or by rail, presumably, to Hoboken, on the railroad that the SUM contemplated.[12]

Then, too, by 1828, the Orange and Sussex Canal project had been transformed to a plan for a railroad instead of a waterway. The Delaware Water Gap-to-Lake Hopatcong road mentioned by Sullivan would also intersect that rail route. An "American gentleman, Mr. Sargent," had invented a railroad that would, Sullivan thought, serve Paterson admirably. It was a system of two parallel lines or tracks of iron bar, fixed on square timbers, which lodged on a line of foundation posts or pilings set in the ground at 12-inch intervals. The posts were to be "somewhat elevated to clear the mud, snow and ice," and hence, Sullivan termed it "The Post Rail Road." "The inequalities of the surface of the ground may be remedied without digging and filling up by the greater or less height of the posts, which keep the Rail Road level, or graduated to the intended slope."[13]

Since this was to be a horse-drawn railroad, one wonders where the horse was to run if the railroad stayed level while the ground did not. However that may have been, *what was to be drawn* over the tracks by the horse was a railroad carriage, invented by Mr. Ross Winans of Sussex, whose mechanical ingenuity in inventing the self-sharpening plough and a method of fulling cloth economically in

**This was the railroad that was surveyed by Beach in 1831, which later became part of the Delaware, Lackawanna and Western Railroad.*

winter by steam was well known. Both Mr. Winans' railroad carriage and Mr. Sargent's post railroad were under contract to John Langdon Sullivan, who was thus enabled "to give these improvements to New Jersey on very moderate and reasonable terms."[14]

But, lest it be imputed to Mr. Sullivan that he had "an interested motive in recommending them," he turned to a more "humane" topic: that of supply, a "supply of good and wholesome water" to New York City from the Great Falls. The Legislature and the Society (SUM) had already approved the plan, notwithstanding its apparent impracticality. There were, after all, rivers, marshes, the trap rock ridge near Hoboken, and the mighty Hudson to be overcome. It would, indeed, be "a costly triumph of civil engineering over natural difficulties to convey the waters that would else run useless in the night, first to a reservoir at Hoboken; whence it may be delivered continually into another reservoir in New York by iron pipes, laid on a bed of stones across the bottom of the Hudson, in the manner and by the means" so invented as "to render certain the fitting of the joints, and their tightness sure and their examination at any time practicable." Such an engineering marval had already been invented — by John Langdon Sullivan.[15]

As a point of interest, Sullivan's idea to pipe water to New York City from the Great Falls by way of Hoboken was picked up during the 1850's by John R. Bartlett, who bought out the SUM for that purpose. Bartlett intended to tunnel the Hudson from Hoboken to Washington Square, in New York's Greenwich Village. He finally abandoned the plan when he realized that he could more profitably sell SUM water to northern New Jersey. The "tunnel" was sold and was completed many years later as the Holland Tunnel, from Hoboken to Christopher Street, New York City.[16]

As for the SUM entering the field of transportation, as the Canal Company threatened to enter the field of manufacturing, Sullivan made some claims based on the Society's charter of 1791. When the Society received its charter, he reported, "the seat of their intended operations was not fixed on," and "the best place for it was not yet ascertained." However, "it was understood that the central situation and relations of this State [New Jersey] in the internal commerce of the Union, gave it a preference, should the streams and water power it might be found to afford, be powerful and durable enough."[17]

From various parts of the new nation had come claims of superior water power, or location, but only the Great Falls of the Passaic "seemed to unite and comprehend so many advantages." The very presence of the Little Falls, four miles above Paterson, added to the desirability of the spot. But the fact that the Passaic River was "navigable for 30 miles above the Falls, with boats drawing 5 feet to Chatham, and beyond with a few locks 50 miles," had apparently been one of the major reasons for the Society's decision to locate its manufacturing city at the Great Falls — Paterson.[18]

Now, according to Sullivan's appraisal, the Society's charter had invested it "with power and possession of the waters of the Passaic, not only as a water power but as corporate owners of the same, giving them the *sole possession* for the purpose, both of mills and manufactories to be established at the Great Falls, and for navigation and feeding of canals leading to and from the principal seat thereof." And when John Langdon Sullivan said "canals," he meant navigable waters along the Pompton River, as well as the Passaic. In fact, construction of any canal through that part of the state watered by the Passaic River was a right reserved to the SUM alone.[19]

He was most particularly definite in claiming that the SUM, and *not* the Canal Company, had rights in Green Pond, since that lake emptied its waters into the Green Pond Brook and the Burnt Meadow Brook, from whence it emptied into the Rockaway River at Wharton, and thus ultimately into the Passaic River below Boonton. In fact, the prospect of having adequate reservoirs to feed the Passaic River at the Great Falls had been "among the primary considerations that fixed the Society's choice, and none were more relied on than that of Green Pond situated among mountains where its depth might be increased without a much greater extension of surface."[20]

As the controversy between the two corporations grew hotter and more public, insults and innuendos grew more frequent and personal. First Sullivan and Colden, then Sullivan and Beach, fought it out in pamphlets and newspapers. Even Macculloch entered the fray: unwillingly, to be sure, but more or less forced to defend his own reputation against Sullivan's allegations.

Writing to the editor of Morristown's *Palladium of Liberty* in December, 1828, several months after Sullivan's Report was published, Macculloch protested being drawn into the controversy. His letter is worth quoting at length because of the light it throws on the growing antagonism between the Canal Company and the SUM and on his relationship to the canal project.[21]

> *Mr. Mann — I have experienced too much satisfaction at my separation from the Morris Canal Company, to entertain any desire of being again involved in its discussions and affairs — accordingly, although quite aware of the fables invented and circulated to my disadvantage, yet knowing the authors and motives, I have suffered these transient attacks to descend unheeded into oblivion. Within three months, two pamphlets distributed gratis by the package throughout the neighboring counties, have made my conduct a theme of very unceremonious comment; I have this moment looked over the pages of a third; how many more are forthcoming, I know not; and under such circumstances, I trust, that without the charge of obstrusion, the public will permit me to give a very brief explanation.*
>
> *In its wisdom, the Legislature of 1824 saw fit, during my absence from Trenton, to change totally, and in every particular, the organization, which I had adopted for the Canal subscription and future direction. The power which I had purposed to retain in Citizens of New-Jersey, was then so arranged as to become absolutely vested on the opposite shore of the Hudson; and from that day up to the present, it has always appeared to me, that the inhabitants of the Canal District were especially interdicted from any influence over the operations of the Company. As for myself, the office conferred upon me* [*President of the Canal Commission*] *was merely nominal; the contracts, the locations, the plans of every sort were secrets uniformly and studiously reserved from me; although President of the operating commissioners, the transactions of that very board came only to my knowledge by the common intel-*

ligence talked throughout the country; and a broad refusal was given to my request of access to some books of the institution. A well grounded jealousy was entertained that I was hostile to the system pursued, and any proposition suggested by me was always certain to meet a prompt and decided negative.

At length in 1826, the Company ran aground; I was called upon to assist in regaining its lost credit; and the reigning power gave me an assurance that in future my advice would be listened to with greater attention. A stoppage of the work at that early period would have left little hope of the canal being ever ultimately accomplished, and this public calamity would also have involved a very serious loss of stock to myself; I therefore acceded to the overture, travelled over the line, drew a Canal report so frequently referred to by Messrs. Sullivan and Colden, and formed plans for new modelling the administration of the Company's affairs. This report was printed under my own eye, in Morristown, but the portion which to me appeared the most important was altered by the board in New York, and a new edition reprinted there. I had distributed some copies of the original among my private friends, and hence arises the discrepancy alluded to by these gentlemen.

The present author interrupts, here, to note the "discrepancy" alluded to. Sullivan's Report quotes from page 17 of Macculloch's July 1, 1826, report to the Canal Company (underscores Sullivan's):

We have located our inclined planes without attempting to secure the ground whereon the mill seats must be erected; we have led our Canal over ground of the most expensive character, blasting rocks and rearing aqueducts with no visible object but to form a chain of the finest water privileges in New-Jersey; yet with such important objects in full view it is hardly credible that not a step has been taken to purchase the land without which the whole operation becomes worthless. We have created at our own expense a chain of several hundred mill seats, forgetting that the only soil on which these establishments can be reared remains yet to be purchased. At the time when our charter was granted some misapprehension hindered the Company from obtaining a right to lands of this description; and while the water belongs to the Company, the soil on which it can be used remains in the hands of proprietors to whom it is worthless.

Thus the State foregoes a vast accession to its manufacturing riches, the Company cannot avail itself of a property which it has created, and the proprietors are deprived of the benefit accruing to their farms by the erection of MANUFACTURING VILLAGES. Some attempt ought be made under the trust clause of our charter to arrange this business with the landlords upon a liberal basis: but should they extort and avail themselves of the vantage position in which they have been inadvertantly placed, the evil may be remedied by applying to the Legislature for leave to take at a fair valuation a narrow strip of ground at the indicated places.

This is perhaps one of the least known but most interesting of intrigues centering on the Morris Canal's first years of existence. The present author has copies of both reports: the one Macculloch claims to have written, and the one he claims was tampered with. They are, indeed, distinctly different. Since Macculloch claims that he had his "original" report printed in Morristown under his own eye, we are safe in saying that the report printed by Jacob Mann of Morristown is the "original." The report printed by William Davis, Jr., of New York, then, must be the so-called "spurious" report. A most curious fact, but true, is that the material quoted by Sullivan in his report to the SUM is simply not in the Davis-printed pamphlet that Macculloch claimed was spurious.

It appears that Macculloch stands condemned by his own evidence. The report he claimed to be his original is the one that contains the incriminating statements that apparently aroused the SUM. In the Mann-printed report, Macculloch had, indeed, promoted the development of mill seats at the Morris Canal's inclined planes. This is not to say, of course, that George Macculloch was involved in any nefarious scheme to destroy Paterson or the SUM. On the contrary, research indicates that Macculloch simply wanted communities along the entire line of the Morris Canal, in particular, communities in Morris County, to enjoy the advantages that were being realized by Paterson with its all-important water power. After all, 13 of the Morris Canal's 23 inclined planes were in Morris County. And as we have seen, Morris County's standard of living had been one of Macculloch's primary concerns when he conceived his dream canal.

What we are witnessing here, at this early period of the state's history, is a conflict born of a legislative blunder. Two corporations: one for manufacturing, one for navigation, had been granted riparian rights in the same lakes and streams. Both charters were granted by lawmakers who should have known better, but who obviously didn't. Small wonder that Macculloch decided early in 1827 that it would be for his "interest and honor to withdraw totally from the concern." If things were going wrong with the canal, and particularly if Paterson were less accommodated by the canal than originally intended, it was "only because the original plans have been abandoned."[22]

All of Sullivan's complaints and accusations, according to Macculloch, seemed to have arisen from "the adoption of the Bloomfield route, which everybody, even including the engineer, who subsequently recommended it, regarded at first as absolutely absurd." Disclaiming any knowledge of the adoption of that route until it was already under excavation, he refused to accept responsibility for the resulting problems. He had, after all, recommended to the board of directors that the route be abandoned, "even after a large expense had been incurred" in its construction. In all fairness, he asked: "Is it just to hold me responsible if water should fail, or be withdrawn from Paterson after my urging a want of water as one of my specific objections to the Bloomfield route?" Again expressing his wishes to wash his hands of the Canal Company-SUM controversy, Macculloch asked "the respectable champion of each institution to cudgel his antagonist any where except across my shoulders."[23]

But Sullivan had raised a very important question: What did the Legislature mean "in taking Green Pond from the Society and giving it to the Morris Canal Company? Did not the Legislature know the Green Pond was a part of all the Passaic waters?" For the Canal Company's charter, Section 11, did state explicitly that it might lawfully "raise the waters in the Green Pond" by means of a dam, and use the "surplus waters thus saved" to the extent that such use was necessary for navigation, and provided the Canal Company pay for damages and losses that should be sustained by owners of lands flowed by such a dam.[24]

For Sullivan, the question of rights in Green Pond was more than academic. To him, the Canal Company charter's coupling of Green Pond with Lake Hopatcong "with a view to blend the power of the corporation over both," was equivalent to giving power over neither. In short, Sullivan said, the whole act was unconstitutional, the entire law was unconstitutional, and therefore void.[25] He recommended that the SUM make immediate application to the Chancellor of the State for an injunction against the Canal Company, prohibiting them from using any where, any part of the Passaic's tributaries.[26]

Sullivan also urged that the SUM "survey, and locate and acquire all the shores, dams and property, at Boonville [Boonton] and Powerville, necessary to the improvement of the natural navigation," allowing the Morris Company the right to cross the Rockaway River at those points by aqueduct. The outlet of Green Pond, he maintained, should be surveyed and located "for the erection of a dam there, to form a reservoir and locks, as originally contemplated in the views of the Society." He recommended, too, that similar works should be constructed at the outlets "of all the headponds and the several branches of the Passaic for the like purposes."

From Little Falls, surveys and plans should be made "to open the natural navigation of the branches of the Passaic." The main dam at the Great Falls, he claimed, should be "raised three feet, and a lock be made from the river into the Great basin of the upper canal." Then, the upper canal should be "extended on the same level and connect with the river at Paterson if the Society should conclude to open the navigation for steam boats up to that place, or to Aquackanonk, if the Society should conclude to open a Canal to the tide at that place, for which," he added, surveys were then proceeding.[27]

The SUM's governor Colt went along with Sullivan's recommendations — at least in part. The SUM did file a Bill in Chancery at the end of 1828, and it was argued during the Spring Term, in January, 1829. Lawyers for the SUM were G. Wood and Philemon Dickerson. Arguing for the Canal Company were D.B. Ogden and Joseph C. Hornblower. In April, 1829, Chancellor Isaac H. Williamson delivered his opinion on the Bill.[28]

Williamson considered it unfortunate that "conflicting claims" had been set up by the respective corporations to the use of the waters of the Passaic River and its tributary streams, claims which had given rise to the present suit.[29] He was "struck with the importance of the case itself," since it "involved questions of great magnitude, affecting the vital interests of the parties, and of great public concern."[30]

The Canal Company's contention was that the SUM had officially gone out of business before 1814 when it failed to exercise the rights granted it by its charter. In brief, the SUM was originally intended to *engage* in manufacturing and it was given that right; as it existed in 1828, it was merely disbursing water power to private manufacurers who bought mill seats that were serviced by the SUM's water supply. As Williamson saw it, the failure to exercise a right did not automatically mean the forfeiture of the right or of the charter which vested it.[31] He considered it "clear, that the Society has a right, as riparian owners, to the flow of the whole of the waters of the Passaic river, in their ancient channel, down to the great falls; and that the Morris Canal and Banking Company cannot divert the waters of the Rockaway, by which the waters of the Passaic at the great falls will be diminished, without encroaching on the rights of the Society."[32]

However, "the Society has no right to the waters of the Great Pond, or any other waters which do not flow into and increase the waters of the Passaic above the great falls." Since the Canal Company claimed that it brought as much water from Lake Hopatcong to Dover and the Rockaway River as it took in and discharged at that point, and again at Powerville, Chancellor Williamson could see no legal point to the Society's argument. As far as he could see, the Canal Company "have a right . . . to acquire, in the manner authorized by their charter, and to use the water of the Passaic, if they can do it without prejudice to the rights of the Society, who can have no serious ground of complaint, unless they sustain an actual injury, which they would do by a diminution of their water power."[33]

In any event, Williamson concluded, "the route of the Morris Canal has long since been known to the Society, large sums of money have been expended [$828,000], and the canal is now in a great state of forwardness. Under these circumstances," he opined, "it would not . . . be a sound exercise of discretion for the Court to interfere, on the ground of apprehended danger." In all good conscience, the Chancellor could not "grant the Injunction, upon assertions, calculations and mere speculative opinions; and where the injury to the defendants [the Canal Company] might be irreparable, if the Court should mistakenly interfere. Where no actual injury as been sustained, but only danger apprehended," Williamson felt, "the defendants must therefore be left to proceed at their peril."[34] In brief, the law would apply to specific evils already committed, not to dangers or threats of future, but uncommitted, crimes.

In any event, the right granted to the Society (even though the right remained unused) of cutting navigable canals couldn't operate "as a restriction upon the power of the legislature to grant canal charters," the Chancellor reasoned, "and any claim set up by the society (SUM), of a right to cut navigable canals wherever and whenever they please, in the state, is extravagant and unfounded. The society has no right to cut a navigable canal but 'for the purpose of transporting goods, wares and merchandise to and from some manufactory by them established in the district.'"[35] Injunction denied.

Almost immediately, the area newspapers began carrying articles that were "pro" and "con" the Chancellor's decision. A person signing himself as "Morris County" decried the fact that Mr. Colt, Governor of the SUM, asserted "that our Legislature have no power to appropriate the waters of the Passaic to any public purpose whatever, and that the exclusive control of them, at and above the village of Paterson, rightly belongs to his society."[36]

Replying to *that* article was that of one signed simply as "Friend":[37]

> *I would ask, who usurps the sovereignty of the people of New Jersey, if the M.C.B. do not? Let the mill owners of the Musconetcong answer — let the farmers of Warren,*

Sussex, Bergen, and Essex, through whose farms this famous canal has been cut, answer; and let those whose lands and quarries have been taken, and whose property has been injured without the legal remuneration, at and from whom that 'inalienable right,' trial by jury, has been wrested by the Morris Canal Company, answer the question.

The way in which the Morris Canal Company answered the question was to order Engineer Beach to go out and buy property contiguous to the outlet(s) of Green Pond and to repair "the works heretofore erected by this company at the outlet of said pond and to cause others to be constructed, if deemed necessary, to establish unequivocally the appropriation by this company to their own use, the waters of said pond together with all the lands contiguous to the same, which the company ought to own, to secure to them the undistributed right of using said pond as a feeder should the same be required."[38]

By December, the purchases had been made. Israel Losey of Dover bought from Judge Lemuel Cobb his 10-acre lot at the outlet of Green Pond. Losey bought in Henry McFarlan's name, paying $100 for the land, and McFarlan, in turn, offered the deed to the Canal Company. It was accepted, and Director McFarlan was reimbursed the $100 plus "all reasonable expenses," with thanks from the directors for his promptness.[39]

At Boonton and Powerville, the Canal Company entered into agreements with William Scott, Canal Commissioner John Scott's brother, who owned and operated the Powerville Forge and Bloomery. Water from the Rockaway River was brought to the head of Plane No. 7 East through a raceway or intake in accordance with a very interesting agreement between the Canal Company and William Scott. Munsell called the agreement "a master stroke of policy," whereby Scott "secured to himself the use of the whole head of water" from the Rockaway River above the Boonton Falls. The dam above the falls was built by the Canal Company, at its own expense, while Scott had the right to use not only the "natural flow" of the Rockaway River, "but also the feed passing through the canal, being bound only to return it into the level below the plane."[40]

The agreement and deed between William Scott and the Morris Canal and Banking Company was partly the outgrowth of a claim for damages filed by Scott, "in consequence of carrying the canal through his property at Boonton."[41] The real reason for the agreement, however, was, in my opinion, to gain "legal" possession of the waters of the Rockaway River at Boonton Falls, against the possible threat of the SUM's injunction.

Other similar agreements had been entered into by the Canal Company and the Ryersons regarding water rights and properties at the outlet of Greenwood Lake. The deeds and agreements are still among the files of the Canal Company at Trenton. Presumably the Canal Company was operating on the old quasi-legal assumption that "possession is nine-tenths of the law," proposing to set itself against the SUM as equal if not superior proprietors of Passaic River tributary waters. Also presumably, the SUM felt itself frustrated at every turn, and so decided to use a new tactic.

Colden's May 1, 1827 *Report to the Directors of the Morris Canal and Banking Company* announced the Company's plans or hopes to borrow money to complete the canal. Stock sales after the scandal of 1826 had been disappointing to say the least, and since they were $10-down-and-$10-a-month installment sales, the amount of money thus far received by the Canal Company had not been enough to pay off its existing debts to contractors and others, much less finish the canal to Powles Hook. Colden was afraid that if such a loan could not be made, thus forcing the Company to proceed on the amount of monies realized on stock subscriptions, progress would be much slower, and completion of the western division of the canal would have to be sacrificed in favor of the section "from the summit level to the City of Jersey."[42]

The State Legislature was accordingly petitioned by the Canal Company for a loan of $500,000, on the ground that the canal was an important state work. But the Legislature was also petitioned by a group signing itself "Inhabitants of Paterson and its vicinity, in the County of Essex, and State of New-Jersey," and objecting to the extension of any credit by the State to the Canal Company. Such a loan to a private company would set a "dangerous precedent," for the Canal Company's charter entitled none but itself control over its "management and concerns." And after all, if the New York capitalists were "unwilling to lend any more money to the Morris Canal," why, the memorialists asked, should the State assume the responsibility?[43]

The political aspects of the canal's attempt to raise money inspired a local poet to write a rather bitter, witty satire. He signed himself merely "Q."[44] The reader will do well to keep in mind that even at this early date, the temperance movement was well underway.

FETCHING THE PUMP — Or, Lobbyism Improved.

Where Legislation holds its session,
(To name the seat were a transgression).
It chanced that late at eve, while walking,
Loud sounded from a tavern — talking.
It seem'd a meeting were held there
Of some unholy kind of prayer.
Enter'd! sole spirit found within
Was whiskey punch — that source of sin —
Its emblematic sweet and sour,
The poison of the social hour;
And every speech and every groan
Ended in "hundreds thousands loan."

As clear'd the fumes — looking around,
Here delegated power was found
Assembled — advocates, cashiers,
Contractors, claimants, engineers —
Great creditors, from York's great city,
And writers in the cause, so witty —
The adjourned lobby all attended,
With bottles, tumblers, Horns, Bills, Bills amended.

Is it thus, my friend, committees meet?
Do humble suitors thus entreat?

"Aye, surely; better give champaigne
Than not the loan essential gain.
Art ignorant, (of flesh thou lump,)
Of this our way to 'fetch the pump?'"

How so? I see no water here,
Nor pump. "The pump of State." 'Tis queer!
"Queer is it? — we boast an excavation
Long as this member of the nation,
Important to the State's existence,

(Tho' this to you I 'own is pretence,)
Petition says 'tis failed of sources,
So serpent-like its thirsty course is;
Its reservoir of funds exhaust,
By reason of its mighty cost,
And all its credit prone and lost,
Like streams which sink to depths profound,
Canalling useless under ground;
Or like a pump, ill made, ill used —
Its bow is bad — from first abused;
Works dry of course, nor yields a drop,
But borrows and pours in at top."

The riddle's solved, its force I see,
Canals which are of water free
With other fluid filled must be!

"Yes, and we know judicial function
On water, straight must lay injunction.
Our engineers by force of mind,
Exceeding erudite — refined,
By calculation more combin'd,
Deem spirit meet, of cheapest kind; —
And though of fire a little risky,
Our loan obtained — we feed with whiskey.
And premium of our known assurance,
In Wall-street — low, for our insurance —
Will easily be had, if we
'Gainst engineers should guarantee.
For some have 'nough of wit and ire
To set Passaic's stream on fire;
And fill'd with water, catch'd, consume
The neighboring labours of the loom.
Besides, 'tis financiered a thing
Likely the western trade to bring,
As all may freely drink who ply —
(No need to fear they'll drink it dry.)
Besides we thus shall aid society
To concentrate all inebriety —
And thus the general good we shall
Promote — by right of our canal;
The right of spirit ours exclusive,
Become good natured — less abusive.

"The improvement here we now begin,
By example prove it is no sin —

As corporations have no soul,
No conscience wakens to control
Our doings, or undoings; all
Is lawful wrong, for our canal.
We're King Canal — our rightful sway
Is felt where'er our whiskey way
O'er land and water tyrannizes,
And every law — of right despises.

"Pretending wrong, we here complain
Is drought of money, as 'twere rain —
Regardless of the proverb wise,
That what's thus done can never rise;
Like water split upon the ground
Can never gather'd be, nor found,
So money's ne'er return'd or found
That's wasteful spent upon the ground —
'Tis neither sown nor planted there,
But buried, no one knows just where.
By granting our demand for loans,
The State for Nature's wrongs atones;
For the State is but a child of nature,
This child ought pay for it, docile creature.

"You see the wherefore and the why
The pump of State's consider'd dry,
Its lower end is in the Treasury.

"To fetch this pump is something new,
But sure as inclin'd planes, 'twill do —
As few can çlearly see what's true —
For planes inclined are soon descended:
'Tis not the question how ascended?
This problem's not for us to solve —
(If it were, we'd further loans resolve) —
Ascent's like paying loan for judgment,
All up hill — and (I'll whisper it) a fudgement;
So Virgil sung, as school-books tell,
Facile's the inclined plane to hell.

"But to return and view the cheerful skies,
There all the mighty toil and labour lies —
For to repay this mighty loan
Were quite impossible — (a groan.)

"But our great plea to prove our merit,
And our best title, too, to credit,
Is, that we're generous ere we're just,

And pay the claimants when we must.
And though 'tis true, beyond negation,
The State should give for education,
Rather than lend this navigation.
Yet we assert great folks are fools,
Who would support our public schools."

Fools, friend! Tho' punch may please the throat
Whiskey ne'er bought a Jersey vote;
The central work that we might boast,
By loans to you were surely lost,
Coal's the best object of its cost." Q

Meanwhile, the engineers continued to squabble in the newspapers, and the issues spilled over into the elections of 1829. An item appeared in the *Paterson Intelligencer* on October 7, 1829, signed, simply: A Stockholder of the Morris Canal Co.:

> *Mr. Editor — Being informed that the ticket selected by the Convention at Orange [for the preparation of the Essex County election slate] in the 3d inst. to be supported at the ensuing election, is composed of persons who are pledged to advocate and sustain the interest of the 'Morris Canal and Banking Company,' to the exclusion of every other local interest that may come in contact with it. My conviction is, that the information is correct, as it is well known that Mr. Amzi Dodd is appointed standing Counsel by the Board of Directors — that Mr. Joseph C. Hornblower is a feed Lawyer by them, and that Mr. John Allen has a very deep interest no one will deny, for as a contractor he has become one of their largest creditors. As for Mr. I.S. Miller,* he is a decided friend of the Canal Company, from what cause I can hardly conjecture — he also may be a creditor by the delinquency of some of the contractors or workmen, in not paying their store accounts, and which he thinks will be paid, provided he in his wisdom, in connection with his colleagues, can prevail*

**This was undoubtedly Isaac Miller, who, with his brothers, kept a general store at Little Falls, where Beattie's Mill is today.*

upon the State to take a portion of the stock, or grant a loan to the Company, which I understand is their great object. If so, I would call upon every elector of the township of Aquackanonk, and especially the people of the town of Paterson, whether they are willing the State should embark on this business, so that the Company then may with impunity encroach upon the vested rights of the "Society for Establishing Useful Manufacturer," [*sic*] — *thereby take from this place the natural flow of the Passaic River, involving the Society in difficulty and law suits, retarding at certain seasons of the year manufacturing operations,* consequently strike a death-blow to the future prosperity of this Town, deprive the poor of support, the mechanic of employment, and the farmer of a market — besides involving the State in a debt, to discharge which the community must be taxed — and all this for the benefit of capitalists of other states.*

I am not an enemy of the Canal, but am opposed to monopoly and any encroachment upon our rights, as a people interested, and deeply interested in the prosperity of the Town of Paterson.

**It is a well known fact, that the manufacturers of this town have this season forwarded to the Governor certificates, under oath, that they were short of water, and it was satisfactorily ascertained that it was occasioned by the Agent of the Canal Company appropriating the waters of the Passaic to the use of the Canal. If so much has been already done by the Company, what may not be expected if the State takes an interest, and the water at the summit level of the Canal should be insufficient? They no doubt will exercise state power, and in case of a protracted law suit with the Society what is to become of our manufactories? Let reason answer.*

"Q" needn't have worried, since the Legislature had already passed an act on February 23, 1829, authorizing the Canal Company to borrow $500,000 — but from another agency. Thus the ground was prepared for the famous "Holland" or "Willink" Loan, or Dutch Mortgage.

As a holder of stock in the Morris Canal and Banking Company, Q might better have worried about the far-reaching implications of

Sullivan's report to the SUM. A railroad whose tracks virtually paralleled the canal across the state to a Pennsylvania coal region as potentially productive as the Mauch Chunk area was a spectre whose shadow was already falling across the towpaths of the Morris Canal.

Still another spectre was raised by Sullivan, with comments about the Canal Company's use of waters from Lake Hopatcong, which normally would have flowed through the Musconetcong River channel, and which would not be diverted, in part, eastward to Paterson's Passaic River system. The owners of mill seats along the Musconetcong would be heard from — as soon as they could find a voice influential enough to speak for them in Trenton.

Meanwhile, the Canal Company would have its hands full, trying to offset the antagonism stirred up by Mr. John L. Sullivan's report on a proposed railroad.

It is tempting to see in the events of the decade 1825-1835 a pattern of agitation for some form of transportation arising from the towns or regions that had once counted heavily on being serviced by the canal. Paterson, slighted by the canal, received the State's second railroad charter, on January 31, 1831. The first charter had gone to the Camden and Amboy Railroad at the same time the Delaware and Raritan Canal was chartered, on February 4, 1830. Elizabethtown, probably disappointed that the Morris Canal did not extend to her borders, chartered the State's third railroad, the Elizabethtown and Somerville Railroad. Citizens from Newark, New Brunswick, and Jersey City threw their efforts behind the New Jersey Railroad and Transportation Company, chartered on March 7, 1832, the State's fourth such act. Sussex County, very much left out of the canal's plans except at Stanhope, chartered the fifth New Jersey railroad the very next day, March 8, 1832: The New Jersey Hudson and Delaware.

Without jumping to conclusions, however tempting, it is a fact that the State's great modern railroads, were prepared for at this very early date.

REFERENCES

1. Isaac H. Williamson, *Law Opinion, In the Case of the Paterson Manufacturing Society Against the Morris Canal and Banking Company* (Paterson, N.J.: 1845), p. 27.
2. *Paterson Intelligencer*, January 31, 1827.
3. *Ibid.*
4. *Report to the Directors of the Morris Canal and Banking Company; Made by the President, Cadwallader D. Colden, May 1st, 1827: With a Statement of the Work Then Done Upon the Canal. A Report from the Chief Engineer, and a Communication from Governor Clinton* (N.Y.: 1827), p. 4 (hereinafter cited as *Report to the Directors, 1827*).
5. *Paterson Intelligencer*, November 14, 1827.
6. *Ibid.*, September 12, 1827.
7. *Palladium of Liberty*, June 21, 1827.
8. Morristown *Patron of Industry*, July 19, 1827.
9. John Langdon Sullivan, *Report on the Origin and Increase of the Paterson Manufactories and the Intended Diversion of Their Waters by the Morris Canal and Banking Company, also on Post Rail Roads, . . .* (Paterson, N.J.: 1828), p. 5 (hereinafter cited as J.L. Sullivan, *Report to the SUM*).
10. *Ibid.*, pp. 43-4.
11. *Ibid.*, pp. 48-9.
12. *Ibid.*, pp. 43, 47, 51.
13. *Ibid.*, p. 55.
14. *Ibid.*, p. 56.
15. *Ibid.*
16. Charles A. Shriner, *Four Chapters of Paterson History* (Paterson, N.J.: 1919), pp. 72-3.
17. J.L. Sullivan, *Report to the SUM*, p. 7.
18. *Ibid.*, pp. 7-8.
19. *Ibid.*, p. 8.
20. *Ibid.*, p. 12.
21. From *Palladium of Liberty*, reprinted in *Paterson Intelligencer*, December 24, 1828.
22. *Ibid.*
23. *Ibid.*
24. J.L. Sullivan, *Report to the SUM*, p. 29.
25. *Ibid.*, p. 30.
26. *Ibid.*, p. 58.
27. *Ibid.*, pp. 58-9.
28. Isaac H. Williamson, *op. cit.*, p. 5.
29. *Ibid.*, p. 9.
30. *Ibid.*, p. 5.
31. *Ibid.*, p. 11.

32. *Ibid.*, p. 18.
33. *Ibid.*, pp. 24-6.
34. *Ibid.*, pp. 26-7.
35. *Ibid.*, p. 19.
36. *Paterson Intelligencer,* January 7, 1829.
37. *Ibid.*, January 14, 1829.
38. Morris Canal and Banking Company *Minutes,* June 25, 1829.
39. *Ibid.*, December 9, 1829.
40. Munsell's *History of Morris County, New Jersey 1739-1882* (N.Y.: 1882), p. 179.
41. Morris Canal and Banking Company *Minutes,* July 21, 1829.
42. *Report to the Directors, 1827,* p. 3.
43. *Paterson Intelligencer,* January 28, 1829.
44. *Ibid.*

8

Getting It All Together

THE State Legislature had already occasioned an even more immediate threat than Sullivan's proposed SUM railroad. On February 19, 1828, one Robert Thomson was authorized "to construct a canal through his land, in the township of Bergen, from Newark bay to New-York bay." Whereas the Morris Canal, so far scheduled to terminate at Newark, was to be 32 feet wide at the surface of its 4-foot depth of water, Mr. Thomson proposed to build his canal at least 40 feet wide, with "at least five feet depth of water at low water." He was to be given three years from the date of the act's passage to complete the canal, the act itself to become null and void if he had not spent $10,000 on the canal within two years.[1]

Mr. Thomson's canal was planned to receive boats with masts, for his authorization specified that he must provide and maintain either a draw or swivel bridge at least 14 feet wide over the public highway leading from the village of Bergen to the Kills and Bergen Point, and that the canal passage *under* the bridge must be at least 32 feet wide. The draw or swivel bridge would be required to admit the "passage of vessels with masts or boats with ladings or structures requiring such open passage."

No mention of this "Thomson Threat" could be found in the Canal Company's records. Yet, in 1828, when the law was passed to authorize Thomson to proceed, the Morris Canal still had friends in Trenton, and it is inconceivable that the Company knew nothing of his plans. Nor is it likely that the news of a rival canal was received without concern. Boats large enough to use Thomson's 40-foot-wide, 5-foot-deep channel would obviously not be using the smaller Morris Canal. But unless the Morris Canal Company intended to run its boats through Mr. Thomson's waterway, it was virtually forced to push for the Jersey City extension of its own canal. So far as is known, Thomson's canal was never built.

While George Macculloch was president of the Canal Commission, the Morris Canal was intended to be terminated at Paterson Landing, at first, then Newark. Macculloch was in favor of Elizabethtown as the terminus of the canal, as we have seen. It was not until Cadwallader D. Colden became president of the Canal Company, late in 1826, that attention focused on Jersey City as the most desirable spot for the canal to end.

The Army Corps of Engineers' report on the canal in 1823 had pointed out that the size of the boats that would run on the Morris Canal would be inadequate for navigation from the Passaic River at Newark, down Newark Bay, around Bergen Point, and up New York Bay in rough weather.[2] An extension of the canal to Jersey City, and a canal basin on the Hudson River, would eliminate that hazardous trip. More to the point, President Colden owned some very choice real estate on the Hudson River at Jersey City.

On November 7, 1827, the State Legislature was petitioned by a "Bergen" group for authority to build the Jersey City extension. A bill was subsequently introduced and engrossed, but was then postponed until the next legislative session. Beach had already surveyed the route from Newark to Jersey City and was able, "by accurate measurement and calculation," to state that the distance "from the Commercial Dock at Newark, to the tide waters of the Hudson, at Jersey City, is only eleven and a half miles." He considered the route "not only practicable," but one that could be "executed with ease" at an "average expense per mile" which would be "considerably less than the average expense of that between the Delaware and the Passaic."[3]

This supplement to the Canal Company's charter was enacted on January 26, 1828.

A canal that was to go to Elizabethtown by way of Newark would require one path; one that was to go to Jersey City would follow quite a different route. Newark businessmen were apparently disturbed at the new plan to make their town just another port along the shipping facilities of a terminal. All or most of Newark's shipping interests at the time were concentrated at the foot of Market Street (today's Commerce Street), where the town maintained docking facilities. If the canal were to stay land-locked when it reached the lower Passaic River — and there are indications that that *was* the Canal Company's intention — it would slight Newark's business district. Essex County lobbyists must have been hard at work in Trenton to protect Newark's interest, for when the Canal Company was authorized to extend their canal to Jersey City, the law contained a guarantee that the canal would accommodate Newark commerce. As of January 26, 1828, the Canal Company was obliged to connect its canal "with the Passaic River at the village of Newark, so that the boats used on said canal, can at all proper times pass from the same into the river, and from the said river into the canal."[4]

Section I of the charter amendment, or supplement, was a modification of the Canal Company's powers, specifying that they might "not take any lands for extending the said Morris Canal . . . until the said lands shall have been surveyed and appraised" according to the terms of the original charter, "and until the said company shall have first paid" the amount of the appraisement to the owners of those lands, unless the owners gave their consent to have the lands taken. This was a change of procedure from the original charter, which had allowed the appropriation of lands by condemnation. The supplement's condition, however, applied only to lands needed for the extension.

Section II of the supplement granted to the Canal Company the right to conduct its canal "through, upon, or along any street, road or public highway" in Newark and Jersey City. However, within the limits of those cities, "the land occupied by said canal, and for its towing path, shall not exceed a space of thirty-two feet." This restriction of size would become a crucial factor in Newark, as we will see in a later chapter.

Section III of the supplement placed on the Canal Company the responsibility of building and maintaining "good and sufficient bridges" over their canal extension.

One route through Newark that Beach had surveyed in 1828 would have run the canal right through Military Common. On February 23, 1828, Newark's citizens had vehemently protested, and the Canal Company's directors bowed to public pressure. Beach was ordered to stop work completely on any part of the canal below the proposed plane into Newark pending a decision on the canal's placement. It would *not* pass through the Common at all, nor through *any* street in Newark, unless three-fourths of the owners of property on said street should consent. A special committee was appointed to look into the Market Street route, which seemed to be the most "eligible," or any other route through town, with the exception of a route that involved the Military Common.[5]

Several proposals emerged from that special committee, each of its members advocating a "favorite" path through Newark.[6] Colonel John Scott, John H. Stephens, and others, represented the so-called "Tracy Route." Charles T. Shipman, and others, were "friends" of two routes: One locating the canal in Market Street; the other, the "Post Office Route," passing the post office on the northeast corner of Broad and Academy streets and entering the Passaic River south of a piece of property owned by Ephraim Beach. Beach, himself, supported a route that would also pass the post office, but enter the river north of his property. Asa Whitehead and Luther Goble introduced the "Franklin Street Route," which would have passed the center of Newark somewhat south of Market Street.

As for the rest of the route, from Newark to Jersey City, the New York *Commercial Advertiser* considered it a "most fortunate circumstance" that the canal would be able to utilize "the old Newark road across the marshes," which had "been abandoned for nearly half a century." This road would "prove an excellent towing path, made already to the hands of the Company."[7] Colden later noted, however, that it was "intended to tow the canal boats across the Passaic and Hackensack rivers by steam boats, or by Teamboats. These boats," he said, were "contemplated to be in connexion with the company, which was incorporated by the legislature of New Jersey last year [1827] to establish a road and ferries between Newark and New York by the old ferries, and will answer the purpose of conveying passengers, and towing the canal boats; and the two companies can at the same time accommodate each other, by uniting to make the bank of the canal both a towing path and road."[8]

These projections about Morris Canal's route were a bit premature, however. The path of the waterway through Newark was far from settled. In fact, it had scarcely been started when Colden's Report of the Joint Commitee (to view the inclined planes) was issued, December 13, 1828. "From the Passaic, the canal ascends, by means of two locks, not yet commenced, to a level passing through the town of Newark, to the highlands immediately above the post office. The ground in the town is believed to be favorable for canalling, requiring but easy earth excavations. The route was represented to us as one that was acceptable to the inhabitants: and, for the purpose of accommodating them further, it is the design of the canal company, by depressing the level of the canal through the town, to render their embankments and bridges, in the street, as low as possible. Above the town, is to be constructed an inclined plane, of seventy feet lift, to which nothing has yet been done."[9]

The canal had been under construction for almost three and a half years when Colden wrote that description. Why hadn't this final leg of the route to tidewater been decided upon? To answer this question, we will have to examine "goings on" in Jersey City, as well as in Newark.

Colden noted in his report to the directors that $777,923.71 had been spent on the canal by December 13, 1828, and that an additional $38,093.89 was then due to the contractors. Moreover, the sum of $284,000 was still needed to complete the canal to Newark, bringing the total estimated expenditures to $1,000,017.60. With the proposed extension of the canal another 11-1/2 miles, from Newark to Jersey City, even *more* money would be required.

In fairness to Beach, his original estimate of $810,000 as the probable cost of building the Morris Canal was made before it was decided to extend the canal beyond Aquackanonk to Newark, and before Jersey City was considered as a terminal point.

At the time of Colden's report in 1828, the Canal Company owed $388,050.50, "including bills in circulation, all loans or other obligations, as well as for work done, and materials furnished or properly acquired for the canal." Another $284,000.00 was considered necessary by the chief engineer to "completely finish the canal and all the necessary works from the Delaware, at Easton, to the Passaic, at Newark." That made a grand total of $672,050.50 needed to wipe the slate clean of old debts and to make the Morris Canal ready to operate profitably.[10] A few weeks later on February 23, 1829, the State Legislature passed an act to authorize the Morris Canal and Banking Company (MC&BCo.) to borrow money and for "other purposes."[11] This started a chain of events unfortunate in the financial life of the Morris Canal, events that overshadowed the apparent, eventual operating success of the canal, itself. Among the "other purposes" of the act was the right of the Canal Company to mortgage its waterway.

Section I of the act entitled the Canal Company to borrow $500,000, "and to issue therefor their post notes, bearing an interest of six per cent per annum, payable to the owners or bona fide holders thereof: *Provided,* that no such post note shall be issued of a less denomination than three hundred dollars." (Post notes are simply promissory notes issued by a bank. They are payable at a fixed time after the date on which they are issued.)

Section II of the Canal Company's charter amendment authorized it (the Company) to put up the canal as collateral security. To guarantee payment of the principal and the 6 percent interest on the half million dollars' worth of post notes it proposed to issue, the Canal Company was authorized to mortgage the canal, "to execute and deliver to any person or persons, body corporate or politic, an assignment, conveyance transfer" of the canal-entire. That meant the "appendages and appurtenances," the "lands, tenements, and premises," and the chartered rights and privileges — everything.

When the law was passed, Colden immediately wrote to Henry De Rham, authorizing that gentleman to negotiate a loan for the

Canal Company in Europe.[12] De Rham, one of New York's largest importers, particularly of French goods, had been the Swiss Consul to New York for many years.[13] While he was using his influence on behalf of the Canal Company, the board of directors called a meeting of the stockholders and directors to organize a drive for subscriptions to the new loan.

Assigned to solicit subscriptions in Newark were Luther Goble, Abraham W. Kinney, and Joel W. Condit; in Morris County, Colonel Joseph Jackson, William Scott, and Abraham Britten; in the Bloomfield area, Israel Crane and Nathaniel Baldwin. New York City was to be canvassed by Dudley Selden, John C. Halsey, Samuel Judd, David Clarkson, *and* the entire board of directors:[14]

C. D. Colden
Lambert Suydam
Robert Carter
Charles T. Shipman
Robert P. Bell
Charles Hoyt
Archibald McIntyre
John Hone, Jr.
Richard M. Lawrence
Isaac A. Johnson
Henry McFarlan
Samuel Jacques
Abraham Ogden
Billopp B. Seaman
Peter M. Ryerson

The esteemed Bank of New York was asked to subscribe to the new loan, the Canal Company "agreeing to take in payment for so much of the subscription the post-notes ($25,000) already held by the bank, and which would soon become due." The subscription was accordingly made, and furthermore, the Directors of the Bank of New York authorized "that the whole amount of $500,000 should be subscribed for, and the security be made equal to that already held by the bank." The Canal Company's indebtedness to the Bank of New York was "paid at maturity by draft on Wm. Willink, Jr., Amsterdam, payable in London, for 8000 lbs. sterling," since the Canal Company had negotiated with Willink for a loan with which to repay their post notes.[15]

That indebtedness to the Bank of New York had been mentioned by Colden in 1828, when he explained that "post-notes of the [Canal] Company in circulation, issued on account of the first loan of

$300,000; which notes, as they shall be issued, are secured by a judgment entered in the Supreme Court of New Jersey, in favor of Charles Wilkes and others, Trustees of these notes,"[16] (Charles Wilkes was president of the Bank of New York at the time, having been affiliated with that institution since 1792. Ill health forced him to leave the bank in 1832 after serving for 48 years.[17] The Bank of New York's headquarters were rebuilt in 1856, one of many New York buildings to be built using the excellent brownstone of the Little Falls quarries in New Jersey, which no doubt was shipped to New York by way of the Morris Canal.)

To repay its debt to the Bank of New York, the Canal Company had authorized Henry De Rham to try to find someone in Europe who would be willing to lend them that money in exchange for a mortgage on the canal. A proposition from Mr. John A. Willink of New York was read into the Canal Company's *Minutes* on September 8, 1829. If Willink undertook to borrow, at an interest of 5 percent, for ten or fifteen years or longer, the sum of $750,000 at the rate of $100 for $95, the Canal Company decided it would allow him personally, a nice commission arrangement: 1 percent for negotiating the loan, 1.5 percent on the amount of the interest for paying the interest that would be due on the loan twice yearly, and 0.25 percent commission for paying off the loan when it came due. To Mr. Abraham Willink of New York, the company would pay $15,000 if he obtained the $750,000 loan on these terms.[18]

The loan *was* obtained, in Holland, and hence was referred to as the Holland Loan or Dutch Mortgage. It was to be repaid in five, equal, annual payments beginning on January 1, 1846. Until that time, 5 percent interest on the loan was to be paid semiannually at Amsterdam, on the first of July and January.[19]

In Europe, "13 millions of guilders were subscribed for this loan in two hours, being more than five times the amount required."[20] The New York *American* reported that the new loan "insured the completion of the canal with the utmost speed," and that the canal would surely be opened from the Delaware to New York before the end of 1830.[21]

By the end of 1829, Shipman's "Post Office Route" had been decided upon. The canal was to go right through the heart of town to the river at the foot of Market Street. Shipman and "friends" were to give the Canal Company title to the lands from the head of the plane to the river, and to lands at the river for a basin just south of Cory's Dock. The titles to those lands cost the Canal Company dearly. A lot on Broad Street owned by stockholder Ephraim Bolles, for example, cost $1600.[22]

Charles T. Shipman was one of Newark's leading leather merchants. Once the canal was completed to the Passaic River, he and stockholder Abraham Kinney established a shipping business there, running a fleet of packet boats from the Canal Dock at Newark to Whitehall in New York. By April, 1831, Shipman & Kinney boasted three packets: The *Palmyra,* the *Bolina,* and the *John Henry.* The following year Shipman added the *Good Intent* to the fleet and Mr. Kinney dropped out of the partnership, whereupon Shipman called his company the "Newark and New York Canal Dock Line."[23]

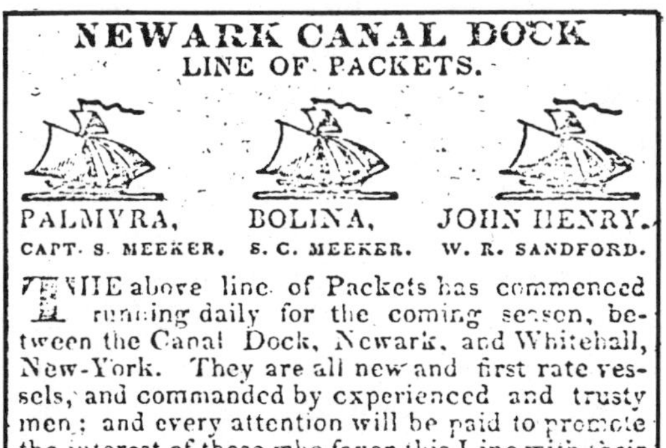

Advertisement for the Newark Canal Dock Line of packets, Shipman & Kinney, proprietors (ETF, Newark Sentinel of Freedom, May 22, 1832).

Once the route was decided upon, Beach was instructed to start work again in Newark, and by the end of September, 1830, was completing the canal from "the lower lock at Newark" to "the Basin at the river, commonly called Shipman's Basin." He had been authorized to put under contract "the whole of that part of the contemplated line of the canal lying between the lower, or Shipman's Basin, at Newark, and the Passaic River at or near the Old Ferry." Beach and Douglass had been asked further to report by October 20, 1830, "on the expense, practicability and advantages of bringing the canal from the Passaic at the old ferry to some point which they are requested to specify on the Hudson near the city of New York."[24] Some of the Canal Company directors had wanted to hold off on the canal's extension until that report was received, but they were a minority, and work was ordered to begin.

The Canal Company felt that it "ought by all means to own as much land at each end of the canal as may be required for the purpose of constructing all such basins, wharves, storehouses, etc., as may be necessary for the convenient transaction of their business."[25] The board further resolved to have "entire control of all basins on the line, and particularly at Newark or at either end."[26]

By March, 1830, "the excavation of the Morris Canal through the town of Newark" was "progressing with great despatch [sic]."[27] Elihu Crane had deeded property to the Canal Company for $2 and the privilege of making a basin at the foot of the plane reaching to Plane Street, on his own lands.[28] Ephraim Beach applied for permission to make both a basin on his property at the river and a lateral cut from the canal to service that basin.[29] He had already given, as a gift, some of his riverfront property to the Company with the condition that it be used as a basin (deed dated January 20, 1830), but whether or not his application referred to the gift property or to another piece of his land, is not known.[30] A basin *was* built on the deeded property, but it had apparently been abandoned by 1854, when the city of Newark proposed to build a road over those lands. The Canal Company then found it necessary to reclaim "the old Canal Basin at Newark to prevent the road from being built."[31]

Another application was received by the Canal Company from a group of Newark citizens represented by Amzi Dodd. They wanted permission to make a lateral cut "from the canal to the New Dock, on condition [that] they procure a repeal of the law requiring the company to lock into the Passaic at Newark."[32] The application was postponed, and it wasn't until April, 1831, that a committee of Canal Company directors went to Newark "to examine the several locations recommended for basins, and for a lock into the Passaic."[33]

The Lehigh Coal and Navigation Company's Board of Managers informed their stockholders that the Morris Canal would be open by August, 1831, according to the predictions of the canal's directors.[34] But in April, 1831, there was not yet a lock into the Passaic River at Newark, much less a terminal at Jersey City. Beach had been instructed to examine the canal's course across the meadows (today's Harrison) and through Bergen Ravine to Jersey City to determine "the best place for basins at the Hudson River."[35] He was, moreover, "to devote his whole time and attention to finishing the work on the line between the Delaware and Passaic rivers," and "not allow anything else to interfere with, prevent or retard the execution of said work with all possible speed."[36] The Lehigh Company must not be disappointed in its expectations.

Still, there was the Jersey City extension effort, and at this point it will be helpful to look at what had been going on in Jersey City.[37]

In 1804, a group of prominent New Yorkers had invested in real estate on the New Jersey side of the Hudson River at a place known as Paulus Hook, or Powles Hook. Their intention was to establish a thriving residential and commercial community within easy commuting distance of New York City. Chief among the group were Jacob Radcliff and Richard Varick, both former mayors of New York, and Anthony Dey, Varick's nephew, son of General Richard Dey, of Preakness, New Jersey, and a prominent and successful New York attorney. Legal advisor for the group was Alexander Hamilton, who was killed a few months later in the Burr-Hamilton duel at Weehawken.

Paulus Hook was owned by the Van Vorst family, specifically by Cornelius Van Vorst, until March 26, 1804, when the property and the ferry rights were deeded to Anthony Dey for an annuity of 6000 "Spanish milled dollars." On April 18, 1804, Dey conveyed the property to Abraham Varick, who in turn conveyed it to Richard

Varick, Jacob Radcliff, and Anthony Dey on April 20, 1804. Perhaps because John Stevens was selling lots in Hoboken at the same time, the Dey group decided to sell parcels of land at Paulus Hook, and on April 16,1804, they advertised such a sale would take place.[38]

The Jersey Company was incorporated on November 10, 1804, and to that corporation, the Dey group (Varick, Radcliff, Dey) conveyed Paulus Hook. The first election of officers was then held on December 10, 1804. The trustees installed were:

Richard Varick, Pres.	Alexander P. McWhorter
Jacob Radcliff	William S. Pennington
James T. Thompson	Isaac H. Williamson
William Halsey	D. B. Ogden, Sec'y.

For 15 years the Associates of the Jersey Company governed Paulus Hook. In 1820, Jersey City was incorporated under the city's first Board of Selectmen (five freeholders, to be elected annually): Dr. John Condit, Samuel Cassedy, Joseph Lyon, John K. Goodman, and John Seaman.[39] Cassedy, Goodman, and Lyon later became actively involved in Morris Canal affairs. Cassedy became Clerk of Bergen County, which then included all of modern Hudson County. Lyon, an Elizabethtown native, was the lessee of the Jersey Associates' new tavern on Grand Street, Jersey City, near Hudson Street. We met him on October 15, 1825, at the Canal Company's groundbreaking ceremony at Lake Hopatcong. A year or so later he became lessee of the ferry at Jersey City.

On November 3, 1804, the Dey Group Associates had sold a block of land to Robert Fulton. To induce him to locate his shipbuilding yard in Jersey City, they made a special arrangement so that Fulton could pay the purchase price of $1000 over a 5-year period without any interest. The Fulton Foundry and works were located on the corner of Green and Morgan Streets, with the dry dock (built by Fulton) in front of the foundry.[40]

In 1809, a Newark group — a Mr. Durand, Elisha Boudinot, and General J. N. Cummings — invested $50,000 to advance the use of steam ferries, and Fulton was asked to build steam engines for two such boats. Actually, Fulton designed the boats, and the New York firm of Charles Brown built them: the *Jersey* was finished July 12,

1812, and the *York* the following year. The Jersey Ferry had consisted of a few rowboats, each manned by two oarsmen. The boats were equipped at times with sails and carried a couple of spare oars. Now it had finally become a steam ferry line.[41]

Apparently C. D. Colden became Fulton's partner in that enterprise, for we find a deed of property and ferry rights from the Associates of the Jersey Company to Fulton and Colden, dated September 3, 1812, for the consideration of $2833.33.[42]

Fulton had married Robert Livingston's daughter, Harriet. After his death in 1815, she remarried and became Mrs. Charles Augustus Dale. In 1825, Mrs. Dale applied to the State Legislature for the right to sell part of her late husband's estate in Jersey City,[43] and was thus authorized to deed to C. D. Colden, Robert Fulton's part of the property that he and Colden had occupied jointly. The ferry rights reverted to the Jersey Associates, and it was the peculiarity of the Associates' charter that those ferry rights could not be enjoyed by any other party in Jersey City, but were exclusively vested in the Associates.[44]

Fulton's widow also sold part of her late husband's property to George and Phineas C. Dummer, New York merchants. Phineas became Jersey City's mayor, and George established a porcelain and glass works at the south side of the canal basin at Jersey City.[45]

Colden, then, owned lands along virtually the entire waterfront of Paulus Hook, extending from Harsimus Cove to Communipaw Cove. And what he didn't own, the Jersey Associates did. James P. Allaire, a "Master Mechanic," took over Fulton's foundry and steam engine works at Morgan and Green Streets on Harsimus Cove, including the dry dock.[46] We will meet Allaire again, when he undertakes a job for the Canal Company.

In September, 1825, the lease of the Jersey Ferry had passed to Colden, Francis B. Ogden, and Samuel Swartwout, for an annual payment of $5000. By October, 1826, Swartwout and Ogden pulled out, and Colden failed to make the ferry a paying proposition, and he surrendered the lease to the Associates.[47] Before Colden died in 1834, he deeded his Jersey City property to John V. B. Van Varick and Archibald McIntyre, both Canal Company directors, on

November 1, 1833. They, in turn, deeded the lands to the Morris Canal and Banking Company on July 17, 1834.[48]

When Colden and the committee of directors and New Jersey legislators toured the canal to view the inclined planes in December, 1828, they started their journey at a 5-acre site directly opposite Courtland Street, New York, "where the river is some yards less than one mile wide, and *immediately adjoining the Jersey Ferry Wharf* (italics mine)."[49] Colden noted then that what the group was inspecting was the proposed site for a canal basin. Yet, almost a year later, on October 20, 1829, the Canal Company's board of directors appointed Abraham Ogden, Robert Carter, and Colden a committee to meet with the Jersey Associates to discuss the canal's Jersey City terminus.[50] On April 6, 1830, director Charles Hoyt and cashier Robert Gilchrist were added to that committee.[51] Obviously something other than a site for a canal basin was under discussion. A listing of events will be helpful in discovering just what that "something" was.

February 24, 1830	Camden & Amboy (C&A) Railroad and Delaware & Raritan Canal charters.
April 6, 1830	Beach ordered to examine canal course across Jersey Meadows and Bergen Ravine.
May 31, 1830	Beach reports on site for canal terminal at Harsimus Cove.
September 12, 1830	Anthony Dey writes Canal Company about "a railroad."*
October 27, 1830	Dey & Associates petition Legislature for a railroad, Newark to Jersey City.
January 1, 1831	Jersey Associates lease their ferry to the New Jersey Railroad and Transportation Company (NJRR&T Co.).
January 21, 1831	Paterson & Hudson River (P&HR) Railroad chartered.

**An educated guess makes this the proposed New Jersey and Atlantic Railroad, devised as a route to rival the Camden & Amboy.*

February 3, 1831	Morris Canal and Banking Company (MC&BCo.) "Junction" Railroad chartered.
February 9, 1831	Elizabethtown & Somerville Railroad chartered.
May 10, 1831	MC&BCo. committee to confer with directors of "Junction" Railroad *re* connecting canal with that railroad.
October 18, 1831	MC&BCo. to prepare petition to Legislature for railroad, Newark to Hudson River.
October 28, 1831	Petition forwarded to Isaac H. Williamson, State Councilman from Essex County, former Governor of New Jersey, and first president of Elizabethtown & Somerville Railroad.
February 18, 1832	MC&BCo. petition withdrawn in favor of application of Jersey City to build the Newark-Hudson River Railroad. The right to hold 50 percent of the stock in that railroad given to MC&BCo.; Beach and Bell, in Trenton lobbying, instructed to act accordingly.
March 7, 1832	NJRR&T Co. chartered, John S. Darcy, first president; Anthony Dey, principal stockholder; Abraham W. Kinney, director, also director of the Canal Company.
March 8, 1832	New Jersey Hudson and Delaware (NJH&D) Railroad chartered to build railroad from Delaware River in Sussex County to the Hudson opposite New York. NJRR&T Co. supporters forced restriction to NJH&D's charters: The road couldn't come closer to Newark than three miles. This was John Rutherford's railroad, the later-named N.J. Midland.

The P&HR Railroad was, of course, the outgrowth of the Canal Company's earlier battle with Paterson's SUM over the use of Passaic River waters. The P&HR had been planned to go into Hoboken by crossing Bergen Ridge on an inclined plane powered by stationary steam engines, or by means of a tunnel through the trap rock. When the NJRR&T Co.'s engineer conducted a survey, he had found a "ravine" about 1-1/2 miles south of the P&HR's proposed spot for negotiating the trap rock ridge. The elevation at the "ravine" was considerably lower than the more northerly part of the ridge. It was proposed that the two railroads use the same pass, the same trackage, to be built by the NJRR&T Co.

The railroads were to terminate at the Jersey Ferry Wharf, precisely where Beach had recommended that the canal basin and outlet lock be located. We now know why negotiations for a canal terminal were necessary: The location had to be changed. The railroads would now occupy the northerly edge of Jersey City on Harsimus Cove; the Canal Company would put its basin at the southerly edge on Communipaw Cove.

The hostility between John Stevens of Hoboken and the Jersey Associates of Jersey City is legendary. The story is too long to be told here, but it must be pointed out that the hostility was rooted in steam and monopoly, and it caused certain "alliances" to be made, keeping the hostility alive and bringing still others into conflict. Very early, the Stevens' had allied themselves with Hoboken, with John Fitch's steamboat patent, and with the C&A Railroad and Delaware & Raritan Canal. Now they had also allied themselves with Paterson, Robert L. Stevens being a stockholder and a director of the P&HR Railroad. The Jersey Associates, on the other hand, had lined up with Jersey City, Fulton, and the Morris Canal, and were now affiliated with the NJRR&T Co. A curious turn of affairs, indeed!

John L. Sullivan had run the first surveys for the P&HR Railroad, and we saw in Chapter 7 how he and Beach had been at odds over the canal's use of Passaic River waters. Now, however, Major McNeill was in charge of the P&HR Railroad, and we saw in Chapter 1 how Beach and McNeill had worked together quite amicably in Catskill, New York. Ephraim Beach was now serving as Chief Engineer of the NJRR&T Co.![52]

The engineers, and, of course, the railroad companies, worked out a compromise whereby the two railroads would form a junction on the west side of the trap rock Bergen Ridge, where the NJRR tracks were to take trains from both railroads into Jersey City. The Jersey Associates had leased their ferry rights there to the NJRR, as we have seen earlier in this chapter, on January 1, 1831, *before* that railroad had been chartered. The P&HR Railroad was to have identical rights in that "junction" railroad. To appease Hoboken, it was decided that the P&HR Railroad would build a branch from Jersey City to that place.

Now, the Canal Company, or rather some of its stockholders and directors, *did* charter a "junction" railroad, supposedly to connect the canal with the P&HR's tracks at Congress (Market) Street in Paterson. The charter was issued on February 3, 1831, to Canal Company director Robert P. Bell, and stockholder Andrew Parsons, among others. This railroad was never built.

Instead, the "junction" between the P&HR and NJ Railroads was effected, and the Canal Company was primarily responsible for that monumental achievement. The P&HR Railroad may have been built by Paterson and the SUM. But, if in fact the Canal Company did buy half the stock of the NJRR&T Co., then that very vital rail connection that was to reach from the Hudson River to New Brunswick was built by Jersey City and the Morris Canal and Banking Company.

This unprecedented cooperation between the Canal Company and the Paterson-SUM faction had not been achieved spontaneously or without earlier antagonism. The Canal Company had fought bitterly to prevent the chartering of the P&HR Railroad. They sent a remonstrance to Trenton on January 9, 1830, protesting that such a road would "serve to defeat the privileges intended to be secured to the stockholders in the Morris Canal and Banking Company." Paterson rejoined by sending its own memorial to the Legislature, accusing the Canal Company of "endangering the very rights of the people" with their attempt to interfere with the projected railroad.[53] The railroad was chartered, January 21, 1831.

Once the engineers got together to work out the logistics of the "joint" or "junction" railroad to be used by the P&HR Railroad and the NJRR&T Co. trains, bitterness subsided and the eye of the storm was reached. In fact, at the P&HR's groundbreaking ceremonies at Paterson on July 4, 1831, the "Officer of the Day" of the celebration proposed a toast: "The Morris Canal, Paterson and Hudson River Rail Road and the Junction Rail Road, united with the Mechanics and Manufacturers, will make the Town of Paterson the Emporium of New Jersey." The toast was followed by six cheers.[54] Patersonians were actually looking forward to the Morris Canal's connection with their railroad! Looking back from today, one wonders why that Junction Railroad was never built. In any case, being dedicated on the Fourth of July, the P&HR couldn't fail, at least according to Alvin Harlow's formula.

In August, 1831, the *Paterson Intelligencer* revealed that the NJRR&T Co.'s surveys had already been made,[55] and by September, the P&HR Railroad announced that it would seek an amendment to its charter to permit a junction with that railroad.[56] Since the NJRR&T Co. hadn't even been chartered yet, these items suggest that cooperation was achieved below the surface and behind the scenes. With Beach and McNeill in the vanguard showing how mutually beneficial planning could yield mutually beneficial results, even Hoboken and Jersey City enjoyed a moment of respite.

That the brief period of calm was but the eye of the storm will be evident later, when the Canal Company and the SUM once again lock horns, and when Hoboken and the NJRR&T Co. become the bitterest of enemies. Meanwhile, during this moment of calm and cooperation, the Morris Canal was at least operating, even if not yet completed to the Hudson River. Boats were plying the waters of the Morris Canal, carrying both freight and passengers from "inland" to the basins at Newark. From Newark it was still necessary to transship freight on steam boats such as the *Newark,* which made two round trips daily between Newark and New York.[57] It was probably the *Newark* that was referred to in the Newark advertisement for the passenger packet *Emeline:*[58]

> *The splendid Packet Boat Emeline, Ephraim Morris, Master, has commenced running from Newark to Pater-*

son. She will leave Newark every morning (Sundays excepted) at half past 7 o'clock, and Paterson every afternoon, at three o'clock. Fare through, 37-1/2 cents: to Bloomfield, 13-3/4 cents. The Emeline will also leave Bloomfield on Monday mornings at 5 o'clock, and arrive in Newark in time for the Steamboat to New York. She will also leave Newark for Bloomfield on Saturday evenings, upon the arrival of the Steamboat from New York.

The Emeline is a large and commodious boat, and every attention will be afforded to passengers.

The scenery on the route is of a romantic and picturesque character, and the attractions in the neighborhood of Paterson, where the boat stops, are too well known to require description. Two Inclined Planes, one at Newark, and the other at Bloomfield, will afford interest to the learned and curious. The patronage of the public is respectfully solicited.

N.B. Passengers on board the Emeline who intend to dine at Paterson, will be conveyed in a Carriage from the Boat to the Passaic Hotel, free of any expense; and again returned to the Boat in the same manner.

In the same newspaper, on July 21, 1831, Joseph O. Ward's advertisement for packet service on the "beautiful *Maria Colden*" announced that she had "lately undergone a thorough repair and been newly painted and fitted up in good style . . . ." Service was to begin on July 22, with the boat running between Newark's Broad Street and Paterson every Tuesday, Thursday, Saturday, and Sunday. Mondays, Wednesdays, and Fridays were reserved for excursion parties, as were all evenings.

By June, 1832, the *Maria Colden,* captained by Mr. Bruen had ceased her Sunday run to Paterson, but a new note had crept into the advertisement: "All those who wish to avail themselves of a trip on the railroad, now in operation between Paterson and Aquackanonk, will find this a good opportunity, as the boat will remain at Paterson long enough for that purpose, and return the same evening."[59]

Perhaps beyond Mead's Basin the canal towns were not blessed with regular packet service, but other benefits accrued to them because of their location on the waterway. Dover was experiencing a real estate boom based on expectations of cheaper coal prices once the canal was in full operation.[60] As already noted, Messrs. Blackwell and McFarlan advertised Dover to be an ideal spot for the location of mechanics and workmen. The same was true of Rockaway, where Colonel Jackson was also promoting the growth of his town. The development of Boonton is legendary. Thomas Gordon remarked in his *Gazetteer of The State of New Jersey* (1834) that grass grew in the streets of the town until the canal was built. And Munsell's *History of Morris County* speculated that if it had not been for the Morris Canal, it is not probable that Boonton would have existed at all. The same statement could no doubt be made about virtually every town the canal intersected.

With the exception of Newark and Jersey City, where building the canal was the slowest because of their special problems, the construction proper was completed between Lake Hopatcong and the Newark Plane by the end of 1830. Niles' *Weekly Register* (January 8, 1831) rang in a new era with these comments:[61]

> *The Morris Canal, as we learn by a letter from a friend, is in operation for the space of fifty miles to Newark, N.J., and boats are constantly passing and repassing. The inclined planes work handsomely, and the writer says that he had passed over several in a pleasure boat, without difficulty — through the perpendicular lifts are from 50 to 100 feet, a boat will pass them in from 10 to 15 minutes. This has had the effect already to reduce the carriage of iron, made at works adjacent to the canal, from five dollars a ton to 50 cents. Such is the result of internal improvements.*

The names of most of the men and companies who contracted to work on the Morris Canal, or who supplied it with material, have been lost, probably forever. Of the few whose names survive, virtually none was ever famous enough to have left his mark in the written records of the times. A handful of notable exceptions, however, is headed by Gouverneur Kemble, one-time owner of "Mount

Pleasant," a famous Newark mansion made more famous by Washington Irving. Irving dubbed the place Cockloft Hall in his *Salmagundi Papers,* as he wrote of his pleasant visits to Newark to enjoy the company of Gouverneur and Peter Kemble and their friends.[62]

Gouverneur Kemble was notable in the business as well as the social and literary worlds, however. He owned the Greenwood Iron Works in New York State and the West Point Foundry Association located at Cold Spring, New York, on the Hudson opposite West Point. Dupuy reports that it was an adjunct of the U.S. Military Academy for defense purposes.[63]

Before 1830, Kemble had established a branch of his foundry and a small machine shop at the foot of Beach Street, New York City, on the Hudson (North) River. There, on May 13, 1829, the West Point Foundry Association assembled and tested the imported English *Stourbridge Lion* for the D&H Railroad.[64] In December, 1830, Kemble's men built the first locomotive in the United States, the *Best Friend of Charleston,* for the Charleston (South Carolina) and Hamburg Railroad.[65] The following year, in August, 1831, the company completed the *DeWitt Clinton* for the Mohawk and Hudson River Railroad.[66] They would continue to make locomotives until about 1835, but not to the exclusion of iron parts for the planes of the Morris Canal.

Blackwell and McFarlan of Dover were already under contract to make iron castings for the Canal Company, but early in 1830, the West Point company offered to supply castings and other iron parts for the Newark and lower Montville planes (No. 9 East).[67] Douglass had insisted that the iron plates for the planes be imported from England,[68] but Messrs. Blackwell and McFarlan and the West Point Foundry each held contracts for the "other" iron parts.*[69]

An example of a contractor's name that is mentioned once or twice (no first names given) but is never identified in Canal Company records is the "Mr. Cooper," with whom the Company intended to contract for a weigh lock. Between May 10, 1831, and March 13,

**Chevalier noted (1839) that the West Point Foundry was the maker of "wheels" for the Canal Company.*

1832, four entries in the Company's *Minutes* refer to "Mr. Cooper" and his proposal to build a weigh lock and erect scales. The last entry locates the weigh lock at Newark. The lock was built, for by 1834 we read that it already needed alteration.[70] Dodd and Baldwin of Bloomfield, builders of the Little Falls Aqueduct and Plane No. 11 East, apparently made the alterations at a cost of $780.89.[71]

A Newark Directory for 1835-6 lists James S. Searing as the weigh master and lock tender at the Collector's Office.[72] The lock, located appropriately, at Lock and Warren Streets, was then numbered No. 15 East; more modern canal maps show it as No. 16 East. The Canal Company authorized the Chief Engineer to buy a lot and erect a house for the "Collector and lock tender of the Newark lift and Weigh Lock, at a cost not exceeding for the lot $400 and for the house $1000."[73]

A canal basin was at the site of that lock, and advertisements offered coal to Paterson customers, delivered, "for fifty cents per ton, on board Canal Boats, at the Weigh Locks, Newark. Wm. C. Gould, Agent."[74] A tremendous rain storm on Saturday, August 5, 1843, swept away the basin at Searing's house, but did not damage either the lock or the plane below it.[75]

In 1847, when the canal and its locks were being enlarged and the inclined planes were being rebuilt, a new weighing device was called for to handle the increased tonnage of the larger boats that were to be built. "Messrs. Fairbanks" were the suppliers of those scales.[76]

Thaddeus Fairbanks of St. Johnsbury, Vermont, devised his first platform scale in 1831, and patented it. Within three years, the Fairbanks scale received a silver medal from the American Institute of New York City — the first award ever to be made by that institution.[77] According to the Fairbanks Genealogy, the Fairbanks Scale Company made weighing machines for everything from prescriptions (measured in portions of grams), to canal boats (measured in tons), to a train of cars.[78] Canal Company records refer to Mr. J. W. Combs, who had been recommended to the job of "weigh master and inspector" in charge of the weigh scales at the Newark Plane, and tell us that the Fairbanks Scale was installed in Plane No. 12 East.[79] Apparently, when the former weigh lock, No. 16 East, was enlarged, it was rebuilt as an ordinary lock.

Later, in July, 1856, a "suitable" scale was to be installed "in the tracks of Plane No. 6 West for the purpose of weighing boats and cargoes," at a cost of $5000.[80] Unfortunately, there is no known record of what company built and installed that device.

The names of other known contractors and suppliers of building materials and machinery parts are contained in Appendix B.

The contractors and suppliers not only helped build and equip a working canal, but also created employment, as well. Beginning in 1830, when the first plane and lock tenders, and the first toll collectors, were appointed to their positions, we find:[81]

Mr. Smith at Plane No. 10 East at $30 per month
Mr. DeHart at Lock No. 14 East at $12 per month
Mr. N. Wilson, Jr. at Mead's Basin }
Mr. A. Parsons at Paterson } 5% of the tolls collected
Mr. Ephraim Morris at Plane No. 11 East at $30 per month.

Both Smith and DeHart were told to find their own assistants, and to pay them out of their own expenses — "if necessary." Mr. DeHart might build his own lock house which the Canal Company was to take in the spring, paying for it or not, as it suited them. In December, 1830, both DeHart and Smith submitted their accounts for service rendered to the Company, and Colden was authorized to settle with them.[82]

According to the *Rules and Regulations for 1830*:[83]

1. The person at each plane or lock was to tend same and collect tolls.
2. No boats were to be allowed to pass a plane or a lock without paying tolls and getting a permit, except boats with special permits issued by the board of directors . . . .
3. Masters [boat captains] were to give an account of their cargo to tenders or collectors at the first lock, plane, or place of collection they came to.
4. Collectors and tenders were to keep an accurate account of all boats, cargos and tolls, and give or send copies of that account to the Canal Company's Secretary at Jersey City every Saturday, with the tolls collected.

5. Tenders and collectors were to inspect cargo to make sure it agreed with the permit issued.
6. A correct measure, counting, or weighing must be made before the boat be allowed to pass, if master and collector disagreed as to weight or quantity of cargo.
7. Toll lists were to be "conspicuously" displayed at every collection point.

The rates of toll ranged from 1-1/2 cents to 4 cents per ton-mile, with mineral coal and iron ore costing the least, and lumber in rafts and oysters (!) the most expensive. Empty boats or, as the regulation states, "other boats, with or without cargo and in addition to any cargo," were charged a rate of 2 cents per mile. Passengers and boats — that is to say — passenger boats paid 10 cents per mile for boat and passengers.

One other regulation required that all boats and rafts must carry a light on their bows at night, and that square-headed or sharp-cornered scows or boats must have a semicircular platform on the bow to defend against injury due to contact. What strange looking craft they must have been!

The Canal Company was considering the propriety of building 50 boats for the use of the company, to be ready for May 1, 1831.[84] Accordingly, an advertisement was placed asking for proposals for contracts to build from 40 to 50 boats of the following description:[85]

Length sixty-five feet in the clear.
Width eight feet from outside to outside.
Depth or height of the sides in the centre 3 ft.
The sides to be of pine plank, in one or more pieces, four inches thick at the bottom; and two and a half inches thick at the top, with a regular taper from bottom to top. The bottom to be flat, or nearly so, to within nine feet of each end, and then to have a sheer of twenty-two inches rise to the stem and stern; the sides to be strait [sic] *to within 9 ft. of the stem and stern, & then to curve to a point at each end; and have a sheer on the gunwale of from fourteen to sixteen inches towards the stem and stern. The planks from where the curves commence on the sides and bottom, to be*

of oak plank one and a half inches thick, to be put on crosswise. The bottom of the boat to be lined with one inch oak or pine boards, to be placed on three fore and aft bearings which are to be two and a half or three inches oak scantling. There are to be three or four cross beams about three by six inches square, to be dovetailed in the top of the sides, for a cabin. — The stern of the boat to be raised so as to give five feet in the clear between the deck and the floor, and come forward about ten feet, with a small window in front.

For the fastenings of the boat, the bottom plank to be spiked on with wrought iron spikes, four and a half inches long, and the ceiling or floor to be nailed on with 20d cut nails; the side plank to be fastened together with not less than ten bolts of proper length in each side, of 5/8 inch round iron, with a head at one end, and a screw nut and washer at the other. The boat to have a competent tiller, rudder- and rudder irons.

The boats must be finished by the first to the 15th of May next, and each boat will be paid for when she is delivered.

By order of the Board. R. Gilchrist, Secretary.

Jersey City, February 21, 1831.

Before this Canal Company advertisement appeared, William Pragnell was already building canal boats. It is not known where he came *from*, but by May 4, 1830, he had "removed to Dover," where he had "commenced the business of building BOATS for the *Morris Canal*." He already had one boat partially built to serve as as sample. It was "built in the most substantial manner, after a model recommended by Professor Douglass, of West Point, Engineer of Inclined Planes; Major Beach, Canal Engineer, and Col. Scott, Canal Commissioner." Anyone wishing to buy one of Pragnell's boats was instructed to "inquire of McFarlan & Ayres, New York, or of Jacob Losey and the Subscriber, at this Place. Dover, May 4, 1830."[86]

After the Canal Company advertised for boats, John Prior offered to build ten of them according to the specifications published in the newspapers. Prior's price was $210 each, but the Board of Directors agreed to have him build the boats provided he accept the price of

$200 each.[87] Within a week, the Company contracted with a Mr. Simpson of Mauch Chunk for the construction of ten boats at $190 each.[88]

Another Canal Company notice appeared in local newspapers, appealing to private boat builders to build craft for use on the canal.[89]

> *From recent reports, received from the Chief Engineer, the Directors of the Morris Canal Company feel themselves warranted in assuring all those who may wish to prepare Boats for the navigation of said canal, that the western division will be completed in all the month of May next, so that the whole line from the Delaware to the Passaic rivers, will then be finished and ready for navigation.*
>
> *The Directors have much pleasure in being able to state further that steady employment will be given to from 50 to 100 boats (in addition to those already built, and now standing) immediately on the completion of the Canal.*
>
> *N.B. The Boats must not be more that* [*sic*] *65 feet in length, and 8 feet 6 inches in breath,*
>
> *By order of the Board,* *R. Gilchrist, Cashier*
> *January 31, 1831*

The Canal Company advertised for individuals to contract for carrying coal. One such public appeal included a bit of the company's "philosophy" or policy about boats in general:[90]

> *The coal will be laden on board the boats by the Lehigh Company, but is to be delivered from the boat at Newark . . . The Company will contract to furnish the boats — the contractor to provide every other requisite, including ropes and towlines, and be at every other expense. Or the contractor may furnish his own boat and bear all expense. Or the Company on good security will advance money, as it may be necessary to enable the contractor to build boats. Or the Company on like security will sell to the contractor boats, at their cost, which are already built.*

The Canal Company promised "to indemnify the owners or captains of boats navigating the canal from all damages in consequence of removing any fence or other obstruction that has or may be placed in or across the canal, by any person or persons other than an authorized agent of the company."[91] That promise was later amended. Boat owners or captains would be indemnified "upon condition that they applied to Mr. Ogden at Paterson or to Mr. Dodd or Mr. Halsey at Newark to defend their suits, and that the defendants used due diligence in conducting the defense."[92]

In 1832, Colden reflected that the boats on the canal during 1831 and 1832 were "very great disappointments." The Company had by then entered into several contracts with boat builders for 100 boats of its own, but "several of the contractors failed altogether to comply with their contracts; and one, who was to have supplied twenty boats, although he built some of them, yet he constructed them so badly that they were worth nothing and could not be accepted." Still, there were an estimated 150 boats already on the canal, and the evidence of a permanent navigation, Colden hoped, would encourage still more private boat builders to put boats in service.[93]

The Canal Company then contracted with boat builders for "not more than 65 boats for the company to be delivered as early in the spring as possible."[94] Moreover, the directors had agreed to sell and deliver coal at Mauch Chunk to persons who loaned money to the company. The price to those lending money was to be "$3.75 per ton to the amount of the loans respectively, provided said loaners furnish their own boats."[95] The board further appointed a committee "to contract for the transportation of coal and to sell any of the company's boats to any person or persons on condition that said boats shall be employed in bringing coal for the company."[96]

Eventually, the company contracted for another 20 to 50 boats for "about $300 each,"[97] and at least three such boats were bought at $350 each from William Pragnell.[98] Levi Barker, still another supplier, sold a new boat to the company for $350.[99] Barker went on to build boats for the Delaware & Hudson Canal. At first he was employed by the Christopher Lane Boat Works, establishing his own boat yard at Hawley in 1849, and remaining in business there for almost 40 years.[100]

A rough draft of the Rates of Toll for 1831 contained some significant changes. Coal and iron ore still cost 1 cent per ton-mile, but a ton was set at 2000 lbs., a short ton. Also in the 1-cent category were sand, manure, clay, gypsum, stone, firewood, passengers (each), and cotton. Costing 1-1/2 cents were hemp, wool, flax, hoops, and hoop poles. Oysters were moved from the 1830 4-cent to the 2-cent rate group. Also in the 2-cent group were "all articles made in New Jersey."[101] However, the "finished" draft of the Rate Schedule was amended to move "cotton, flax and wool" and "all articles manufactured in New Jersey" to the 3-cent category.[102] Still later, cotton was moved again — to the 1-1/2 cent category.[103]

"Passengers" enjoyed a category of their own, at 1 cent per mile each, in contrast to the 1830 Regulations which had set the toll at 10 cents per mile for passenger boat and each passenger. In 1831, all boats paid a toll of 2 cents per mile.

Spiritous liquors and wine in casks were shipped at the rate of 3 cents per mile. Colden had proposed that no spiritous liquors be sold on the canal, but the proposal was "postponed" by the board of directors.[104] It was never brought up again. Years earlier, a gentleman signing himself "Aristides" had written to Morristown's *Palladium of Liberty*:[105]

> *. . . How much we would gain by the carriage of our cider to New York, instead of being obliged to distill it into a poison; a poison which is beginning to be made so copiously at New Orleans, that we must in a very few years be undersold and driven out of every market. In a short period spirits must fall in price, and if we cannot send our cider to market by water, we may cut down half our orchards, or what is worse, be obliged to drink all our whiskey ourselves . . . .*

Aristides knew what he was talking about. New Jersey's cider was famous even during Colonial days. The Swedish traveler, Peter Kalm, noted that it, along with New York's, was the best cider in America. Stills had been imported into Morris County as early as 1773, and it is estimated that one acre of apple orchard ultimately yielded 125 gallons of "Jersey Lightning," or apple jack, which in 1794, sold for $.75 per gallon.[106] In short, "spiritous liquors" were a very important item of New Jersey trade, the temperance movement notwithstanding.

Of course, the most important items to be transported were coal and iron ore. At the outset of 1831, the Lehigh Coal and Navigation Company reported to its stockholders that large manufacturing establishments along the Morris Canal's Eastern Division were awaiting coal deliveries. Besides those industries at Newark, Bloomfield, Paterson, and Dover, there were "fifty-four iron furnaces and forges already in operation," and "thirty-nine forges and furnaces in the neighborhood which have ceased to operate for want of fuel; these will undoubtedly be revived, when they can be supplied with Lehigh Coal." Anticipating the opening of the Morris Canal to coal traffic, the Lehigh Company had "agreed upon the terms of a contract for the sale of fifty thousand tons of coal to be delivered at Mauch Chunk, and to pass by the route of the Morris Canal." Completion of the Morris Canal was expected early in 1831, after some of the "mechanical work at a few of the inclined planes" was corrected. Even with the existing defects, "the planes caused little or no detention, as the boats had completed the passage on them in nearly the same time that the horses walked the same distance."[107]

By early August, 1831, Major Douglass reported that the inclined planes were all finished and that they were "ready for navigation."[108] And none too soon, for as the Lehigh Company report to its stockholders had been pleased to note: "The actual commencement of the Delaware and Raritan Canal and Railroad, and of the Camden and Amboy Railroad" would open "a new and extensive market" for Lehigh Coal, "and in connection with the Morris Canal, will afford ample means of reaching the tide waters of New York on the most reasonable terms."[109] The "eye of the storm" was passing, and with it the era of relative calm.

The Mauch Chunk *Courier* announced in August, 1831, that the first boats laden with Lehigh Coal had left for Newark.[110] Whatever happened to those boats *en route* is something of a mystery. It wasn't until late in October that the Newark *Sentinel of Freedom* reported "thirteen boats from Easton, laden with Lehigh coal and flour, arrived at the inclined plane, at this place yesterday morning. These we believe are the first fruits of the season from the extreme of the canal. We hope that a 'good harvest' may yet be in reserve."[111] The mystery is anything but solved by the Canal Company's *own* report that "the first cargo of coal, via the Morris Canal, reached Newark on the 29th day of October, 1831."[112] This coal must have been carried by the "several boats" that were reported by the Easton *Whig* on October 25, 1831, to have "left the Delaware for New York, via the Morris Canal," "last week," ascending "the first plane very handsomely." October 25th that year was a Tuesday. The reference to "last week" means that boats loaded with coal bound for Newark left South Easton one day between October 17 and 22. Why boats would have taken a full seven days to reach Newark is indeed a mystery.

The Paterson paper noted in passing on November 2, 1831, that "several boats loaded with Lehigh coal, passed by this place last week from Easton, through the Morris Canal." Since the *Paterson Intelligencer* was published weekly on Wednesdays, "last week" would have meat any day from October 24 through 29. A very long trip, if these were the same boats that had left Mauch Chunk in August!

Coal was needed desperately to benefit the poor as well as for industry. The Paterson editor was moved to remark:[113]

> *We should be glad to see a supply of coal arrive from this or any other source, which would bring down the price of wood to its proper level. We do not recollect to have known such high prices to be asked for wood in our market in any previous season. Fuel is an article of prime necessity, and a few shillings difference in a load of wood, will amount to a considerable item in the expenditures of a poor family during a long and cold winter.*

Despite that editor's concern, there was little demand for coal during 1831![114]

> *The shipments of 1830 greatly exceeded the demand, and vast quantities of Lackawanna, Lehigh and Schuylkill coal lay over in different markets, for which, at the opening of the season of 1831, there was neither any domestic demand, nor any manufacturing demand, nor even any export demand. This spread a panic among many of the*

Schuylkill Colliers, and they totally suspended their operations through the whole season of 1831.

A few other coal suppliers had had to reduce their price to $4 in Philadelphia in order to sell off their old stock.

The Canal Company, acting virtually as coal agent for the Lehigh Coal and Navigation Company, was under contract to that company for 50,000 tons of coal annually. In turn, the Morris Company contracted to deliver Lehigh coal to local coal dealers. One such contract, with Mr. Job Crane of Elizabethtown, backfired. Crane demanded damages because the Canal Company had failed to deliver on their contract for 80 tons of coal, delivered at Elizabethtown at $6.25 a ton. The Canal Company offered to settle with Mr. Crane for $100 as compensation for his disappointment, if he would give a receipt in full for all claims and damages, *and* surrender the company's obligation to furnish 80 tons of coal.[115]

Josiah White of the Lehigh Company had calculated that on a canal such as the Morris, 640,000 tons a year could be transported.[116] The price of Lehigh coal to the Canal Company was set at $3 a ton, free of tolls to Easton. The Company expected to make a considerable profit on that contract. But if it could not honor its own contracts with dealers such as Mr. Job Crane of Elizabethtown, the Morris Canal would be in serious difficulty. The year of 1832 would prove to be crucial for the Canal Company.

REFERENCES

1. *Laws of New Jersey,* 1828, p. 97.
2. *Report of the Commissioners Appointed by the Legislature of the State of New Jersey, For the Purpose of Exploring the Route of a Canal to Unite the River Delaware, Near Easton, with the Passaic, Near Newark, with Accompanying Documents* (Morristown, 1823). "Report of the Army Corps of Engineers," appended, pp. 16-17.
3. *Paterson Intelligencer,* May 30, 1827.
4. *Laws of New Jersey,* 1828.
5. Morris Canal and Banking Company *Minutes,* December 1, 1829 (hereinafter cited as MC&BCo. *Minutes*).
6. *Ibid.,* December 5, 12, and 19, 1829.
7. *Paterson Intelligencer,* September 5, 1827.

8. *Report of the Joint Committee of Council and Assembly Appointed to View The Morris Canal and Inclined Planes* (N.Y.: 1828), p. 4 (hereinafter cited as *Report of the Joint Committee, 1828).*
9. *Ibid.*
10. *Ibid.*, pp. 15-6.
11. *Laws of New Jersey,* 1829.
12. MC&BCo. *Minutes,* April 28, 1829.
13. Robert G. Albion, *The Rise of New York Port* (N.Y.: 1939), pp. 65, 211, 329.
14. MC&BCo. *Minutes,* May 12, 1829.
15. Henry W. Domett, *History of the Bank of New York: 1784-1884,* 4th edition (Cambridge, Mass.: 1885), p. 79.
16. *Report of the Joint Committee, 1828,* p. 15.
17. Domett, *op. cit.*, p. 82, 92.
18. MC&BCo. *Minutes,* September 15, 1829.
19. *Ibid.*, March 27, 1830.
20. *Paterson Intelligencer,* March 17, 1830.
21. *Ibid.*, July 14, 1830.
22. Deed in Morris Canal and Banking Company *Docket Files* (hereinafter cited as MC&BCo. *Docket Files*).
23. *Newark Sentinel of Freedom,* April 12, 1831 and June 2, 1832.
24. MC&BCo. *Minutes,* September 28, 1830.
25. *Ibid.*, December 1, 1829.
26. *Ibid.*, December 19, 1829.
27. *Paterson Intelligencer,* March 17, 1830.
28. MC&BCo. *Docket Files,* agreement dated January 9, 1830.
29. MC&BCo. *Minutes,* June 15, 1830.
30. MC&BCo. *Docket Files,* deed dated January 20, 1830.
31. MC&BCo. *Minutes,* November 18, 1854.
32. *Ibid.*, April 13, 1830.
33. *Ibid.*, April 19, 1831.
34. Lehigh Coal and Navigation Company Annual Report to the Stockholders, January 11, 1830, p. 7.
35. MC&BCo. *Minutes,* April 6, 1830.
36. *Ibid.*, May 5, 1830.
37. Material about the Associates of the New Jersey Company was derived from a variety of sources: Cadwallader D. Colden, *Life of Robert Fulton* (N.Y.: 1817); Charles H. Winfield, *History of the County of Hudson, New Jersey, From Its Earliest Settlement to the Present Time* (N.Y.: 1874) (hereinafter cited as Winfield, *History of Hudson*); *History of Jersey City, New Jersey: A Record of Its Early Settlement and Corporate Progress,* compiled for the (Jersey City) *Evening Journal* by Alexander McLean (Jersey City, 1895) (hereinafter cited as *Evening Journal History of Jersey City*).
38. Winfield, *History of Hudson,* pp. 249-253.
39. *Ibid.*, p. 287.
40. *Evening Journal History of Jersey City,* pp. 28-9.

41. Winfield, *History of Hudson*, pp. 249-253.
42. MC&BCo. *Docket Files*, Lehigh Valley Railroad Company statement *re* ownership of Jersey City Basin property.
43. *Paterson Intelligencer*, November 23, and December 20, 1825.
44. *Laws of New Jersey*, 1825, p. 18.
45. MC&BCo. *Docket Files*, deed dated January 11, 1825.
46. Albion, *op. cit.*, p. 149.
47. Winfield, *History of Hudson*, p. 149.
48. MC&BCo. *Docket Files.*
49. *Report of the Joint Committee, 1828*, p. 3.
50. MC&BCo. *Minutes*, October 20, 1829.
51. *Ibid.*, April 6, 1830.
52. *Paterson Intelligencer*, March 13, 1833.
53. *Ibid.*, January 20, 1830.
54. *Ibid.*, July 7, 1831.
55. *Ibid.*, August 17, 1831.
56. *Ibid.*, September 21, 1831.
57. Newark *Daily Advertiser*, September 30, 1831.
58. *Ibid.*, June 6, 1831.
59. *Ibid.*, June 18, 1831.
60. *Palladium of Liberty*, June 25, 1828.
61. Niles' *Weekly Register*, January 8, 1831.
62. *A History of the City of Newark . . .*, Vol. I (N.Y.: 1913), pp. 431-2.
63. James M. Ransom, *Vanishing Ironworks of the Ramapos* (New Brunswick, N.J.: 1966), p. 220.
64. *The Story of Anthracite* (N.Y.: 1932), p. 86.
65. Stewart H. Holbrook, *The Story of American Railroads* (N.Y.: 1947), p. 23.
66. Agnes C. Laut, *The Romance of the Rails* (N.Y.: 1936), p. 99.
67. MC&BCo. *Minutes*, February 26, 1830.
68. *Ibid.*, November 3, 1829.
69. *Ibid.*, February 26, and July 6, 1830.
70. *Ibid.*, December 26, 1834.
71. *Ibid.*, May 7, 1835.
72. *Directory of Newark, N.J.*, 1835-6, p. 79.
73. MC&BCo. *Minutes*, March 30, 1837.
74. *Paterson Intelligencer*, September 14, 1841.
75. Newark *Daily Advertiser*, August 7, 1843.
76. MC&BCo. *Minutes*, March 26, 1847.
77. Mildred Jailer, "The Museum in St. Johnsbury," *The Royle Forum*, No. 135, June 15, 1971, p. 14.
78. *Fairbanks Genealogy*, n.p., n.d., p. 332.
79. MC&BCo. *Minutes*, June 20, 1846.
80. *Ibid.*, July 21, 1856.
81. *Ibid.*, October 5, 1830.

82. *Ibid.*, December 29, 1830.
83. *Ibid.*, September 28, 1830.
84. *Ibid.*, December 29, 1830.
85. Newark *Daily Advertiser*, February 21, 1831.
86. *Paterson Intelligencer*, May 19, 1830.
87. MC&BCo. *Minutes*, March 1, 1831.
88. *Ibid.*, March 8, 1831.
89. *Paterson Intelligencer*, March 9, 1831.
90. Easton *Whig*, January 24, 1832.
91. MC&BCo. *Minutes*, May 31, 1831.
92. *Ibid.*, July 1, 1831.
93. *Report of a Joint Committee of the Directors and Stockholders of the Morris Canal and Banking Company, on the Affairs of the Company, with a Communication from the Directors Who Were of the Joint Committee, to the Stockholders*, August 23, 1832, in *American Railroad Journal*, September 29, 1832, p. 627 (hereinafter cited as *Report of a Joint Committee, 1832*).
94. MC&BCo. *Minutes*, January 12, 1832.
95. *Ibid.*, March 6, 1832.
96. *Ibid.*, March 13, 1832.
97. *Ibid.*, September 4, 1832.
98. *Ibid.*, October 2, 1832.
99. *Ibid.*, November 6, 1832.
100. Manville B. Wakefield, *Coal Boats to Tidewater* (Grahamsville, N.Y.: 1971), p. 62.
101. MC&BCo. *Minutes*, February 18, 1831.
102. *Ibid.*, February 24, 1831.
103. *Ibid.*, July 1, 1831.
104. *Ibid.*, March 8, 1831.
105. *Palladium of Liberty*, September 5, 1822.
106. Peter O. Wacker, *The Musconetcong Valley of New Jersey: A Historical Geography* (New Brunswick, N.J.: 1968), p. 58.
107. Lehigh Coal and Navigation Annual Report to the Stockholders, 1831, p. 8 (hereinafter cited as LC&NCo. Report, 1831).
108. MC&BCo. *Minutes*, August 9, 1831.
109. LC&NCo. Report, 1831, p. 4.
110. James P. Snell's *History of Sussex and Warren Counties, New Jersey* (Philadelphia: 1882), p. 486.
111. *Paterson Intelligencer*, October 26, 1831.
112. *Report to the Stockholders of the Morris Canal and Banking Company* (N.Y.: April 1, 1861), p. 17.
113. *Paterson Intelligencer*, November 2, 1831.
114. *Ibid.*, December 28, 1831.
115. MC&BCo. *Minutes*, December 20, 1831.
116. *Report of a Joint Committee, 1832*, p. 627.

9

A Walk in The Water

WATER was let into the canal early in April, 1832. Navigation between the Delaware and the Passaic was to be possible at last.[1] Well — almost possible. A breach near Paterson had held up canal traffic for "a few days," so that it was April 16 when the first boat of the 1832 season descended Plane No. 11 East at Newark.[2] Boats began "plying daily along the line, freighted with the various products of the country through which it passes." The Canal Company was expected to transport "large quantities of Lehigh coal during the summer," making it available at more moderate prices. One newspaper editor hoped that "the experience of last winter," the shortage of 1830, would "induce an early provision for the demands of next winter."[3]

The *Argo* had arrived in Newark from Easton on May 16, 1832, its coal being "the first cargo which has come down the canal this season."[4] The *Argo's* coal was consigned to Jonathan Cory at Newark's Commercial Dock. On May 18, the *Paul Jones* and the *Tecumseh* arrived with coal from Mauch Chunk, consigned also to Cory.[5]

However, it was the arrival of the *Walk in the Water* that was to

ARRIVED LAST WEEK.

Canal Boat *Argo*, Capt. Perry, from Easton, with Lehigh Coal to J. Cory.

Canal Boat *Paul Jones*, Capt. Conner, from Mauch Chunk, with Lehigh Coal to J. Cory.

Canal Boat Tecumseh, Capt. Sherman, from Mauch Chunk, with Lehigh Coal to J. Cory.

ALSO—a number of Boats with sundries.

N. B. The above COAL is course and first quality, for sale at the Dock by J. CORY.

Newark, May 22, 1832.

ARRIVED—Canal boat *Walk in the Water*, from Mauch Chunk, with Lehigh Coal, to Stephens & Condit. This is the first boat laden with Leghigh Coal, that has entered the tide waters of the Passaic, through the Morris Canal.

Announcement of boats arriving at Newark via the Morris Canal. Walk in the Water was the first boat to enter tidewater (ETF, Newark Sentinel of Freedom, May 22, 1832).

be the most significant of all. She arrived on May 19, 1832,* with a consignment of coal for Stephens & Condit. She not only *arrived;* she was the first boat to enter the tidewaters of the Passaic River "through the Morris Canal."[6] The outlet lock was finally completed.

Colden himself rode into Newark on one of a squadron of coal boats on May 23, 1832, passing the Bloomfield Plane at 10:00 A.M. By that time Plane No. 11 East had "already passed eighteen boats, and there were at least as many more, mostly coal boats, between that and Mead's Basin, which would pass in the course of the day." As far as President Colden was concerned, the canal was "now in

**Shaw's History of Essex and Hudson Counties placed the date as "May 19 or 20," but May 20, 1832, was a Sunday, and it is unlikely that freight boats would have been operating on the Sabbath.*

perfect order, from one end to the other."[7] The great expectations were about to be realized.

The canal was, of course, new; the banks were not yet settled and firm; the mechanisms were still subject to malfunction. The canal banks between Little Falls and Paterson had given way in May, 1829, and had been repaired.[8] The Pompton Aqueduct had suffered a breach in September, 1831.[9] A breach in the canal at Boonton,[10] and a "disarrangement of the plane at Bloomfield" in June[11] and in July, 1832, had been quickly repaired. On July 18, seven boats with wood and ore and two boats with merchandise cleared the plane at Newark.[12]

"It must be admitted," Colden said, "that the breaches and accidents which have interrupted the navigation of the canal during this, the first year of its operations, have exceeded our expectations, and have greatly diminished the amount of tolls which it was calculated would have been received; and at the same time, these occurrences have increased the expenses for the year [1831-2] above the estimate; and yet it is certain that there have been fewer breaches on the Morris Canal than has [sic] occurred on many other canals in the incipiency of their operations. More than three hundred breaches happened on the Erie canal . . . before it had been navigated a year."[13] But the Morris' planes were the big problem.

Thirteen of the inclined planes had been supplied with sprocket chain of American manufacture of such inferior quality that the chains broke continually "after a very little use." Eight planes using English-made chain had given excellent performances with no breakage; hence, the Canal Company intended to supply all planes with the English chain to convince everyone of the superiority of planes over locks.[14] To quell the fears of boatmen in going over the planes, with the danger of breaking chains, and as an additional attraction, the Canal Company decided to indemnify the boatmen against loss. The Company advertised that it would furnish "good and sufficient chains" for every inclined plane:[15]

> *But in order to afford perfect security to all persons now engaged, or that may hereafter embark in the boating business on the Morris Canal, it is hereby further resolved, that this Company will indemnify all persons whose boats*

or property may receive injury while passing the inclined planes of the Morris Canal, or which may be occasioned by the breaking of chain, or any part of the machinery of a plane. Provided that such injury is not occasioned by the misconduct of those who may have charge of the Boat — and provided also, that the master, owner, or persons having charge of the Boat, shall conform in all things to the printed regulations for the navigation of the Canal.

Perhaps the most famous accident on an inclined plane of the Morris Canal involved the *Electa** on the Boonton Plane in May, 1832:[16]

A boat called the 'Electa,' of Rockaway, owned by Colonel Joseph Jackson, and partly loaded with merchant iron, was on its way to Newark, having on board the captain, his wife and two children. Just as the boat passed the summit the chain broke, and the car with the boat ran down with great velocity, striking the water with such force as to throw an immense wave over the towing path, which carried the boat with it down an embankment from 15 to 20 ft in height and landed it on the rocks below, amid some trees standing there, but fortunately without striking any. People hastened to the boat to ascertain the fate of those on board. On opening the cabin door, the wife with her two children, was found sitting there rather composedly, and uninjured. When told what had happened, she seemed surprised, and said she "thought the boat went down very swift, but supposed that was the way the thing worked."

Upon receiving a report of the accident, the Canal Company directors asked Colden to write to Jackson for an interview about "the accident involving the Electa on the Boonton plane." However, Colden was to exercise caution and not "admit in any shape the liability of the company to pay damages in case of accidents of this nature."[17] This incident took place seven months before the Canal Company decided to indemnify boatmen and boats against such accidents.

**Jackson had named the boat for his second wife, Electa Beach, daughter of Enoch Beach. This was Electa's second marriage, as well; she was the widow of Silas Dickerson.*

Accidents continued to happen, however, and John Righter put in a claim for his ruined boat, the *Aqueduct*, caused by the breaking of a chain, also in May. He was paid $130 in damages.[18] Another accident took place at Plane No. 6 West, at Port Colden in 1846. Captain Frederick Fliger, master of "C. F. Randolph & Company's scow No. 122, was passing this plane loaded with iron ore for Newark. When going up the plane the chain broke and the scow ran down, killing one woman and three children and breaking up the scow. The Captain put in his claim to the Canal Company for fifty dollars to cover his personal property, which was allowed."[19]

One newspaper reported another instance during the 1850s, in which a family of five recently arrived German immigrants took passage west on a returning ore boat at Jersey City. The boat took on ore en route, and all went well until the boat reached Port Colden, Plane No. 6 West. There, as the boat was descending the plane, just past the brow, the sprocket chain snapped under the strain. As the boat hurtled down the plane, the cargo of ore shifted and crashed through the side of the boat. Seven persons — the five immigrants and two boatmen — were thus carried to unmarked graves. "The story of the incident alone remains; the names of the victims . . . perished with them."[20]

These last two stories are similar, and could have been reports of the same incident. In any event, it appears that the story about the family of German immigants took place on July 6, 1847: "While a boat was descending the plane, the chain broke, when the boat was precipitated into the water with such violence as to dash the cabin to pieces. In the cabin was a family of German emigants, who had been living a year or two in Paterson. Of these unfortunate people, the father, mother, and three of the children were drowned at the time of the accident, or have died of their injuries since."[21] The Canal Company authorized the president to donate money to Mr. ____ (name not filled in), "the husband and father of the individuals who recently lost their lives in consequence of a canal boat running down Plane No. 6 West."[22]

Canal Company *Minutes* indicate that C. F. Randolph had contracted for and purchased "70 scow boats, all of which he expected would be on the canal by or about the first of August next," and his proposal to operate those boats for the company was approved on June 8, 1847. He conceivably could have had scow No. 122 on the canal in time for the accident on July 6, 1847, or perhaps was already operating boats previous to the agreement mentioned in the minutes for June 8 of that year. The second story, as reported by the Washington (N.J.) *Star*, is suspect, since sprocket chain was no longer being used on Plane No. 6 West during the 1850s. As we will see, it was in January, 1848, that twisted wire rope, made by John A. Roebling, was introduced at that plane.

Many canal boats were apparently rather prosaically numbered, as was Captain Fliger's No. 122. Private boats, and perhaps nicknamed company boats, boasted names of wives, famous men, politicians, military heroes, and even such humorous appellations as *Lager Bier*. One boat, the *Sarsey Fanny,* was subsequently immortalized in a Harper's *New Monthly Magazine* article in 1860. We have already met the boats *Walk in the Water, Argo, Tecumseh, Emiline, Maria Colden, Electa, Aqueduct, Dover,* and others. Colonel John Scott, Canal Commissioner during the first few years of the canal's history, had a "splendid packet boat" called the *Jersey*. He loaned it to a Pennsylvania military cadet drill group, the Fencibles, for an excursion over the canal from Easton to Newark, even supplying free refreshments. In appreciation, the Fencibles had had a "neat silver goblet" made for Scott by Mr. Lownes, of Chestnut Street, Philadelphia.[23]

The Canal Company wanted to show its gratitude to Scott, too, for his "long and faithful service." Accordingly, the directors resolved that Scott's boat, the *Jersey,* "so long as he owns her and whenever she is employed on the canal as a pleasure boat for the transportation of passengers only and not for hire, and the Colonel shall be with her, shall pass on the Canal free of tolls."[24] Still later, the Canal Company bought the *Jersey,* and had her brought to Newark to be used in the "service of the company."[25] Thus, the Canal Company had its first packet boat in regular service for monthly inspection tours of the canal.

The *Maria Colden* was still operating in May, June, and July, 1832, but in August, she was taken out of regular service and became available exclusively for excursions.[26]

When Colden reported to the directors on the affairs of the Company in September, 1832, there had already been spent $1,859,724.88 on actual construction. Boats had cost another $13,742.00. Freight on coal (transportation) amounted to $3,781.25. By the time he added in the bills receivable, amounts due from "sundry" individuals, contingent expenses, and so on, the canal had cost $2,104,413.69. The debts of the company amounted to $1,120,117.75, of which $750,000 was owed on the Holland Loan. Still, Colden felt that the company was in excellent condition and that the stockholders had every reason to be confident. Any reports in conflict with his report were "as untrue as a report that was once current, that the Canal had not a sufficient supply of water, or a report afterwards circulated that the Mauch Chunk mines were exhausted."[27]

He explained that the Canal Company still held 4000 of its own shares of stock, and that 2000 of those had been pledged to the Lehigh Coal and Navigation Company as collateral security for coal purchased of them, while some of the remaining 2000 shares had been pledged as security for other loans. If the Company's debts could be discharged, Colden felt, "the connexion of the canal and bank" would be of the greatest advantage to the bank." The canal would be "a better safety fund than any that has yet been devised, because everyone who holds a bill of the bank will have pledged for its payment a property which has cost near two millions of dollars." However, it was in the Company's power to execute trusts that Colden put *his* trust, for "the trust property is by the charter entirely separated from the bank and the canal, and is not liable for the debts of the company."[28]

Yet, the board of directors were considering asking Willink (of the Holland Loan) for yet another $250,000 loan to complete the canal to Jersey City![29]

Financial problems were decidedly not solved. When Colden was reelected president of the Canal Company, it was at the reduced salary of $1500 yearly. Previous reductions in 1828 had placed the president's salary at $2000 a year, with an additional $1000, including expenses, for his position of President of the Canal Committee. At the same time, Chief Engineer Beach's salary had been reduced to $2000, as was the Canal Commissioner's (John Scott).

These reductions were not a reflection on the ability, fidelity, or integrity of the officers, but were simply a necessity, forced upon the directors by pressure from stockholders who were outraged by the high salaries of the Company's officers.[30]

Beach, for one, eventually objected to his reduced salary.[31] But an economy drive was necessary, and instead of a salary adjustment, Beach was asked to give accounts of all employees under him, to see if expenses "could, with propriety, be reduced."[32] Even contractors were being paid for their work with the company's post-notes,[33] and the cashier had been instructed to write to all persons who owed money to the company "for stock or otherwise," telling them that unless their debts were settled on or before October 15, 1829, the accounts would be placed in the hands of the board's attorneys for collection. Of course, the board thought it would be wise to review a list of those debtors, to see if there were any persons "who it would be advisable not to sue."[34] Beach, obviously, was in this category. He owned 70 shares of stock, and owed installments on all of them. Since the board was, at the time, considering giving Major Douglass $4000 yearly to Beach's $2000, they voted to give Beach a $500 raise *if* he paid up the back installments on his stock.[35]

The economy drive had accompanied the Canal Company's attempt to amend its charter so that it might borrow still *more* money. Ostensibly, the new amendment was sought on the grounds of a need to clarify the procedure by which appraisers (of property for the canal) were chosen. Section II of the amendment, however, had given the Morris Company the right "from time to time," to borrow "all such sum and sums of money as the Board of Directors" considered "necessary and proper." The act had passed the Legislature on January 30, 1830, and a new $100,000 loan had been sought, with shares to be given as security. These were among the remaining 2000 shares, still owned by the Company, that had been pledged for the loan.[36]

In 1832, however, anticipating a full and successful boating season, the board slowed the economy drive long enough to permit the hiring of three superintendents for the line of the canal, at $1000 each per annum, or not to exceed $3000 for all three. The amount to be paid each superintendant was to be apportioned by a committee formed for that purpose.[37]

Lorenzo A. Sykes had been hired as Resident Engineer on August 10, 1831, at a salary of $2000.[38] But at the outset of 1832, Sykes and Canal Commissioner Scott, also the Vice President of the Company, were told that their offices had been abolished.[39] By February, the salary of the Chairman of the Canal Committee was also abolished.[40] With these three moves, the Canal Company cut expenses by $5000 a year.

Contractors who normally worked for the Canal Company, Mason & Pratt, were, at this point, working on the New Jersey Railroad. The firm applied to the Company for the sum of $5000, but the board resolved to advance no more than $3000 at any time.[41] The board's action suggests a "rob-Peter-to-pay-Paul" situation: Cut back on canal expenses in order to subsidize a railroad.

Tolls for the 1832 boating season were set with an eye to competition, i.e., the Delaware Division Canal. With that canal in operation, the Lehigh Company had its choice of routes for transporting its coal. Care had to be taken not to set the tolls for the Morris Canal higher than those on the Delaware Division.[42] When boats began to move over the Morris Canal, the Newark newspapers reported their arrivals and departures, and made mention of the cargoes carried, as well as their origins and destinations:[41]

ARRIVED AT NEWARK

Tuesday, June 5	*—13 boats from Easton, Washington, Stanhope, Dover, and Boonton.*
CLEARED	*—4 boats for Easton and 11 for Greenwich, Anderson, Stanhope, etc., with iron, coal, and mer'dize.*
Wednesday, June 6	*—8 boats from Easton, Stanhope, Dover, and Boonton, with produce, firewood, iron and iron ore.*
CLEARED	*—3 boats for Mauch Chunk, and 7 for Washington, Stanhope, etc., with merchandize, coal, pig iron, steel, etc.*
Thursday, June 7	*—10 boats from Stanhope, Dover, Rockaway and Boonton, with planks, stoves, etc.*

CLEARED	*—8 boats with merchandize, coal, lumber, etc., for Stanhope, Dover, Rockaway and Boonton.*
Friday, June 8	*—6 boats from Stanhope, Andover, Boonton and Rockaway, with wood, iron ore, etc.*
CLEARED	*—3 boats for Mauch Chunk, and 7 for Boonton, Rockaway, Washington, etc.*
Saturday, June 9	*—7 boats from Stanhope, Rockaway, Dover and Boonton.*
CLEARED	*—2 boats to Easton, 11 to Boonton, Rockaway, Dover, etc.*

Monday, June 11 — Cleared — Eleven boats to Dover, Stanhope, Rockaway, Boonton, etc., etc., with merchandize, coal, etc. —

Arrived — Twelve boats from the above places, laden with iron, fire-wood, etc., etc.

Interestingly, this listing indicates more local, or "way," traffic than through traffic. Moreover, virtually every "clearance" back-shipped coal to points on the canal that were much closer to the source of Lehigh coal than Newark. The coal from Newark consisted of Lehigh and other coal, and had come into that city by way of Newark Bay or by backshipping Lehigh coal. Even this early back-shipping was a general practice — a practice carried out later at the expense of inland coal users who paid the extra freight costs.

The remainder of the 1832 boating season was relatively eventless. The Canal Company authorized Robert P. Bell (for the moment, Superintendent of the Western Division of the canal) to buy "six more horses and carts."[44] To buy *more* livestock implies that earlier purchases had been made, although this entry was the first of its kind in Canal Company records.

A committee of directors plus President Colden was authorized to contract for the building of a dry dock at Newark, although the location was not specified. Eventually, an agreement was made with James Searing, of the weigh lock (Lock No. 16 East), to erect a dry dock, indicating that the dock was probably built at the basin, there, at "Searing's" lock.[45] In fact, the Index to Property of the Morris Canal and Banking Company shows land parcel No. 121 to have been "Searing's Dock Property."

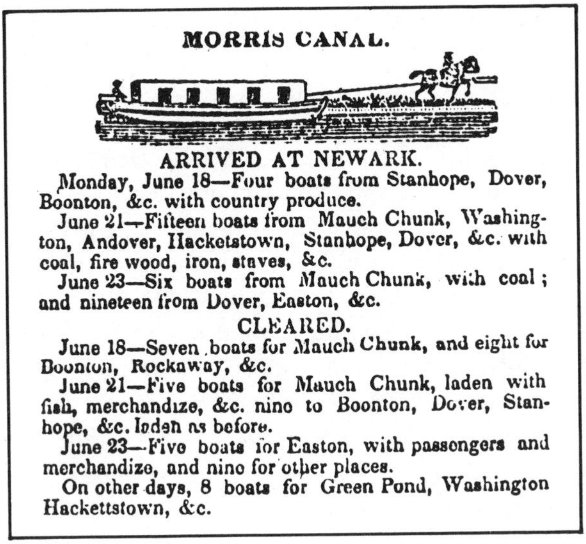

MORRIS CANAL.

ARRIVED AT NEWARK.

Monday, June 18—Four boats from Stanhope, Dover, Boonton, &c. with country produce.

June 21—Fifteen boats from Mauch Chunk, Washington, Andover, Hacketstown, Stanhope, Dover, &c. with coal, fire wood, iron, staves, &c.

June 23—Six boats from Mauch Chunk, with coal; and nineteen from Dover, Easton, &c.

CLEARED.

June 18—Seven boats for Mauch Chunk, and eight for Boonton, Rockaway, &c.

June 21—Five boats for Mauch Chunk, laden with fish, merchandize, &c. nine to Boonton, Dover, Stanhope, &c. laden as before.

June 23—Five boats for Easton, with passengers and merchandize, and nine for other places.

On other days, 8 boats for Green Pond, Washington Hackettstown, &c.

Announcement of boats arriving and clearing at Newark (ETF, Newark Sentinel of Freedom, June 12, 1832).

In October, Cashier Robert Gilchrist was asked to try to make arrangements with the owners of the ferry at Jersey City "for a free passage across the Ferry for all the officers and directors" of the Morris Canal Company.[46] There is no entry telling of the outcome of that arrangement, but given the amicable relations between the Canal Company and the New Jersey Railroad, then the "owners" of the ferry, no doubt free passage between New York and Jersey City was arranged.

Before the year was out, the Canal Company was offered the chance to buy the "Saw Mill Lot" in Drakesville, a lot which included the saw mill, the improvements, and all the water rights and privileges, for the sum of $800. The offer was made by Mr. F. Kerr through his agent, Judge Drake, of Drakesville. Accordingly, the Company bought Tract 488-1/2, a parcel squeezed between lands once belonging to Ebenezer Woodruff and Levi Carnes. In fact, on Canal Company maps, it appears that "Tract 488-1/2" could have been "two-plus" acres, at or very near Emmans Road, which crossed Plane No. 3 East.[47]

In 1832 Frances Trollope published her *Domestic Manners of the Americans,* describing the American part of her journey from England to Cincinnati and thence to Washington, D.C., Baltimore, Philadelphia, New York, and New Jersey. All of those travels had taken place between November 4, 1827, and the Spring of 1831. The publication of her book brought to a wide readership the accomplishments of the Morris Canal, of which she wrote:[48]

> *We spent a delightful day in New Jersey, in visiting, with a most agreeable party, the inclined planes, which are used instead of locks on the Morris Canal.*
>
> *This is a very interesting work; it is one among a thousand which prove the people of America to be the most enterprising in the world. I was informed that this important canal, which connects the waters of the Hudson and the Delaware, is a hundred miles long, and in this distance overcomes a variation of level amounting to sixteen hundred feet. Of this, fourteen hundred are achieved by inclined planes. The planes average about sixty feet of perpendicular lift each, and are to support about forty tons. The time consumed in passing them is twelve minutes for one hundred feet of perpendicular rise. The expense is less than a third of what locks would be for surmounting the same rise. If we set about any more canals* [*in England*], *this may be worth attending to.*
>
> *The Morris Canal is certainly an extraordinary work; it not only varies its level sixteen hundred feet, but at one point runs along the side of a mountain at thirty feet above the tops of the highest buildings in the town of Paterson, below; at another, it crosses the falls of the Passaic, in a stone aqueduct, sixty feet above the water in the river. This noble work in a great degree owes its existence to the patriotic and scientific energy of Mr. Cadwallader Colden.*
>
> *There is no point in the national character of the Americans which commands so much respect as the boldness and energy with which public works are undertaken*

and carried through. Nothing stops them if a profitable result can be fairly hoped for. It is this which has made cities spring up amidst the forests with such inconceivable rapidity; and could they once be thoroughly persuaded that any point of the ocean had a hoard of dollars beneath it, I have not the slightest doubt that in about eighteen months we should see a snug covered rail-road leading direct to the spot.

"As profitable result fairly hoped for," as Mrs. Trollope said, may have been the motive for the Canal Company's decision "to sell the whole or any part of the 4000 shares held by them of the capital stock thereof, . . . at the par value of $200 the share."[49] The number of shares to be sold was set at 2500, and a notice of the sale was accordingly put in three newspapers: one in Morristown, Newark, and New York City. The shares were to sell for $200 each for a net increase in the Company's capital stock of $250,000, and the terms of sale specified $1.00 per share to be paid at the time of subscribing. If all 2500 shares were not subscribed on or before January 28 (1833), they were to be offered to existing subscribers, or shareholders.[50] The *Minutes* of the Company reveal that President C. D. Colden bought them all — all 2500 shares. Aside from this rather interesting maneuver of Colden's, the year 1833 started innocuously enough. In January, Director Enoch Green recommended that John Steel, Jr. be hired as General Superintendent of Planes.[51] However, it was Ephraim Morris who was appointed a superintendent of planes at $1000 a year, with the Canal Committee given the power to give him a $100 raise at its discretion.[52]

On that same day (May 1, 1833) Roswell B. Mason was made Resident Engineer, replacing Lorenzo A. Sykes, and Colonel John Scott retired from the service of the Company to assume the presidency of Dover's Union Bank.[53] On that same day, too, the Paterson papers announced that the Morris Canal was "open for navigation"[54] for 1833 with boats having been "running for several days past."

There were requests to have basins made, too. A Mr. Howell had asked for "a slip or basin" for his accommodation "on the towpath side of the canal, at or near Greenwich."[55] Another application for a basin "near Hackettstown" was granted, but the board of directors specified that the "whole work" was to be "done under the direction of the superintendent at that place."[56] The application had been submitted by Mr. W. Dusenberry. The place "near Hackettstown" was Port Colden, known to area residents as "Dusenberry's Folly."

Many claims for damages and threats of lawsuits marred an otherwise quiet year. Samuel Farrand, who owned two pieces of property at the canal (156 and 158 on Weir Maps) in Bloomfield, claimed "damages alleged to have been sustained by him in taking materials from his land to repair breaches," and in broken fences.[57] The Company settled with Farrand for $100.[58]

Because of its unique charter, the Canal Company had been guilty of many offenses. The kinds of offenses are revealed in the MC&BCo. *Docket Files,* in an undated, unsigned, annotated list of complaints, claims, and comments. Beginning in the east with Newark:

1. *Aaron Johnson for damage to a well at New-Ark — Settled.*
2. *Ephraim Beach for a ditch to drain land near New-Ark.*
3. *Col. Ward for a bridge near New-Ark.*
4. *Farrand wants a trunk under canal and many other things — settled.*
5. *John Morrell at Paterson wants a bridge or would sell to the company the piece of land separated from his farm at a fair price. — "Remark" — M. Morrell purchased this property sometime after the canal was completed. If he bought of Robert Carrick, it was appraised and has been paid for. If of Dan'l Holsman, it was assigned.*
6. *Silas Cook. We took of him 70/100 of an acre of land when he assigned without charge — and the appraisers allowed him $100 for damages to cider and gristmill, which was tendered to him and refused — this gentleman's claims you have all heard. All I can say is that he has been a uniform and steady friend of the company, and altho I do not think he has been injured by the canal, yet I should be glad to see the co. give him $2 or even $300 in addition to the amount awarded him to settle all differences and make him happy.*

7. *Lot for a house at Plane No. 8 East. Done.*
8. *Look at a house and lot at Lock No. 12 East which has been offered to the company for $500.* [*This was Boonton's Lock No. 13 East.*]
9. *Widow Randolph's right of dower in the Brooklyn property — settled.*
10. *Cary and Hamel sawmill near Drakesville. They obtained a judgment against the company for $339.08 which has been paid lately.*
11. *Dr. Woodruff — Bridge at Drakesville near Plane No. 3 East.*
12. *Job Burt's claim for damages at Plane No. 2 East. We took from Mr. Burt 83/100 of an acre for which the appraisers awarded $4.15. This amount has been paid to Mr. Burt.*
13. *Crane, Shaver, and Styer — At Hackettstown — about water, I cannot think but the easiest way to settle these difficulties will be to purchase the property and dispose of it subject to the interrupting action by* [*of*] *our works. But the slightest intimation of such intention on the part of the company must be carefully avoided, or it will cost us dear.*
14. *Unsettled claims at the Pond* [*Lake Hopatcong*]

Joseph Sharp	*$ 45.57*	
Widow Ludlow, wife of Timothy Southard	*$510.20*	*Offered payment 50 times if they would make a sufficient title.*
Robert Henderson	*$ 30.27*	*To be paid, as is said, to Jos. Dickerson*
John Henderson	*$154.57*	*Mr. H. owes the Co. $950 for stock and property is mtgd.*
Robt. R. Livingston	*$ 36.99*	*Both claimed by Mr. Shipman*
Thos. Cadwallader	*$129.72*	
Proprietors of East Jersey	*$ 41.22*	
Chas. F. Randolph	*$ 8.74*	
David B. Hurd	*$200.54*	*Owes on 3 shares — $570.00*

Trowbridge &
Oliver *$ 9.72*

Some of the above claimants allege that more land has been taken than is contained in the survey. To all such, with whom settlements have been made, we have promised to allow them the same price per acre when they shall prove an error in our survey, and to all others we have made the same offer, that is we would now pay agreeably to our survey and if upon being again measured it should appear we have taken a greater quantity we would pay for it at the same price.

15. *Dr. Cook — Hackettstown — damages to a farm in Greenwich. Sold by him since the canal was made.*
16. *Dan'l Hultziser, near Stewartsville, for damages.*
17. *Adam Van Delin near Washington.*
18. *John Fulmer — at Stewartsville — water.*
19. *Mr. Shipman at Plane No. 9 West.*
20. *Mr. Reese*
21. *Mr. Bowers*
22. *Dr. Cooper, Easton, damages to meadow by leakage.*
23. *Henry Winters, Greenwich, the same — settled.*
24. *Paul Feit*
25. *Jacob Sharp*
26. *John Leisenring — Mauch Chunk — This man is entitled to something, but not much.* [*John Leisenring was later superintendent and engineer of the Lehigh & Susquehanna Railroad, keeping that position until the summer of 1868.*]

? *Matthias Kitchell, Rockaway. Look at this man's claims, but don't pay him money. He owes for stock.*
? *Abraham Woolston*
? *Christopher Hagerty.*

The appropriately numbered item 13 on the list gives a clue of things to come. John L. Sullivan had already touched on it in his 1828 Report to the SUM: The rights of the millers along the Musconetcong River to the undiminished flow of those waters. We will be returning to this subject before the chapter ends.

Among the other pedestrian affairs of the Canal Company during the year 1833, there appeared certain noticeable trends — ominous trends. There was no denying that all the canals had reduced freight prices.[59]

> *Just after the War of 1812, it cost as much to ship goods 30 miles overland as it did to ship them from Great Britain to the United States. Before the era of canals, a coal mine and an iron deposit existing within ten miles of each other were both useless, because the cost of transporting either the iron or the coal overland was prohibitive. By wagon, or primitive roads, freight costs were 30 cents a ton mile. Turnpikes cut the cost of shipping approximately in half. Canals produced even greater economies. The Erie Canal, for example, reduced freight costs from Buffalo to New York from $100 to $15 a ton.*

The ominous countertrend, however, was increasing demands for higher wages. The Canal Company experienced this trend during 1833 through two requests for raises. William Wilson, who had replaced Robert P. Bell (resigned, October 2, 1832) as Superintendent of the Western Division, asked for more money. He'd been promoted in title only, and was still being paid at his old rate.[60] A second request for more income came from William Carlisle, a plane tender, who claimed "additional compensation for living of $5 a month," on the ground that Robert P. Bell, while Superintendent of the Western Division, had promised him that raise.[61]

In his capacity as Canal Company director, and especially as a member of the all-important Canal Committee, Bell had been in a position to pick up parcels of land, many on the Western Division, for the canal's navigational usage. However, several parcels of land on both sides of the summit level were obviously not bought with navigation in mind.[62] It will be instructive to review the little that is known of Bell's career with the Canal Company.

Robert P. Bell resigned from the Canal Company's board of directors under conditions that are at best puzzling. In December, 1831, "several stockholders residing in Philadelphia" had requested the appointment of Mr. Robert Earp to the board. Director Charles Hoyt offered to resign, but his offer was refused. Bell then offered to resign and was accepted, and Earp was appointed.[63] By August 8,

1832, Earp, too, had resigned. Two months later, Bell resigned as Superintendent of the Western Division of the canal.[64]

In 1833, no longer connected with the Canal Company's direction, Bell built and operated a store in Stanhope. His clerk was Andrew A. Smalley, who later became Stanhope's second Postmaster. First to serve in that position had been Gamaliel Bartlett, whom we met earlier as a contractor for some of the Morris Canal's western planes.[65]

Beginning in 1830, Bell began to accumulate significant amounts of acreage, strategically situated in and about the canal, particularly at Stanhope. On September 2, 1830, from T. J. Speer of Bloomfield, Bell bought 1-18/100 acres of land occupied by the canal and its banks. Later this property was conveyed to Augustus W. Bell, who, in turn, conveyed the land to Andrew Smalley. Then from Joseph Cutler he bought a piece of land 50 by 200 feet along the towpath at the top of Plane No. 1 East, on the Summit Level. This parcel, too, was later conveyed to Augustus W. Bell who, in turn, conveyed the property to Andrew Smalley.

On August 1, 1831, Bell bought 5-1/2 acres, on both sides of the canal at Port Colden. This property, too, was deeded to Augustus W. Bell who in turn conveyed it to Smalley. This piece of property, however, did eventually find its way to the Canal Company, when Smalley deeded the land to the Canal Company on August 15, 1857. In turn, the Company conveyed the property to Alexander P. Berthoud, Canal Superintendent after the Civil War.

Also at Port Colden, Bell bought 32 acres of land from John G. Parke, on August 1, 1832. One-half acre was next to Lock No. 6 West, containing the waste water from the lock. As might be expected, this property was deeded to Augustus W. Bell, who, in turn, conveyed it to Smalley, etc., etc., on June 2, 1856. All that is currently known of Augustus W. Bell is that he once owned a building at the corner of Park Place and Bank Street, Morristown. The building was occupied in 1812 by the State Bank, which continued in business "a number of years."[66] All that is known of John G. Parke is that he was murdered on May 1, 1843, in what was Warren County's most sensational crime — the Castner Murders.

There is a curious document among Canal Company Records, in Docket File No. 518, that reveals Bell's real estate activities. It is entitled: "Location of Tracts with Remarks: Schedule, Location, and Description of Various Lands in Sussex County. R. P. Bell and Others." The document itself is undated, but internal evidence suggests 1838 as the time of its compilation.

One thing that the Schedule shows is that Robert P. Bell and his wife, Charlotte, were then residents of Stanhope, for they mortgaged a piece of property to an Isaac Dennis on September 9, 1836, the property described as "situate near the dwelling house occupied by R. P. Bell at the date of the deed." Several other old deeds had shown the Bells to have been residents of the City and County of New York. Among the other parcels or tracts of land Bell bought or gave mortgages on was the so-called Hude Tract, which once had belonged to Silas Dickerson.

Bell bought the 116.98-acre tract on May 20, 1834, for $8000, whereupon he began to operate the gristmill on the site, and the iron works. The works were incorporated under the name Morris & Sussex Manufacturing Company. The forges and nail factory operated on water from the Musconetcong River, which, of course, originated in Lake Hopatcong. When the canal diverted some of that water to its own use, a conflict between Bell and the Canal Company was inevitable. There were claims and counterclaims until finally the Canal Company board of directors agreed to put the issues in front of a pair of referees or arbitrators, William Pennington and Chief Justice Joseph C. Hornblower.[67]

Perhaps as a result of that arbitration, Bell, and his wife, Charlotte, deeded to the Canal Company 3000 square feet of land at Plane No. 1 West and an equal amount at the head of Plane No. 2 West. Both deeds were executed on April 3, 1837. Mention of these deeds on the undated schedule of Bell's properties helps to date that document, but it is an entry in the Canal Company's *Minutes* of December 7, 1838, that suggests that the list was compiled to help the Canal Company's cause in litigation with Bell. On that date, the board of directors decided that it would not be "expedient to make any offer to compromise to Mr. Bell, or to attempt to purchase his property at Stanhope, but allow the case to go to the jury." A statement at the end of that schedule of Bell's properties indicates that there were "no judgments on record in the Clerk's Office of Sussex County against Robert P. Bell."

It was characteristic of the Canal Company to buy out owners of property who were or might be a source of problems to the Company or its canal. The document might have been compiled to determine just how much property Mr. Bell owned that the Company might be interested in buying. However, in 1840, Bell signed an agreement with the Canal Company. He agreed "to release, exonerate, and discharge" the Company from his claims "for or by reason of any injury done by them or their agents or persons in their employ to the water privileges" on his "lands situate in the Counties of Sussex and Morris, and for or by reason of any proceedings had or done by them or their agents touching or purchasing" his "lands and real estate, situate in the said two counties above mentioned. It is understood that the suit or suits now pending against the Morris Canal and Banking Company prosecuted by me [Bell], are to be discontinued." The agreement was dated April 28, 1840.[68]

The following year, on June 17, 1841, one Andrew A. Smalley of Byram Township, deeded a parcel of land to the Stanhope Iron Company, described as "abutted and bounded" in part by lands owned by or lately owned by Robert P. Bell.

The career of Robert P. Bell after 1840 is a blank. He had bought a house and 1-1/2 acres of land in Newton from Sussex County Sheriff, Uzal Haggerty, on August 1, 1834, but whether or not he ever lived there is unknown. He had been appointed a "Justice" of Sussex County on March 19, 1839.[69] Whether or not he pursued a legal career after that is also unknown.

Andrew A. Smalley, however, is less a blank. He had been Bell's partner in the Stanhope store, and had apparently been a partner in Bell's real estate operations. He, too, must have been a resident of Stanhope, for he was a trustee of the Stanhope Methodist Episcopal Church. That he lived in Sussex County, precisely in Byram Township, at the time is evident by notations in Snell's *History of Sussex and Warren Counties.* He was, for example, an incorporater and director of the Sussex County Mutual Life Insurance Company, and a commissioner of the Sussex Mine Railroad when it was organized in 1848.[70] In 1858, Smalley became a State Senator, serving three terms, 1858, 1859, and 1860. At the same time, he was the owner of the T. V. Johnson steamboat line at Newark.[71]

Thomas Vail (T.V.) Johnson, owned a packet towing and freighting line, at least as early as 1838, at Newark's Commercial Dock at the foot of today's Commerce Street.[72] Smalley bought the company in 1857 and continued to operate with the steamer *Jamaica* and the barge *Charles Mayo*. In 1862, the Johnson line was merged with that of Stephens and Condit, with Smalley putting up $2000 plus his boats. Smalley stayed on as superintendent of the line until 1866, when he sold out his interest in it.[73]

The interests, as well as the influence, of the Morris Canal grew rapidly once navigation and commerce were firmly established. One of Canal Commissioner John Scott's last official acts for the Company, for example, was to negotiate an agreement with William A. Wood of Woodport, on Lake Hopatcong. The "agreement", or "contract," specified the location and the dimensions of a navigable canal to connect "Brookland Pond" (Hopatcong) with the "Little Pond."[74] As his brother, Freeman "Supercargo" Wood, had done in Dover, William A. Wood opened a store at Woodport in 1831. Judging from the benefits that he *and* his in-laws (the Hurd family of Hurdtown) would derive from the presence of the Morris Canal at the outlet of Lake Hopatcong, it is not likely that Scott had much difficulty in getting Wood to enter into an agreement that would extend navigation from the Lake to his store on Union Turnpike (today's Route 15) and beyond, to the Hurd's iron mine, forge, and sawmill.

Wood's agreement with the Canal Company was to make and open a navigable channel from the head of Brookland Pond to the Little Pond, near the store and dwelling house "where Wood then resided. The course of the channel was to be in or near the bed of the stream of water passing from the Little Pond to the head of Brooklyn Pond, the bottom of the cut or channel to be at least 4′3″ below the top of the gates in the stand lock at the dam crossing the outlet of Brookland Pond and of sufficient width to pass any boats that can navigate the Morris Canal . . . and so as to enable any boats navigating the Morris Canal to pass and repass freely out of one pond into the other." The agreement was dated January 2, 1832.

For the construction of this channel "from the head of Brookland Pond to his storehouse at the [Union] Turnpike," Wood was paid a total of $2220. Scott's notes, the last of which was dated "Dover, July 25, 1832," were all canceled, indicating that the navigable cut was indeed made, and that the two ponds were thus connected. The subject needs more research, but it looks at this point as if Lake Hopatcong, as Brookland Pond, was enlarged to include the Little Pond by way of this connective canal built by William A. Wood, plus the raising of the Lakes' waters by a Canal Company dam. The new higher level of water would have flowed the lower lands between the two ponds, covering the connecting canal, effectually creating the larger single lake, Hopatcong.

Still others would benefit generally from the presence of the canal at Lake Hopatcong. Dr. Samuel Fowler of Franklin had written to Scott before the close of 1830, to point out the revenues to be derived from navigation on the lake.[75] Dr. Fowler stood to realize a bit of revenue himself, since zinc and Franklinite ores, plus the products of his Franklin Furnace, would be accommodated by lake-going canal boats. One thinks automatically, here, of Macculloch's expectations in 1826, of using the "chain of ponds" that could be utilized to extend navigation northward past Franklin to meet the "Lackawaxen Canal."[76]

Scott's note to Wood, dated July 25, 1832, was payable in three months. By then Scott had been named president of Dover's new Union Bank. In order to give full time to his new position, he resigned as Canal Commissioner on May 1, 1833. Other persons associated in some way with the Morris Canal and serving on the Union Bank's board of directors were William Scott, Jacob Hurd, Israel Losey, and Freeman Wood. It was this bank, according to Munsell, that was largely owned by Anson G. Phelps, the wealthy New York iron merchant, whose son later formed the Phelps-Dodge Corporation together with William E. Dodge.[77]

Scott's resignation left Israel Crane the only one of the original ten incorporators of the Morris Canal still in service on the board of directors. After 11 years of what appear to have been devoted service to the Canal Company, Mr. Crane received a slap on the wrist. He was instructed, along with Mr. Pentz and Charles T. Shipman, to attend meetings of the board or else resign. "Neglect of duty" was the charge.[78] What Crane was preoccupied with is unknown, but he resigned in January, 1834.[79]

Charles T. Shipman's preoccupation was no doubt his shipping line, not to mention his leather business. However, he was also busy building a large sawmill "immediately below the waste weir or feeder near the inclined plane No. 1 East." The board was displeased and resolved to serve notice of their disapprobation on Mr. Shipman.[80] What he was doing was establishing Shipmanport, or Shippenport, or Shippingport, as it was variously called throughout the years. There, presumably, was where he built boats, for that is the legend that persists.

However, waters that were to be used by the Canal Company for navigation could not be used for any manufactory of any kind, not even by Director Charles T. Shipman. There is no entry in the Company's *Minutes* explaining Shipman's resignation. Nor does the list of directors for 1834 contain his name. Perhaps Shipman was chastised by the board of directors for using that waste water at Plane No. 1 East, for the Canal Company directors had already applied to the Legislature for an amendment to its charter to enable them to use canal water for manufacturing.[81]

In addition to the water-power application, the Canal Company had also asked for an act to create a transportation company. That application had been dismissed from the Legislature, and Colden, afraid that the water-power application would also be ousted, went to Trenton to lobby. He wanted "to represent fairly to the Legislature of this state, the wishes of the board and to aid in the accomplishment of the objects of this board now pending in the legislature."[82] Colden enlisted the aid of an impressive battery of legal guns to go to Trenton on behalf of the charter amendment. General John S. Darcy, Abraham W. Kinney, Ashbel Cory, Abraham Britten, William Scott, and Amzi Dodd — giants all — went off to Trenton to use their considerable influence on behalf of the Canal Company.[83]

Predictably, Patersonians objected strenuously to the Canal Company's latest bid for increased powers. A letter from "A Friend to the Millers and Manufactures" appeared in the *Paterson Intelligencer* on February 6, 1833, to war Paterson of what the Canal Company was up to in Trenton. The "Friend" could scarcely contain his sarcasm as he explained:

> *if their stock was at par, the stock held by the company would enable them to complete the canal to Jersey City, and carry on banking operations, as had been contemplated when the charter was granted; that it would raise the value of their stock, if the legislature would grant them a privilege, which they say will be a great advantage to the public, and not occasion the least sacrifice of the people's property or interest! — Then they cautiously disclose — that the privilege desired is a repeal of so much of the 27th section of their charter, as prohibits them from applying any water taken into the canal, to any other purpose than mere navigation. They modestly profess not to desire liberty to take any more water than they are now authorized to take, and express confident hopes of success in their application, because it is owing to the great exertion of the company that New-Jersey may claim to have a canal that overcomes a greater elevation than any other in the world! After this lofty climax, they conclude by praying, that this section of their charter may be so modified as to permit them to apply to useful purposes, the water which now runs from the feed gates and waste wiers of the canal.*

Paterson rallied and held a public meeting on Monday evening, February 11, 1833, at the Passaic Hotel. Wasting little time, and few words, the audience resolved that since they had felt the power of the Canal Company over the waters of the Rockaway during the drought of 1830, they had no wish to witness that power again. The consensus was to appoint a delegation to go to Trenton "to promote the views of this meeting." Appointed were: Judge Reynolds, Messrs. E.B.D. Ogden, Robert Morrell, John Vail, and B.W. Vandervoort.[84]

In Trenton, Colden and Amzi Dodd, Esq. (from Essex) appeared on behalf of the Canal Company; Messrs. Ogden and Speer represented "certain mill-owners and citizens of Paterson;" while Garret D.

Wall, Esq., took up the cause of the SUM. A correspondent to the Paterson paper noted that the Canal Company's application was particularly important, especially to the mill-owners along the Musconetcong, the Rockaway, and the Passaic.[85]

The act did *not* pass the legislature, but the movement did not expire. It would reappear shortly, as would the millers along the Musconetcong River, and Garrett D. Wall, and the whole cast of characters.

For the general public, however, the high spot of 1833 was June 14, when President Andrew Jackson visited Newark. He was touring the eastern portion of the country and had expressed interest in seeing at first hand the working of the Morris Canal plane at Newark. He reportedly rode from Jersey City to Newark on his magnificent horse. After traveling the turnpike, he was met at Bridge Street by a procession of citizens who escorted him to the plane. He witnessed "the passage of two boats over the inclined plane," and then "the cavalcade moved to the Military Common."[86]

Despite the early warnings and suggestions of Josiah White to make the Morris Canal the same size as his Lehigh Canal, and despite Fulton's advice in his *Treatise,* the fact is that the Morris Canal had been built too small. Fulton had said that all future canals should be of the same dimensions so that where they meet, boats could pass from one to the other.[87] As it was, the huge boats in use on the Lehigh Canal could not negotiate the locks or planes of the Morris Canal. At the time (1833) Morris Canal boats could carry 16 tons, but Lehigh boats could carry about 70 tons. Morris Canal boats either had to make the trip up to Mauch Chunk to be loaded with coal, and pay all the requisite tolls to the Lehigh Company, or else they could pick up coal at South Easton, at a much higher price per ton. An alternative was to have coal transferred from the Lehigh's large boats to the Morris' smaller ones. For a time, the Canal Company contracted with the firm of Everitt & Hubbs, who brought "coal from Mauch Chunk to Easton in large boats, there to be put on board the Morris Canal's boats."[88]

None of these methods was economical, although all were used. At Mauch Chunk, the Morris Canal could load coal for $3.03 per ton; at South Easton, where coal first had to be unloaded by Lehigh boats, or those of Everitt & Hubbs, coal cost $3.75 a ton.[89] If Morris' boats went up to Mauch Chunk and back, they paid tolls. If the boats only went as far as South Easton, they paid tolls, too: ferry tolls for the boats to cross the Delaware River; tolls for the horses and drivers to use the Easton toll bridge; tolls for the use of the Easton outlet lock.

Eventually, Colden requested that the Lehigh Company direct its boats "to pass into the Delaware River to supply our [Morris'] boats with coal in order to save the toll charged at the Delaware Lock."[90] Still later, the Canal Company sent a representative to Harrisburg, Pennsylvania, to try to make arrangements with that state's Canal Commission whereby Morris Canal boats might "be exempted from the payment of 75¢ for passing in and out of the outlet lock at Easton."[91] Then the Canal Company discontinued the practice of having boatmen pay their own ferriage across the Delaware River and picked up that expense itself.[92]

The Canal Company decided to pay the tolls "on the Easton Bridge for all such horses and boys as are or may be engaged in the transportation of coal for the company until some convenient means are provided by the company for passing said horses and boys across the Delaware River." Robert P. Bell, the superintendent of the Western Division, and Director Enoch Green were empowered by the board of directors "to make arrangements with the owners of the Easton Bridge for a free passage to all horses and boys in the employ of the company or engaged in navigating the Morris Canal." Moreover, Mr. Bell was instructed to take any necessary means to have constructed as soon as possible some fixture calculated to facilitate the passage of boats across the Delaware at Easton."[93]

Early in May, 1832, the canal boat *Hopatcong* descended into the Delaware River on Plane No. 11 West and somehow made it across the river to the Lehigh outlet lock on her own. Encouraged by this success, and no doubt trying to escape paying the toll for ferriage, another canal boat captain refused an offer by a Pennsylvania boatman to pilot his boat across the "Lehigh Basin." He tried to cross the Delaware by "pushing and paddling" with a couple of hoop poles, but the river current caught the boat and swept it over the Phillipsburg Falls. The boat narrowly escaped shipwreck on the rocks on the Jersey side of the river.[94]

These incidents undoubtedly inspired the Canal Company's May 9 instructions to Bell, and probably initiated the development of the "cable ferry" at Phillipsburg. A cable strung between the New Jersey and the Pennsylvania banks of the river, and suspended at a height that would permit log rafts and other craft to pass under it on their way downstream, gave canal boats a fixed point which kept them from being swept down the river by the side currents that propelled them across. A towboat was attached to the cable by means of two ropes — a short rope at the bow; a longer rope at the stern. Canal boats were lashed to the towboat, and together they were propelled by side currents from one side of the river to the other. The practice was much like "tacking" in a sail boat, but using water currents instead of wind.

The Lehigh Coal and Navigation Company advertised for "proposals for boating coal from Mauch Chunk to Newark, by the Lehigh and Morris Canals, and from Easton to Newark by the Morris Canal," with contractors furnishing boats and all the necessary appliances." Boats would be loaded at Mauch Chunk by the Lehigh Company; at Easton by the contractors.[95] At the same time, the Lehigh Company was given authorization "to transport coal on the Morris Canal for tolls already established, provided that transportation for their company doesn't interfere with the loading of Morris Canal coal at Mauch Chunk or its transportation."[96]

The Canal Company began to look for "a suitable place for depositing coal at South Easton, Phillipsburg, or elsewhere."[97] In brief, Morris was trying to find a home of its own instead of using the Lehigh Company's facilities at South Easton. The Company's toll collection offices were at this time on the Pennsylvania side of the Delaware River. Edward A. Reeder, Easton-born, but later a resident of Stanhope, entered the employ of the Canal Company at the toll collector's office in Easton, where he remained in service for several years.[98]

Proposals to build boat yards were received by the Canal Company from "Mr. Sherwood," who applied for a lease of a lot of land at Port Delaware for a boat yard,"[99] and from Messrs. Savitte & Davis. To the latter, the Company leased for 10 years, at an annual rent of $100, "80 ft. front on the canal by 200 ft. deep of the land owned by the company at Phillipsburg on the South side of the Rock, or that side nearest to E&R Green's" so that they might make a boat yard.[100]

"Port Delaware" and "Phillipsburg" were used interchangeably until the area around Plane No. 11 West actually became a "port," i.e., a place for loading and unloading cargo. The spot became such a port in 1835, and from then on, the Toll Collector's office was at Port Delaware, Mr. J. Kerr, Collector. It appears unlikely, however, that a basin, as such, or coal-storage facility was yet located at Port Delaware. Rather, boatmen were induced to contract for a boating season by the prospect of picking up their boats already fully loaded at the head of "the plane at Port Delaware opposite Easton."*[101] Many of the larger or more important basins were not built until the 1837 season, among them, Dover, Bloomfield, and Mr. Dusenberry's own Port Colden.[102] Port Murray had also been established in this general period, named for the Canal Company president at that time, James B. Murray.

Meanwhile, something obviously had to be done to bring down the costs of coal and of transportation on the Morris Canal. Transshipment of cargo was a stopgap measure necessitated by the inferior dimensions of the canal; it certainly wasn't an efficient or economical way to run a canal company. The Delaware Division Canal was ready for navigation before the end of 1832, after extensive repairs; and as operating canal ready to receive Lehigh coal, it posed a certain kind of threat to the Morris Company. Delaware Canal boats, although smaller than the Lehigh's, were still capable of 67 tons burden, or almost three times what a Morris boat could then carry.[103]

Clearly, enlargement of the Morris Canal was called for, and as quickly as possible. But first it was essential that the canal's extension to Jersey City be finished so that transshipment of cargo would no longer be necessary on the *eastern* end of the waterway.

**This advertisement was published by William C. Dusenberry of Port Colden, who had bought 25 new boats from the Canal Company.*

MORRIS CANAL.

PERSON desirous of RENTING BOATS for the purpose of transporting Coal for the Company, are hereby informed that the undersigned will attend at DOVER on MONDAY and TUESDAY next, the 10th and 11th inst. to make contracts.— After that period, application for boats may be made to J. KERR, Collector, Port Delaware, or to the subscriber at Newark.

Boats will be rented for the Coal trade ONLY.

a5 3tis S. JONES.

Advertisement for boatmen to rent Morris Canal Company boats (ETF, Newark Daily Advertiser, April 5, 1835).

REFERENCES

1. Newark *Daily Advertiser*, April 9, 1832.
2. *Ibid.*, April 17, 1832.
3. *Paterson Intelligencer*, May 2, 1832.
4. *Ibid.*, May 17, 1832.
5. Newark *Sentinel of Freedom*, May 22, 1832.
6. *Ibid.*
7. *Paterson Intelligencer*, May 30, 1832.
8. Morris Canal and Banking Company *Minutes*, May 12, 1829 (hereinafter cited as MC&BCo. *Minutes*).
9. *Ibid.*, September 20, 1831.
10. Newark *Sentinel of Freedom*, July 10, 1832.
11. MC&BCo. *Minutes*, June 12, 1832.
12. Newark *Daily Advertiser*, July 18, 1832.
13. *Report of a Joint Committee of the Directors and Stockholders of the Morris Canal and Banking Company, on the Affairs of the Company, with a Communication from the Directors who were of the Joint Committee, to the Stock-*

holders, August 23, 1832, in *American Railroad Journal,* September 29, 1832, p. 627 (hereinafter cited as *Report of a Joint Committee, 1832*).

14. *Ibid.*
15. MC&BCo. *Minutes,* December 17, 1832, and Newark *Sentinel of Freedom,* December 28, 1832.
16. Munsell's *History of Morris County, New Jersey: 1739-1882* (N.Y.: 1882), p. 179 (hereinafter cited as Munsell's *Morris*).
17. MC&BCo. *Minutes,* May 10, 1832.
18. *Ibid.,* July 31, 1833.
19. Cornelius C. Vermeule, Jr., *Final Report of Consulting and Directing Engineer, June 29, 1929* (East Orange, N.J.: 1929), p. 59.
20. Washington (N.J.) *Star,* September 28, 1934.
21. *Paterson Intelligencer,* July 14, 1847.
22. MC&BCo. *Minutes,* August 12, 1847.
23. *Paterson Intelligencer,* August 22, 1832.
24. MC&BCo. *Minutes,* May 1, 1833.
25. *Ibid.,* September 24, 1834.
26. Newark *Daily Advertiser,* August 3, 1832.
27. *Report of a Joint Committee, 1832,* p. 626.
28. *Ibid.,* p. 627.
29. MC&BCo. *Minutes,* February 10, 1832.
30. *Ibid.,* August 28, 1828.
31. *Ibid.,* April 21, 1829.
32. *Ibid.,* May 12, 1829.
33. *Ibid.,* September 8, 1829.
34. *Ibid.,* September 15, 1829.
35. *Ibid.,* February 26, 1830.
36. *Report of a Joint Committee, 1832,* p. 626.
37. MC&BCo. *Minutes,* January 18, 1832.
38. *Ibid.,* August 10, 1831.
39. *Ibid.,* January 18, 1832.
40. *Ibid.,* February 1, 1832.
41. *Ibid.,* November 12, 1832.
42. *Ibid.,* February 1, 1832.
43. Newark *Sentinel of Freedom,* June 12, 1832.
44. MC&BCo. *Minutes,* July 3, 1832.
45. *Ibid.,* March 29, 1833.
46. *Ibid.,* October 12, 1832.
47. Morris Canal and Banking Company *Docket Files* (hereinafter cited as MC&BCo. *Docket Files*).
48. Frances Trollope, *Domestic Manners of the Americans* (London: 1927), pp. 307-8.
49. MC&BCo. *Minutes,* December 10, 1832.
50. *Ibid.,* December 27, 1832.

51. *Ibid.*, January 29, 1833.
52. *Ibid.*, May 1, 1833.
53. *Ibid.*
54. *Paterson Intelligencer,* May 1, 1833.
55. MC&BCo. *Minutes,* January 29, 1833.
56. *Ibid.*, February 18, 1833.
57. *Ibid.*, March 6, 1833.
58. *Ibid.*, January 8, 1834.
59. E.A.J. Johnson and Herman E. Krooss, *The Origins and Development of the American Economy* (N.Y.: 1953), p. 345.
60. MC&BCo. *Minutes,* January 29, 1833.
61. *Ibid.*, February 5, 1833.
62. MC&BCo. *Docket Files.*
63. MC&BCo. *Minutes,* December 20, 1831.
64. *Ibid.*, October 2, 1832.
65. James P. Snell's *History of Sussex and Warren Counties, New Jersey* (Philadelphia: 1881), p. 467 (hereinafter cited as Snell's *Sussex*).
66. Munsell's *Morris,* p. 162.
67. MC&BCo. *Minutes,* February 22, 1836.
68. MC&BCo. *Docket Files.*
69. Snell's *Sussex,* p. 161.
70. *Ibid.*, pp. 230, 167, 465.
71. William H. Shaw, *History of Essex and Hudson Counties, New Jersey* (Philadelphia: 1884), pp. 189-190 (hereinafter cited as Shaw's *Essex*).
72. *Paterson Intelligencer,* June 13, 1838.
73. Shaw's *Essex, loc. cit.*
74. MC&BCo. *Docket Files.*
75. MC&BCo. *Minutes,* December 30, 1830.
76. *Annual Report Upon the Morris Canal for the Year Ending July 1, 1826* (N.Y.: 1826), p. 14.
77. Munsell's *Morris,* p. 315.
78. MC&BCo. *Minutes,* August 7, 1833.
79. *Ibid.*, January 15, 1834.
80. *Ibid.*, October 2, 1833.
81. *Ibid.*, January 11, 1833.
82. *Ibid.*, January 29, 1833.
83. *Ibid.*, February 5, 1833.
84. *Paterson Intelligencer,* February 13, 1833.
85. *Ibid.*, February 20, 1833.
86. *A History of the City of Newark, New Jersey* (N.Y.: 1913), Vol. I, pp. 495-6.
87. Cadwallader D. Colden, *The Life of Robert Fulton* (N.Y.: 1817), p. 334.
88. MC&BCo. *Minutes,* April 14, 1834.
89. *Ibid.*, March 29, 1833.
90. *Ibid.*, July 18, 1834.

91. *Ibid.*, January 1, 1835.
92. *Ibid.*, July 18, 1834.
93. *Ibid.*, May 9, 1832.
94. Newark *Daily Advertiser*, May 31, 1832.
95. Newark *Sentinel of Freedom*, June 21, 1832.
96. MC&BCo. *Minutes*, June 12, 1832.
97. *Ibid.*, July 3, 1832.
98. Snell's *Sussex*, p. 472.
99. MC&BCo. *Minutes*, January 1, 1834.
100. *Ibid.*, September 24, 1834.
101. *Palladium of Liberty*, June 28, 1836.
102. MC&BCo. *Minutes*, March 1, 1838.
103. *The Story of Anthracite* (N.Y.: 1932), p. 76.

10

Growing Pains

AFTER years of apparent stalling, the canal's extension to Jersey City suddenly assumed a position of paramount importance. A sense of urgency fills the Canal Company's records throughout 1835 and 1836, although the cause of that urgency is nowhere explained. Nor is there any explanation of why the canal's extension had not been completed at least a year or two earlier, for the law authorizing the extension to Jersey City had been in effect since January, 1828. We have seen how a railroad from Newark to Jersey City had been authorized in 1832 and the extent to which the Canal Company was involved in those plans. Perhaps it should not surprise us to learn that by the end of 1832 the Canal Company might *not* have wanted to complete the Jersey City extension at all. Messrs. Brooks, McIntyre, and Varick were appointed a committee "to take into consideration the propriety of continuing the canal between Jersey City and Newark."[1]

Yet we saw that Beach had run surveys for that extension in 1827, and that early in 1830 Beach and Douglass had been instructed to report on the canal's route from Newark to Jersey City.[2] According to Beach, Douglass' time had been "extensively occupied with other engagements in addition to his professional duties," so that it was "uncertain when his attention can be expected to this subject." Beach, himself, had experienced ill health, which delayed his report until May 31, 1830. However, Beach took "the liberty of submitting the following observations subject nevertheless to the examination of Major Douglass or any other Engineer to whom the Board may think proper to refer same."[3]

Enlarging upon the "hurried report" he'd made on May 12, 1827, Beach explained that "the marsh through which the excavation must be made appears more spongy and permeable and the bottom much softer and more yielding," while "the ridges through which the excavations must be made, instead of gravelly loam (as was represented) are composed of fine sandy soil and quite springy." There were, after all, no precedents in the United States for making canals across salt marshes except at the mouth of South River (the Raritan Canal — abandoned before it was completed) and on the Delaware and Chesapeake (Saint George's Meadow — completed, but very expensive).

Descriptions of excavating the Erie Canal through the Montezuma Swamp probably also applied to the Morris' extension from Newark to Jersey City through the tidal marshes of the Jersey Meadows.[4] The Meadows, being tidal, could not be drained so that the canal prism might be lined with clay. Yet, unless the prism *was* "puddled," it is difficult to imagine how the spongy soil of vegetable mold, as described by Beach, could have held water. That soil, like that of the Montezuma Swamp, probably required the special sharp-edged spades devised by the Irish "bog trotters" for use in cutting peat, and used in digging the Erie Canal.

Beach's plan was to construct one 8-foot lift lock at the start of the marsh on the west side of the Passaic River and three tide locks: one on the east side of the Passaic River, one on the east bank of the Hackensack River, and one at the confluence of the canal with the Hudson. The canal would be filled by tidal water between the Hackensack and Hudson rivers, as it would be between the Passaic and Hackensack Rivers.

Beach thought that the canal basin at the Hudson River should "be located with a view to future improvement of Jersey City by extending Hudson [Street] northward," and that "Hudson Street should form the pier between the Basin and River." Then, communication between the basin and the river "should be formed by placing the lock across Hudson Street, at or near the intersection of Wayne Street."[5] We have seen how and why this plan was changed.

It appeared that only Prior's Mill on Mill Creek would be affected by the canal's extension. That creek "being a tide creek which can be filled or discharged at either end," presented no insurmountable difficulties. It operated only on an outgoing tide, and since both "ends" of the creek lay in tidewater (the Hudson), one end could be filled "from excavation of the canal on each side," so that the canal could be "continued across it entire without increased expense. This plan is no doubt the best and, upon the supposition that the mill can be purchased at a fair price (or a better water power commuted from the canal to compensate for the damages — which may be done without injury), is the cheapest that can be adopted for crossing that stream."[6]

Prior's Mill stood on a hillside at about the southwest corner of Fremont and Wayne Streets. To this bit of upland the Americans had retreated when the British invaded New York. Mill Creek, whose waters worked the mill's machinery, emptied on the southern end into Communipaw Cove. A circular knoll at the mouth was surrounded on three sides by salt marsh and on the east by Communipaw Cove. The spot was known variously as the "Point of Rocks," "The Creek of the Woods," "Jan Delacher's Point," and "Mill Creek Point." The name "Jan DeLacher," or "The Laugher," supposedly immortalized "the jovial disposition of Jan Evertsen Bout, the Point's first occupant" about 1638.[7] The canal was to pass far to the south of the mill, at the Point.

The Jersey Associates had originally planned to make Mill Creek into a tidal canal, so that in effect, Jersey City would have been an island — a mini-Manhattan — "bordered by piers and docks and surrounded by navigable water."[8] The Canal Company's plan to fill in Mill Creek, as we shall see later, did not materialize, but other plans involving the creek later brought the combined wrath of Jersey City and the Central Railroad of New Jersey on the Company.

Acting on Beach's report, the board of directors began considering offers from various contractors for the construction of the Jersey City extension. In 1830, a Mr. Toughts had offered to excavate the marsh between Newark and Jersey City at 21¢ the cubic yard. He offered further to take half the amount of his price in Canal Company stock at par value, and would give satisfactory security for the performance of any contract he might make.[9] When excavation did get underway, however, Mr. Toughts' name was not among the contractors who were put to work. Instead, the firm of Mason & Pratt, contractors for several of the inclined planes, including Nos. 9 and 10 West, were rehired for the Newark-to-Jersey City segment.[10]

At first it was thought that the Jersey City extension should "be made of the same capacity as the Erie Canal . . . 28 feet at bottom, 40 feet at top water line, and 4 feet deep."[11] However, when the final surveys, maps, and calculations were made, they were ordered to be made "on the supposition that the canal between Newark and Jersey City will be of the same size as the canal now made connecting the River Delaware with the Passaic."[12]

The 11.5-mile extension from Newark to the Hudson River was divided into half-mile work sections, beginning with number 99. The following chart was compiled from Company *Minutes* of May 23 and June 8, 1834, to show to whom the work was contracted:

Section	**Contractor**
99, 100, 101, 102 (West of Passaic River)	O'Brian Pursell & Co.
103, 104 (Between Passaic and Hackensack Rivers)	Hill & Vreeland
105, 106, 107 (East of Hackensack River)	George Taylor
108, 109	Capt. M. Andrews
110, 111	Patrick Garrett
112, 113, 114	Green & Hall
115, 116, 117, 118	Arnold Mason & Co.
119(?) Basin Wall (North)	Samuel Roberts & Co.

Once again a committee of Canal Company directors met with the Jersey Associates and the Newark and Paterson Railroads to negotiate about "suitable basins, wharves, etc., for the accommodation of the Morris Canal at Jersey City."[13] We saw in Chapter 8 how Beach had recommended that the canal basin be located in

Harsimus Cove with an outlet lock into the Hudson River at Wayne and Hudson Streets. We saw, too, how that location was rendered impracticable by the projected P&HR and NJRR&T Co. railroads which would be using the Jersey City Ferry wharf. These new negotiations of 1834 determined the site for the canal basin: at "the southern part of Jersey City, as laid out by the Engineer, Captain Beach."[14] At first glance, "as laid out by" Beach could refer either to the canal basin *or* to the city itself. As it turns out, it refers to both, since in 1827, when he was conducting his preliminary surveys for the canal's extension, Beach had been employed by the Jersey Associates to lay out those streets of Jersey City that had not been laid out on Mangin's map of 1804.[15] Many of the streets were "paper streets," existing only on the map at the time, since they were projected to be built on land to be reclaimed from the waters of New York Bay.

An August 1 deadline had been set by the board of directors for the canal's completion; no delays could be tolerated.[16] Beach was instructed to get ready "all surveys, maps, and calculations necessary to be submitted to the Chief Justice prior to the application to him for the appointment of appraisers."[17]

With his deadline only two months away, Beach was told to get tough with the contractors to finish their sections on schedule. Visiting the line of the canal to determine the causes of delay, Beach met everywhere the same "cry" he'd met before: "Impossible to get men." Only George Taylor (sections 105, 106, 107) felt he could finish on time. Captain Andrews (sections 108, 109) doubted that men enough could be hired to make the August 1 date. Samuel Roberts flatly said, "Impossible." Even September 1 would be doubtful because enough men and teams couldn't be gotten at the prices he'd contracted for in May. He was paying his men 75¢ a day for labor. If the Canal Company would pay the difference, he could try offering $1.25 a day to attract more men. But even then, he wouldn't finish his sections until September.[18]

Beach definitely did not favor the Canal Company footing the bill for an almost 100-percent increase in wages. What, he asked, would "be the effect? Raise the wages on the large jobs and the small ones must raise too or abandon. Hence a great sacrifice to the company or ruin to the contractors, but the company must sustain or the contractors will abandon." The work could be done by September 1, with money enough to pay the men, most likely, but "since canal boats are successfully towed by steam with their cargoes to New York," he felt that "the object will not justify the sacrifice."

As for any aspersions cast upon his ability or his character, Beach was quick to answer Canal Company cashier, Robert Gilchrist: "From your long acquaintance with me, you will know sir that on the question of performance, if it comes within the range of possibilities, the word *cannot* never belonged to my vocabulary. You also know that I have not been in the habit of promising what I could not perform . . . . I shall not cease to urge on the contractors."

Even with urging, work proceeded slowly. By the end of 1835 there was still no "basin," as such. Samuel Roberts had contracted in 1835 to build a wall "on the north side,"[19] but there is no indication as to just where that wall was built. In fact, in 1836 the Canal Company was still trying to find out if they had a right to the submerged lands on the southern edge of Paulus Hook. The Canal Committee was asked to ascertain and report to the board of directors "whether the Canal Company now have such a title to the water lots between Bergen and South Streets, from Washington Street east to Hudson Street as gives them the right to fill up said lots and extend the area of the basin south of Quay Street."[20]

There is evidence to suggest that the area from Henderson Street east to Van Vorst Street was used as a "natural basin," for Communipaw Cove at that point formed a gentle indentation in the shoreline. S. Baldwin, Esq., Civil Engineer, was engaged in 1836 to survey and report on the proposed basin area.[21] His findings were not recorded, but in 1837 the Canal Company was still trying to finalize its plans for a basin, so we can assume that Mr. Baldwin's plan was not used. The Company wanted "a plan for the permanent regulation of the natural basin adjoining Crane's property near the termination of the canal at Jersey City."[22] The plan was to include "one or more places of landing and deposit at that place, so as to embrace a portion of the present natural basin." Arrangements were to be made with the owners of property at that place at a cost not exceeding $2000.[23]

Completion of the tidewater basin was obviously reserved to the future.

With or without the Jersey City extension, or a completed basin, the main canal opened for business on March 17, 1834, and boats began to arrive at and clear Newark regularly.[24] New freight lines sprang into existence, boasting covered boats built with an eye to keeping cargoes dry as canal boats were towed across Newark Bay and up into New York Harbor. A combine formed by the firms of Thomas Christe, Whitehall Dock, New York; Stephens & Condit Company, Newark; and Charles Rodenbaugh, Easton advertised that they would begin service between New York and Easton during July, 1834. Boats would leave Newark and Easton on Tuesdays and Fridays during the season, with freight-forwarding service available for goods bound for Bethlehem, Allentown, and Mauch Chunk. Warehouses were provided for goods that required storage.[25]

Rodenbaugh, who was also a Phillipsburg store owner, was permitted to tender "I.O.U.'s" to toll collectors in lieu of tolls. This special arrangement also applied to the New Jersey Iron Company of Boonton, the E&R Green Company of Phillipsburg, the Murdoch Company, and Charles T. Shipman.[26]

Newark became a Port of Entry on June 30, 1834, as did Jersey City, and the prospects for a profitable boating season were bright. Newspapers reported large numbers of boats "actively employed" on the canal, with many more "building at various points along its banks."[27] Newark was said to have shipped 24,000 tons of coal to New York during 1834, either by sloop or schooner.[28] Lehigh Coal and Navigation Company figures for 1834 indicate that 27,000 tons of coal were transferred to the Morris Canal. Only 13,000 tons had been transferred in 1833.

A Newark Directory published in 1836 claimed that Newark's becoming a Port of Entry had brought "a corresponding increase in the trade with the city of New York, and the northern and eastern ports, in which trade nearly 100 vessels are constantly employed, several of them being exclusively engaged in the coal trade, as connected with the Morris Canal." The directory also noted "that the fuel for the consumption of the town is chiefly supplied by this canal."[29]

The Lehigh Coal Company was advertising its coal, both hard and soft. As for that company's hard coal, the advertisement claimed, "it is sufficient to state that for parlor grates, stoves, and culinary uses, no coal is more generally approved in this place [Philadelphia]; while as fuel for steam engines, for malting, heating brewers' and dyers' coppers, nail works, foundries, and rolling and slitting mills, it has decidedly the preference on account of its great durability and cleanliness to the extent of more than two thirds of the aggregate consumption of anthracite coal for those purposes in Philadelphia and its vicinity. The Lehigh *soft* coal ignites more easily and burns more freely than the *hard,* and on that account is preferred by many consumers."[30]

Coal could be bought from the Lehigh Coal and Navigation Company's agent at Newark, C.H. Barker:* Hard Coal, $5.25 per long ton (2240 lbs); Soft Coal, $4.75. Coal would be raked at Newark and "delivered on board of vessels for 25 cents per ton in addition to the prices named above for that place." Of course, coal could also be bought at "Greenwich, two and a half miles from Easton," for as little as $3.80 (hard) and $3.30 (soft). At Dover, forty-eight miles from Easton, hard coal sold for $4.65, soft coal for $4.15. Prices at other points along the Morris Canal were proportionate to the distances from Easton.[31]

Those prices, of course, didn't apply to the Canal Company, which contracted for quantities much larger than those quoted for by the Lehigh Company. Even at the end of the 1834 boating season, the board of directors authorized President Murray to order 4000 tons of coal, to be picked up at South Easton by November 15. The Canal Company then would try to transport that amount to Newark before the season ended, "before the company's boats were stopped by the ice on the line of the Morris Canal." The Company assumed responsibility for delivery of that coal, in case ice did stop the boats.[32]

There had always been some doubt as to the legality of the Canal Company's ownership and operation of boats on its own canal. In the opinion of Thomas Nesbitt McCarter, a Canal Company

**These prices stood for quantities not less than three boat loads or of 50 tons, and under 1000 tons. Larger quantities brought lower prices.*

Counsel, the charter had very specifically constrained the Company to owning and operating a canal, nothing else.[33] In what appears to have been an effort to avoid litigation, the Company had bought canal boats, but then had rented them to independent boatmen,[34] offering a $9 premium to "any master for performing three trips, Newark to Mauch Chunk, within thirty days, with not less than 18 tons of coal each trip."[35] A further premium of $2 was offered to "every boatman who shall perform a full trip between Mauch Chunk and Newark in ten days carrying not less than 18 tons of coal."[36]

In order to put more boats on the canal, the Canal Company began shopping around for contractors who would make the best boats for the least money. A Mr. Reilley of Rochester, New York, contracted for 12 boats at $350 each, payable on delivery of the boats at the Long Wharf, Jersey City.[37] William R. Quick was paid $375 for the canal boat *Barry*.[38] Not all boats were that expensive, but a few cost even more. Thirty-two boats ordered for delivery on or before May 1, 1834, had price tags ranging from $325 to $380:[39]

Number of Boats	Where Built	Price
5	Montville	$325 to $335 each
3	Dover	
3	Drakesville	
1	Stanhope	
12	Easton, Pa.	
2	Lehigh Gap, Pa.	
2	Weisport, Pa.	
2	Mauch Chunk, Pa.	
2	Newark	$380

The firm of Everitt, Hubbs & Company had contracted to transport 40,000 tons of coal for the Canal Company, free of tolls on their boats and cargoes.[40] The Canal Company was to supply Messrs. Everitt and Hubbs "with at least 80 boats."[41]

The Canal Company seems to have been so desperate for boats that they proposed to hire a portion of the Delaware and Hudson Canal's boats.[42] That canal company's president Wurts responded that he would be willing to sell boats to the Morris Canal,[43] an offer that the Morris Company declined.[44] But they even tried to hire boats and boatmen from Pennsylvania's Union Canal.[45] The outcome of the latter attempt is not recorded. Boats from the Union Canal *did* operate on the Delaware Division Canal, however, and the possibility of Union Canal boats on the Morris Canal is not ruled out.[46]

Reporting on the 1834 boating season, President Murray acknowledged that it was "the want of boats alone" that had prevented a substantial increase in business. "The boats in operation were principally procured by means of private contributions among the directors and a few of the stockholders in February last (1834)."[47]

It was no coincidence, then, that on February 8, 1834, the State Legislature had passed "An Act to Incorporate the Morris Canal Transportation Company." The incorporators were Samuel Swartwout, John V.B. Varick, Samuel R. Brooks, Charles T. Shipman, James B. Murray, Samuel C. Jaques, Henry Yates, Billopp B. Seaman, Robert Gilchrist, Samuel Cassedy, John M. Cornelison, Peter M. Ryerson, and their associates.[48]

Of the incorporators, Swartwout, Varick, Brooks, Murray and Yates were simultaneously directors of the Morris Canal and Banking Company. Shipman, Jacques, and Ryerson had in the recent past filled the same positions with the Canal Company. Robert Gilchrist, of course, was the Company cashier. Samuel Cassedy, Clerk of Jersey City and Clerk of Bergen County later, was at the time the Canal Company's bookkeeper. Thus, the only name listed among the incorporators who had no obvious connection with the Morris Canal was John M. Cornelison.

It is neither difficult to imagine, nor unfair to assume, that the "private contributions" came from these men for the boats that ran on the Morris Canal from 1834 for several years hence.

The Morris Canal Transportation Company's charter gave the company the right to buy and sell coal and to transport "persons or passengers, and all manner of things, through or upon the Morris Canal." (Section 1.) The charter was to be in effect for 20 years, expiring in 1854. During that time the Transportation Company was *not* to be permitted "to erect on the real estate" purchased for its expressed purposes, "any kind of water works or machinery, so as to require any of the waters of said canal, and shall not use any of the waters of said canal, for any purpose whatever, except navigation." Moreover, the company was to navigate the Morris Canal "according to the Canal Company's rules, regulations, and toll schedules of the Canal Company." (Section 4.) Thus the Canal Company was to own and maintain a navigable waterway, while the Transportation Company was to operate boats on that waterway. But not *all* of the boats, since some at least were rented by the Canal Company to private masters.

Such boats as had been ordered during 1834 had not been completed "until a large portion of the season had passed away." However, 1835 would be a better year, for "besides upwards of 200 effective boats now on the canal, and ready to begin with the opening of the next season, 100 more will be supplied from the contracts for building, which have been made since the close of the last."[49]

At the outset of 1834, the board of directors had considered purchasing the steamboat *Connecticut* for use as a towboat,[50] and authorized a committee to look into the price of that "or any other boat suitable for the transportation of coal from Newark to New York."[51] That subject was not mentioned again until 1835, when the Canal Committee was authorized to *hire* a steam towboat to tow canal boats from Newark to New York in quantities of 15,000 tons.[52]

The steamboat *Experiment,* captained by Master H. Battle, was advertised in the Newark newspapers during 1835:[53] "Freight, Passage and Tow Boat between New-York and Newark . . . The owners of this Boat intend running her between New York and Newark every day during the season. She will leave the wharf at the foot of Morris street, (N.R.) (North River), New York, at 6 o'clock A.M. and Newark at 12 M. every day, Sundays excepted. For freight, passage, or the price of towing, apply to the Captain on board. N.B. — Boats navigating the Morris Canal will be towed to and from New-York in preference to all others, and on the most reasonable terms."

The advertisement tells us that the *Experiment* began its season on June 2, 1835, but doesn't identify the "owners." The Canal Company's *Minutes* show that Mr. James P. Allaire* was paid $419.35

**James P. Allaire was a master machinist, who acquired Fulton's steam engine works at Jersey City. He later moved the works to the East River, New York, where he also had an extensive iron works in the Corlear's Hook area.* [Robert Albion, *The Rise of New York Port* (N.Y.: 1939), p. 149.]

"for repairing the steamboat Experiment."[54] Then, the Canal Committee was authorized to dispose of her on July 10, 1835, and Gilchrist accordingly advertised the boat for sale.[55] In this roundabout way, we thus find that the Canal Company was the *Experiment's* owner.

The Lehigh Canal had long been using its own steamboat to tow canal boats between Philadelphia and Bristol at the lower end of the Delaware Division Canal.[56] Before long, the *Car of Commerce,* a steam vessel especially designed for canal navigation, would be passing through the Delaware and Raritan Canal itself, en route from New York to Philadelphia.[57] It would not be until 1856, however, that any kind of steam boat would be run upon the Morris Canal.

A very significant trend in thinking began in 1834, when the Canal Company board of directors authorized Resident Engineer Roswell B. Mason to build a house for the plane tender at the head of Plane No. 8 East at Montville.[58] The materials for building the house had been removed from the Rockaway Plane (No. 6 East) and taken to Montville so that a house could be built "at the latter instead of the former place."[59] By mid-1834, the Company authorized the Canal Committee "to cause lock and plane houses and offices for collectors to be erected at such places on the canal as they may deem necessary and also to procure lands for the plane and lock tenders . . ."[60]

Housing for Canal Company employees was discussed as early as 1830, at least for lock tenders.[61] However, it can hardly be said to have been Company policy to provide living quarters for its help. Recall Mr. DeHart's instructions to build himself, or have built, a house at Lock No. 14 East, Beavertown. At the end of the season, DeHart was to submit his account, the Company to pay for the house, or not, as it saw fit.

Nevertheless, and understandably, the issue had come up from time to time. Early in 1833, the Canal Committee had been authorized to "contract for construction of as many lock and plane houses as they may think proper, at such price as they may think reasonable."[62] That authorization yielded one plane tender's house, that at the lower Montville Plane, No. 9 East. William Gladson, the builder, was paid $80 for building a basin at Montville, and $350 for building the tender's house.[63] A deed on file with the Canal

Company's records shows that Gladson conveyed 1/12 acre to the Canal Company on August 9, 1833, presumably for the land on which either the plane tender's house stood, or the basin — another point of policy. The Company had resolved that it should own all basin facilities in order to exercise control over them.

Still another plane tender's house was ordered built during 1834. The Canal Committee visited the line of the canal early in September, and had agreed then with Mr. George Perry to buy "one-half acre of land at the head of Plane No. 5 West, on the south side for the purpose of erecting thereon a house for the plane tender for the sum of $20."[64]

By 1859, Canal Company policy regarding housing for employees had developed, apparently, as "extensive repairs and additions to houses occupied by plane and lock tenders" swelled the expenditures for that year. Nevertheless, the work had been done "because the use of such houses constitutes part of the pay of the men who occupy them."[65] By the time the Civil War had ended, the Canal Company considered employee housing "necessary for the proper accommodation of the men near to their work, tending to ensure at those places, the services of such men as are needed for the efficient operation of the canal."[66] This implies that not all lock and plane tenders yet had housing, but that some lived in their own homes at some distance from work. However, cheap labor was becoming more and more scarce and harder to hold on to, and concessions had to be made if the Company expected to keep its faithful employees — or its employees faithful.

Evidence of beginning labor problems appeared in Beach's 1834 report to Gilchrist on the progress then being made on the Jersey City extension. The contractor's suggestion to almost double his men's wages to meet the Company's deadline is a case in point. But labor was not the biggest problem facing the Canal Company while the extension to Jersey City was still being built. Despite President Murray's glowing report on the success of the inclined planes, there were still defects inherent in the basic mechanical principles of the planes. Murray claimed that experiments had "satisfied the Board, that the introduction of lock planes instead of summit planes, and a slight alteration in the machinery, involving together not more than $2000 additional expenses on each plane" would improve their performance.[67] Even as they were then operating, the planes were veritable tributes to Yankee ingenuity, as Mrs. Trollope had suggested. "Loaded boats are seen creeping up the sides of mountains, hitherto deemed scarcely accessible by the usual modes of travel, affording proud specimens of the untiring spirit of American enterprise and ingenuity."[68] Regardless of this high praise, the inclined planes still needed the most attention.

The substitution of lock planes for the planes that had summit designs, as referred to by Murray, was an outgrowth of experiments begun as early as 1832. Ephraim Morris had been authorized by the board "to procure materials for altering the Boonton Plane to a lock plane, and that he may get the iron work at Boonton if he shall think it best."[69] Douglass had recommended that the Boonton plane design of Colonel John Scott be adopted for the planes that remained to be built after the initial experimental period. Accordingly he had built the Newark Plane on the plan of a summit plane, and then had had all planes from Rockaway west built along those lines. Morris continued to convert the system's summit planes to his design for lock planes. Three such planes, unfortunately unidentified, were ordered altered in September, 1834, but a subsequent change in Company plans reduced that number to one.[70]

Throughout 1834, suppliers of parts for the inclined planes were kept busy. Some illegible handwriting in the Canal Company's *Minutes* refers to "rope" for the inclined planes to be supplied by Durpes (or Dupres) May & Company, who had also suggested "some alterations" to be made in the planes.[71] More illegible handwriting "identifies" Jacob Alyen or Algen or Alger* of Newark as a supplier of castings for the inclined planes.[72]

New English chain was purchased for Plane No. 4 West, while the Peru Iron Company, of 32 South Street, New York City, and the Berback Company, Brooklyn, furnished American *inch* chain.[73] Peru Iron Company also supplied one-inch chain for Plane No. 1 East, while A.W. Kinney of Newark made back-chain for the

**This person was later identified as Jacob Alyea, a partner of Abraham W. Kinney and others in the Newark Malleable Iron Mfg. Co.*

planes.[74] Chain was apparently still a major cause of concern. Accidents must have continued to happen despite Colden's earlier prediction that once all the planes were supplied with imported English chain the problem would be solved. The firm of Davis & Brooks (assumed to be Directors Thomas B. Davis and Samuel R. Brooks) was denied payment for chain sold to the Canal Company because it had proved defective.[75] The firm of Samuel Hicks & Son continued to supply the Canal Company, but with imported cast iron chains for the inclined planes and rolled railway iron.[76]

Rope continued to be tried, and a flat rope ordered from Mr. W. Jackson of Philadelphia proved unsuitable.[77] Still another order of one-inch rope, "of sufficient length to suit the Newark Plane," was given, but no mention is made of the supplier.[78]

According to the minutes for May 12, 1835, the firm of C. Logan & Company, Liverpool, England, received an order for:

2 chains	600′ long—7/8″	All welded on the sides without swivels. The links not to exceed 3-1/2″ width.
1 chain	700′ long—7/8″	
3 chains	800′ long—7/8″	
1 chain	2000′ long	(1/2″ back chains)

It is generally accepted among canal buffs that the planes of the Morris Canal were altered during 1835-36. Sufficient examples have been included here to show that alterations on an experimental basis (at Boonton) had begun as early as 1832. For the month of September, 1835, Roswell B. Mason submitted his account for expenses and "materials for altering the inclined planes amounting to $4,700 which was ordered to be paid."[79]

The large-scale alterations of 1835-36 got underway in October 1835, when the Canal Company engaged P.M. Ryerson to superintend the contemplated alterations of planes," at a fee of $1500.[80] By February, 1836, Henry McFarlan tendered his bill for $14,458 worth of castings that had already been "furnished in altering the planes."[81]

Three anomalies appear in the records for this time period. First, there is an enigmatic note in the Morris Canal Company Docket File 486 explaining that, in 1833, Lock No. 1 East had been on the lands of E.B. Woodruff of Drakesville. The lock had been removed "when the plane was lengthened." This, of course, would have reduced the number of locks on the Eastern Division.

The second anomaly, however, explains the first, for William Scott deeded to the Canal Company a piece of land 66 feet wide by 100 feet long above the Powerville Basin. This land, according to a notation on the deed, was the location of a new lock, No. 9 East (Docket File 425).[82] Thus the number of locks on the Eastern Division remained the same.

The third anomaly involves the canal's depth. The only mention given to the increased depth of water in the canal was made by new President McLane in a report to the stockholders (June 17, 1836): "The Morris Canal, with its increased depth of water and enlarged and strengthened planes, offers the cheapest and most convenient transportation to New York" for Pennsylvania's "immense coal trade."[83]

McLane was elected in 1836 to succeed James B. Murray, who had filled the office as president of the Company since the death of C.D. Colden on February 7, 1834,[84] and had been elected in his own right in 1835.

McLane's report to the stockholders in June, 1836, actually covered the year 1835, and the work that had been accomplished under Murray's administration of the Canal Company. He took great care to explain to the stockholders that the inclined planes had been "reconstructed so as to afford greater power and security; and are now capable of passing boats carrying 35 ton cargo, or, including the weight of the boat, 45 tons without risk."[85] The new tonnage was made possible by widening the planes by two feet, as well as by making the machinery stronger and safer.

At least *two* boatmen would have been shocked to hear that there was no longer any risk while traveling the Morris Canal's inclined planes. On July 20, 1836, an accident occurred at Plane No. 4 West, at Waterloo:[86] "While one boat was descending, and another boat, ascending, the inclined plane at Old Andover, the shaft of the water wheel gave way, and the machinery being thus freed from all restriction, whirled impetuously, in a ratio corresponding to the speed of the descending boat — which plunged into the basin, with prodigious force, and instantly sunk; while the ascending boat rushed with equal velocity against the lock at the top of the plane and carried it away. It will take some time and much money, to repair the damage. No person was injured, that we heard of."

This type of accident was precisely what had made the board of directors commission Major Douglass and Ephraim Morris "to have made at Philadelphia or elsewhere, a break [sic] or machine for stopping the cars on the inclined planes in their descent in case of accident." Either engineer, or both, was to have the mechanism made "on their own plan, not to exceed $100." They were ordered to report on the cost of having 23 such "breaks" made, "if they should be wanted."[87] Morris later reported on the subject of "breaks or stops for the cars," but it is not recorded whether or not this mechanism was adopted for use.[88] We do know, however, that Morris received a patent in 1838 for the first American-made railroad braking mechanism. At that time he was still employed by the Canal Company, as he would be until 1843.

Morris' planes were the source of interest to visiting engineers. The Scotsman David Stevenson left a complimentary description of "several of the planes on the canal near Newark," which he visited in the company of Major Douglass. Stevenson's book was published in 1838, but its descriptions seem to be of an earlier date. What he saw, he described as lock planes, indicating that it was Morris' design that was then in use. What Major Douglass was doing there at that time remains a mystery to the author.[89]

The boats then in use on the canal, according to Stevenson, were[90]

> *8-1/2 feet in breadth of beam, from 60 to 80 feet in length, and from twenty-five to thirty tons burden. The greatest weight ever drawn up the planes is about fifty tons.* [*The boat car*] *consists of a strongly made wooden crib or cradle . . . on which the boat rests, supported on two iron waggons running on four wheels . . . the cars run on plate-rails laid on the inclined planes, and are raised and lowered by means of machinery driven by water wheels . . . . The railway, on which the car runs, extends along the bottom of the canal for a short distance from the lower extremity of the plane; when a boat is to be raised, the car is lowered into the water, and the boat being floated over it, is made*

fast to the part of the framework which projects above the gunwale . . . . The machinery is then put in motion; and the car bearing the boat, is drawn by a chain to the top of the inclined plane, at which there is a lock for its reception.

The lock is furnished with gates at both extremities; after the car has entered it, the gates next the top of the inclined plane are closed, and those next the canal being opened, the water flows in and floats the boat off the car, when she proceeds on her way. Her place is supplied by a boat travelling in the opposite direction, which enters the lock, and the gates next the canal being closed, and the water run off, she grounds on the car. The gates next the plane are then opened, the car is gently lowered to the bottom when it enters the water, and the boat is again floated.

The second foreign visitor to view Morris' planes was Michel Chevalier, who recorded the fact that the water wheel then in use at Plane No. 9 West was 26 feet in diameter and 8 feet wide![91]

These enlarged and strengthened planes gave investors new reason to buy Canal Company stock, which was now selling at $142 a share, with subsequent asking prices of $145. "But a short time since, this stock was sold for less than $40 per share."[92]

President Murray wrote to the Canal Company stockholders:[93]

It is indeed scarcely credible that such desponding prophecies [of those who speak ill of the canal] should have come from the lips of some of those who at its commencement were willing to subscribe largely to the stock, and being unsuccessful in their application, have from that day opposed or sought to destroy it. It is true that the fluctuations of the tariff have greatly depressed the iron trade, and have checked the prosperity of Dover and other extensive works; but nevertheless our returns show that the tonnage in manufactures of iron during the last year exceed that of the preceding by more than fifty percent, and should the extensive experiments now in progress, under the direction of Dr. Nott, succeed in enabling*

**Nott, as we saw earlier, was president of Union College, and also headed a firm that made patented stoves that would burn anthracite coal. Now he was experimenting with ore furnaces.*

manufacturers to smelt iron ore by means of anthracite coal . . . it is difficult to predict the extent of the advantages which will be secured to the Morris Canal. . . .

Only two breaches had occurred during the last season, "not costing to repair them more than $2000; while the limestone district, heretofore the most vexatious, has not experienced a single accident."[94] In fact, if there "were anything wanting to insure confidence in our prospects, it is to be found in the progressive increase of the Lehigh, Schuylkill, and Union Canals, the two first of which depend almost exclusively on coal for their income."[95]

"The capacity of the Morris Canal," Murray continued, "has been fully proved during the last season, when in one day there were passed over the Boonton plane, without any extraordinary exertion, 97 boats. The period is not far distant when it will be required to employ as much tonnage as its capacity will allow, which, with its present supply of water, is estimated by Benjamin Wright, Esq., and other competent judges, to be 650 boats."[96]

It would be desirable for the Canal Company to pay off its debts, however. Murray continued:[97]

A canal of 100 miles in length, the only one connecting the great commercial emporium of the Western world with a boundless interior, free of debt, and every year adding to its resources, a property that can suffer neither by death, revolution, fire, or flood, it would soon become the repository of the wealthy, enjoy the confidence of the public, and be sought after as an investment with an avidity proportioned to the infatuation that has heretofore looked upon it as valueless.

We [*the directors*] *feel no apprehension for the future; the night of delusion and despondency has passed away; onward is our watchword, and it rests with our fellow-citizens to say how rapid shall be the march.*

We are reminded that this encomium is, after all, a report to the stockholders, an all-out effort to induce even more stockholders to come into the fold. What Murray failed to mention was the fact that the Morris Canal seemed to depend almost exclusively on loans for its income. A January, 1835, act of the Legislature had authorized the company to increase its capital another million dollars, and the Morris took on the appearance of temporary prosperity. Across the Hudson, Wall Street speculators were playing bull- and bear-baiting games with the Morris' floating stock.

In February, 1835, inspired, perhaps, by the glowing report Murray had just issued about the Morris Canal's prospects, the Wall Street stock manipulators attempted to corner the market with contracts to buy Canal Company stock. The New York brokers, at that point, "voted 28 to 24 not to recognize Morris Canal stock contracts and to rescind all the stock engagements."[98] Holders of contracts to buy stock were then called upon to pay their debts. Of course they couldn't, and so faced ruin. On top of all the stock-manipulations credited in the past to the Morris Canal's Wall Street promotors, this one earned for the company the deprecation: "Curses on Morris."[99]

Still, things could have been worse, as Murray pointed out.[100]

> *Applications have already been made, and in some instances, charters obtained, for lateral canals and railroads to connect the Morris Canal with the rich mineral and agricultural districts, replete with water power, along its borders.*

The Morris Canal — as yet — had no reason to consider lateral canals and railroads as threats. A branch canal to connect Bottle Hill (Madison) with the Morris Canal at Montville was looked on favorably by the Canal Company's directors.[101] A charter was subsequently given to the "Bottle Hill and Montville Canal Company," on February 26, 1834.[102] The proposed canal was to commence "at or near the village of Bottle Hill, in the Township of Chatham, in the county of Morris, running through the townships of Hanover and Pequannock, to the Morris Canal, so as to intersect the same" near "the lowest inclined plane at Montville." Moreover, the Bottle Hill to Morris Canal was to be permitted, "by means of a raceway or canal, to connect the waters of the Rockaway river from a point at or near the old Boonton Works, below the Boonton falls, with the said canal, in that section thereof, which will be between said river and the aforesaid termination of said canal." However, waters taken *from* the Rockaway must be returned either *to* the Rockaway, or into the

Passaic or any of its tributaries above the Great Falls at Paterson. Under no circumstances might the waters taken at Boonton be mingled with the waters of the Morris Canal.

As for lateral or connecting railroads, the Morris and Essex (M&E) Railroad was chartered to be built "from Orange in the county of Essex through the townships of Livingston and Hanover to some point on the Morris Canal in the county of Morris."[103] Instead of Orange, however, the incorporators chose an Essex County connection with the NJRR&TCo. at Newark. Chartered on January 29, 1835, the M&E was to receive as much help from Ephraim Beach as the NJRR&TCo. had, three years earlier.

In May, 1836, Beach was in Morristown exploring a "projected railroad from this town to Carpenter's Point, where it is proposed to connect it with the Lake Erie Railroad. At several points he took an elevation, and at no spot did he find it over 60 feet to the mile — thus rendering it certain that the road will be constructed before many years. We learn that surveyors will soon commence operations on the line."[104] Carpenter's Point, no longer on modern maps, is now known as Port Jervis,[105] very near the D&H Canal and Railroad and the Erie Railroad.

Murray ended his *1835 Report to the Stockholders* on a triumphant note.[106]

> *The sufficiency of the supply of water for all present purposes, has been fully tested during a season of unusual drought; and while our neighbors have frequently been totally dry, we have the satisfaction to say, that not a day has been lost along the line of this canal for want of water. Should, however, its increasing business render it important to secure a further supply, a reservoir equal to our present principal source, is within our reach. The procurement of such additional means will become important should the Company succeed in obtaining legislative authority to lease or sell the use of the water now passing from one level to the other.* [*In fact, that authority*] *might be the source of great profit to the Company.*

Besides the 1500-foot fall from the Summit, "a line of two miles in length, where the canal runs by the side of the Passaic at Newark, might be studded with mills and factories, approachable by the river on the one side, and fed by the canal on the other."[107]

The ink on Murray's *1835 Report to the Stockholders* was not dry before Ephraim Beach received a letter from Robert Gilchrist, Canal Company Cashier, requesting a "certified description" of the canal, particularly as to its capacity. When at the end of a week Beach had not replied, Gilchrist wrote again on February 26, 1835:[108]

> *Dear Sir,*
>
> *I have been anxiously expecting to receive the Certificate respecting the capacity of the Canal, etc. It is much wanted, and I am desired by Mr. Brooks to say he thinks you might attend to all small matters of this kind which in his opinion can be done in an hour.*
>
> *Do my dear sir send in this paper as soon as you possibly can. I believe it is of some importance to him in some negotiations now going on — and it would be a pity that he should want that or any other paper or thing to aid him he has done much and is doing all he can — Send it tomorrow if you can.*

Beach's reply was dated February 24, 1835, and had apparently been on its way while Gilchrist's second letter was traveling to Beach. Beach's health had been the cause of his tardy answer:[109]

> *I deem it proper to state that the location of the Morris Canal and all its works, together with their plan of construction, were made by me and executed under my immediate supervision and direction with the exception of some of the Inclined Planes, which although not under my immediate direction, were constantly subject to my observation and inspection. Nor would I derogate any from the services rendered me by the valuable young men who aided me during the progress of the work. And although the work was executed at a time when situation of the Company required the most strict adherence to economy, permanence and durability were subjects of the first consideration.*

At this time, Beach reported that the canal from the Passaic at Newark to the Delaware at Easton was about 90.5 miles long. The prism was 32 feet wide at top water line, 20 feet wide at bottom, with 4 feet of water. All culverts and aqueducts were made of "stone masonry laid in water cement, except in cases where there was not sufficient space between the bottom of the canal to admit of stone arches, in which case wooden trunks or aqueducts were constructed supported by permanent stone piers and abutments." The 6 stone dams along the line of the canal had been built of stone masonry. Wooden trunks served both as aqueducts and waste weirs.

Twenty-four lift locks overcame a total of 213 feet: 156 feet in the 17 eastern locks; 57 feet in the 7 western locks. All had been made of stone laid in water cement. Five guard locks had been placed in 5 of 6 dams. One tide lock at Newark, made of wood, overcame a maximum of 10 feet at lowest tide.

Of the 23 inclined planes, 12 on the Eastern Division overcame 748 feet of elevation, and 11 on the Western Division overcame 691 feet — a total of 1439 feet. "Although Inclined Planes are said to be yet in experiment, experience upon the Morris Canal has so far tested that experiment as to place their utility beyond a doubt in facilitating transportation over large elevations not only for boats with their cargoes of the size of those of the Morris Canal but for boats with their cargoes of twice their size."

The aqueduct at Little Falls was given special praise by Beach, who described it as a huge stone arch of eighty feet span, "with two small arches, one on each side of about ten feet span each for the accommodation of the water power of the river." The work had been expensive, but the structure was "bold in its design," and Beach was sure that the arch would be "a durable and splendid monument of the persevering energies of the company and of the fidelity and skill of the contractors who constructed it."

As for the capacity of the canal to handle tonnage, he felt that, given a sufficient number of boats and an ample supply of water, there was no doubt that "the resources of the country" would supply business to the Canal Company even at low rates of toll.

While he knew of no mill owners in the canal's vicinity who "had any just cause of complaint" over the use of water, Beach felt that it would be "very desirable" to increase the Canal Company's water supply. On the Western Division, there were several spots for reservoirs "between Hackettstown and Stewardstown [sic]," while the Eastern Division had one spot which would offer not only an extensive reservoir, but would afford the extension of "navigation into a different district by means of a feeder." Without naming the potential reservoir, Beach was not doubt referring here to Greenwood Lake.

The size of the boats then in use on the canal, plus the "facilities of passing the elevations," determined the capacity of the Morris Canal. Locks 9 feet wide and 75 feet long obviously regulated the size of the boat to "about 8-1/2 beam and seventy feet in length, competent to carry upon the Canal a cargo of thirty tons" maximum. Beach preferred to base his calculations on the average 25-ton cargo, that "being the load usually contemplated for Morris Canal boats." He calculated that 384,000 tons would be carried east to New York during a 240-day season. Return freight west to points along the canal would probably amount to 50,000 tons, for a total anticipated tonnage of 434,000 annually.

All this, however, depended upon "a competent supply of boats," and Beach estimated that "at least 800 boats" would be necessary to carry that much tonnage during a 16-hour day, 240-day boating season. A 24-hour day, 240-day season would require 1000 boats for through trade, plus another 100 or 125 more "for the intermediate business." At those rates, Beach reasoned that income would be:

For Transporting 384,000 tons 90 miles @ 1¢ per ton-mile	$345,600
Return freight 48,000 tons 90 miles @ 1-1/2¢ per ton-mile	55,800
Intermediate freight 65,000 tons 40 miles @ 1-1/2¢ per ton-mile	39,000
	$440,400

With the completion of the canal's extension to Jersey City, Beach expected that "probably ten percent at least" could be added to the annual receipts for a total of $484,440. That should pay a "large dividend" on the stockholders' investment, and he was sure that all the canal's discouragements" would "yet reap a rich reward for their untiring perseverance."

What Ephraim Beach had told Robert Gilchrist and Samuel R. Brooks, both directors of the newly formed Morris Canal Transportation Company, was exactly what they needed to know in order to organize their new enterprise for maximum operating efficiency. Since early 1833, the Canal Company had been angling and lobbying for this extension of its powers. The new transportation company would be quite an enlargement, indeed.

REFERENCES

1. Morris Canal and Banking Company *Minutes*, December 4, 1832 (hereinafter cited as MC&BCo. *Minutes*).
2. *Ibid.*, April 6, 1830.
3. Morris Canal and Banking Company *Docket Files* (hereinafter cited as MC&BCo. *Docket Files*), letter from Ephraim Beach to the board of directors, May 31, 1830.
4. Alvin Harlow, *Old Towpaths* (N.Y.: 1926), pp. 54, 298.
5. MC&BCo. *Docket Files*, letter from Ephraim Beach to the board of directors, May 31, 1830.
6. *Ibid.*
7. Charles H. Winfield, *History of the County of Hudson, New Jersey, From Its Earliest Settlement to the Present Time* (N.Y.: 1874), pp. 331-7.
8. *History of Jersey City, New Jersey: A Record of Its Early Settlement and Corporate Progress*, compiled for the (Jersey City) *Evening Journal* by Alexander McLean (Jersey City: 1895), p. 26.
9. MC&BCo. *Minutes*, July 20, 1830.
10. *Ibid.*, March 26, 1834 and April 8, 1834.
11. *Ibid.*, May 9, 1834.
12. *Ibid.*, May 28, 1834.
13. *Ibid.*, March 26, 1834.
14. *Ibid.*, July 10, 1834.
15. *Paterson Intelligencer*, Jan. 20, 1830.
16. MC&BCo. *Minutes*, May 13 and 15, 1834.
17. *Ibid.*, May 28, 1834.
18. MC&BCo. *Docket Files*, report of Ephraim Beach to the board of directors, June 8, 1834.
19. MC&BCo. *Minutes*, November 3, 1835.
20. *Ibid.*, March 15, 1836.
21. *Ibid.*, March 31, 1836.
22. *Ibid.*, March 10, 1837.

23. *Ibid.*, March 14, 1837.
24. *Paterson Intelligencer,* March 12, 1834.
25. Newark *Sentinel of Freedom,* July 15, 1834.
26. MC&BCo. *Minutes,* March 20, 1834.
27. *Paterson Intelligencer,* May 28, 1834.
28. Robert G. Albion, *The Rise of New York Port* (N.Y.: 1939), p. 137.
29. *Directory of Newark, N.J.,* 1835-36, p. 23.
30. Newark *Sentinel of Freedom,* May 7, 1834.
31. *Ibid.*
32. MC&BCo. *Minutes,* November 6, 1834.
33. MC&BCo. *Minutes,* April 1, 1867.
34. Newark *Daily Advertiser,* April 5, 1837.
35. MC&BCo. *Minutes,* September 13, 1834.
36. *Ibid.*, September 18, 1834.
37. *Ibid.*, April 8, 1834.
38. *Ibid.*, August 19, 1834.
39. *Ibid.*, March 20, 1834.
40. *Ibid.*, April 2, 1834.
41. *Ibid.*, March 30, 1834.
42. *Ibid.*, July 2, 1834.
43. *Ibid.*, July 22, 1834.
44. *Ibid.*, July 23, 1834.
45. *Ibid.*, July 8, 1834.
46. Easton *Centinel,* August 1, 1834.
47. *Report of the President and Directors of the Morris Canal and Banking Company, to the Stockholders* (N.Y.: 1835), p. 4 (hereinafter cited as *Report to the Stockholders, 1835*).
48. *Laws of New Jersey,* 1834.
49. *Report to the Stockholders, 1835,* p. 4.
50. MC&BCo. *Minutes,* January 1, 1834.
51. *Ibid.*, January 28, 1834.
52. *Ibid.*, March 18, 1835.
53. Newark *Daily Advertiser,* June 12, 1835.
54. MC&BCo. *Minutes,* May 7, 1835.
55. Newark *Daily Advertiser,* July 16, 1835.
56. *Paterson Intelligencer,* March 6, 1833.
57. Newark *Daily Advertiser,* May 3, 1837.
58. MC&BCo. *Minutes,* April 30, 1834.
59. *Ibid.*, January 1, 1834.
60. *Ibid.*, June 4, 1834.
61. *Ibid.*, December 29, 1830.
62. *Ibid.*, January 29, 1833.
63. *Ibid.*, May 1, 1833.
64. *Ibid.*, September 13, 1834.

65. *Report to the Stockholders of the Morris Canal and Banking Company, April 1, 1861* (N.Y.: 1861), p. 14.
66. *Report of the President and Directors of the Morris Canal and Banking Company to the Stockholders* (Jersey City: 1867), pp. 6-7.
67. *Report to the Stockholders, 1835,* p. 9.
68. *Ibid.*
69. MC&BCo. *Minutes,* November 12, 1832.
70. *Ibid.,* October 7, 1834.
71. *Ibid.,* April 14, 1834.
72. *Ibid.,* May 9, 1834.
73. *Ibid.*
74. *Ibid.,* July 2, 1834.
75. *Ibid.,* July 8, 1834.
76. *Ibid.,* September 10, 1835.
77. *Ibid.,* August 19, 1834.
78. *Ibid.,* October 20, 1834.
79. *Ibid.,* October 1, 1835.
80. *Ibid.,* October 31, 1835.
81. *Ibid.,* February 1, 1836.
82. MC&BCo. *Docket Files,* deed dated April 12, 1837.
83. *Report of the President and Directors of the Morris Canal and Banking Company, to the Stockholders* (N.Y.: 1836), p. 17 (hereinafter cited as *Report of the President, 1836*).
84. *Paterson Intelligencer,* February 12, 1834.
85. *Report of the President, 1836,* p. 14.
86. *Paterson Intelligencer,* July 27, 1836.
87. MC&BCo. *Minutes,* February 17, 1832.
88. *Ibid.,* April 10, 1832.
89. David Stevenson, *Sketch of the Civil Engineering of North America* (London: 1838), pp. 127-8.
90. *Ibid.,* pp. 128-9.
91. Michel Chevalier, *Histoire et Description de Voies de Communication aux Etats Unis* (Paris: 1840), n.p.
92. *Paterson Intelligencer,* January 14, 1835.
93. *Report to the Stockholders, 1835,* pp. 3, 7-8.
94. *Ibid.,* p. 8.
95. *Ibid.,* p. 10.
96. *Ibid.,* pp. 10-11.
97. *Ibid.,* pp. 13-14.
98. Earl J. Heydinger, "Railroads of the Lehigh Coal and Navigation Company: Group VI, the Little Schuylkill . . . ," *Railway and Locomotive Historical Society Bulletin, No. 108,* April, 1863, pp. 28-9.
99. Wheaton J. Lane, *From Indian Trail to Iron Horse: Travel and Transportation in New Jersey, 1620-1860* (Princeton, N.J.: 1939), p. 236.

100. *Report to the Stockholders, 1835,* p. 6.
101. MC&BCo. *Minutes,* January 15, 1834.
102. *Laws of New Jersey,* 1834, p. 129.
103. *Jerseyman,* October 15, 1836.
104. *Paterson Intelligencer,* May 25, 1836.
105. Manville B. Wakefield, *Coal Boats to Tidewater* (Grahamsville, N.Y.: 1971), p. 3.
106. *Report to the Stockholders, 1835,* pp. 9-10.
107. *Ibid.,* p. 10.
108. MC&BCo. *Docket Files.*
109. *Ibid.*

11

The Enlargement Myth

THE "enlargement" of the Morris Canal is a myth. Enlargement was continual — at least until 1860 — taking many and varied forms, the first of which was the obvious one of length, as when, in 1826, the canal was lengthened beyond Aquackanonk, its earliest planned terminus, to reach Newark. The 1828 supplement to the Canal Company's charter authorized further lengthening of the canal to Jersey City. William Wood's construction of a navigable cut at Lake Hopatcong is also probably best categorized as an enlargement of the canal's length.

Another amendatory act in 1829 enabled the Company to borrow half-a-million dollars for an enlargement of capital, which was further enlarged by an act of 1830 that made possible the $750,000 "Dutch" loan. Then the act of January, 1835, had allowed the Company to enlarge its capital stock by an additional 10,000 shares ($1,000,000).

But enlargement also took the form of "extended" authority, when an act of 1832 gave the Canal Company the right to protect its own waterway and punish those who obstructed or damaged it. And there *were* obstructions in the canal in those earliest of days. The Company had promised "to indemnify the owners or captains of boats navigating the canal from all damages in consequence of removing any fence or other obstruction that has or may be placed in or across the canal, by any person or persons other than an authorized agent of the company."[1] That promise was conditional, however. Boatmen were required to use the professional legal services of either "Mr. Ogden at Paterson," or "Mr. Dodd or Mr. Halsey at Newark" in order to enjoy the Canal Company's protection.[2]

This act of 1832 enabled the Company to order Martin Hulsizer at Plane No. 8 West to get his eel weirs out of the plane's raceway. Roswell Mason was then instructed to see to it that all similar contraptions were removed from the canal "or in any raceway on the line, or at the outlet of the Great Lake."[3]

There were "enlargements" in the Canal Company's responsibilities as well. The Town of Newark, for example, filed a bill against the Morris Canal Company "for neglecting to repair its bridges."[4] The Company rose to the occasion, however, perhaps because there was some advantage to itself to do so. The board of directors decided "that the space between the Rail Road Bridge and the Bridge on the East Side of Broad Street in the City of Newark, be bridged" at the Canal Company's own expense.[5]

Employee housing, already discussed in Chapter 10, was another form of enlargement. Still another was in dockage facilities. The Company bought a piece of land from C. Van Emburgh, whose Newark property was adjacent to the Newark Dock Association. The Company agreed to let Van Emburgh contract for the dock, to be completed by July 1, 1837, at a cost of $10 per running-foot front. Roswell B. Mason was to have the final say on the quality of workmanship. The dock was to occupy a spot "from the southerly end of Newark Dock Association, south 300 feet at such a distance from the shore as will give at least 8 feet water at ordinary low tide and as nearly parallel with the Newark Dock Association's wharf, and a return bulkhead to connect with present upland; the whole to be at least two feet above ordinary high water mark."[6]

The canal's capacity had grown too, as we have seen, with the enlarged and strengthened planes. By mid-1836 there were "400 boats upon the canal," with many others being constructed for use "in the course of the season."[7] Boats cost more money by then, and the board of directors resolved to "limit the price to be paid in any future contracts for boats to $400 for each boat."[8] Of course, that limitation didn't apply to the new packet boat that was to be built later for the use of the Company's directors and guests. "The sum of $1000" was appropriated to build that packet "in conformity to the views of a Committee."[9]

Still another form of enlargement was in privileges, as in the momentarily unsuccessful attempt to use canal water for manufacturing. Canal Company President Murray had let the stockholders know that that was the goal of certain petitions before the State Legislature.

The *least obvious,* but not the least important, form of enlargement lay in the area of influence, or scope. Consider (1) the implications of the Morris Canal Transportation Company; (2) the enlarged scope implied by the term "lateral canals and railroads"; (3) the meaning of the new spirit of peace and harmony between the Morris Canal Company and the residents of Paterson inaugurated by their joint venture of building *the* railroad from Bergen Junction to Jersey City — the railroad that the Paterson and Hudson River (P&HR) Railroad Company would use jointly with the New Jersey Railroad and Transportation Company (NJRR&TCo.). But most of all, consider what it must have meant to the Canal Company to have as its new president Louis McLane, with his backing of the Second Bank of the United States. *That* was an enlargement of influence.

While Roswell L. Colt of Paterson's SUM and Cadwallader D. Colden of the Morris Canal Company had had a history of harsh words and harsher dealings, the appearance of virtually a whole new cast of characters promised a whole new interpretation of the whole new concept of mutually beneficial cooperation. This drama would contain no protagonists or antagonists, but simply a chorus.

We have seen the Morris Canal Company become the Morris Canal and Banking Company, much to George P. Macculloch's distress. Then we saw it become the Morris Canal, Banking, and Transportation Company. Now we shall see it try to become the Morris Canal, Banking, Transportation, Manufacturing *and* Investment Company. What is more, we shall see that these aspirations will be fostered by virtually every existing railroad in the state, including the Camden & Amboy (C&A), and some "foreign" railroads.

Fraternal twins had been born that February, 1830, day in Trenton, as the State Legislature engendered both the Delaware and Raritan Canal *and* the C&A Railroad. Within a year, the two companies had reached the point of mutual destruction. In an effort to head off disaster for the twins, their legislative parents passed the so-called "Marriage Act," creating the Joint Companies (February 15, 1831). By means of that act, both companies were protected from each other and from other railroads being built in their bailiwick. In return for this saving legislation, the State of New Jersey received 1000 shares of stock from each of its offspring. No railroad, from that time on, could be built across the state between Philadelphia and New York. By eliminating the competition between the canal and the railroad, the State ushered in an era of cooperation. However, it had also made of itself, plus the Joint Companies, partners in a virtual monopoly of transportation.

I submit that the same spirit of cooperation prevailed in northern New Jersey between the Morris Canal and the various small railroads that were springing into existence between 1832 and 1836. However, there was a difference. This northern cartel gave birth to itself and would maintain itself without the dubious benefit of state intervention. This entire concept must be examined, step-by-step, so that its validity may be seen. This was the most subtle "enlargement" of them all.

When the Canal Company had petitioned the legislature for an amendment to its charter so that it might use the potential water power of its canal, it also asked that its board of directors be increased from 15 to 23, to be appointed for life. The petition also asked for the right to establish a branch of its bank in Newark.[10] The petition was denied.

An amended petition was then forwarded to Trenton, asking that the number of directors be increased from 15 to 23 and grouped in "classes" to serve different lengths of time, so that *all* the directors would not have to be elected annually by the stockholders. The petition *still* asked for the right to use surplus canal waters for manufacturing, but it now included a request to take Greenwood Lake into the canal by way of a navigable feeder and the Pompton River. Tolls, of course, would be collected for navigation and freight on this feeder canal. This petition yielded results, and the charter was amended on March 5, 1836.

The charter supplement, or amendment, provided for this new leadership first of all by increasing the number of directors to 23 men. These men were to be divided into five "classes," each of which would serve a different length of time:

Class	No. of Directors	Term
1	3	1
2	5	2
3	5	3
4	5	4
5	5	5

Thus only one class of directors, three members of the board, would come up for annual reelection. The other classes would serve 2-, 3-, 4-, and 5-year terms so that a certain amount of stability and permanence was given to the governance of Canal Company affairs. It was the board members themselves, however, with their varied interests, who gave a complementary dynamism to the Company at the same time. The "enlargements" of the Company's scope that were to take place during this period are a case in point.

Contingent upon the passage of that law was an agreement between the Morris Canal and Banking Company (MC&BCo.) and the SUM, which had already been signed on February 17, 1836, only a few days before the Canal Company's petition became law. The agreement concerned the use of the waters of the Passaic River and all of its tributaries. And here can be seen the new spirit of peace and harmony, of mutually beneficial cooperation at its best.

The last time the SUM had learned that the Company intended to use SUM-owned water to establish mill seats along the line of the canal, John L. Sullivan was employed as the SUM's champion. The two companies ran head-long into each other — an irresistible force meeting an immovable object. But by 1836, however, an almost complete change in the cast of characters gave an almost complete change of direction to the drama.

The new spirit of harmony between the SUM and the Canal Company is a bit less shocking when we learn that Philemon Dickerson, Esq., was both the counsel for the SUM and a director of the MC&BCo. Moreover, he had been the first president of, and was a stockholder in, the P&HR Railroad. On January 12, 1836, the Company's board of directors received a letter from Mr. Dickerson, containing a proposal from the SUM which suggested a compromise over the use of Passaic River waters. The letter was referred to Chief Engineer Beach for his professional opinion as to whether or not such a compromise would be detrimental in any way to the Morris Canal.

In what may well have been his last official act for the Canal Company as Chief Engineer, Beach reported on the SUM's proposal with characteristic thoroughness.[11] Since the canal had been started, he had been in the habit of taking measurements of the Passaic River and its tributaries, particularly during dry seasons, to furnish himself with data about any effects canal operations might have on water power, particularly at Paterson. He had found that the most accurate readings could be made at Little Falls, about 4 or 5 miles above Paterson. There, "upon the reef at the head of the Messrs. Miller's Pond," he measured the lowest water in 1827. According to his gauge, there was at that time

passing over the reef in 24 hours	*25,166,016 c. ft.*
A 22 feet aperture under 30 inches head will discharge in 24 hrs.	*23,760,000 c. ft.,*
leaving to supply absorption + evaporation and to guard against more excessive drought	*1,406,016 c. ft.,*

which I think none too much taking into consideration the liability of streams to fail as the country becomes improved. 23 feet aperture will discharge 24,840,000 c. feet; 24 feet, 25,920,000. I see no material objection to discharging the 2 square feet of water from the Basin upon the top of the hill —

A 2 feet aperture under 30 inches head will in 24 hours discharge	*2,160,000 c. ft.*
To supply the Canal from Paterson to Newark for absorption + evaporation 75 feet per mile per minute	*1,512,000 c. ft.*

Amt. to supply lockage once in 10 minutes 16 hours per day — *648,000 c. ft.*
The 2 feet for the Society absorption evaporation + lockage — *4,320,000 c. ft.*
To pass which through the Canal will create a current of about 1/3 of a mile per hour.

Supposing that 22 feet would be constantly supplied from the natural streams, Beach estimated that creating a 3-square-mile reservoir at Long Pond (Greenwood Lake) would yield another supply of water of about 836,252,000 feet. That water should allow Paterson's manufacturers to continue operating for 120 days through an ordinary dry season. Adding this to the 6,969,600 feet supplied throughout the rest of the year by the ordinary flow of the area's streams, the "flow from the summit to Paterson will be 2/3 mile per hour and from Paterson to Newark 1/2 mile per hour, which will be well compensated for by the water power it will afford."

To repeat: This January 16, 1836 document may well have been Beach's last "official" report for the Canal Company. When the charter amendment was passed on March 5, 1836, the Legislature named a new board of directors, who then held their own election on March 30. When all the appointments had been made, Beach's name was conspicuously absent from the list of employees; Roswell B. Mason had been appointed Chief Engineer.[12] Ephraim Beach had already gone on to begin building the Morris and Essex (M&E) Railroad.

Meanwhile, armed with the Chief Engineer's report, the board of directors authorized the president to conclude the agreement with the SUM. That agreement, signed on February 17, 1836, was to be contingent upon the passage of the act enabling the Canal Company to enlarge its water supply system by taking in Greenwood Lake as a reservoir.[13] The SUM-MC&BCo. agreement specified that "the Morris Canal and Banking Company was to discharge water from their Canal, through a basin to be formed by the SUM and by them kept watertight, in the natural ravine, on the Rocky Hill adjoining the Canal near the Falls."

Anticipating the benefits to be derived from the additional supply that would be forthcoming for their mills at Paterson when Greenwood Lake's waters were conducted to them by way of the Morris Canal, the SUM immediately set out to comply with the terms of that agreement.[14]

Note that it was the SUM's proposal that resulted in this agreement. Note, too, that the agreement was proposed and finalized *before* the Canal Company's charter amendment was passed on March 5, 1836, and in fact was to become effective only upon passage of that amendment.

The SUM built a new 8-foot-high watertight masonry dam across the Passaic River, approximately where the present Spruce Street (or Wayne Avenue) bridge stands. The dam was intended to create more pondage for feed water which was to be taken into the SUM's raceways. It would supercede the old 2-foot-high wooden dam the Society had built.

Earlier the SUM had a feed system of two raceways, later known as the "Middle" and "Lower" races or canals. In 1829, after losing its first battle with the Canal Company over the use of the Rockaway River, the SUM had built a third, or "Upper," raceway.

According to the SUM-MC&BCo. agreement, the SUM was to make its Middle Raceway watertight, so as to maintain in it a specified level of water at all times. The waste gates of that raceway were to be so adjusted that "the water should stand in the same not less than thirty inches above a monument set therein; that being the ordinary height of the water in that race. That they should so regulate the gates to the mills on that race, that not more than twenty-two square feet of water should be drawn from the same, under a pressure or head of not more than thirty inches, measured from the center of the apertures to the surface of the water."

A lock gate, or gauge gate, was to be placed at the end of the natural ravine on Rocky Hill, about 43 feet distant from the canal. In the lock gate was to be installed "an aperture or gauge as required in the contract for the measurement of the said three feet of water." The end of the ravine had originally been filled up by the Canal Company, and the towing path had been laid across the ravine, atop the fill. Now, the ravine was cleared out, in accordance with the February, 1836 agreement, and a towpath-bridge was built to span the mouth of the natural cut so that water could flow uninterruptedly "from the canal through the ravine over the lands of the Society to said gate or gauge."

Rocky Hill is today's Mulberry Hill, at the intersection of Grand Street and Stony Road. The new upper basin built by the SUM was at the foot of the hill which holds today's Stony Road Reservoir. Water from the canal filled the basin, from which it flowed into the Passaic River above the new 8-foot dam, to be taken into the SUM's upper raceway. As a result of this anticipated increase in water power, the SUM was to lay out 12 to 15 new mill sites, all of which would be supplied, in part, by water from Greenwood Lake, by way of the Morris Canal.[15]

Upper and Middle Raceways of the SUM, Paterson. Stony Road Reservoir, filled by water from the Morris Canal, is atop the hill at the left (Report on Water, Vol. III of the State Geologist Report, 1894).

Aside from supplying the SUM's raceways with water, the Canal Company may have been the supplier of water for domestic purposes in Paterson. The SUM had apparently announced plans to make arrangements with the Canal Company to furnish the town of Paterson "with a supply of water for domestic and other purposes." Water was to be piped from the canal "or from the upper basin, near the river."[16] Whether or not the arrangement was ever made is a subject for further research.

The spot where the Morris Canal and the SUM's upper basin connected is clearly noted on the C.S. Van Wagoner map of Paterson, dated May 1, 1840. At that point the two companies were destined to clash again in 1845, as we will see. However, the map does not help to explain how the canal boat *Sally Ann* could have gotten into the Passaic River in August, 1857, when she was swept over the SUM's dam. Fortunately for the *Sally Ann,* the water level in the river that year was not high enough to carry her over the Great Falls.[17] According to some notes of the late 1800s, left by Paterson's former mayor, William Gledhill, a wing gate from the Upper SUM raceway into the river did exist. At least one boat, captained by a Mr. Parker, actually started from the upper canal, went through the wing gates, traveled up the Passaic, and returned with a load of stone from Little Falls. Mr. Gledhill, however, mentioned no date for the event. Still, it is possible that it was through this wing gate that the *Sally Ann* gained access to the Passaic River.

The Canal Company's charter amendment provided for yet another form of enlargement — that of privilege. The Company had asked for the right to engage in manufacturing with the use of canal water, but the Legislature had granted only the right to lease canal water to *others* who would then be permitted to engage in manufacturing. "The privilege of farming out the water of the canal between Boonton . . . and Newark, on the Passaic, for manufacturing purposes" was now built into the charter. An "extensive manufactory" had already been established at a plane at Newark, employing "not less than 200 hands," and paying water-rent to the Canal Company. The charter supplement authorized a further enlargement. The Canal Company's capital stock was to increase to 6000 shares to complete the Jersey City extension and finance the Pompton Feeder.[18]

At the outset of 1835, the Canal Company directors had asked their Canal Committee to report on "the expediency of procuring immediately the right to take and use the waters of the Long Pond (Greenwood Lake)."[19] The committee thought it would be most expedient, and another committee composed of Company President Murray, Henry W. Hicks, John Haggerty, and Washington Irving visited Greenwood Lake "to report on the expediency of introducing the waters of said pond into the Morris Canal by means of a feeder from Pompton near Judge Ryerson's."[20]

Among the iron industries that would be better served by a navigable feeder up to Pompton Falls were six forges and furnaces, from Pompton up to Charlottenburg, and nine mines. They were all owned by Peter M. Ryerson and his brothers, Jacob M. and Martin J. The Ryersons had also inherited their late father's lucrative interest in manufacturing hoops, poles, and barrel staves. It was said that they employed "thousands of men" altogether.[21] Peter M. Ryerson, Jr. had been born in 1798. He was but 29 years old when he became a Canal Company director in 1827. His multifaceted business interests were characteristic of the men who would now, in 1836, help hold the reins of the Canal Company.

An earlier desire of Peter M. Ryerson to locate his manufacturing interests closer to a source of reliable transportation had caused him to become one of the incorporators of the Dundee Manufacturing Company in 1832. The story of that company is not really relevant to our story except in that it illustrates Ryerson's need for transportation facilities. The reader will recall that Beach's very early report on what route the canal should follow on the Eastern Division had focused on a navigable feeder at Pompton. When it was not built, Ryerson undoubtedly decided to move at least his non-iron industry to the Passaic River where sloops could transport his products. His interest in transportation for his goods also led Ryerson to become an investor in the P&HR Railroad, and he was one of its directors.

The prospect of a canal to Pompton Falls led Ryerson to rebuild his furnace in 1837. He introduced the hot-blast charcoal method there, but when it proved unprofitable, he converted the furnace to a rolling mill in 1839.[22]

The Ryerson brothers deeded lands at the outlet of Long Pond to the Canal Company for a new dam on March 27, 1837. The only exception to the deed was that Jacob Ryerson was to be entitled to rebuild the old dam to service his Long Pond Iron Works. He was also to be entitled to take as much water from the lake as he needed to operate the iron works. If in the future the lake was no longer to be used "to save water," his rights in those waters and the waters themselves were to be "expressly reserved for J.M. Ryerson, or his heirs and assigns," but not for his successors in business, however.[23]

Even with the Pompton Feeder, Peter M. Ryerson was apparently not successful in his iron business. He was hit hard by the various iron tariffs that were plaguing the American iron industry in general. The Tariff of 1847 was to affect all of his enterprises; the Panic of 1857 was to destroy them. He was then compelled to abandon his businesses and surrender his property to his various auditors. He had already conveyed his interest in the Dundee Company to the SUM on May 22, 1850.[24]

Ransom claimed that it was through Peter M. Ryerson's efforts that the Morris Canal feeder was built.[25] Canal Company records tend to support that claim, for Ryerson was appointed a superintendent of construction, particularly for the rebuilding of the inclined planes. It is almost a certainty that he supervised the construction of the feeder canal that would enable him to ship Ryerson products and to receive fuel for his iron works.

The new dam at Greenwood Lake raised the water level eight feet so that ample water could be conducted into the Pompton Feeder. No descriptions of the actual construction of the feeder are known to exist, but George W. Keppler left a brief, undocumented commentary on what it must have been like:[26]

> *Within the six-mile extent of feeder canal, two dams and 3500 feet of guard bank had to be constructed. The experience gained on the Jersey City extension proved helpful in digging the feeder through the Pompton River flats, which were often flooded during high water. The banks of the canal as it ran along the edge of the river had*

to be built high enough to prevent the river from entering the canal when its waters were swollen. Most of the construction consisted of throwing up parallel banks across the flats and reinforcing the outsides of those banks with heavy timbers and stones. The channel was puddled with heavy clay to prevent seepage, and the many smaller streams that criss-crossed the flats were diverted and channeled under the canal by means of stone culverts.

At the head of the feeder canal, where the artificial channel met the impounded river at the dam, a guard lock was installed to allow boats to enter and leave the river on their way to and from the Pompton ironworks. The guard lock also helped regulate the amount of water let into the feeder canal. The whole canal extension was begun in 1836 and took a little more than a year to complete.

Ryerson, of course, was not the only beneficiary of the Canal Company's new feeder. The SUM's mills were to receive more water from Greenwood Lake via the canal at Paterson, as we saw earlier. Paterson's prospects, too, had begun to look very promising.

One enthusiastic supporter of the Canal Company-SUM agreement for the use of Passaic River waters had written that within 12 to 17 miles of New York, there would now be water power enough to drive "150 to 200 mills." What, he asked, "are the prospects of Paterson? Who can believe that the vast resources which are here concentrated are always to slumber in inactivity? In what part of this country is there open such a wide field to make permanent investment?"[27]

Ryerson was not the only one of the 23 new board of director members to see those opportunities to make "permanent investments" because of the new charter amendment. Andrew Parsons, the Canal Company's first toll collector at Paterson, and owner of a canal basin at about Clay and Marshall Streets there, was quick to see the advantages of the new arrangement. He applied for permission to use canal water to manufacture bricks at Paterson.[28] He also wrote to the Canal Company to suggest that the MC&BCo. open a branch bank at Paterson. The Canal Company's effort to establish such a branch at Newark had failed, but Parsons wanted the people of Paterson to "apply to the legislature for an extension of the Banking Capital of the Morris Canal $500,000 to be used exclusively for the benefit of this place." If Patersonians did *that,* Parsons wanted to know if the Canal Company "would agree to have a Branch at Paterson."[29]

Clearly, expectations of great progress and profit were engendered by the Canal Company's charter amendment, while its new leadership had infused confidence in its investors.

When James B. Murray wrote his 1835 report to the stockholders of the Canal Company, he had remarked that applications had "already been made, and, in some instances, charters obtained, for lateral canals and railroads to connect the Morris Canal with the rich mineral and agricultural districts, replete with water power, along its borders."[30] The implications of that comment are now becoming clearer. The ramifications are immense.

In January, 1834, a Mr. Bishop had written to the Canal Company about "a canal to intersect the Morris Canal at or near Montville." The letter was read to the board of directors, who instructed Cashier Gilchrist to inform Mr. Bishop that they were "favorable to the project."[31] The Bottle Hill and Montville Canal was chartered on February 26, 1834. The incorporators of that company were:[32] Jacob Wilson, director of the Dover Union Bank; Noadiah P. Thomas, a cotton manufacturer whose works were located near the Eden Mill on the Whippany River; Walter Kirkpatrick, son-in-law of Lemuel Cobb; Benjamin L. Condict, part owner of a bark mill and tannery in Montville; and William Britten, a director of the State Bank at Morris and soon-to-be director of the Morris and Essex (M&E) Railroad. Their canal was to connect Madison (Bottle Hill) with the Lower Montville Plane (No. 9 East), by way of Hanover and Pequannock Townships. It was never built.

Another lateral canal was proposed by Samuel Swartwout, a Canal Company director in 1834 and 1835. He had asked for a loan from the Company's bank and pledged some of his property at Hoboken as security. He also proposed a branch of the Morris Canal to that place.[33] That canal was never built, and Swartwout went on to become involved in an "unsavory scandal" when, as an appointee of President Jackson to the New York Customs House, he "diverted more than a million dollars to some of his private ventures," then skipped to Europe when he was discovered.[34]

Over the years, Swartwout had been a business partner of C.D. Colden, William Bayard, Joseph G. Swift, Henry Eckford, David B. Ogden, Charles Kinsey and others. Most of these men had been Canal Company directors or supporters, and several of them, as we saw, had also become involved in unsavory financial dealings. It would be interesting to know whether or not Swartwout's proposed canal branch to Hoboken was one of the private ventures to which he had planned to divert Customs House funds. Swartwout was also involved in the Bergen County Railroad, chartered in February 17, 1836, to connect Hoboken with Hackensack. Among the incorporators with Swartwout were James B. Murray and James A. Stevens.

The Orange and Hanover Railroad was also chartered during this era, on February 21, 1837. Among the incorporators were Colonel Joseph Jackson and Calvin Howell, assistant to Ezekiel Kitchell in building the Canal Company's first experimental inclined plane at Rockaway. This railroad was supposed to connect with the Morris Canal at Rockaway and with the M&E Railroad at Orange.

On the other side of the State, the town of Belvidere was anticipating a connection with the Morris Canal. Like Morristown, Belvidere had been left out of the main stream of things when the chosen canal route failed to accommodate its citizens. The reader will recall the proposed branch canal that was being surveyed to Belvidere, but like the Morristown area's "Bottle Hill" canal, it was never built. As Morristown sought a rail connection with Newark, Belvidere sought one with Trenton. Thus, the M&E and the Belvidere Delaware (Bel Del) railroads were chartered on January 29, 1835, and March 2, 1836, respectively. The M&E was connected with tidewater by way of an agreement to use the tracks of the NJRR&TCo. from Newark to Jersey City. The Bel Del was to have a connection with the Lackawanna Valley coal fields by way of a railroad that had been contemplated in Pennsylvania since 1826, but just hadn't been able to get itself built. This is no forced analogy between Morristown and Belvidere. As we will see, the efforts of both of those towns gave birth to the all-powerful Delaware, Lackawanna and Western

(DL&W) Railroad. And here we must digress for the sake of clarity. Interestingly enough, Ephraim Beach is central to the story.

Henry W. Drinker and William Henry were two of Pennsylvania's most enterprising and farsighted promoters of internal improvements. These gentlemen had hired Ephraim Beach in 1831 to conduct a survey for the Susquehanna & Delaware Canal and Railroad Company (Chapter 2). Over the intervening years, plans for "Drinker's Railroad" were refined so that instead of merely connecting Pittston on the Susquehanna River with the Delaware River somewhere near the Water Gap, the railroad was also to extend northward to the New York State line, where a connection with the New York and Erie Railroad was supposed to be built. In fact, this extension had been a condition of Drinker's receiving financial backing from Wall Street financiers.[35] That was in 1833, and the Pennsylvania Legislature passed a charter for such a railroad, called "the 'Susquehanna Railroad' Act of Assembly of 1833." This railroad, with an outlet at the Delaware River north of Belvidere, interested and involved Dr. George Green, one of Belvidere's most prominent citizens. Green is credited with having originated the Bel Del Railroad. It was in this railroad, too, that Drinker and Henry tried to involve the MC&BCo. in August, 1836. At that time, Henry "met and made arrangements with the Morris Canal Board of Directors to raise $150,000 on stock subscriptions to commence the road." However, Drinker's efforts to raise "100,000 pounds sterling in England" failed, and construction was postponed.[36]

In May, 1836, Ephraim Beach was in Morristown, already working on the M&E Railroad, when he explored a projected railroad route from Morristown to Carpenter's Point (today's Port Jervis), where it was "proposed to connect it with the Lake Erie Railroad."[37]

Drinker's projected railroad was postponed, and the Morristown-Carpenter's Point Railroad was not undertaken. What *was* built was the M&E, followed shortly by the Leggett's Gap Railroad, another Drinker project; and yet another, the Delaware and Cobb's Gap Railroad. The latter two railroads merged in 1847 to become the DL&W Railroad. And when the Warren Railroad was built to connect the Delaware Water Gap with the Central Railroad of New Jersey at Hampton, New Jersey, intersecting the Morris Canal at

Washington in 1856, the DL&W immediately absorbed it into its own system in 1857.

This Warren Railroad also would intersect the M&E Railroad at Washington when the latter was extended from Dover to Phillipsburg about 1860. Eventually, in 1868, the M&E, too, would be absorbed into the DL&W Railroad.

All of this was made possible by the Scranton brothers: George W., Selden T., and Charles; and James I. Blair. The Scranton's bought Slocum Hollow, Pennsylvania, with all its coal and ore, in 1840. Then they bought Drinker's charter for the Leggett's Gap Railroad in 1847. They had already bought William Henry's lease of the Oxford Furnace in 1839, just a little west of Port Colden, a little north of Washington, and a little south of Belvidere.[38]

Belvidere had secured a charter for the Belvidere Delaware (Bel Del) Railroad on March 2, 1836. This road was supposed to connect the Delaware Water Gap with Belvidere and with the feeder of the Delaware and Raritan Canal, or with the city of Trenton. A month earlier, on February 9, Belvidere had gotten a charter for the Port Colden-Belvidere Railroad. It is no coincidence that William Henry's name appears among the incorporators of that company, or among those of the Port Colden-Morristown Railroad, incorporated on February 28, 1837.

This railroad was supposed to start at "some point on the Morris Canal at or within two miles of Port Colden" and connect with the M&E Railroad "at or within four miles of Morristown." Other incorporators of the Port Colden-Morristown Railroad were Caleb H. Valentine, soon-to-be-director of the Canal Company; William C. Dusenberry, builder of Port Colden's basin; John S. Darcy, president of the NJRR&TCo.; and Ephraim Marsh, president-to-be of the Morris Canal Company.

Still another charter was obtained at this point, for the Morristown and Easton Railroad, March 9, 1837. This road was to begin at the M&E Railroad in Morristown, run by way of Mendham and Chester to Schooley's Mountain, and, generally following the Musconetcong River through Asbury in Warren County, reach the Delaware River opposite Easton. This was apparently a preliminary attempt to extend the M&E Railroad to Phillipsburg. That extension was not accomplished until 1861, however.

Note that Schooley's Mountain was to be serviced by this proposed Morristown and Easton Railroad. Small wonder that Ephraim Marsh, owner of Heath House on that mountain, was one of the incorporators. Others were:

James Vanderpool	Robert C. Stephens
Samuel Searing	David W. Neighbour
Joel W. Condit	Gershom Rusling
Stephen D. Day	Robert S. Kennedy
Abraham Britten	Enoch Green
William N. Wood	Thomas G. Stewart
John Marsh	William B. McCullough
Nathan A. Cooper	George Creveling
Lawrence Hager	Jacob W. Neighbour
David W. Dellicker	

These 20 men represented the whole swath of land and towns that would have been traversed by such a railroad.

The Port Colden-Belvidere Railroad with the Morristown-Port Colden Railroad would have given New York a rail connection with the Lackawanna Valley; the Morristown-Easton Railroad would have given New York the same connection with the Lehigh Valley; Morristown already had the M&E Railroad with its NJRR&TCo. outlet to tidewater. Thus, all these early charters seem to have been attempts to establish through-routes. The role of the Morris Canal in these plans is all too obvious: Only consider the role played by the engineers of the Canal Company in surveying and building these early railroads. We saw Beach as Drinker's choice to survey his original route in 1831, as Chief Engineer of the NJRR&TCo. in 1833, and again as engineer of the M&E Railroad. It was doubtless his position of influence with the NJRR&TCo. that made it possible for the M&E to connect with that road, thus gaining for it an outlet at the Hudson River at Jersey City.

Roswell B. Mason, Canal Company engineer, Beach's assistant, was given permission by his board of directors to devote "two days to the concerns of the Morris and Essex Railroad Company — provided he can do so without prejudice" to the interests of the Canal Company.[39] Following Beach, Mason became the NJRR&TCo.'s

Chief Engineer and, at the same time, was a Canal Company engineer. When he left the NJRR&TCo. to undertake the building of the Illinois Central Railroad, Lorenzo A. Sykes, yet another Canal Company assistant engineer, replaced him.

There was, to be sure, excellent reason for the Canal Company to cooperate with these early railroads, for when its charter had been issued, the State Legislature had stricken the clause that would "prohibit the construction of railroads, as well as canals, within a certain distance."[40] But at this point in the Canal Company's history, the view is that even "cooperation" is not precise or adequate to describe this canal-railroad combine. More intensive research into the Port Colden-Belvidere and other railroads would doubtlessly reveal the pervading hand of Morris Canal Company engineers. The Canal Company's interest both in the Port Colden Railroad and in Drinker's Railroad, which it was intended to meet at the Delaware River, is nowhere more clearly stated than in the annual report to the stockholders for the year 1836. In that report, the road is described as intended to connect Port Colden, "immediately on the margin of the canal, 85 miles from New York," with Belvidere, "on the river Delaware, a distance of only 12 miles." When built, the railroad would then "connect with the others authorized by the legislature of Pennsylvania, leading from the Delaware at Belvidere, to the Susquehanna River, into the anthracite and bituminous coal mines of Pennsylvania." Coal from these mines was intended for rail delivery to Port Colden, then shipment by canal boat to New York.[41]

Now, the Port Colden-Belvidere Railroad alone could not have effected that connection. The Bel Del Railroad was undoubtedly included in this Canal Company inference. Associated with Dr. George Green in the Bel Del Railroad were Enoch Green, once a Canal Company director, and John M. Sherrerd, a known Canal Company stockholder. It will be instructive, here, as it has been before, to see who the incorporators of the Bel Del were:[42]

John Kinney, Jr. Jeremy Mackey George Green Peter Kleinhaus John M. Sherrerd	These were 5 of the 11-man board of directors of the Belvidere Bank.

Nelson Jay — Warren County freeholder
Adam Butts — President, Belvidere Delaware Bridge Co.
Enoch Green — Phillipsburg — Green's Mill
Thomas L. Woodruff — known only to have been from Trenton
William L. Happock
Philip Fine — Justice of the Peace; later a Judge
Benjamin Riegel — Riegel Lumbering Co. (later Riegel Paper), Riegelsville
Charles J. Ihrie — Phillipsburg; Judge; Justice of the Peace; later State Senator
Hart Johnson
George Carpenter
William Green — Phillipsburg — Green's Mill?

<table>
<tr><td>Cornelius Ludlow
Daniel Van Syckle
Henry Quinn
John Lilly
Wilson Bray
Joseph Titus</td><td>These men were undoubtedly from Hunterdon and/or Mercer County.</td></tr>
</table>

The interesting fact about the Belvidere Bank's interest in this railroad is that James I. Blair, who would later build the Warren Railroad, and who was so closely associated with the Scranton brothers and the DL&W Railroad, was the virtual owner of the bank. Even more intriguing is that Garret D. Wall, a director of the Canal Company, was on the bank's board of directors.[43]

A look at the board of directors of the Morris Canal Company for 1836 will confirm, if nothing else, the theme of this chapter: A new era of mutually beneficial cooperation among businesses, railroads, and the Morris Canal equalled an era of "enlargement" on the grandest scale possible.

New York

Louis McLane	Edwin Lord
John S. Crary	Joseph L. Joseph
George Griswold	Daniel Jackson
John Haggerty	Henry W. Hicks
Samuel R. Brooks	James B. Murray

Washington Irving
Henry Yates
Stephen Whitney
Jonathan Goodhue

New Jersey

Garret D. Wall
James Parker
Philemon Dickerson
John S. Darcy
Peter M. Ryerson
Christian B. Zabriskie
Isaac H. Williamson
John Travers

Pennsylvania

John Moss

Of the New Yorkers, President Louis McLane had recently ended terms as Jackson's Secretary of the Treasury (1831-33) and Secretary of State (1833-34). In financial circles, he was said to be second only to Nicholas Biddle of the Bank of the United States. Whether he was affiliated at this time with any railroad is not known, but when he resigned the Morris Canal's presidency, he did so to become president of the Baltimore and Ohio Railroad.

Edwin Lord was said to have had connections with the Erie Railroad, then in its beginning-construction phase. Perhaps he was related to Eleazor Lord, first president of the Erie, or to D. N. Lord, an Erie director who was later to become a Canal Company director.

Stephen Whitney was a director of the Erie Railroad. As a shipper, he had made his fortune smuggling cotton through the British Blockades during the War of 1812, becoming one of New York's first millionaires. Together with William B. Astor, John Jacob's son and former Canal Company director, Whitney founded New York's first Merchants Exchange, the infant capital of capitalism — Wall Street. Also in Wall Street was Joseph L. Joseph, with offices at Wall and Hanover Streets. He was to lose everything in the financial crisis known as the Panic of 1837.

John S. Crary was a partner in a very large New York commercial house which also had offices in Liverpool. John Haggerty owned one of New York's two largest auction houses. (The other was owned by John Hone, former Canal Company director, and brother of New York's Mayor Phillip Hone, first president of the Delaware & Hudson Canal Company.) Henry W. Hicks represented the firm of

Samuel Hicks & Sons, wealthy shipping magnates. Washington Irving, a friend of Phillip Hone, of Eleazor Lord, first president of the New York & Erie Railroad, and of Macculloch, was also the brother of William Irving, prominent merchant on New York's East River.

Jonathan Goodhue had very close business relations with the Baring Brothers, international financiers, who made it possible, by a loan, for Goodhue and Company to buy the celebrated and very lucrative Black Ball shipping line. Daniel Jackson was a P&HRR stockholder.

George Griswold was part owner of the Black X Line of ships. The marriages of his daughters: Matilda to Frederick T. Frelinghuysen, and another to John Clive Green, would later bring the sons-in-law into the circle of Canal Company directors. Green would eventually become one of the owners of the Michigan Central Railroad, while Frelinghuysen would enter the United States Senate.

Henry Yates, James B. Murray, and Samuel R. Brooks are not at present known to the author except in their earlier directorships in the Canal Company.

John Moss, the Philadelphia, Pennsylvania, director of 1836, is not known to this author, but an educated guess would make him a representative of one of the several railroads that were being built to carry anthracite to the Morris Canal.

Of the Jerseymen, we have seen Peter M. Ryerson's interests in both the P&HR Railroad and the Dundee Manufacturing Company. Also affiliated with the P&HR was Philemon Dickerson, as we also have seen. Dickerson would resign from the Canal Company within the year, however, to become governor of New Jersey. A former governor of New Jersey, Isaac H. Williamson was the first president of the Elizabeth and Somerville Railroad, predecessor of the Central Railroad of New Jersey. John S. Darcy was president of the NJRR&TCo. from 1832-3 until his death in 1863. Christian B. Zabriskie was a New Jersey State Councilman (Senator) from Bergen for two terms, 1822-24 and 1834-36 — both periods crucial to the Morris Canal.

John Travers, with James B. Murray and Thomas E. (B?) Davis, all Canal Company directors, formed the Bergen Port Company.[44]

As "proprietors of a tract of land at Constable's Point" (Bayonne), the men proposed to establish a town with "a safe and convenient port." They also intended to erect "buildings for manufacturing" and to engage in manufacturing *in* those buildings. Among the powers of the Bergen Port Company were the rights to make a canal to intersect the Morris Canal and to build a railroad to intersect the NJRR, "to make a communication by railroad through Bergen Port to the Kills."

In considering this era of cooperation at this point, an enigmatic comment made by Eleanor Morton, the author of Josiah White's biography, begins to make sense. According to Morton, Josiah White received word that the Morris Canal now had "a new shipping place at Perth Amboy."[45] At first reading, the statement makes little sense or impact. Then, we come to realize that Garret D. Wall and James Parker, both of the C&A Railroad, were Canal Company directors at this time. Moreover, Parker was a native and one-time mayor of Perth Amboy, who had also served as Collector of that Port during 1829 and 1830. At this time, however, Parker was a member of the U.S. House of Representatives (1833-1837). Surely there was "cooperation."

If a more convincing argument is needed, consider the fact that the NJRR&TCo., originally built from Jersey City to Newark, was extending southward through Elizabeth at the time. By January, 1836, tracks had reached Rahway, and in July, 1836, they were completed to the north bank of the Raritan River, with every prospect of shortly connecting with the C&A Railroad at Perth Amboy.[46] And we have already met John S. Darcy, Canal Company director, president of the NJRR&TCo. "Cooperation"?

Garret D. Wall is one of the most interesting of all the 1836-37 Canal Company directors. We saw earlier in this chapter that Wall had represented the SUM *against* the Canal Company, when the SUM tried to prevent the legislature from allowing the Morris Canal Company to use canal water for manufacturing.[47] Now, in 1836, here he was, working *for* the Canal Company. In trying to determine "why," Wall's very deep involvement in the affairs of Belvidere, New Jersey, came to light.

When Warren County was formed in 1825, the Legislature appointed a three-man commission, among them Nathaniel Saxton, to decide where to locate the seat of that county. As an inducement to choose Belvidere for that honor, Wall offered to donate a parcel of land for the new county buildings, for a public square, and for churches around the square. It worked. He had acquired the lands immediately prior to his deeding them to Belvidere. Among the lands he *kept,* however, were all the business portion and "all of the water power within the town of Belvidere."[48] It seems a certainty that Wall was interested in the development of Belvidere's manufacturing potential. Although his name is not among the incorporators of the proposed Belvidere railroads discussed in this chapter, we may fairly assume that he was interested in those transportation facilities.

When we consider the developments at Belvidere; the SUM's plans to expand at Paterson; Ryerson's expectations for his iron works due to the Pompton Feeder; the prospect of "Bergen Port," to be established by Messrs. Travers, Murray, and Davis, etc., we begin to realize the magnitude of the "enlargement" that was brought about by the Canal Company's charter amendment of March 5, 1836. When we consider developments in Pennsylvania, we are convinced that Morris Canal was caught up in a whirlwind drive to expand the economy of the entire developing nation.

When President Louis McLane issued his 1836 report to the stockholders he could well afford to be enthusiastic. He announced that the Canal Company had "increased the contract with the Lehigh Company, for the transportation of 60,000 instead of 40,000 tons of coal during the current year [1836]. Indeed, at no period since the canal has been in existence, has the company transacted so much business, or presented such brilliant prospects, as it has the present season."[49]

So good had the season been, in fact, the Canal Company declared a dividend of three percent for the first half of 1836.[50] As the 1837 season approached, President McLane announced that the board of directors had notified "the holders of the debt in Holland, that both the principal and interest" on that mortgage loan would be paid on July 1, 1837, so that the Morris Canal would be "released from incumbrance."[51]

McLane explained to the stockholders how it had become apparent during the 1836 season "that the augmenting concerns of the company, growing out of the active employment of the other powers of the charter, would not permit either the executive officers, or the directors, to give the requisite attention to the canal. The successful management of such a work calls for more than a skillful police and general supervision on the line. It requires, also, constant and judicious observation and attention, *in developing the resources of the country which the work penetrates; and unremitting exertions, not only to make known its advantages, but to multiply and extend its connections, in order that it may enjoy a proper share of the trade for which there will always be so much competition* (italics mine)."[52]

We have seen how it was proposed to develop the resources of the country penetrated by the canal. We have seen, too, how the Canal Company planned to multiply and extend its connections in New Jersey, and into Pennsylvania (by way of the "Drinker" Railroad) and New York State and beyond (by way of the Erie Railroad). But there were other developments afoot, too, that would have a significant impact on the Morris Canal's future.

The Beaver Meadow Railroad, which had been chartered in 1830, was under construction from the Beaver Meadow Coal Fields supposedly to Easton. Josiah White prevented the completion of that road beyond Parryville on the Lehigh River, however. At one point he armed his men with muskets, as did Mr. Ingham, president of the Beaver Meadow, in anticipation of armed conflict over the location of the latter's railroad.[53] By 1836, however, the Beaver Meadow was transporting anthracite from its coal field to the Lehigh Navigation at Parryville, where it was loaded on canal boats for shipment to South Easton.

The Hazleton Railroad, chartered in 1836, was built to connect the Hazleton Coal Field with the Beaver Meadow road at Weatherly.

In the Schuylkill coal region, other anthracite railroads were being built, too. The Philadelphia & Reading, chartered in 1833 to connect those two cities, was frustrated by the Pennsylvania Legislature in its attempt to reach Pottsville. When the Little Schuylkill Railroad (chartered 1828) was authorized to extend its Port Clinton-to-

Tamaqua road to Reading, the Philadelphia & Reading Company convinced the Little Schuylkill Company to relinquish that privilege to them, thus adding another twenty miles to their trackage toward Pottsville. The Philadelphia & Reading then carried the Little Schuylkill's coal freight from Port Clinton to Port Richmond, on the Delaware River.

Josiah White began building his Lehigh & Susquehanna (L&S) Railroad in 1837, reportedly as "a link in the railroad system between Philadelphia and Buffalo," although his immediate aim was to connect White Haven, on the upper reaches of his Lehigh Navigation, with Wilkes Barre, on the North Branch of the Susquehanna River.

George W. Edwards, a Philadelphia & Reading man, bitterly attacked White because of his L&S Railroad. The coal that would be tapped by the L&S, Edwards claimed, should be transported by the "natural and easy outlet" of the Little Schuylkill — and hence, the Philadelphia & Reading.[54] Since the Philadelphia & Reading was being hampered in its attempt to expand trackage by the state's lawmakers, Edwards' hostility is understandable — especially if White's plan *was* to reach Buffalo — for the Reading had big plans, too.

Anticipating the story by some years, it is interesting to note that the Beaver Meadow and Hazleton Railroads were later absorbed by the Lehigh Valley Railroad, while Josiah White's L&S became a vital part of the Central Railroad of New Jersey. Incidentally, Edwards later became a Canal Company director.

Still another Pennsylvania railroad being built at this time was the Sunbury & Erie, chartered on April 3, 1837, and destined to have an impact on the Morris Canal. This railroad was to start at Sunbury, go through Williamsport, and eventually reach Erie, Pennsylvania, that state's only port on Lake Erie. By starting at Sunbury, the railroad would be connected, or "merged," with both the North and West Branches of the Pennsylvania Canal system. Other railroads connecting with the Sunbury & Erie at Williamsport were the proposed Pittsburgh & Susquehanna and the Williamsport & Elmira roads.

In 1836, the Bank of the United States had established a branch at Erie, Pennsylvania. President of that bank, Nicholas Biddle, was to be the president of the Sunbury & Erie Railroad.[55] Edward R.

Biddle, Nicholas' nephew, became a Canal Company director just one month before the Sunbury & Erie and Pittsburgh & Susquehanna were chartered.

In view of all these things, i.e., the expansions and enlargements in the economy in general, and particularly in the coal trade, the Canal Company's board of directors felt it was their duty, "either to reorganize the company, so as to place the canal and its business under the management of a distinct board (which would have materially augmented the expense), or *to combine with it other connections, whose interests would lead them to cooperate in promoting and extending the trade* (italics mine). After full deliberation, the board determined to adopt the latter expedient."[56] The decision was made on December 24, 1836.[57] Thus, the president and directors[58]

> *entered into an agreement with the Little Schuylkill and Susquehanna Rail-Road Company, whereby, in consideration of the right of using the canal, and enjoying all the water power and real estate attached or appurtenant to it, — so far as by the charter they may lawfully do, — that company have agreed to purchase of the Morris Canal and Banking Company all their old boats at a valuation by appraisers mutually chosen, and the new boats at the cost; have guaranteed all the expenses of working the canal, and keeping it and the appurtenances in repair; and have agreed to pay, for the term of five years from the 1st day of December last, a net income of six per centum per annum upon the entire expenditure made, and to be made, upon the canal and its appurtenances, and upon the cost of the real estate.*
>
> *The contract, also, secures to the Little Schuylkill and Susquehanna Company, the option of extending the time for a further term of five years at the same rate; and afterwards, for a like term of five years, at the rate of eight per cent per annum. The exercise of the franchise conferred by the charter remains, in the meantime, in this company.*
>
> *By this arrangement, it is believed, there has been secured an annual revenue from the canal and its works*

equal to four and a half per cent upon the capital of $4,100,000.

The Little Schuylkill and Susquehanna Rail-Road Company is incorporated by the state of Pennsylvania, with authority to construct a rail-road from Williamsport to the rail-road of the Beaver Meadow Company, in the state of Pennsylvania.

It has the authority, also, to hold five thousand acres of coal lands, of which a considerable portion has already been purchased. It may be presumed that a chief motive by which this company may have been influenced in forming the contract, herein explained, is to acquire at all times certain and ample means of transportation for the produce of the extensive mines; and the directors are understood to possess full confidence in their ability, from the various means at their command, to render the contract immediately profitable.

The capital of the company may be extended to the sum of two millions of dollars, of which, it is understood, an amount has already been subscribed, by the Bank of the United States and responsible individuals, sufficient to insure the early completion of the rail-road and the seasonable development of the mines . . . .

The Morris Canal will be connected, by way of the Lehigh Canal and the Beaver Meadow Rail-road, with the Little Schuylkill and Susquehanna Rail-road, extending to Williamsport, a distance of eighty-eight miles.

In connection with these continuous routes, the projected rail-roads to Elmira, Pittsburgh, and Erie, will open the nearest and most direct communication from the bituminous coal field to Jersey City and the city of New York . . . .

From Williamsport the rail-road to Pittsburgh will be two hundred and seventeen miles, and to Erie two hundred miles; making the total distance from New York to the former place [*Pittsburgh*], *four hundred and sixty seven miles, and to Erie four hundred and fifty miles.*

It has been a *long* way from Paterson and Long Pond with the Canal Company's charter amendment of March 5, 1836. The enlargement contemplated is truly staggering, and the cooperative effort to have made such an enlargement possible is just as overwhelming to imagine.

Canal Company president Louis McLane knew what he was talking about when he explained to the stockholders about "the trade for which there will always be so much competition." The very prospect of trade had brought about a proliferation of canals and railroads designed to feed them. The hope of expanding trade brought about the kinds of combinations or "mergers" of canals and railroads that we have described here. The desire to monopolize trade would bring about the eventual destruction of the canals and the supremacy of the railroads. But that was still in the future by many years.

Louis McLane's confident and enthusiastic report to the stockholders was written on March 6, 1837. Four days later, on March 10, the Bank of the United States suspended payments. The Morris Canal would soon follow suit. The Panic of 1837 had begun.

REFERENCES

1. Morris Canal and Banking Company *Minutes,* May 31, 1831 (hereinafter cited as MC&BCo. *Minutes*).
2. *Ibid.,* July 1, 1831.
3. *Ibid.,* September 13, 1834.
4. *Paterson Intelligencer,* July 22, 1835.
5. MC&BCo. *Minutes,* November 17, 1836.
6. Morris Canal and Banking Company *Docket Files* (hereinafter cited as MC&BCo. *Docket Files),* "Specifications for a Dock," dated December 30, 1836.
7. *Report of the President and Directors of the Morris Canal and Banking Company, to the Stockholders* (N.Y.: 1836), p. 19 (hereinafter cited as *Report of the President, 1836).*
8. MC&BCo. *Minutes,* November 17, 1836.
9. *Ibid.,* March 14, 1839.
10. *Paterson Intelligencer,* February 3, 1836.
11. MC&BCo. *Docket Files,* report of Ephraim Beach to the board of directors, dated January 16, 1836.
12. MC&BCo. *Minutes,* March 20, 1836.

13. *Ibid.,* January 16, 1836.
14. The following several paragraphs have been excerpted from the *Paterson Intelligencer,* September 3, 1845.
15. *Paterson Intelligencer,* January 25, 1837.
16. *Ibid.,* May 22, 1839.
17. Paterson *Daily Guardian,* August 12, 1857.
18. *Report of the President, 1836,* pp. 7-8.
19. MC&BCo. *Minutes,* January 29, 1835.
20. *Ibid.,* June 10, 1835.
21. William W. Scott, *History of Passaic and Its Environs* (N.Y.: 1922), p. 99.
22. James W. Ransom, *Vanishing Ironworks of the Ramapos* (N.Y.: 1966), pp. 114-117.
23. MC&BCo. *Docket Files,* #C53.
24. Scott, *op. cit.,* p. 267.
25. Ransom, *op. cit.,* p. 104.
26. G.W. Keppler Collection, unpublished manuscript, "Goodman Novel," pp. 197-8.
27. *Paterson Intelligencer,* April 6, 1836.
28. MC&BCo. *Minutes,* May 30, 1836.
29. *Ibid.,* October 29, 1836.
30. *Report of the President and Directors of the Morris Canal and Banking Company, to the Stockholders* (N.Y.: 1835), p. 6.
31. MC&BCo. *Minutes,* January 15, 1834.
32. *History of Morris County, New Jersey: 1734-1882* (N.Y.: 1882), pp. 162, 223, 227, 250, 315.
33. MC&BCo. *Minutes,* November 2, 1837.
34. Robert G. Albion, *The Rise of New York Port* (N.Y.: 1939), pp. 226-7.
35. H. Hollister, *History of the Lackawanna Valley* (N.Y.: 1869), pp. 384-5.
36. *Ibid.,* p. 386.
37. *Jerseyman,* May 25, 1836.
38. James P. Snell, *History of Sussex and Warren Counties, New Jersey* (Philadelphia: 1881), pp. 67, 613-14 (hereinafter cited as Snell's *Sussex*).
39. MC&BCo. *Minutes,* January 29, 1835.
40. *Paterson Intelligencer,* January 27, 1829.
41. *Report of the President, 1836,* p. 16.
42. Snell's *Sussex,* identifications derived from various pages.
43. Snell's *Sussex,* p. 545.
44. *Laws of New Jersey,* 1837, p. 389.
45. Eleanor Morton, *Josiah White: Prince of Pioneers* (N.Y.: 1946), p. 222.
46. Wheaton J. Lane, *From Indian Trail to Iron Horse: Travel and Transportation in New Jersey, 1620-1860* (Princeton, N.J.: 1939), p. 314.
47. *Paterson Intelligencer,* February 20, 1833.
48. Snell's *Sussex,* pp. 532-3.
49. *Report of the President, 1836,* p. 19.
50. *Paterson Intelligencer,* July 20, 1836.

51. *Report of the President and Directors of the Morris Canal and Banking Company, to the Stockholders* (N.Y.: 1837), p. 4 (hereinafter cited as *Report of the President, 1837).*
52. *Ibid.,* pp. 4-5.
53. *The Story of Anthracite* (N.Y.: 1932), pp. 99-101.
54. Morton, *op. cit.,* p. 223.
55. George H. Burgess and Miles C. Kennedy, *Centennial History of the Pennsylvania Railroad Company: 1846-1946* (Philadelphia: 1949), n.p.
56. *Report of the President, 1837,* p. 5.
57. MC&BCo. *Minutes,* December 24, 1836.
58. *Report of the President, 1837,* pp. 5-8.

12

Sowing the Wind

TO understand many of the events in the Canal Company's history from 1837 to 1844, it is essential to come to grips with the Bank of the United States, and with Andrew Jackson's "war" on that bank.

The bank was originally Alexander Hamilton's idea, and was to have been to this country what the Bank of England was to that country.[1] Its charter was granted in 1791 and was to expire in 1811, just before the outbreak of the War of 1812, and just when the nation needed the "stable currency and the machinery for floating loans that it alone could provide."[2] The bank was rechartered for 20 more years in 1817, however, and was thus due to dissolve in 1837. Nicholas Biddle, the bank's president, was the uncle of Edward R. Biddle, who became a Morris Canal Company director in March, 1837.

Before the old charter died out, Clay and Webster submitted a bill to Congress to recharter the Bank, and in 1832 the bill was passed by both Houses, only to be vetoed by Jackson. The Bank, Jackson declared, was un-American, undemocratic, and unconstitutional, because more than a quarter of its stock was held by foreigners. What wasn't held by foreigners was held by the very rich. Jackson worried about "the dangers which 'might flow from such a concentration of power in the hands of a few men irresponsible to the people.'" Most all, he "deplored the requests of rich men to be made 'richer by act of Congress." '[3]

As a result of a Supreme Court case in 1819, the bank was assured of a tax-free monopoly status — one staunchly opposed by President Jackson. He interpreted his reelection in 1832 as a mandate against the Bank of the United States. Rather than sit back and wait for the Bank's charter to expire in 1837, Jackson tried to withdraw the deposits of the United States Government from the Bank — about 12 million dollars.[4] In that action he was not supported by the Constitution, since only the Secretary of the Treasury could make such a withdrawal. Louis McClane was that Secretary, and he refused to act on Jackson's order. McClane was taken from his position and made Secretary of State.

But in June of 1836, Jackson agreed to distribute the surplus funds of the National Treasury to the states in the form of loans. Accordingly, a Congressional act provided for the distribution of the surplus: Whatever money in excess of 5 million dollars was in the Treasury on January 1, 1837, was to be apportioned among the states in accordance with their representation in the electoral college, and paid over to them during the year in four quarterly installments. Dazzled by this prospect of receiving these generous subsidies, many states incurred debts so large and so unwarranted by their taxable resources for repayment that the bubble *had* to burst. Moreover, many states also lent their credit freely in aid of dubious private enterprises. Unfortunately, they found a ready market for their securities among foreign investors, who were misled into thinking that the credit of the individual states was as good as that of the United States Government.[5]

English exporters of goods to this country were the first to observe that the phenomenal prosperity of their customers here might have a disastrous ending. By 1836 the balance of trade was running strongly against the United States. This unfavorable trade balance caused American drafts and bills of exchange to sell at a discount in England, and the Bank of England promptly raised its discount rates. Exporters refused new credits to American customers and demanded immediate settlement of old accounts. American importers, on the other hand, had to make immediate collection on their accounts to pay the foreign debts. Since only specie would be accepted by the foreigners, the American banks were deluged with requests for hard money. Of course they couldn't possibly fill the requests, and by May, 1837, every bank in the United States had suspended specie payment.

What followed was a rash of bank failures, an epidemic of bankruptcies as business after business, unable to collect its debts, was forced to close down. This was true in England as well as in the United States. When English cotton mills closed down, the cotton planters in the South and West lost their best market. When the South and West lost their buying power, the manufacturing Northeast lost *its* best market, and so on. Jackson, fortunately for him, had left office a few weeks before the panic broke. Biddle, meanwhile, had received a charter from Pennsylvania, and he established his Second Bank of the United States in Philadelphia and began establishing branches.[7]

Jackson's actions may not have been the *direct* cause of the Panic of 1837, but they were certainly responsible for the incredible boom and overexpansion that preceded and precipitated the Panic. His war against Biddle, and the apparent unfairness of his tactics, certainly seemed to have wrought changes in Biddle's banking philosophy. The Bank of the United States, although called by some a monopoly, had always enjoyed the reputation of being restrained and conservative.[8] The Second Bank of the United States was to be anything but restrained and conservative, for as Nicholas Biddle was reorganizing his empire in Pennsylvania, his nephew, Edward R. Biddle, was embarking on his own career with the Morris Canal and Banking Company (MC&BCo.).

The Panic of 1837 had had a head start within the Canal Company, itself. Robert Gilchrist, cashier since 1826, "defalcated" with $121,229.47 in Canal Company funds.[9] Before Gilchrist resigned in January, 1837, the Company had succeeded in recovering "$3000 in cash," plus "sundry conveyances, duly made and recorded, of certain lots at Jersey City and Newark," evaluated at $80,000. The new lessees, the Little Schuylkill and Susquehanna (LS&S) Railroad Company, agreed to take the "sundry" lands conveyed by Gilchrist and to accept and pay $10,000 for his bond. This left a total of $28,229.47 for Gilchrist to repay the Canal Company. Thomas A.

Alexander was chosen to fill the position of cashier. Meanwhile, the Canal Company had to make good the loss out of its earnings for the last half of 1836. Even after doing that, however, the Company was expected to be able to declare a 4-percent dividend on its stock for the same period.[10]

At about this time, Louis McLane resigned from the Canal Company to become president of the Baltimore and Ohio Railroad. The board of directors declined to accept his resignation on January 31, 1837, however, and when the annual election of directors and officers came up on March 9, he was reelected president of the Company.[11] McLane again tendered his resignation on June 14, 1837. The board accepted McLane's resignation this time, and Henry Yates was chosen to be president *pro tem* until a suitable successor for McLane could be chosen. Joseph L. Joseph resigned as director, too, and Samuel L. Southard was nominated to fill that vacancy.[12]

What followed over the course of the next two weeks is an indication of how the Canal Company would be run for the next four years. Southard was elected a director on June 22, and, at the same time, he and Edward R. Biddle were nominated for the presidency and vice presidency. Southard declined to serve. Since the bylaws did not provide for a vice president, they were altered accordingly, and on June 30, 1837, Biddle and Edwin Lord were elected president and vice president, respectively.[13] Southard would not be elected president until the following year.

The 1837 election of Canal Company officers added three new Philadelphia directors: Biddle, Thomas C. Cadwallader, and Washington Jackson; and two new New York directors: Simeon Draper, Jr., and Edwin Post. New Jersey directors Parker, Darcy, and Dickerson were not reelected, nor were New York's Crary and Haggerty.[14] The board was taking on a new aspect, with this shift in membership reflecting the shift in interests discussed in this chapter. The effects of the Panic of 1837 were not felt in the Canal Company's board room as they were being felt in the economy at large. Throughout 1837 and 1838, the Company had vast financial dealings with a varied array of individuals and institutions that belied the economic condition of the commercial world.

In a "Report of the Special Committee of the House of Assembly, on the Embarrassments of the Country," read before that House on November 7, 1837, A.C.M. Pennington painted a grim picture, indeed. When Jackson had left the White House on March 3, 1837, his farewell speech congratulated the American people on their "prosperity and happiness," and congratulated himself for "his humble efforts to restore the constitutional currency of gold and silver." Now, Pennington complained, the country was in a "calamitous condition." Everywhere could be seen "the most touching exhibitions of embarrassment and distress," and the "melancholy wrecks of our former prosperity." Within weeks, "as if by enchantment, credit perished; confidence was destroyed; business in all its branches prostrated; the arm of enterprise paralyzed."[15]

In the following list, the reader will see no signs of the "universal ruins and bankruptcy" described by Pennington. No attempt is made or intended to interpret these Canal Company dealings. They are noted merely to illustrate the scope of the Company's activities during the Panic of 1837. All items are from the Company's *Minutes;* most of them reflect the apparent solvency of the Canal Company's bank.

Abraham R. Lawrence, Pres., New York & Harlem River Railroad: loan, October 6, 1836

Commissioners of the State of Texas: bonds, November 2, 1837

Farmers Loan Co., Mobile, Alabama: notes, March 11, 1837

Long Island Railroad: loan, September 7 and October 1, 1838

Mr. Hudson: loan, with Pea Patch Is. as security, November 1, 1838

Mr. Colin Mitchell: Appalachicola stock, March 9 and April 12, 1838

Agricultural Bank of Natchez: bonds, April 30, 1838

Mr. Judson & Co., Mobile, Alabama: real estate, June 7, 1838

Bank of Mobile, Alabama: stock, September 7, 1838

Heading this list, perhaps, should be a letter read into the *Minutes* of October 27, 1836, in which Isaac Coe proposed "that the Bank and

Branches in Indiana" were "to do business with" the Morris Canal Bank, and asked an "allowance of interest on their deposits." This was followed by an October 29, 1836, entry which recorded instructions to Cashier Thomas A. Alexander: "Resolved that the Cashier, under the advice and direction of Messrs. Griswold, Joseph, and Brooks, be authorized to pay to Isaac Coe, Esq., Commissioner, the cash payment which will be due on the 1st of November on the contract made with him for Indiana 5% bonds."

According to one source, the Canal Company gave to the State of Indiana a second mortgage on the Morris Canal in exchange for $3,000,000 worth of bonds.[16] Another source set the figure at $1,000,000.[17] Bonds were bought from the new State of Michigan, too. Indiana and Michigan bonds were then sold or hypothecated by the Canal Company before they were paid for.[18] Mr. Coe's complicity consisted of selling Indiana's bonds to the Canal Company (of which he was a stockholder) for 96 cents on the dollar, while he credited the State of Indiana with but 88 cents.[19]

The point the author makes here is that these Canal Company dealings in 1837 and 1838, while the rest of the country was economically depressed, indicate that the Panic of 1837 did not immediately affect the Morris Canal's plans for expansion — in particular, plans for the LS&S Railroad.

Construction on that railroad had begun in 1837, but even before it was completed, those new lessees of the Morris Canal operated the canal boats they had bought. When navigation closed at the end of the 1837 boating season, on December 14, coal tonnage clearing the Canal Company Collector's Office at Easton had included "18,385 tons, which was transported by 3,646 boats, in the service of the Little Schuylkill and Susquehanna Railroad Company, averaging 23 tons each."[20] (Obviously, the item means 3,646 *boat loads,* not 3,646 distinctly separate boats.)

Coal was definitely reaching the Morris Canal from the LS&S Railroad during 1838. Paterson's Andrew Parsons, alone, received enough coal from that company to run up a bill of $29,938.24, with interest. When the railroad company wrote to protest to the Canal Company and to ask for payment, the board of directors resolved to insist that Parsons pay his bill.[21]

By early 1839, the Canal Company proposed that the contract between it and the LS&S be annulled, in exchange for $200,000 worth of LS&S stock, or for $120,000 worth of stock plus the LS&S's coal lands. Work already done on the railroad would be paid for by the Canal Company, and half of the work necessary to complete the road would be paid for by the Canal Company and the Bank of the United States, with the LS&S Company footing the remaining bill.[22] That proposal was accepted by the LS&S, and the agreement was signed in May, 1839.[23] Then, estimates were sought to determine how much it would cost to complete the railroad. Those estimates are not given in the Canal Company's *Minutes,* but the engineer who prepared the report was identified as "S.W. Roberts" — perhaps Solomon White Roberts, Josiah White's nephew.[24]

A "Lehigh Branch" of the LS&S was eventually built in 1840. During that year, the Company offered to lease to miners coal lands in the Linder Gap area, and then offered special rate advantages for mined coal that was shipped by way of the Morris Canal.[25]

The coal lands of the LS&S Railroad were conveyed to the State of Michigan, presumably in exchange for the bonds the Canal Company had "bought" from that state. When construction on the LS&S Railroad came to a halt in 1841, it is said that Indiana found itself the holder of "13,600 shares of LS&S stock par $230,000, $100,000 in Beaver Meadow Railroad stock and 75,000 shares in the Baltimore and Ohio Railroad," plus a worthless second mortgage of $1,250,000 on the Morris Canal.[26] Indiana also was said to hold "nearly 10,000 shares of Morris stock."[27] Former president McLane was now president of the Baltimore and Ohio.

Through President Biddle, the Canal Company had contracted to buy "a majority of the stock in the Beaver Meadow Railroad and Coal Company."[28] And, of course, the Company also owned LS&S stock — $136,000 worth, according to the Annual Report of 1837.

Biddle had also negotiated the Canal Company's contract with the State of Michigan for a $5,000 loan. A provision of that contract was that an advance of 80 percent on that money be realized by the Canal Company.[29] Henry Yates, alone of the 23-member board of Canal Company's directors, cast the single dissenting vote in accepting that contract. Now, 80 percent of $5 million is *four* million, but the records are unfortunately silent on what Biddle wanted to do with that sum.

As already stated, the MC&BCo. continued to expand during the early part of the Panic of 1837. The Company's suspension of payment on its notes was done only to conserve resources, not because the Company had none to pay with. As noted earlier, by May, 1837, every bank in the nation had suspended payment — many for the same purpose of conserving cash supplies.

Meanwhile, it was "business as usual" on the canal itself. Rates of toll for 1837 were established:[30] new lower cargo rates for bricks, charcoal and mineral coal, firewood, flour, grain and grindstones, iron ore, lumber and lime, salt and seeds, whiskey, and window glass. All were to be shipped at one cent per ton-mile, *provided* they were to travel at least 80 miles on the canal. Shorter distances were to cost one and a half cents. The ton was no longer "short," or 2000 pounds, but "long," or 2240 pounds, as it was measured on the Lehigh Canal.

All boats were to travel free if they carried cargo of at least 20 tons. Boats carrying less than that were to be charged three cents per ton-mile, as were boats of less than 60 feet in length. Lumber and timber in rafts were prohibited without written permission. Even with permission, they were to be charged three cents per ton-mile. Emphasis appears to have been placed on through traffic and on larger cargoes.

The Chief Engineer was to have general charge of the whole canal, with a superintendent to assist him. Two supervisors of the inclined planes were to be appointed, one east and one west of the summit. Moreover, four supervisors of the canal itself were to be appointed to be responsible for work gangs, two such crews east and west of the summit.[31]

A basin was to be built at Port Colden, as William C. Dusenberry had requested.[32] Dusenberry already had a fleet of boats operating under his name, as his earlier advertisement of June 28, 1836, indicates:[33]

BOATMEN

The subscribers wish to employ hands and mules sufficient to run 150 boats on the Morris Canal for the season. The boats will be furnished already loaded at the summit of the *plane at Port Delaware opposite Easton. Good wages will be paid and no detention to be used in unloading at Newark. None of the wages will be held back for security of performance as the price paid will be sufficient inducement for persons running boats to hold on the season. All the subscriber wants is good hands and they shall be paid promptly. Industrious men can earn by the above arrangement from $60 to $70 per month for themselves, boy, and horse. The canal is now in permanent order. Applications to be made!!*

William C. Dusenberry
and Co.
Port Colden

Of course, he was a bit premature in describing the canal as being in "permanent order." He had been premature, too, in ordering 18 new canal boats from the Canal Company to swell his fleet. He requested to be relieved of his contract for those new boats, and the board of directors decided to credit his account to the amount of the boats' fair evaluation, according to the terms of their contract with the LS&S Railroad Company.[34]

The Canal Company fully expected to begin enlarging the canal's dimensions during 1837-8.[35] It should have been a good time to employ men for that task, since the Panic of 1837 had thrown many out of work. However, the Lehigh Coal and Navigation Company was also seeking laborers to work with its contractors on the Lehigh "Navigation." That company advertised for "3000 laborers, 300 masons, and 300 carpenters," offering them steady employment, and 20 percent higher wages than before, in "one of the healthiest sections of the country."[36] Men were probably also needed to work on the Lehigh and Susquehanna Railroad, which had just been put under construction in 1837.

Recommendations to enlarge the Morris Canal were submitted in a report by the Canal Committee for the board of directors' consideration. The canal banks were to be raised to hold 5 feet of water, the work "to be done under the superintendence and direction of the engineer of the lessees." Chains strong enough to lift boats of 54-ton burden were to be ordered: 5 from England; 1 to be made of Pure [Peru?] iron; 1 to be made of Catawissa Iron. The rails of the inclined planes were to be made stronger so that they could be used on the ground or on stone only. Experiments to that effect were to be conducted at the "Easton Plane," (Plane No. 11 West) to determine the effect of not using timber under the rails.[37]

The "experiments" at Plane No. 11 West would eventually lead to the introduction about 1841-1845 of forged "Tee" rails, as a substitute for the old strap iron rails.[38] After the "Tee" rails were put into service, the use of timbers between the earth and the rail was discontinued.

The extension of the canal to Jersey City and the Little Basin, familiar to today's canal buffs, was not yet completed. The board had authorized the construction of "a tide lock through Bergen Street at Jersey City," however, and had directed the engineer to report "a plan for a basin at the termination thereof."[39]

Water was actually let out of the Morris Canal at the end of 1837 so that enlargement of the canal could proceed during the winter months.[40] The width of the tide lock at the Jersey City terminus had been set at 11 feet so that larger boats could have access to tidewater.[41] When the new tide lock at Newark was to be built, it was decided that that lock, too, should conform to the dimension of the new lock at Jersey City.[42] With larger locks and strengthened planes, not to mention larger boats, it was hoped that the canal could successfully carry Lehigh Navigation boats.

Two final recommendations in this report concerned bitter complaints from mill owners along the Musconetcong River and Lopatcong Creek. As to the latter complaint, the Engineer was directed to make the proper survey, plans, and estimates to enable the board to decide on the propriety of relocating and reconstructing the canal from the foot of Plane No. 10 West, so as to avoid all difficulty in relation to the mills of the Messrs. Green.[43]

The "Messrs. Green" had written to Ephraim Beach on July 27, 1836, about their agreement with the Canal Company:[44]

Dear Sir:

As the board of directors and different agents of the Morris Canal and Banking Company are frequently changing, we have deemed it expedient to address you upon the subject of the arrangement which was made by us with yourself as engineer of said canal at the time of the location of their works, in order that we may have the benefit of your recollections upon the subject as a matter of reference hereafter in case it may ever be necessary to use them. We have subjoined a statement embracing what we believe to have been the arrangement and if your recollections coincide with ours you will confer a favor by giving us a certificate to that effect, and also by depositing a similar statement with the Company. Our recollections are as follows: The Morris Canal were to enter upon the Valley of the Lopatcong Creek (which is the stream upon which our mills are situated) at the foot of Plane No. 9 West and were to occupy its bed and interlock with its waters until it reached the dam at our place, from said dam to the plane at the Delaware River. The canal was to be so constructed for a 5 foot depth of water, the top water line to be so arranged as to flow 6 inches over the top of the dam aforesaid, by which means the mutual use of the water of said level by both parties would afford the opportunity of drawing that level 6 inches below the top of said dam and of retaining on said level the water of the stream at such times as the mill was not in operation. It being also understood that in the practical operation of the canal, there should always be as much water brought over Plane No. 9 into the valley of said stream as would be required to pass boats over Plane No. 11 and that the company would always keep the banks and other works tight and in good order between the last lock on the canal [then, No. 7 West, later No. 10 West] and the river plane, so as to prevent any leakage, as well as always to keep the level of top water up to what is before specified.

And further, that the said company should be permitted to pass thru a garden which was situated a short distance below our buildings, and to have the necessary quantity of ground for the canal and towing path free of charge upon condition of permitting us to use the tow path from our stable near the road to the back of our store house, as well as to the extent of our ground as far as the waste weir, with waggons or horses, whenever we might find it necessary to do so.

Very respectfully yours,

C. & R. Green

Beach certified that he'd agreed to this agreement with the Messrs. Green and that it had been proper to do so, "inasmuch as it secured to them no privilege which they did not previously enjoy, and imposed no obligation on the company which they must not encounter under any other circumstances."[45] Later, Canal Company president Daniel Tyler would object to Beach's certification on the grounds that he probably had not *been* Canal Company engineer at the time. Tyler also objected that Beach had acted more "as if he represented Messrs. Green's interest than the Company's."[46]

Charles Green had complained earlier of an excessive flow of water from the Canal-Creek combination in October, 1833, because the waste weir or culvert there had been too small to carry surplus water away. His claim for damages in the amount of $244.87-1/2 was paid by the Canal Company.[47]

As to the other complaint, the users of the Musconetcong River waters had made numerous similar claims over the years. The last recommendation of the report on enlargement had been that president Southard and vice president Biddle buy Colonel Nathaniel Saxton's lands or water rights in the Musconetcong River.[48]

Saxton had become the agent of and mouthpiece for both himself and the several landholders in the Musconetcong Valley who had damage claims against the Canal Company, as noted in a handwritten Company document.[49]

1. The widow of Henry Bidleman, Green's Bridge (flour and saw mill)
2. John Fulmer, bark mill, tanning mill, clover mill, hide-breaking mill, and pump for tan yard
3. Samuel Stewart } farms damaged by backflowing
4. John Stewart }
5. Andrew Best, farm above H'town, at lock and dam
6. John Trace [sp?], lands overflowed

7. Richard Baggott, orchard damaged; clover fields, too
8. Wm. and Jas. Young, fencing damaged, canal cut off access to fields
9. N. Saxton — lands occupied by canal; *Where the company use and have the benefit of the bed of the creek as ground whereon to raise a reservoir for the use of the canal. I can perceive no reason why they should not pay for it, it does not otherwise belong to them. Call it water or call it land, they are bound to pay for it if they take it. N.S. bought it and paid for it by the acre, 24 yrs. ago, and he would now use it for his own benefit or sell it by the acre if it was not occupied by the Canal Co. 12 or 14 yrs. ago he was offered 12 (?)* acre for a farm in his valley including the creek.*

Lumber for a bridge and for the gates of two locks (Locks No. 4 West and 5 West) was supplied by Nathaniel Saxton, at whose sawmill the wood was prepared. A bridge between the two locks was also built of Saxton-prepared lumber, as was a culvert under the canal below the field of the Messrs. Young.

The Canal Company's contractors with about 100 hands had taken possession of Saxton's grounds (without consent), cutting from his woods enough saplings to build log houses or shanties and stables — one of them 50 feet long — to accommodate and lodge the workmen and horses. The crews lived there two or three years. During that time, they burned Saxton's wood for cooking and for warmth, and when the contractors, such as Jerry O'Connell, were unable to pay for the services of their men, they broke and ran away. The Canal Company had promised to everyone along the line of the canal, including Saxton, that it would pay all claims, but it didn't. Now the owners wanted compensation, plus interest.

These early claims were for damages, lands flowed, materials appropriated, and the like. Later claims were for the diversion of waters from the Musconetcong River to the extent that the millers could not use their waterwheels. The movement to protest the use of the river water which originated in Lake Musconetcong had had an earlier start. John L. Sullivan's 1828 report on the P&HR

**Handwriting is illegible; the symbol looks as much like Pounds as Dollars.*

Railroad for the SUM had mentioned the Canal Company's effect on the Musconetcong millers.

Early in 1835, the State Legislature received a petition from "citizens of Warren and Morris, complaining of the Morris Canal taking the waters of the Musconetcong." The petition was presented to a committee appointed to investigate and report on the complaints.[50] Apparently no satisfactory adjustments were made in Canal Company policy or actions, for two years later the millers of the Hackettstown area of the Musconetcong were again demanding that their claims be honored. President James B. Murray and director Peter M. Ryerson were authorized to make "some arrangement" with the owners of the water power on the line of the canal near Hackettstown, "by purchase or otherwise."[51] By October, the millers were still demanding "satisfaction for water taken from the Musconetcong River above their town" at Saxton's Falls.[52] The reader will recall the undated document, already cited, in which the anonymous compiler of the report advised that these mills in the vicinity of Hackettstown be bought by the Canal Company.

Saxton himself had proposed that a waste gate at Saxton's Falls be left open permanently so that water sufficient for two wheels could flow through. He also proposed to sell his entire property to the Canal Company at cost (plus interest) and cancel, thereby, all his claims against the Company.[53] By May 1, 1828, the sale was effected, and the Canal Company owned Saxton's three tracts of land on the Musconetcong River, including his saw mill, and two adjoining tracts.[54]

The Canal Company had agreed to put a waste weir in the canal at or near the Guard Lock (No. 5 West) at Saxton Falls to supply one wheel during eight weeks of the year and not less than two wheels during the rest of the year. Owners of the Hackettstown area mills were invited to make offers to sell their mills to the Canal Company by not later than February 20, 1836. The Canal Company would have the option of buying those mills within 30 days. If the Canal Company bought the mills, however, the waste weir proposal would become null and void. The Company's acceptance of offers to buy the mills would "probably depend upon the condition that the application now pending before the legislature should succeed."[55]

The application had passed the legislature on March 5, 1836, permitting the Canal Company to take Greenwood Lake "and other waters" into the Company's water supply system. The "other waters" referred to here were: Saxton's Falls in Warren County; Cranberry Lake, Punk Horn Brook, and Bear Swamp in Sussex; Mount Pleasant Brook in Morris. All these were expected to be taken into the canal as feeders or reservoirs.[56]

Green Pond had been reconsidered as a possible reservoir in early 1834, when Moses Phillips wrote to the Canal Company on that subject.[57] The directors declined to act on his proposal for the time being. Another possible reservoir spot was at Drakesville.[58] Still another was at Mount Pleasant, on the 105-acre, D.B. Hurd tract. President Southard estimated that about two-thirds of that tract would be flowed by the contemplated reservoir. The owner asked $3000 for the land, but Southard was sure he could and should buy it at $1500. Moreover, he thought the tract above Hurd's, where the "Middle Forge" was located should be bought for $3000 to $5000, although the owners of the Middle Forge and another mill on the property had asked $8000 for that tract. Southard felt clearly that the Canal Company would require that land if a reservoir were to be made, but the trick was to buy it cheaply enough.[59]

The Middle Forge, mentioned above as part of the Mount Pleasant tract, was at this time owned by Moses Phillips, of Orange County, New York. Phillips had long been a Canal Company stockholder, as we saw when, in 1829, he was a member of a committee to obtain stock subscriptions.[60] According to Munsell, Phillips' forge was also known as the "Aetna."*[61] Apparently, Southard's idea of buying the Mount Pleasant tract for an additional reservoir was based on an earlier Canal Company plan to buy Mount Pleasant in 1836.[62]

The canal clearly needed more water. The number of complaints from the millers along the Musconetcong River were evidence of that. The agreement with the SUM about the mutual use of Greenwood Lake waters had brought temporary peace to the Passaic Valley, and the Company hoped that the latest enlargements of the

**It was this site that was bought by the U.S. Government in 1880 for a powder storage depot. Today it is known as Picatinny Arsenal.*

canal's reservoir system would silence the Musconetcong Valley, too, so that the enlargement could proceed.

Once the Canal Company's application to the State Legislature was passed in March, 1836, the board of directors then drew up an agreement to effect an amicable settlement of the differences between the mill owners and the Morris Canal Company. The Company proposed that:

1. The new charter supplement determine the amount of water to be discharged at Saxton's Falls into the river;
2. The Company be at liberty to build a mill at or near the dam;
3. The quantity of water necessary for the mills be determined by competent scientific persons mutually satisfactory to both the Company and the millers.

If this proposal was acceptable to the millers, it would apply only to the future and would not interfere with unsettled claims already submitted for damages previously sustained.

Another, even older grievance was brewing at Broadway over the use of Brass Castle Creek, where the creek flowed through the orchard of Adam Wandling. Wandling's agreement with the Canal Company was to allow the engineers to dig a canal 6 feet wide and a "sufficient distance" through his meadow and orchard to connect with the Creek. The Company was to have the use of the water except for what was "a reasonable supply for the wants of the immediate neighbors below the said Wandlings." Wandling was given $200 as payment in full for those privileges and "for the damages hitherto sustained to his barnyard, by his having to build a fence around his barnyard, and for all past damages to his grain fields due to leakage and for the present [1832] season, but not to prohibit him from suing for future damages due to leakage."[63] Incidentally, Wandling later became a director of the Warren Railroad.

Another agreement for the use of Brass Castle Creek concerned William M. Warne, who had sold a large parcel of land to Henry Winters of Broadway, but had kept the water rights for himself. These water rights Warne then "leased" to the Canal Company for $250 annual rent.[64] The sale to Winters had taken place on January 19, 1833.

When the Canal Company failed to pay its annual water rent for

1836 and 1837, Warne wrote to President Samuel L. Southard on January 4, 1838. Getting no answer to his letter, Warne caught up with Southard aboard the Canal Company's boat on its way to Easton, and Southard promised a reply when he returned to Port Colden. When Warne still got no reply, he went to Port Colden to look for Mr. Southard. Canal Company records do not indicate whether or not Warne's claim was honored; however, in 1845, when the canal was being enlarged, Warne got an injunction to prevent the Company from taking water from his part of Brass Castle Creek.[65]

Oddly enough, Warne had once offered to sell to the Canal Company all of his rights in the creek's waters.[66] That the sale was not effected is obvious, although in 1835 the Company still considered buying Brass Castle Creek, if expedient. At the same time (1835) the Company had considered buying lands "about the head and branches of Lubbers Run for the purpose of creating additional reservoirs."[67] The charter amendment which was to provide for an additional reservoir was still a year away.

Robert C. Stephens also made many claims against the Canal Company. Stephens occupied a gristmill seat on the Musconetcong River, but since he occupied a seat in the State Legislature, he was "handled" a little differently from other, less influential millers. His claims were settled for $525 on March 13, 1839, about a year before the Canal Company's decision to submit surveys, maps, and engineering data to Chief Justice Joseph C. Hornblower to establish and prove the need for still more reservoirs.[68] Stephens had already become an incorporator of the proposed Morristown and Easton Railroad, chartered on March 9, 1837. With him in that act of incorporation was Ephraim Marsh, among others.

Before 1838 was over, the Canal Company would once again apply to the State Legislature for "an increase of the Capital Stock of the Morris Canal and Banking Company, and also for the privileges of using the waters on a part or the whole of the line of the Morris Canal, for Manufacturing and other purposes, and for other privileges."[69] What "other purposes" and "other privileges" are meant will soon become apparent.

Meanwhile, the 1838 Rates of Toll reflected another change in Canal Company policy. Whereas boats had been allowed to use the canal free of charge if they carried 20 or more tons of cargo, the 1838 rates specified a toll of 2 cents per mile on each boat, regardless of what was carried. Passengers, who had ridden for 1 cent per mile in 1837, would now be able to ride for 1/2 cent per mile. All other rates were the same as those in 1837.[70]

Despite the financial problems plaguing the Canal Company's directors during 1838, the waterway itself enjoyed a relatively successful boating season. A July 4 celebration at Little Falls was swelled by the attendance of persons from the Paterson area, who were able to board Andrew Parsons' canal packet boat at his basin at 8 A.M. and at 1 P.M.[71] Parsons also ran his freight boats from Paterson to Newark, where forwarding was done by T.V. Johnson's "Daily Line of Packets to New York."[72]

Newark citizens began to complain in mid-1838 about boats on the canal on Sundays, in "violation of the Sabbath," and the Canal Company received a letter from Newark Justice of the Peace, D.R. Crane, in protest.[73] By October, 1838, President Southard wrote to say that nothing could be done "in relation to the boats not running on the Sabbath day this season," and promised to do something about it for the next season.[74]

By mid-season, 1838, the Newark *Daily Advertiser* announced: "The Morris Canal is now in prime order, and appears to be doing a brisk business . . . . The improvements in progress west from Mauch Chunk must ere long extend the communication to the rich valley of the Susquehanna, and its dependencies. As the business of the country revives, the benefits of this route must become more and more manifest."[75] The "improvements" referred to the LS&S Railroad, or to the Lehigh and Susquehanna (L&S) Railroad, or both.

A few snags in the smooth texture of canal operations occurred that year, however.[76]

> *One of the abutments of the aqueduct of the Morris Canal across the Pompton river near Mead's Basin gave way on Wednesday afternoon* [*August 1, 1838*] *and an empty canal boat which was passing at the time was caught about midway on the end of the aqueduct and was broken nearly in the middle. The helmsman, seeing the*

abutment giving way, saved himself by running to the forward part of the boat.

From 75 to 100 hands, besides a large number of teams, were employed in repairing the breach, which was to be completed by the following Saturday. Meanwhile, "a temporary stop gate was erected between the Pompton feeder, nearby, and the aqueduct, and the canal between the feeder and Newark immediately filled, so that this section of the canal is in full operation."[77]

Then, too, there was the minor "snag" of counterfeit $5.00 MC&BCo. notes, found to be circulating in Flatbush, Long Island. The notes were made "from an old plate which differs materially from the old one now in use."[78] Years earlier, the Delaware and Hudson Canal Bank's notes had also been tampered with "and altered from one to five dollars, and were then circulated."[79]

A sad note was struck in 1838 by the death of General William Colfax on Friday, September 7.[80]

> *He was buried with the honors of war, by a detachment from General Godwin's Brigade, composed of the Union Cadets, General Godwin's Guards, with their respective Bands of Music, commanded by Captains Conger and Allen. The military ceremonies were conducted by General Godwin, assisted by Colonel Garrison. The solemn and affecting Music of the Bands, the precision of the exercises, and admirable appearance of the military, with a procession more than three-fourths of a mile, gave to the scene an imposing character seldom surpassed.*

The troops from Paterson had traveled to the General's funeral by way of the canal and the newly completed Pompton Feeder. In a gesture of respect for the General, tolls in the amount of $14.14 collected from the troops were returned to them by the Canal Company's board of directors.[81]

A dredge for cleaning the bottom of the canal, its basins, and feeders, was to be contracted for during 1838. With an eye to deepening the prism to hold 5 feet of water, the board considered a proposal for a "mud machine" that was forwarded by N.L. Griswold, director George Griswold's brother and business partner.[82]

By the end of the 1838 boating season, the Canal Company presented its petition for "other privileges" to the Legislature. Henry Van Nest, Washington (New Jersey) grain merchant and grocer, made the presentation. When the substance of the petition was publicized, it drew a scathing attack from the *Paterson Intelligencer*:[83] "Among the causes which have contributed to bring odium upon chartered companies, is the grasping avarice and selfishness which characterize these soulless corporations." Among these corporations one of the worst, because most powerful, was the MC&BCo. "Not satisfied with the grants already conferred upon them," this Company had "come before the Legislature with the modest request to be allowed to increase their banking capital thirty millions of dollars. Of these thirty millions, they ask the state to subscribe ten millions, and pay for it by issuing state bonds, at 5 per cent interest." In view of these "odious" designs of the Canal Company, "the people cannot be too vigilant — for, as Jeremy Bentham well observes, the evil in legislation is, that the private interests of the legislators are too often in favor of a course injurious to the public, and in a conflict of interests, the latter generally goes the wall."

The author who penned that attack was not overstating his case, at all.

The Canal Company had presented its reasons for petitioning the Legislature for these "other" privileges in a 21-page pamphlet entitled: "Reasons For Granting to the Morris Canal and Banking Company An Increase of Banking Capital and the Loan of the State Bonds." The pamphlet is unsigned, and is unfortunately undated, but internal evidence and newspaper references to the Canal Company's petition indicate a probable date of early 1839.

The argument of the pamphlet is persuasive, citing 18 out of the nation's total of 26 states that were already issuing state bonds "to promote their internal prosperity and welfare." Seven of the 26 states, Massachusetts, New York, Pennsylvania, Maryland, Virginia, Ohio, and Michigan, had issued bonds "with a direct design to construct and promote internal improvements."[84]

Using the comparative growth rates of the states' populations as a standard, the pamphlet's author made his point: New York and Pennsylvania led the field in population growth. He attributed that growth to the fact that those states had "been the great theatres of the influence of internal improvements, constructed with means acquired by the issue of their state bonds." New Jersey, without a policy of issuing bonds, was far down on the list of population growth rates. "New York on the one side and Pennsylvania on the other, had been developing their gigantic resources." It was, he argued, "important, nay indispensable to their plans that they should establish communications with each other, to do which, they must pass through New Jersey."

The interests of these two great states had demanded and had been responsible for constructing, up to the time of this petition, "four great thoroughfares entirely through our state; two canals and two railroads; and to these chiefly is New Jersey indebted for her mediocre advancement."

"We ought to have had three commercial cities" in New Jersey, one with the same harbour and privileges as the city of New York, another with the same as Philadelphia, and a third at Amboy. "We should have had our own shipping and commercial connection with all the southern and western states, supplying them with goods, and receiving their cotton, rice, tobacco, pork, grain and flour in return. We should have had our long lines of warehouses, loaded down with domestic and foreign merchandise and produce, our custom house, and our European packets. But it is believed, that Jersey City was not even a port of entry until two years since [1837]."

The argument continues that it was not too late yet to have all these things. "But they will never come, if extensive capital be not first procured, and naturalized, to act as the lever to raise up these weighty and long neglected interests, and to serve as the seed of other and indigenous capital."

And so it went, on and on, piling reason upon reason, for 21 pages Apparently it didn't take the legislature one-tenth the time to refuse the request as had been taken to read it. According to the *Paterson Intelligencer* of October 7, 1840, the Canal Company's petition for an increase in its Capital Stock, and for the right to use canal water for "manufacturing and other purposes, and for other privileges," was "rejected in a most summary manner," and was "dismissed from the files of the House."

If 1838 ended with a storm of protest over the Canal Company's intention to use canal water for manufacturing, the year 1839 opened with a storm of unprecedent violence — from Nature. On January 26, 1839, "about 112 feet of the roof and upper story" of the Rogers Locomotive Works in Paterson was "blown off and carried by the gale into the adjoining yard," severely injuring a machinist named Crawford. The same storm flooded the cellars of several homes in Jersey City and damaged many wharves, while taking off the chimney and part of the roof of the Morris Canal Company's banking house on Green Street.[85]

Again, early in 1839, spring freshets hit Paterson with the worst flooding that city had had since 1810. The canal had given way under the excessive water in three different places between Little Falls and Paterson, with repairs expected to take two weeks.[86]

Possibly from the same storm system, the New Village property of Jacob Melick, on the opposite side of the state, was damaged by a breach in the towpath bank. The break was repaired, and the Canal Company awarded Melick $150. He thought he deserved more, however. From 1826 to January 15, 1838, Melick had written damages to his property, due to the presence of the canal, totaling $889.37-1/2. His letter to Canal Company president Samuel L. Southard, dated February 8, 1838, is even more interesting for its outward appearance than for its contents. The sheet of paper on which Melick wrote his claim was folded in such a way as to make its own envelope. On the "envelope" is a return address which reads: "Jacob Melick P.M., New Village, N.J., Feby, 8th." In the upper right corner is the word "Free," from which we might assume that Mr. Melick, P.M., was the Post Master, and that he didn't have to pay for his postage. What we would *like* to assume is that mail was carried by canal boat. However, despite the existence of many such folded letters in Canal Company files, that must remain only an assumption. (Sheets of paper thus folded were considered "letters" until the development of the separate envelope. The English practice of attaching stamps was not considered for adoption in this country until 1842.[87])

Samuel L. Southard resigned the presidency of the Canal Company at the beginning of 1839, withdrawing "from a lucrative station, that he may be able to devote himself the more exclusively to the public service in the Senate of the United States."[88] Mr. Southard actually had requested of the Canal Company's directors not complete withdrawal, but a change in his status with the Company — a change from being its president to being its attorney. Far from resigning in order to be "able to devote himself the more exclusively to the public service" as a U.S. Senator, he proposed to stay on with the Canal Company at an annual salary of $4000, plus his house in Jersey City and, of course, expenses.[89]

Edward R. Biddle ascended to the presidency once Southard resigned to become a U.S. Senator. When the 1839 elections were held, Peter M. Ryerson and Washington Irving were reelected to Class 5 directorships, as were Jacob Hurd, Alexander Lardner, and Robert Bacot. Bacot was former Cashier Robert Gilchrist's son-in-law.[90]

For the 1839 boating season, the Canal Company contracted with the Beaver Meadow Railroad and Coal Company for 50,000 tons of anthracite. The Morris Company was to supply the boats to transport that coal. Delivered to Jersey City, the coal was to cost $1.40 a ton.[91] In Paterson, the new coal yard of Westervelt and Geroe opened for business on the canal bank, boasting not only Beaver Meadow coal, but Hazleton as well, plus a "cargo of Virginia coal, for Blacksmiths."[92] With all that coal available, it is ironic that an unidentified black woman had been found frozen to death near the canal on January 24, 1839.[93]

Now that the anesthetic effects of the Panic of 1837 were beginning to wear off, President Biddle anticipated a lively increase in tonnage. He asked the board of directors for the authority to mortgage the canal to someone or some agency who would enlarge it to handle boats of 54-ton burden. Immediately upon being so authorized, Biddle announced that arrangements had already been made to mortgage the canal to the State of Indiana.[94]

The Morris Canal, of course, was already mortgaged to the Dutch firm of Wilhelm Willink. What Indiana received was a second mortgage — in exchange for a loan of a quarter million dollars. The State of New Jersey had refused to grant the Canal Company its request for an increase in banking capital. Note that it was banking capital sought. The State of Indiana was not so wise as her older sister-state.

REFERENCES

1. John D. Hicks, *The Federal Union: A History of the United States to 1865* (N.Y.: 1937), p. 222.
2. *Ibid.*, p. 303.
3. *Ibid.*, p. 423.
4. *Ibid.*, p. 428.
5. *Ibid.*, pp. 431-2.
6. *Ibid.*, p. 433.
7. *Ibid.*
8. *Ibid.*
9. Morris Canal and Banking Company *Minutes*, January 26, 1837 (hereinafter cited as MC&BCo. *Minutes*).
10. *Ibid.*, March 6, 1837.
11. *Ibid.*, January 31 and March 9, 1837.
12. *Ibid.*, June 15, 1837.
13. *Ibid.*, June 15, 22, and 30, 1837.
14. *Ibid.*, March 6, 1837.
15. A.C.M. Pennington, *Report of the Special Committee of the House of Assembly, on the Embarrassments of the Country* (Trenton, N.J.: 1837), p. 3.
16. Earl J. Heydinger, "Little Schuylkill & Susquehanna Addenda," *Railway and Locomotive Historical Society, Bulletin No. 109*, October, 1963, p. 28.
17. Weaton J. Lane, *From Indian Trial to Iron Horse: Travel and Transportation in New Jersey, 1620-1860* (Princeton, N.J.: 1939), p. 237.
18. *Ibid.*
19. Heydinger, *op. cit.*, p. 29.
20. *Paterson Intelligencer*, January 3, 1837.
21. MC&BCo. *Minutes*, July 20, 1838.
22. *Ibid.*, April 25, 1839.
23. *Ibid.*, May 2, 1839.
24. *Ibid.*, June 3, 1839.
25. Heydinger, *op. cit.*, *loc. cit.* p. 29.
26. *Ibid.*
27. Lane, *op. cit.*, p. 238.
28. MC&BCo. *Minutes*, October 1, 1838.
29. *Ibid.*, April 1, 1838.
30. *Ibid.*, January 26, 1837.
31. *Ibid.*
32. *Ibid.*, January 19, 1837.
33. *Palladium of Liberty*, June 28, 1836.
34. MC&BCo. *Minutes*, January 19, 1837.
35. *Paterson Intelligencer*, January 3, 1838.
36. *Ibid.*, May 10, 1837.
37. MC&BCo. *Minutes*, August 21, 1837.

38. Cornelius C. Vermeule, Jr., *Final Report of Consulting and Directing Engineer, June 29, 1929* (E. Orange, N.J.: 1929), opposite p. 76.
39. MC&BCo. *Minutes,* August 21, 1837.
40. *Paterson Intelligencer,* December 27, 1837.
41. MC&BCo. *Minutes,* November 25, 1837.
42. *Ibid.,* October 1, 1838.
43. *Ibid.*
44. Morris Canal and Banking Company *Docket Files* (hereinafter cited as MC&BCo. *Docket Files*), Docket #647.
45. *Ibid.,* Letter from Ephraim Beach to the board of directors, Newark, September 27, 1837.
46. *Ibid.*
47. MC&BCo. *Minutes,* October 13, 1833.
48. *Ibid.,* August 21, 1837.
49. MC&BCo. *Docket Files,* Docket #525-1/2.
50. *Paterson Intelligencer,* March 4, 1835.
51. MC&BCo. *Minutes,* February 12, 1836.
52. *Ibid.,* October 6, 1836.
53. MC&BCo. *Docket Files,* letter from James B. Murray to Louis McLane, May 28, 1835.
54. *Ibid.,* deed dated May 1, 1838.
55. *Ibid.,* proposal to the Hackettstown Millers, by Louis McLane and Peter B. Shafer, February 11, 1836.
56. *Paterson Intelligencer,* January 17, 1838.
57. MC&BCo. *Minutes,* April 30, 1834.
58. *Ibid.,* January 19, 1837.
59. MC&BCo. *Docket Files,* letter from Samuel L. Southard to Edward R. Biddle, Morristown, December 23, 1838.
60. MC&BCo. *Minutes,* May 12, 1829.
61. Munsell's *History of Morris County, New Jersey, 1739-1882* (N.Y.: 1882), p. 45.
62. MC&BCo. *Minutes,* October 6, 1836.
63. MC&BCo. *Docket Files,* Docket #595, dated June 16, 1832.
64. *Ibid.,* Deed dated June 19, 1832.
65. *Ibid.,* All documents referred to in this paragraph are in Docket #595.
66. MC&BCo. *Minutes,* March 6, 1833.
67. *Ibid.,* March 11, 1835.
68. MC&BCo. *Docket Files.*
69. *Paterson Intelligencer,* November 27, 1838.
70. MC&BCo. *Minutes,* March 15, 1838.
71. *Paterson Intelligencer,* July 4, 1838.
72. *Ibid.,* August 8, 1838.
73. MC&BCo. *Minutes,* July 13, 1838.
74. *Ibid.,* October 1, 1838.
75. *Paterson Intelligencer,* August 1, 1838.

76. *Ibid.*, August 8, 1838.
77. *Ibid.*
78. *Ibid.*, August 22, 1838.
79. *Ibid.*, April 12, 1826.
80. *Ibid.*. September 12, 1838.
81. MC&BCo. *Minutes,* October 1, 1838.
82. *Ibid.*
83. *Paterson Intelligencer,* February 20, 1839.
84. "Reasons for Granting to the Morris Canal and Banking Company An Increase of Banking Capital and the Loan of State Bonds," pp. 1, 4, 7. This pamphlet bears no date and no author's name.
85. *Paterson Intelligencer,* January 30, 1839.
86. *Ibid.*, April 17, 1839.
87. Eleanor Morton, *Josiah White: Prince of Pioneers* (N.Y.: 1946), p. 242.
88. *Paterson Intelligencer,* January 30, 1839.
89. MC&BCo. *Minutes,* January 15, 1839.
90. William H. Shaw, *History of Essex and Hudson Counties, New Jersey* (Philadelphia: 1884), p. 1171.
91. MC&BCo. *Minutes,* March 29, 1839.
92. *Paterson Intelligencer,* June 12, 1839.
93. *Ibid.*, January 30, 1839.
94. MC&BCo. *Minutes,* October 6, 1840.

13

Reaping the Whirlwind

WHEN the Canal Company's board of directors decided to push for completing the Morris Canal's enlargement, they met with Josiah White late in 1840 "to promise that their locks would be enlarged, at last — to 11 feet in width and 95 feet in length." The tide locks at Newark and Jersey City had already been enlarged. We have seen that Solomon White Roberts, another of Josiah White's engineers, had worked on estimates of the cost of completing the Little Schuylkill and Susquehanna (LS&S) Railroad. Considering the amount of "mutually beneficial cooperation" still evident among the builders of Morris Canal's "connecting works," it is not surprising to learn that a Lehigh Navigation engineer would be supervising the Morris Canal's initial enlargement. It was, after all, very advantageous to the Lehigh Company to have the Morris Canal made fit to receive its boats. Thus it was that in December, 1840, Josiah White's Chief Engineer, Edwin A. Douglass, was appointed Chief Engineer of the Morris Canal to superintend the canal's enlargement.[2]

Douglass had been a stockholder in the Morris Canal, possibly from its beginning, and is known to have held five shares of stock in 1831. His early professional affiliation, however, had been with the Delaware and Raritan Canal. He had superintended the building of that canal's western section and the lower part of its feeder.[3] From the Delaware and Raritan Canal, Douglass moved into Josiah White's Lehigh Valley about 1835-1836, and became Chief Engineer of the Lehigh Navigation. He is said to have kept that position for the rest of his life.[4]

According to a Philadelphia *Ledger* story reprinted in a Paterson newspaper, the Morris Canal was making arrangements for a "speedy enlargement," to enable it "to pass boats carrying 55 tons." An "efficient engineer," unfortunately unidentified, "has been engaged, and the work of the enlargement has been actually commenced, and will be prosecuted to completion without delay." Means are to "be provided by gentlemen who are large holders in the present stock." A line of freight boats was to be run from New York to White Haven by canal. The freight would then go to Wilkes Barre by way of the Lehigh and Susquehanna (L&S) Railroad. The *Ledger* pointed out that Philadelphians were apt to lose the trade from the interior to the Morris Canal unless steps were taken to enlarge the locks on the Delaware Division Canal. Baltimore was already "diverting considerable trade" from Philadelphia; only the future would tell if New York would "take the balance."[5]

The enlargement of the Morris Canal's locks, undertaken during the winter of 1840-41, was "to be immediately resumed and perfected, so that the navigation will be opened by the first of July. The work is to be paid for in post-notes, which, by a resolution of the Directors of the Bank, are to be receivable at par, in all debts due to the Bank, and for the purchase of coal transported on the Canal."[6]

Early in 1841, the *New York Times* [cited in the *Paterson Intelligencer*] was "happy to learn that the Morris Canal, through its whole length, 102 miles, has suffered no damage by the late freshets, which have been so destructive in many parts of the country." More than 1000 men were reported working at the enlargement, and the *Times* hoped that once Lehigh boats were able to navigate the Morris Canal, New York would get its coal "without the expense and loss attending re-shipment," hopefully by "next spring."[7]

The *Times'* optimism was a little premature, however. The Morris Canal was unscathed by the 1841 floods, but Josiah White's Grand Canal was heavily damaged by the raging Lehigh River. Work on the

Morris Canal's enlargement was understandably suspended. White would have needed Douglass to help pick up the pieces and put them back together again.[8] But White needed more than an engineer; he needed money for repairs to the Lehigh Canal, and he wrote to the Canal Company in March, 1841, asking the directors to raise funds for his reconstruction task.[9]

Shortly before the suspension of work on the Morris Canal's enlargement at the close of 1840, Canal Company President Biddle had issued a special report to the stockholders on the financial problems and future prospects of the canal.[10] He expected transportation costs to be cut by thirty-five to forty percent as soon as the "scow boats" in use on the Lehigh Canal, with their 54-ton carrying capacity, could navigate the Morris Canal. In fact, Biddle expected the Morris Canal's tonnage to double, with a corresponding increase in receipts. Stockholders would thus receive dividends on their investments.[11]

It was true, Biddle admitted, that the Canal Company's credit had been impaired "for more than a year past," and that the price of Canal Company stock was "greatly depressed." The Panic of 1837 had occurred "at a time when a large indebtedness existed to the Company." Then, too, there had been "the unexpected deprivation of a large amount of funds in the hands" of the Canal Company's London bankers, causing the Canal Company to relinquish "about seven hundred and fifty thousand dollars of active capital."*[12]

The improvements to the Canal — "feeders, reservoirs, piers, coal landings, basins, and permanent structures" — had cost about $850,000 since 1835. Another $800,000 had been spent, or invested, "in internal improvements connected with the canal," but Biddle expected that those improvements would "prove highly beneficial" in the future to expanding and extending the canal's trade. Another reason for the Company's financially embarrassed condition was the "transactions in state bonds in 1838, and 1839, which the sudden and heavy decline in the value of those securities in the latter part of 1839, disabled the Company from completing the payments for, punctually

This is possibly the reason for not paying off the Holland Loan, as McLane had promised the stockholders would be done.

at the stipulated periods." However, the prospects were actually good, despite the embarrassed condition of the Company's finances: there had been, in 1840, "occurrences," the character of which would "make an entire revolution in the prospects of the Company."[13]

Imported iron was still cheaper than that made domestically because of "very heavy duties imposed on iron, except railway iron," and because "iron being made in Great Britain was being manufactured more cheaply with bituminous coal instead of charcoal, as in this country." Charcoal had been very expensive to produce, while bituminous coal could be "used in its natural state, or after being reduced with comparatively little expense to the form of coke." At last in this country, "after long and preserving efforts," a way had been found to use anthracite coal in making iron. A method similar to that used in Britain had been "successfully and practically accomplished within the present year [1840]."[14]

In the Lehigh Valley, David Thomas, a Welshman imported by Josiah White, had been experimenting with the use of anthracite coal in smelting iron ore. He was using a new smelting process invented by his countryman, George Crane. A "Crane" furnace had been built at Mauch Chunk, and on August 8, 1840, Thomas sent to Philadelphia the first run of iron made with Pennsylvania anthracite coal and American iron ore.[15] The furnace was named the Lehigh Crane Company in honor of Crane. Here, suddenly, was a new potential market for New Jersey iron ore. Morris Canal boats could, once the improvements were done, carry ore up the Lehigh Canal to that furnace — and all the others that were bound to be built on the Crane-Thomas principles.

That development, plus the connecting works available to the Canal Company, was to work the "revolution" Biddle referred to. Coal would flow through all those connecting works: The Beaver Meadow and LS&S Railroads and the Lehigh Canal; plus the proposed Leggett's Gap and Belvidere-Port Colden Railroads, into the Morris Canal to be delivered to New Jersey's iron works. There, "iron would be manufactured and passed through the remainder of the line of the Canal to New York." As if that "occurrence" didn't justify renewed confidence in the Morris Canal, Biddle reminded the stockholders that the canal's inclined planes afforded "admirable sites for furnaces." Each plane, he claimed, provided enough water power for two furnaces, which could be approached by canal both at the head and base.[16]

These prospects were reason enough to keep working on the canal's enlargement. And, indeed, throughout 1841 enlargement work did continue, despite the destruction of part of the Lehigh Canal. In mid-1841, Vice President Edwin Lord informed the Newark *Daily Advertiser* that the work of enlargement was three-fourths done, and that the Canal Company's engineer felt that the work could be "completed to Newark within 30 days," or just about the time when the whole line of navigation "would be open from Mauch Chunk."[17] Then, as much as 4000 tons of coal per month would be delivered for sale at Newark and Jersey City.[18]

At Newark, the Canal Company's coal agent, George S. Wills,* would be ready to do business as soon as coal arrived.[19] However, one of the locks at Newark was not yet completed by October, and the Canal Company directors authorized a maximum of $450 to be spent on putting it in working order.[20] In anticipation of being able to use the canal again, Anthony Dey of New York wrote to the Canal Company "in relation to the transportation of merchandise on the canal." Seldon T. Scranton of Oxford Furnace, New Jersey, also wrote, asking whether the Canal Company's post notes "would be received for tolls whether they had matured or not." The answer was affirmative.[21]

Up to the time of Lord's letter (July 3, 1841), $67,063-worth of the Company's post notes had been disbursed in payment for work done.[22] There was apparently some doubt that those post-notes were legal, and the question was referred to Mr. Southard, Canal Company Counsel, for his opinion.[23] Southard reported that he considered the issue of post notes under the denomination of $300 as perfectly legal.[24]

As the opening day for navigation approached, sections of the canal were filled with water as they were finished and ready. Boats were to be "loaned" west of Stanhope, and enlargements would

**This could be "Mills," since the handwriting in the Minutes is not clear.*

continue "as far as funds would allow."[25] It is not entirely clear what "loaned" means in this context, but the canal was filled from the Delaware to Stanhope before the end of October, 1841. The *Paterson Intelligencer* on October 27 lamented the fact that the section between Stanhope and Boonton would probably not be completed until the following spring.

There was excellent reason to finish the Stanhope-to-Phillipsburg section first, at least from Biddle's point of view. For one thing, there were the new anthracite smelters in Pennsylvania to service. There was an even more excellent reason, however. In 1841, on April 5 (according to Ransom), the Stanhope Iron Company went into blast. Conflicting dates are given by other sources. There, at Stanhope, Edwin Post is credited with having successfully made the first iron in this state using anthracite coal.[26] (It is more likely, for reasons shown later, that C.S. Dickerson should get that credit.) Edwin Post was president of the Stanhope Iron Company, and Samuel R. Brooks was his associate. Both men were Canal Company directors. With them in their adventure, as principal stockholders, were Canal Company President, Edward R. Biddle, and Vice President, Edwin Lord. Messrs. Biddle and Lord had bought the iron works with Canal Company funds. If the reader thinks such a maneuver should be investigated, rest assured that the Canal Company's board of directors did too.

By October, 1841, the board of directors had resolved:[27]

> *That considering the position assumed by Messrs. E.R. Biddle and E. Lord towards the other members of the board, precluding the possibility of any harmonious co-operation and taking into view the facts which have recently come to the knowledge of the board in regard to the appropriation of large amounts to the individual use of these gentlemen — it is deemed indispensable that they should resign their situations as Directors of this Company.*

Neither Biddle nor Lord intended to withdraw voluntarily from the Canal Company. The board therefore resolved on October 15 that the men were to be removed from the offices of president and vice president of the MC&BCo. A committee to investigate the facts behind the alleged official mismanagement of company funds submitted its report. That report revealed the lengths Biddle and Lord had gone to bring the Canal Company to ruin.

The committee announced in the January 10, 1842, report "with great reluctance" that their findings were "most unfavorable to the prospect of any resuscitation so far as the interest of the present shareholders are concerned and that this state of things has been produced by great mismanagement and misapplication of funds on the part of the Executive Officers to whom their disbursement has been but too blindly confided."

The Canal Company's bank had "never recovered from the transactions of the years 1835 and 1837, when the prospects of large profits caused by dealing in foreign and domestic exchanges entailed upon it losses which have proved to be the primary cause of its present condition. Its agency for the sales of the bonds of the states of Indiana and Michigan furnished the means of postponing the fatal effects of that period, but they have come at last with a force increased in the proportion to the delay."

The death blow, in short, had been delivered by the president and vice president of the Canal Company. In what has to be the understatement of 1842, the committee reported:[28]

> *Messrs. E.R. Biddle, the late President, and Edwin Lord, the late Vice President, have had a personal interest in several purchases of mineral and other lands in Pennsylvania and New Jersey, on which they have made extensive improvements for their private benefit, with the funds of this Company, and that they have otherwise appropriated them to their own use and to that of their immediate connexions* [sic] *in a way which calls for the strongest censure of the Board.*
>
> [*We are*] *constrained here to say a few words in explanation of what may appear to be unaccountable that such a course of mismanagement should have been so long going on without an earlier decisive interference on the part of the directors. It must be borne in mind that the great balance of power lay with the shareholders in Philadelphia, who confided it in their proxies exclusively to the late President, Mr. E.R. Biddle. He was placed in that position*

by their interference although known at the time to be a large debtor to the Bank, and as the former directors retired from the board simultaneously with the ebbing tide of its pecuniary prosperity produced principally under the management of its finances in 1836 and 1837, the present board, most of whom are new members holding but a small interest in the stock and have not been sufficiently conversant with the course of proceeding to take these ulterior measures to wrest from the Executive Officers the unlimited powers with which their predecessors had invested them. It has been done at last, but too late to prevent the wreck which now presents itself to our view.

The committee then recommended protective measures be taken to protect the circulation of funds and deposits as quickly as possible and a decision be made how to treat those whose "large abstractions" had been the principal cause of the present crisis. The committee felt there was a need for the directors to be increasingly vigilant over its executive officers.

What were these large abstractions made by Messrs. Biddle and Lord which so devastated the Canal Company? According to the Committee's report, "the principal accounts which constitute the real debtor items are the following:[29]

The Montour Iron Works *$174,986.10*

These works are situated at Danville and WilkesBarre in Pennsylvania, and are stated to belong principally to Messrs. Biddle Chambers & Co., a firm in which it is understood Messrs. E.R. Biddle and E. Lord are or were partners. In exchange for the large amount of money thus advanced, the Iron Co. have given obligations for $227,500 (which makes an apparent balance in their favor) payable in iron at the market price in several installments of which $200,000 have been assigned to the State of Indiana and $27,500 to the State of Michigan as collateral security. Of*

**Thomas Chambers was a Canal Company director in 1841.*

this amount, $66,000 have become due but nothing has been paid and it is said that there are mortgages on the property (having a lien prior to this Company) to an amount equal to its value.

The advances were made in and after the month of December, 1839, by direction of Mr. Biddle and without any authority of the Board, he stating to the cashier at the time that it was a mere temporary arrangement for his personal convenience, and would soon be repaid. It was accordingly in the first instance placed to the individual debit of Mr. Biddle, but on the 7th August, 1840, was transferred by his direction to that of the Montour Iron Works.

Biddle's Montour Iron Company was described in the June, 1862, *Harper's New Monthly Magazine* as being, "next to the Cambria Iron Works, the largest in the State of Pennsylvania."[30] The works reprocessed old railroad strap-iron rails and rerolled them for use as rails again.

The committee's report continued with Biddle's and Lord's illicit deeds.

The Stanhope Iron Company $122,594.91

To represent this debt the Morris Canal Company received obligations for the delivery of iron (similar to those given by the Montour Iron Company) for $277,500, of which $177,500 are assigned to the State of Michigan and $100,000 to the State of Indiana.

The Stanhope Iron Company was chartered by the State of New Jersey with an original capital of $50,000 and afterwards increased to $250,000 and the principal shareholders were Messrs. Biddle and Lord, who have paid nothing in on the score of capital except for the sum obtained from the Morris Canal Company and there are original mortgages in other hands on the property to the amount of $36,000, which have a preference over any other claim and some of which are now in progress of foreclosure.

The original capital of $50,000 is represented by the following notes held by the Morris Canal Company —

Edwin Lord, dated 11 Aug. 1840	*$25,000*
Edward R. Biddle, dated 11 Aug. 1840	*10,000*
C.S. Dickerson (note prepared, but he declined to sign it)	*15,000*
	$50,000

These notes are made payable in 1846 with interest semi-annually, but no payment whatever has been made either on account of these promissory notes or the obligations for iron issued by the Company.

The advances thus made commenced in June, 1840 and terminated in February, 1841, but appear to be wholly unauthorized by the Board.

C.S. Dickerson *$17,678.00*

This appears on the books as a private debt or over-drawing by Mr. Dickerson but is understood to have been advanced by order of Mr. Biddle for the purchase of property near Dover and for making experiments in the use of anthracite coal in making iron. Mr. Dickerson is quite willing to transfer the property to this Company on being released from all claims by them.*

No authority of the Board is to be found for this transaction.

Mount Hope Mining Company *$46,939.06*

Incorporated by the State of New Jersey with a capital of $50,000 and purchased of the then shareholders by Mr. Biddle with the funds of the Morris Canal Company but without the knowledge of the Board.

This Company, as in the case of the Montour Iron Co., appears to be a creditor in consequence of being credited with a bond and mortgage on the property for $50,000 which was made in favor of the Morris Canal Company, assigned to the State of Michigan as a collateral under their guaranty. It consists of 800 acres of land in Morris County with a deposit of iron ore, but without improve-

*This mention of Dickerson's experiments suggests that he, not Edwin Post, developed the Stanhope anthracite iron-making process.

ments of any value, and it is not believed to be worth one-fourth of the amount of the mortgage thus given.

Edwin Lord's Collateral Account *$8,091.30*

The principal debit in this account arises from a loan made to Mr. Lord on the 13th May, 1841, of $5000 in Indiana Bonds under Mr. Biddle's authority on an assignment of Hazelton coal, stock and other securities of little value, and with Mr. Lord's written agreement to return the bonds in kind, which although repeatedly required have not been complied with.

Edwin Lord's Private Account *$10,328.00*

Mr. Lord has drawn this out from time to time and claimed a credit against it for salary as vice president at the rate of $6000 per annum, which Mr. Biddle ordered to be passed in the Books but was declined by the cashier unless directed by the Board.

It appears that Mr. Lord acted as vice president in the first instance by request of Mr. Biddle who promised to divide his own salary with him.

In October, 1840, Mr. Lord was elected vice president by the Board at which time the salary of each of the officers was $6000 per annum but in February, 1841, they were reduced to $4000 each.

Beavertown Property at Saw Mill *$7,861.61*

[*Assigned to the State of Michigan*]

Tamanend Coal Property *$80,900.61*

[*Consisted of 100 acres in Luserne County, Pa., received from the LS&S Railroad Co.*] *in consideration of releasing them from the lease of the canal which they had taken for a term of years. It is assigned as security to the State of Michigan.*

Canal Boats *$131,249.00*

[*Assigned to State of Indiana as security*]

There were other items on this list compiled by the committee; these are merely the more costly and, to the author, more interesting ones. These items alone account for more than $600,000 of Morris

Canal monies misappropriated by Biddle and Lord. Small wonder that the committee concluded its report on such a gloomy note:

> *The entire prostration of credit and reputed disappointments sustained by the public in relation to the financial affairs of this company leaves us no room to hope that any renewal subscription could be obtained to rebuild it in its present state of dilapidation.*
>
> *Proceedings have already been commenced for the foreclosure of the mortgage held in Holland involving the canal and its franchise under which it is most probable it will pass into the holders of that mortgage.*
>
> *In the meantime, the period appears to have arrived which has been anxiously waited for by the friends of this undertaking. It is a point now conceded that the planes are capable of conveying boats of 50 tons burthen and a small additional expense will enlarge to that extent the capacity of the whole work while the increasing consumption of coal and the various manufacturing establishments which have sprung up in its vicinity must gradually increase its trade.*
>
> *Be the consequences therefore what they may to individuals, the great public interest of the region through which the canal passes as well as of the city of New York, its point of termination, will be greatly benefitted by its completion.*

The report ends with the recommendation that the list of company employees be examined with an eye to economy, so that an "immediate reduction" could be made "both in number and compensation." How, one wonders, could the Company's losses be recovered by cutting the $15 monthly salary of a lock tender?

Bank Officers named were:

Isaac Gibson, Cashier	$2500 yrly	
Wm. A. Ellery, Assoc. Cashier	1500	
I.W. Kay, Bookkeeper	1000	
S.L. Southard, Counsellor	2000	
		$7000

Canal Officers named were:

Wilson Knott, Superintendent	$2500	
D. McCormick, Supv. @ $75 mo.	600	
Wm. Groff, Supv. @ $65 mo.	520	
Peter Hopper, Supv. @ $65 mo.	520	
W. Kerr, Collector @ $50 mo. (Easton)	400	
B.S. Shultz, Collector @ $75 mo. (Newark)	600	
		$5140

Plane and Lock Tenders, paid monthly, but only during the boating season:

23 Plane Tenders @	$42.50
18 Lock Tenders @	$15.00
1 Lock Tender @	$25.00
2 Lock Tenders @	$30.00
1 Lock Tender @	$40.00

Now we can see the extent of Biddle's financial wizardry. We can understand, too, how Indiana ended up with a worthless second mortgage on the Morris Canal and with stock in both the Beaver Meadow and LS&S railroads. And we can see how Biddle and Lord created the conditions that would destroy the Company.

This committee report, although dated November 3, 1841, was not entered in the Canal Company's *Minutes* until January 10, 1842. Why that was so must, of course, be a matter of conjecture.

It is inconceivable that the board of directors did not know what Biddle and Lord were doing with Canal Company funds. Samuel R. Brooks and Edwin Post, both directors, were associated with them in the Stanhope Iron Company. The notes drawn by Biddle and Lord were dated August 11, 1840 — almost a year and a half before the committee was appointed to investigate their activities. Then, too, there is the August 25, 1841, agreement between the Canal Company and the Iron Company, duly and fully recorded in the *Minutes* during a board of directors' meeting. By that agreement, "the Stanhope Iron Company, late the property of Robert P. Bell," was authorized to use canal water "from the head of Plane No. 1 West," which the Canal Company promised to keep supplied from the Summit Level. The Iron Company agreed to return the water into the canal.

Included in the agreement had been the Iron Company's right to take canal water to operate the Stanhope gristmill, located on the canal above Plane No. 2 West.[31]

In short, when Biddle wrote his December 31, 1840, "Special Report to the Stockholders," in which he remarked that the canal's inclined planes would provide "admirable sites for furnaces," he had the controlling interest in the Stanhope Iron Company, which was situated on one of those "admirable sites" above Plane No. 2 West, Stanhope — surely with Canal Company knowledge.[32]

For the use of the Iron Company, water could be taken out of the canal "into a spur constructed by them and leading from there southerly and terminating at or near to a wheel house erected by said Iron Company," the water to be returned to the canal at the foot of Plane No. 2 West.[33] Here, then, was the origin of the so-called "Singer Spur" at Stanhope. This little canal branch once fed the Stanhope Iron Company, then the Sussex Iron Company, then the Musconetcong Iron Works, and later the Singer Manufacturing Company.

Edwin Post was not the only person to switch allegiance from the Morris Canal Company to the Sussex Iron Company. At age 19, Edward A. Reeder had entered the employ of the Canal Company when the canal first opened to traffic (1831), working in the toll collector's office at Easton. In 1846, Reeder moved to Stanhope, where he entered the Sussex Iron Company's employ as bookkeeper. Later he became secretary of that iron works. Still later he would become a staunch lobbyist in the State Legislature for the M&E Railroad.[34]

In 1846, the iron company installed a steam engine to replace the water-powered engines formerly used in conjunction with canal water. In 1853, Edwin Post was said to have conducted experiments with franklinite ore, trying to extract zinc as well as iron from it. According to Snell, resulting explosions and fires destroyed everything and the company lay idle until 1864, when the Musconetcong Iron Company bought the works and rebuilt them.[35] The Paterson *Daily Guardian* puts the date of destruction at 1852, with "the explosion of the gas reservoir connected with the experimental furnace No. 4." Furnaces 1, 2, and 3 were all supplied with steam power in 1845, according to the *Guardian*, and were operated in order to process magnetic ore from the Irondale mines, "8 miles distant to the east." No. 4 furnace had been built to experiment on the franklinite ore.[36]

Regardless of the date of destruction, Edwin Post, president, and Edward A. Reeder, secretary and superintendent, were still at the helm of the Sussex Iron Company in 1856. The author possesses copies of several letters that passed between that company and the Cooper-Hewitt Company, addressed to or signed by Post and Reeder.[37] A December 24, 1856, letter is interesting because it names the mines from which ores were shipped to Cooper-Hewitt, and identifies Morris Canal boats 128, 1253, and 1309, and one named the *Roving Irishman.* Ores were taken from the Hilts and Osborn mines in Roxbury and Mount Olive Townships, respectively. The letters, disappointingly, list no ore from the Mount Hope mines owned by Biddle until 1855.[38]

Despite the apparent success in smelting iron ores with anthracite coal at Stanhope, which should have reduced the cost of locally produced iron, Edwin Post joined other local makers of iron in a protest about the tariff system on imported iron. If the duties on imported iron goods were to be lowered, it "must inevitably destroy the existing establishments, and effectively prevent the further extension of the iron manufactures in this country." The iron men "asked for no monopoly in this business," but despite the improvements they had introduced in the making of iron, and the resulting lower costs to consumers thereby, they still found it impossible "to compete with the surplus production of the foreign manufacturers." Indeed, "the great experiment of manufacturing iron in all its branches with anthracite as fuel, which has thus far promised to be entirely successful, and which before all other means, under a proper discrimination of duties, would tend to revive and resuscitate the iron manufacture of New Jersey, must be effectually crushed by the operation of the present tariff law."[39]

The group were to present a memorial to their representatives in Congress. Noted in the paper along with Post in this protest were M. John Ryerson, Samuel B. Beach, Jr., Peter P. Brown, Samuel F. Righter, Joel C. Post, John M. Eddy, Calvin A. Kanouse, Charles

McFarlan, John Warren, Elijah D. Scott, Ebenezer F. Smith, Whitfield H. Hurd, Guy M. Hinchman, Colonel Stansbury, George Vail, Joseph Jackson, Sydney Ford, Chilion F. DeCamp, and Henry Brevoort.

The year 1841 had closed with the Canal Company's notice that proposals would be received until January 5, 1842, "for letting the Morris Canal, for the Season of boating during the year 1842. The lessees will be required to put the canal in a condition for navigation at their own expense, and either pay a stated sum by way of lease, or to stipulate for a division of proceeds over and above the actual expenditures."[40] The "Lehigh Coal Company" was the first lessee, according to one newspaper.[41]

At Buffalo, New York, the Erie Canal netted tolls in the amount of $11,689 during the first eight days of the 1842 boating season.[42] During its first eight days of 1842, the Morris Canal faced foreclosure and bankruptcy. Governor Pennington, who was also the State's Chancellor, "appointed Receivers of the Morris Canal and Banking Company, viz. Messrs. Dudley S. Gregory of Jersey City, Elias B.D. Ogden of Paterson, and Stephen Britten of Elizabethtown."[43]

The receivers gave public notice on February 9, 1842, in the *Paterson Intelligencer* that "all persons along the line of the Canal and elsewhere are forbid taking, selling, or converting to their uses, any of the property or effects of the Company, and from purchasing the same from any of the former agents or managers of the company, and also from making payments to any persons but the Receivers, or their lawfully constituted agents, under such penalty as may fall upon them."

Gregory, who was Jersey City's first Mayor, and part owner of that city's first steel company, had also been a Canal Company director since 1838. Whatever his reasons, Gregory declined to serve as receiver, and A.C.M. Pennington was appointed in his place.[44]

The first notice about the probable lessee of the canal had named the Lehigh Coal Company. It was Lewis S. Coryell, however, who is credited with having leased the canal from the receivers. By April, his workmen were "busily engaged in making the necessary repairs to enable them to commence navigation by the first of May next."[45]

The newspapers speak of Coryell as a Philadelphian, although he had been born in Round Valley, Hunterdon County, New Jersey. Earlier, he had worked at "improving" the Delaware River above tidewater, then had superintended the "mechanical work" on the Delaware Division Canal. He was also a director in "several railroad, coal and other companies."[46] An educated guess makes him a close friend of Josiah White.[47] If he was, as was said, "extensively engaged in the coal business," one of the coal companies he was involved with was no doubt the Lehigh.[48] In any case, with Coryell as lessee and also friend of Josiah White, it was as if White *had* leased the Morris Canal.

Coryell's son, Martin, an engineer, and also a Delaware Division Canal alumnus, was put in charge of the Morris Canal "under John Cryder representing the English bondholders." Later he would work with Ashbel Welsh in building the Bel Del Railroad from Trenton up to Phillipsburg.[49]

This interesting article, picked up from the London *Spectator* and run in the Jersey City *Advertiser,* throws a little light on the "English bondholders" mentioned above:[50]

> *Some error appears to exist about the object of Mr. R. Chrichton Wiley's voyage to New York. That gentleman has recently sailed on the 'Columbia,' not as generally believed as general agent for enforcing claims of English holders of American securities, but as agent of English bondholders of Morris Canal and Banking Co. to watch the sale of the canal which will be put to auction in New York in consequence of the foreclosure of the mortgage of 750,000 dollars held by some Dutch capitalists upon the canal. The English holders have no chance of payment except by purchasing the canal and carrying it on for its own benefit.*
>
> *Shortly after the issue of the bonds of the English loan, a change took place in the directors of the company, and the whole management of its affairs got into the hands of Mr. E.R. Biddle, the president, and Mr. Lord, his partner*

and vice president, who contrived to lend to themselves and their co-speculators in iron and coal works in New Jersey and Pennsylvania, a large portion of the canal funds, nearly $300,000. It would appear that the whole proceeds have been applied to the benefit of Messrs. Biddle and Lord by the most shameful fraud. Mr. Biddle is the nephew of the celebrated Nicholas Biddle, the late manager of the United States Bank, and proceedings which have destroyed the Morris Canal Co. were a miniature of those by which the United States Bank has been utterly destroyed.

Whatever ill could be said about the conduct of the Canal Company's former managers, however, one newspaper admitted that "the canal itself is a work of almost vital importance to the country through which it passes, and an important convenience to our state."[51]

Despite what the Canal Company had earlier told Oxford Furnace, newspapers were now warning that the post notes issued by the now defunct Canal Company would *not* be "receivable for tolls on the canal."[52] That was not acceptable to many holders of those post notes, however, for the face of those notes carried the promise that they *would* be receivable for tolls, and the disappointed holders proposed to test the worth of that promise. "Those notes were taken in good faith by the farmers, mechanics, merchants, and laborers, and rather than be swindled out of their hard earned but now worthless money, an attempt will be made to compel its redemption as promised."[53]

Instead of rejoicing over the prospect of the lessees reopening the "improved" canal, Newark's town fathers were threatening to fill in the canal in their city unless the bridges were "made less dangerous to life and limb than their present delapidated state."[54] The threat was not carried out, of course, and water was let into the canal to the weigh lock at Newark in early September, 1842, with the "town level" to be filled shortly thereafter.[55]

The bridges apparently continued to be unsafe, for almost a year later an elderly man fell to his death from the canal bridge at the foot of Market Street. The bridge was described as "about 12 feet high," and "without any railing." It was one of the bridges that Newark had complained about "some months ago," but which had remained unfixed.[56]

Newark's councilmen weren't the only ones with complaints about the Morris Canal. Creditors by the scores wanted to be paid the monies that were owed them by the Canal Company. On January 5, 1843, the creditors met at the hotel of N.B. Luce, in Morristown, to decide how to proceed. The group included "the bill and post note holders, contractors, laborers, and particularly those creditors who became such on account of the late enlargement of the Canal." The 54 attendees each represented debts of $40,000 or more. A committee of five was selected to represent the creditors, and also "to investigate the several liens alleged to exist against the company, particularly that called the Holland mortgage." The committee consisted of James M. Porter, Daniel K. Allen, Jacob Hurd, Charles J. Ihrie, and former Canal Company director, Andrew Parsons. To a man, the creditors thought that "common justice requires that they should be paid what is honestly due to them."[57]

At a subsequent meeting of the creditors, it was decided that since a bill had already been filed in Chancery by the firm of Wilhelm Willink to recover its $750,000 loan, the creditors would institute proceedings in the Equity side of the Circuit Court of the United States for the District of New Jersey, to recover the creditors' monies. A counsel would have to be hired, of course, and to pay him, a special committee of three was appointed "to collect from the creditors 1 percent upon their respective claims." That committee consisted of Solin Chapin of Easton, S. Hayward of Stanhope, and B. Shultz, of Newark.[58] Mr. Shultz, as noted earlier, was the Canal Company's toll collector at the Newark weigh lock. Solon Chapin, we shall see later, was to be the contractor who rebuilt Planes 9, 10, and 11 West.[59]

As if there weren't enough old problems for the canal, 1843 proved a record year for new ones. There was navigation — sporadically, at least. Up to July 1, 1843, "1210 boats passed Port Delaware with 4,880* tons of coal," and the canal was said to be

**This figure is undoubtedly a misprint. If it were accurate, it would have meant only 4 tons of coal per boat. Probably 48,800 tons was meant, which would have placed a more reasonable 40 tons on each boat.*

"in good order, with every indication of a thriving business season."[60] The ink on that newspaper was barely dry when the *Intelligencer* on July 19 reported a breach in the canal at Port Delaware. The water "broke through the embankment at the foot of Mt. Ida, and emptied itself into the Delaware."

The "thriving business season" was further hampered by a "turn-out," or strike, of boatmen on the Lehigh Canal. Boats lay idle at South Easton, their masters having "gone home and entered into other employment."[61] One source claims that the strike had begun in the spring of 1843, that miners and raftmen had joined in, with the boatmen from the Schuylkill Canal later swelling the ranks.[62] Not to be daunted by a strike among workmen, the Lehigh Coal and Navigation Company drew the water out of its canals, ostensibly to make repairs. It was expected that the canal would be filled again on Tuesday, August 22, 1843, and that boats would once again be on their way. There had been "a settlement of the differences with the Beaver Meadow boatmen, but the Lehigh boatmen continue to stand out."[63] The Commonwealth of Pennsylvania actually filed bills against and indicted "the individuals arrested for a conspiracy in obstructing the navigation of the Mauch Chunk Canal for riot."[64] It was doubtful that the indictments would result in a trial, however, since an agreement had been reached, and business was to resume immediately. The strike caused the price of freight shipment to advance 10 cents per ton-mile, presumably to afford raises for the strikers.[65]

While that strike was in progress, a severe rain storm dumped over 15 inches of rain in 12 hours over most of the Northeast. Paterson, Little Falls, Wagaraw (Hawthorne), Oldham (North Haledon), Newark, Belleville, New Brunswick, Princeton, and Bound Brook in New Jersey were particularly hard hit. New York had heavy rain, but no damage was reported. "Philadelphia suffered severely, and according to the papers of that city, the loss is immense. A number of bridges in that neighborhood were destroyed, buildings blown down, and seven or eight lives lost."[66] An "extensive breach" in the Morris Canal bank about a mile south of Paterson caused damage to roads and bridges.

Between Paterson and Little Falls there were four very serious breaches in the space of one mile, ranging from about 30 to 250 feet in width. The Little Falls Turnpike was extensively damaged by the canal water as it rushed out, and stone and gravel were deposited in corn and buckwheat fields along the way. One "small breach in the canal a short distance above the stony road" was fortunately kept from becoming a large breach because of a pile of lumber lying on the towpath. The water carried the lumber into the gap, thus providing some reinforcement and holding back some of the water.

Below Paterson there were several other breaches in the canal's banks, and the weigh-lock basin at Newark "gave way, and the flood carried away the basin near the wharves, and thus the canal was soon emptied. The damage to the canal is estimated at $2000. Prompt measures have been taken to repair it."[67] Even Dame Nature seemed to be conspiring to keep the Canal from operating successfully.

Kicking a man when he is down has never been considered sporting, but on top of all Morris Canal's troubles, the city of Newark was again threatening to fill in the canal "where it crosses River-street in that city, sufficiently to make it safe for those who may travel this street, unless the Receivers of the Company shall make the necessary repairs to the bridge, within a period of ten days." The receivers had "refused to put any repairs on this bridge" because "this part of the canal" was not leased![68]

These problems about bridges eventually forced the Canal Company to seek yet another charter supplement. On March 19, 1845, the State Legislature passed an act to allow "the proper officers of any city, township, or county of this state, and the several turnpike companies therein, whose road passes over the said canal, may, for a proper consideration to be paid, contract with the said Morris Canal and Banking Company for the maintenance and repair of said bridges," by the aforesaid agencies or officers. Once built, those bridges must "be kept and maintained as other public bridges," by whoever, or whichever agency, built them.

According to the Company's *Minutes,* out of this act later emerged an agreement between the Company and the city of Newark, "in reference to bridges, etc., in the city of Newark." It is the "etc." that is interesting here because it refers to the Center Market which was later to be built in the space over the canal in Market Street. A tentative agreement about the market was reached in March of 1847 between the city and the Company, but the Canal Company directors wanted it understood that, when it was built, there be an "abundance of light" under it.[69]

Meanwhile, canal business proceeded. Apparently, Coryell had leased the Morris Canal for only 1842, for in 1844, the Easton *Argus* announced that Messrs. Mills & Sykes, "the enterprising lessees of last year" [1843] had once again "rented the Morris Canal from the Receivers of the Company," and had intended "to put the whole line in operation for the ensuing season."[70]

This area needs further research, but another educated guess is that "Sykes" was Lorenzo A. Sykes, former Morris Canal engineer who had gone on to work for the NJRR&TCo. In 1862, a Mr. Sykes worked on the D&H Canal,[71] and one of the D&H's locomotives was named the "Major Sykes."[72] Mills (and this is a less-educated guess) could have been John B. Mills, who had worked on the Erie Canal, and who had worked on the 1824 survey for the D&H Canal.[73]

The 1843 boating season, under Messrs. Sykes and Mills, closed on December 6. The "old" MC&BCo. was in its death throes. In October, 1843, Chancellor William Pennington was asked to decide which of the many creditors had a prior right to be paid. His decision was that Wilhelm Willink, holder of the first mortgage on the canal, should be satisfied first. Willink's cause had been argued by Asa Whitehead, while Benjamin Williamson had represented the State of Indiana in its claims.

In rendering his decision, Chancellor Pennington pointed out that Indiana had known of the Morris Canal's first mortgage when it entered into its agreement with the Canal Company for a second mortgage. If there was any money left after Willink was satisfied, he decided, it would "be appropriated to the payment of the mortgage" held by Indiana. Indiana appealed Pennington's decision, but it was upheld by the Court of Appeals.[74] The canal must be sold to satisfy its creditors.

And so it was that on October 21, 1844, the Morris Canal was auctioned off for $1,000,000. That amount was used to pay Mr. Willink

$908,403.37, including the interest on the original loan of $750,000. The remaining $91,596.63 was paid to Indiana.[75]

On October 1, 1844, just before the canal was sold, several members of the New Jersey bar were asked to give an opinion as to "what rights and privileges will be vested in the purchasers" of the Canal. Their collective opinion was that the sale of the Morris Canal would "pass the whole of the mortgaged premises specified in and covered by the said mortgage, and that the purchaser will acquire all the chartered rights and privileges of the said 'The Morris Canal and Banking Company' . . . . The powers and privileges possessed by the company under the charter will reside prefectly and entirely in the purchaser." The new owners, however, "would be discharged from all claims by, or liabilities to, the creditors and stockholders of the old company, and such creditors and stockholders would be deprived of all claims and liens upon the canal or the chartered privileges of the company." The new owners might keep the name "Morris Canal and Banking Company," or they would be permitted to adopt "another and different name."[76] The signers of that opinion were P.D. Vroom, Henry W. Green, B. Williamson, Asa Whitehead, Frederick T. Frelinghuysen, David B. Ogden, and J. Hoyt.

Ira C. Whitehead, Master in Chancery, brought the gavel down on the old Morris Canal and Banking Company on October 21, 1844. Benjamin Williamson, Asa Whitehead, and John J. Bryant, "all of the county of Essex," had bid "one million of dollars; and no other person bidding as much," they became the new owners of the Morris Canal.

The "Morris Canal and Banking Company of 1844" was thus created.

REFERENCES

1. Eleanor Morton, *Josiah White: Prince of Pioneers* (N.Y.: 1946), p. 220.
2. Morris Canal and Banking Company *Minutes,* December 31, 1840 (hereinafter cited as MC&BCo. *Minutes*).
3. James P. Snell, *History of Hunterdon and Somerset Counties, New Jersey* (Philadelphia: 1881), p. 109 (hereinafter cited as Snell's *Hunterdon*).
4. Morton, *op. cit.,* pp. 205, 228.
5. *Paterson Intelligencer,* January 13, 1841.
6. *Ibid.,* March 31, 1841.
7. *Ibid.,* January 27, 1841.
8. MC&BCo. *Minutes,* February 18, 1841.
9. *Ibid.,* March 25, 1841.
10. "Special Report to the Stockholders of the Morris Canal and Banking Company," December 31, 1840.
11. *Ibid.,* pp. 9, 11.
12. *Ibid.,* pp. 3, 4.
13. *Ibid.,* pp. 4, 5.
14. *Ibid.,* p. 7.
15. Morton, *op. cit.,* pp. 216, 218, 226.
16. *Ibid.,* pp. 8, 9.
17. Newark *Daily Advertiser,* July 3, 1841.
18. *Paterson Intelligencer,* July 18, 1841.
19. MC&BCo. *Minutes,* August 24, 1841.
20. *Ibid.,* October 15, 1841.
21. *Ibid.,* September 24, 1841.
22. *Paterson Intelligencer,* July 18, 1841.
23. MC&BCo. *Minutes,* September 24, 1841.
24. *Ibid.,* October 7, 1841.
25. *Ibid.,* October 15, 1841.
26. James P. Snell, *History of Sussex and Warren Counties, New Jersey* (Philadelphia: 1881), p. 467 (hereinafter cited as Snell's *Sussex*).
27. MC&BCo. *Minutes,* October 14, 1841.
28. *Ibid.,* January 10, 1842.
29. *Ibid.*
30. "The Catawissa Railroad," *Harper's New Monthly Magazine,* Vol. 25, June to November, 1862, p. 27.
31. *Ibid.,* August 25, 1841.
32. "Special Report," December 31, 1840, p. 9.
33. MC&BCo. *Minutes,* August 25, 1841.
34. Snell's *Sussex,* p. 472.
35. *Ibid.,* p. 467.
36. Paterson *Daily Guardian,* August 3, 1859.
37. Abram S. Hewitt Papers.

38. Munsell's *History of Morris County, New Jersey: 1739-1882* (N.Y.: 1882), pp. 55-6 (hereinafter cited as Munsell's *Morris*).
39. *Paterson Intelligencer,* January 12, 1842.
40. *Ibid.,* December 29, 1841.
41. *Ibid.,* January 26, 1842.
42. New York *Observer,* May 7, 1842.
43. *Paterson Intelligencer,* January 26, 1842.
44. Jersey City *Advertiser and Hudson Republican,* February 22, 1842.
45. *Ibid.*, April 19, 1842.
46. Snell's *Hunterdon,* p. 292.
47. Morton, *op. cit.,* p. 182.
48. *Paterson Intelligencer,* May 25, 1852.
49. Snell's *Hunterdon,* pp. 292-3.
50. Jersey City *Advertiser and Hudson Republican,* May 10, 1842.
51. *Ibid.,* May 17, 1842.
52. *Paterson Intelligencer,* April 20, 1842.
53. *Ibid.,* May 25, 1842.
54. Jersey City *Advertiser and Hudson Republican,* August 2, 1842.
55. Newark *Daily Advertiser,* September 6, 1842.
56. *Paterson Intelligencer,* June 4, 1843.
57. *Ibid.,* January 11, 1843.
58. *Ibid.,* March 8, 1843.
59. MC&BCo. *Minutes,* July 11, 1850.
60. *Paterson Intelligencer,* July 12, 1843.
61. *Ibid.,* August 2, 1843.
62. Morton, *p. cit.,* pp. 243-4.
63. *Paterson Intelligencer,* August 16, 1843.
64. *Ibid.,* September 6, 1843.
65. Newark *Daily Advertiser,* August 25, 1843.
66. *Paterson Intelligencer,* August 9, 1843.
67. *Ibid.*
68. *Ibid.,* August 16, 1843.
69. MC&BCo. *Minutes,* March 26, 1847.
70. *Paterson Intelligencer,* February 21, 1844.
71. Manville B. Wakefield, *Coal Boats to Tidewater* (Grahamsville, N.Y.: 1971), p. 89.
72. *The Story of Anthracite* (N.Y.: 1932), p. 106.
73. *Ibid.,* pp. 63-4.
74. *Report of the Morris Canal Investigating Committee on the Abandonment of Navigation on the Morris Canal* (Trenton, N.J.: 1903), pp. 58-60.
75. *Ibid.,* p. 70.
76. *Ibid.,* pp. 73-4.

14

A Brand New Start

SOME historians have claimed that just before the canal was sold, Wall Street speculators had bid Morris Canal stock up in order to unload their shares. "This they did in the belief that the floating supply was small and that the receivers had no right to sell treasury stock. The latter, however, acting under directions from the Chancellor, sold a portion of the shares held by them."[1] The speculators questioned the legality of the receivers' sale of stock, but it was ajudged legitimate, and they had to resort to new methods of recouping their losses. They bought in the Dutch loan at a 20-percent discount ($600,000 instead of $750,000), and paid the state of Michigan $3000 for the "various assets which had been given as security for the money owed that state," meanwhile disposing of their own Morris shares in the stock market.[2]

The newly reorganized Canal Company then "bought in the claims of the Wall Street clique at a large profit to the latter. In the reorganization an assessment of $25 a share" was called for.[3]

The real "owners" of the MC&BCo. of 1844 were not Williamson, Whitehead, and Bryant. The men who had put up the money for the sale were:[4]

John Rankin	1 part	S.R. Brooks	1 part
J.W. Hayne	1 part	J.J. Coddington	1 part
J.C. Green	1 part	J. Hoyt	1 part
George Griswold	1 part	N.G. Bucknor	4 parts
C. Clark	1 part		12 parts

On November 30, 1844, the new company elected 23 men to its board of directors.

Thomas Tileston	Benjamin Williamson
Joseph B. Varnum	Peter McMartin
Jophet Bishop	Frederick T. Frelinghuysen
Chester Clark	John Strader, Jr.
George Griswold	James J. Scofield
Jonathan J. Coddington	John J. Bryant
John Rankin	Henry McFarlan
Jacob Little	Elijah Scott
Samuel R. Brooks	John A. Willink
Daniel Tyler	John C. Green
Woodward Haven	Zebedee Cook, Jr.
Joseph N. Alsop	

Benjamin Williamson was chosen president at that first election. He declined the position, however, and Daniel Tyler was then elected first president of the MC&BCo. of 1844.[5] One of the first official acts of the new company was to announce that the Morris Canal was to be enlarged.[6]

Daniel Tyler was just the man for the job.[7] Tyler had resigned West Point in May, 1834, when President Jackson refused to give him a well-deserved promotion. In 1835, as president of "an iron and coal company in Lycoming County, Pennsylvania,"* he began to build the "first coke, hot-blast furnace ever built in America, and after a hard struggle of two years succeeded in making pig iron, using bituminous coal for the fuel." Tyler left that iron and coal

**Williamsport is in Lycoming County, Pennsylvania. The reader will remember from Chapter 11 that all the railroads planned for Elmira, Erie, and Pittsburgh were to have their origin there, and that the LS&S Railroad would ultimately connect the Morris Canal with Williamsport.*

company when it failed because of "the great fall in the price of the manufactured article," and became president of the Norwich and Worcester Railroad in 1840.

The Canal Company, according to Tyler's biographer, "had fallen into the hands of New York capitalists owning large stock interests in the Norwich and Worcester Railroad." Once Tyler put that road in good and profitable order, he was "invited to become the President and Engineer" of the Morris Canal Company. At the time Tyler accepted that post, the locks of the canal had been but "partially rebuilt to pass the Lehigh Canal Company's boats carrying seventy-five tons of coal." He saw at once that if the carrying capacity of the Lehigh boats "could be given to the canal, it would constitute the only direct route from the Lehigh mines to the city of New York." First, however, the inclined planes would have to be redesigned and rebuilt.

Tyler's "great problem" to be solved was that of water supply. This he did, "by diminishing the consumption of water, enlarging the prism of the canal's waterway, and by an ingenious use of 'Summit Planes' with railroad tracks upon which cars were moved by the power of turbine wheels."

The new president first experimented with a new type of plane at Montville and, satisfied that the mechanism would work, decided to use the same type of construction at Port Colden.[8] That plane proved more successful than the one at Montville[9], although both planes had been able to pass "boats carrying 60 tons of freight, and with less water than was used on the old planes to pass boats carrying 22 tons."[10]

Ready to undertake a "very material enlargement," the Canal Company advertised to receive proposals for the work.[11] The re-building of the planes was ordered for the summer of 1845, with two planes to be completed by June 1, 1846.[12] Boats of 60-ton burden were to be bought, and a contract for 100 boats "of a size adapted to navigate the Morris Canal when enlarged" was given to W.H. Phillips of the Millport Chemming Company.[13]

Contracts were also let for canal sections to be enlarged. According to one report, "The labor used was largely Irish, as were the con-tractors. The contractors received about one dollar a day for laborers and for superintendents $1.65. The laborers were boarded in shanties along the line of work at the cost of $2.00 per week, with meals at $.09 each. The contractors received from $.085 to $.15 per cubic yard for earth excavation, $.41 per cubic yard for solid rock, and $.18 per cubic yard for loose rock."[14]

In April, 1845, 1200 men were at work along the canal, hoping to have it ready for the new 60-ton boats by June 1.[15] The Company announced that 130 of the old Morris Canal boats "lying in the Morris and Lehigh Canal" were to be "sold at Public Auction, to the highest bidder, on the following days, at the following places, viz: All those boats in the canal from Newark to Little Falls, at the house of Mrs. Chandler in the city of Newark, on Monday, the 14th of April; all the boats between Little Falls and Stanhope, at the house of Jacob Hurd, in the town of Dover, on Wednesday, the 16th day of April; and all the boats from Stanhope to Mauch Chunk, at the house of Peter Carlin, in Phillipsburg, on Friday, the 18th of April." The notice was signed by "Wilson Knott, Agent."[16] Before the canal was sold, Knott, together with A.L. Downes, had leased the Morris Canal boats from Indiana, to which they had been assigned by Biddle.[17] Now, President Daniel Tyler was authorized by the board of directors to sell or "let" any or all of the company's boats.[18]

Some sections of the canal were completed in June, 1845, and filled with water. The entire canal prism was expected to be enlarged "early" in July, 1845.[19] Water was being let into the rest of the canal from Lake Hopatcong about mid-July, "but an accident occurred which prevented it from filling the channel" and delayed the canal's opening.[20] 1845 was a year of near-drought, and water supply was a problem. The problem was further compounded by the Stanhope Iron Company's use of the waters of Lake Hopatcong and the Musconetcong River "for manufacturing purposes without restriction."[21] Much more about that problem will be forthcoming, however.

The canal prism was originally "intended to be 32 feet on water line, 20 feet on bottom, with 4 feet of water." When Tyler took over the task of enlarging, he found "that the slopes had washed in so as to make the canal 36 feet on water line, 15 feet on bottom, and 3-1/2 feet deep on the average, and in some places, the canal did not exceed 8 feet on bottom, and 3 feet of water."[22] The enlarged canal was now "42 feet wide on the water line, 25 feet on the bottom, and 5 feet of water (equal in every respect to the Delaware Division of the Pennsylvania Canal)."[23] It is probably from this time that the practice of lining the banks with stonework, or rip-rap, was begun to prevent erosion from water currents.

Not only the canal prism had been in bad condition when Tyler became Canal Company president and engineer. The 23 inclined planes "were found in the worst possible order, having been built of wood, and some ten years ago." During the 1845 boating season, $12,892.35 had been spent on repairs to the old planes, but they could not "by any possibility, be repaired to stand beyond another season." Even if they could be repaired, the old planes could not "be made or modified to take over a boat with more than 25 tons of freight." The new planes did in fact "take over boats with 60 tons freight," and were "built of imperishable materials." Tyler's contribution to the design of Morris Canal's inclined planes was to substitute iron parts where wooden ones had been used. Two planes had already been scheduled to receive Tyler's improvement. The one at Port Colden was completed; the one at Montville was almost completed; and the Newark Plane was to be rebuilt in 1846.[24] Also to be rebuilt was the Mountain View Aqueduct.[25]

After the canal enlargement was complete, Tyler was sure that "freighting can be done as cheap on it as on any other canal of like capacity in the United States."[26] Even the newspapers seemed to agree with that. Regardless of what the Canal Company had been in the hands of stock speculators, if the current work to "make the canal what it is capable of being made, it must eventually be as productive as any canal in the country." The Delaware and Hudson Canal, was at that time "opposing the construction of the New York and Erie Railroad" and seemed to have lost sight of "a much greater rival than the Erie Railroad can possible be." That rival was "the Morris Canal."[27]

Old controversies and new tempers were to flare up between the Morris Canal and SUM during the enlargement of the canal. As Tyler had mentioned in his report to the directors, 1845 had been a dry year. In an apparent effort to get more water out of the canal at Paterson as the 1836 agreement had allowed, Roswell Colt, himself, was said to have gone up to Garret Mountain on July 18, 1845, and "hoisted" the canal's waste gates at Stony Road. Mr. Spence, the Canal Company's gate keeper "immediately put them down again, and spiked them down." Mr. Colt, it was reported, had sworn that he would hoist the gates again the next day, "even if he goes to the Penitentiary" for it.[28]

President Daniel Tyler received this report from canal engineer James Moore while he was in Washington, New Jersey, inspecting the work being done on the Port Colden Plane.* He advised Moore to take no action other than shutting the gates, but to let the law decide between parties. If Mr. Colt should tear out the Canal Company's gates, Moore was to tell Peter Hopper, one of the canal superintendents, to "shunt off the entire feed at Long Pond [Greenwood Lake]."[29]

Later, on August 8, 1845, Moore reported to Tyler, who was then in Montville inspecting the plane, that Colt had again taken action. As Spence told the story:[30] "At about noon a party appeared, in number about ten, headed by Edward Dickerson, Esq., Clerk of the District Court of the U.S., and son of the Hon. Philemon Dickerson, the Judge of that Court.

"Mr. E. Dickerson proclaimed that he was the head of the party, and was employed by the Society for the Establishment of Useful Manufactures to destroy the gate, and let the water out of the canal; and as proof of this fact, he waved a paper, which he said was a Bond of Indemnity from the Society, to hold him harmless for the outrage he was about to commit." Later, at Paterson's Franklin House, Mr. Dickerson took from his pocket a six-barrel pistol — a Paterson-made Colt six-shooter, no doubt — and put it in a pistol case which had been behind the bar. He was said to have told the bartender that "he had fired of all the charges after he had left the ground" where the canal gates were, "for the purpose of seeing how near he

It is not possible to identify Moore precisely. However a Lt. James Moore served with Alexander P. Berthoud during the Civil War, as did Williams I. Powers. Both Berthoud and Powers would become Canal Company Superintendents. A Col. James Moore surveyed the Elizabethtown & Somerville Road, forerunner of the Central Railroad of New Jersey, of which he later was chief engineer.

could come." Since there was water going over the SUM's dam at the time, Moore could see no reason or "motive but that of malice for the violence committed."[31]

That, at least, was the Canal Company's side of the story, as told to the board of directors by Tyler in a "special report." The report was published in some of the area newspapers. The editors of Paterson's *Intelligencer* expressed the hope that the same newspapers would be fair and give equal coverage to the SUM's side of the story. To lead the way, the *Intelligencer's* issue on September 3, 1845, was almost entirely devoted to the SUM's story.

In Chapter 8 we saw how the era of peace and harmony between the Canal Company and the SUM had been ushered in by the Paterson and Hudson Railroad (P&HR) — New Jersey Railroad and Transportation Company — Canal Company agreement. In Chapter 11 we saw how the two corporations wedded their individual interests in the Stony Road Reservoir project. Now, the honeymoon was over. Roswell Colt's statement about the "violence" on Garret Mountain was a step-by-step description of how the "marriage" had come about, and how it had suddenly dissolved.[32]

> *In pursuance of the* [*1836*] *agreement, the Society built the dam across the Passaic, and performed all the covenants on their part, at an expense of more than ten thousand dollars. The Canal Co. availed themselves of the reservoirs on the headwaters of the Passaic, and diverted some of its tributaries. They made a small canal from their canal to the waters of the Society, in which the Society put a gate about thirty feet from the canal bank, with an iron aperture inserted, through which but three square feet of water could be discharged. The Canal Co. then tore away their canal bank and suffered the waters of the canal to flow to this gate.**

The Society's statement was published as a one-page "Extra," in which Roswell B. Colt denied that there had been any personal violence, "nor was more force used than was necessary for removing the unlawful obstruction to the passage of the water," despite the

**This "small canal" can be seen on Van Wagoner's 1840 Map of Paterson.*

Canal Company's claims about the six-shooter wielded by Mr. Dickerson.

Colt explained how the Canal Company had filled in the mouth of the ravine at Stony Road so that canal water could no longer flow into the SUM's Stony Road Reservoir, or "Upper Basin." Since reduced water meant diminished power for Paterson's mills, the Society had ordered the solid embankment torn down. The new managers of the Canal Company of 1844 showed by their actions that they intended to repudiate the compromise agreement of 1836. Cold would show them by *his* actions that he did not intend to stand for any further encroachments on the SUM's rights by this Canal Company. Once again we see what a change in the cast of characters can do to a drama.

Upon first hearing of the July incident at the canal waste gate, Colt had sent "three distinct messages, through directors of the Company, to their President," requesting an interview to settle their differences. The requests were ignored. Instead, Canal Company employees had begun filling the mouth of that ravine with earth and stone. Despite repeated attempts to communicate with Canal Company President Tyler, Colt was unsuccessful in having the work stopped at Stony Road. The pistol incident of August 8, 1845, made the Canal Company cease building, but building resumed a day or two later.

Then, on Monday, August 11, two of the Canal Company Counselors were "dispatched" to Schooley's Mountain, where Ephraim Marsh, soon-to-be-elected Canal Company president, kept Heath House, a "watering place in Morris County." Chancellor Vroom was visiting there at the time. According to Colt, those two counselors presented no Bill of Complaint nor request for injunction against the SUM. However, that same day, again according to Colt, Canal Company cashier John J. Bryant and another counselor had scooted off and somewhat illegally gotten an injunction from a Master in Chancery at Trenton. That injunction was served on the SUM on August 12, 1845, the same day that the SUM asked for an injunction against the Canal Company!

Chancellor Vroom's decision on the Bill in Chancery was handed down on December 17, 1845. It is worth examining, as it was interpreted by the newspapers:[33]

1. The right of the Canal Company to drop the waters of the Canal into the Passaic, and to take an equal quantity out is res adjudicata [*That is, a matter already decided by a competent court*].

2. Supposing the agreement [*of 1836*] *made after the date of the Dutch mortgage* [*1832*], *between the companies,* [*to be*] *binding on the present Canal Company, what is the agreement?*

If for a perpetual flow of 3 square feet, the evidence shows that enforcing the agreement would ruin the Canal, thus working to them irreparable mischief. Under such circumstances, the Court would not grant an injunction against the Canal Company.

3. If the case were clear of all other difficulties, the Chancellor is not satisfied that the agreement calls for a constant flow of three square feet. He expresses no opinion as to its meaning.

4. If the agreement be what the Society allege, should Injunction issue?

> *1. Is the Canal Co. bound by the agreement? If yes. —*
>
> *2. Then the Society have not fulfilled the agreement on their part.*
>
> *3.* [*There is*] *no sufficient evidence that the Society have not sufficient water if properly husbanded to supply 22 square feet.*

If the agreement is what the Society allege, they must not permit waste of water. If it is impracticable for them to prevent the escape of the water, then it is impracticable for them to comply with the agreement.

5. If the operation of filling the enlarged Canal in a dry season is the cause of the deficiency of water at Paterson, the cause has ceased and there is no need of an injunction. I am, however, by no means clear that filling the Canal occasioned the deficiency.

There is no ground upon which the injunction against the Society can be dissolved. It follows that the Injunction against the Society, already granted, must continue.

The storm of protest that raged in response to Vroom's decision is too lengthy and detailed to go into here. The reader is referred to the *Paterson Intelligencer* of December 24, 1845, for a full-page rebuttal. A brief excerpt from that rebuttal conveys the editor's opinion that Vroom's decision was one "which would be discreditable to the merest Attorney; it displays no reach or grasp of thought, no familiarity with just and sound general principals, and none of the clear, distinct, and firm impressions which such a case will make upon a logical, discriminating mind."[34]

This was the same Peter D. Vroom who had signed his name to the October 1, 1844 "Opinion," stating that the "New" Morris Canal Company would fall heir to *all* of the "Old" company's rights, powers, and privileges. Apparently none of the new Canal Company's managers felt that the duties and responsibilities of the "Old" company went along with the privileges conferred by that sale.

The controversy between the two corporations over water had become a source of bitter mockery to Patersonians. That famous September 3, 1845 issue of the *Paterson Intelligencer* also carried two items reflecting that bitterness:

School Exercises

Question: John! What is a Sieve?
Answer: A compound of small stones and sand, such as composes the banks of the Morris Canal.

Question: Why is the Morris Canal like an old miser?
Answer: Because it keeps all it gets, and gets all it can.

Question: What is a lyre?
Answer: I don't know, s-s-sir. Ask President Tyler.

The second item, also bitterly satirical, concerned a topic which would also blossom into a controversy, although not of the magnitude as that of the water supply. This topic was the Canal Company's post notes, with which the costs of the enlargement were to be paid.

"Morris Canal Post Notes. — We would advise those who run boats on the Morris Canal, to provide themselves with Post Notes to pay their toll. The Company say they still have banking powers; of course then they will redeem. 'Gentlemen, walk up to the Captain's office and settle.'"

Chapter 13 has shown that the farmers, mechanics, laborers, contractors, and creditors of the Canal Company had planned to test the promise of payment contained in those "promissory notes." A town meeting was held at Peter Archdeacon's Cottage on the Cliffs, symbolically perhaps, since that inn overlooked the SUM's dam and the Great Falls. The entire town was urged to attend the meeting "for the purpose of making application to the Legislature for a law to protect the people of Paterson against the impositions of the Morris Canal Company."[35]

Paterson, primed for any future controversy, was now going "to compel the Canal Company to receive their post notes in payment for tolls." The people of Paterson still had faith that the State Legislature "will do what is right in the premises." The best thing that governing body could do, however, "would be to appoint commissioners to fill it up. The public, we think, would justify such a proceeding at this time, on the general principle that it was a nuisance. At this day the idea of keeping the thing in operation for the sake of its benefits to the public, appears perfectly ridiculous."[36]

The State of New Jersey *did* bring suit against the Canal Company during the 1846 March Term "on indictment for a nuisance," at which time, a summons was "duly issued."[37] However, the summons was "returned, not served, and an affadavit having been made by the Sheriff of Passaic County, that he hath made diligent inquiry, and cannot ascertain the name of any President, Secretary, Cashier or Director, resident in this County, on whom to serve said process," a public notice was inserted in the *Paterson Intelligencer* for six weeks. The purpose was to notify the Canal Company that it was ordered to make an appearance, or plead to the indictment!

Note that the sheriff could find not a single Canal Company officer resident in *Passaic* County. It is an amazing fact that of the 23 directors for 1846, 11 were New Yorkers, one was a Philadelphian, one was from Connecticut, one from Massachusetts, three were from Jersey City, two from Elizabeth, one was from Newark, and three were from Morris County! Passaic County was not represented at all. Note, too, that up until now Patersonians had complained — bitterly enough — about the Canal Company, but that now they were urging the canal's abandonment!

During the understandably intense controversy at Paterson, the Canal Company advocates were sending letters to the newspapers. These letters were signed by "A Tenant," the inference being that the author was a tenant under the SUM's water-lease system. One of these letters suggested that Paterson ought to adopt the use of steam-engines to power its mills instead of being so totally dependent on the waters of the Passaic River. *That* suggestion brought about a response almost as violent as that about the Canal Company's "high-handedness" at Garret Mountain!

The Canal Company had claimed — or more accurately the "Tenant" had claimed — that the enlarged canal would make coal deliveries so easy and prices so cheap that it would be worth Paterson's while to consider the conversion from water power to coal-generated steam power. Replying to that suggestion, an SUM advocate, signing himself "Vindex," wrote that only the Canal Company could suggest such a thing, since "it is well known that to produce an amount of power by steam, equal to that given by one foot of water, the wages of the hands to attend the engine would be as much as the whole rent given for the water . . . . When steam power becomes as cheap as water power, then our mill owners and operatives, and all depending upon them, will move to Jersey City, or Gloucester Point, and lease *'that coal'* to be consumed by those who are unable to get there."[38]

"A Tenant Under the Society" had also replied to the Canal Company's suggestion by asking: "Does not the company perceive, that if coal can be made to operate mills cheaper than water, Paterson is ruined? The operators would not certainly pay freight to Paterson and back again, upon goods and raw material, when they might take the coal at Jersey City or Philadelphia cheaper than at Paterson. If coal were cheaper than water, no man would be mad enough to drive a spindle at any distance from the market." Moreover, the "Tenant" was sure that "if steam is to be used, it will be applied anywhere rather than at Paterson."[39] And of course it was.

It is surely no coincidence that at this same time the Canal Company convinced Edwin Post to install a steam engine of at least 120 horsepower at the Stanhope Iron Works. In a way, the Canal Company had had just as much trouble over the water supply on its Western Division as it had on the Eastern. In a document dated December 16, 1845, Post and Samuel R. Brooks agreed to have their new steam engine built and ready for operation by February 15, 1846, under a penalty of $100 a day to be paid to the Canal Company for failure to comply. The Canal Company's part of the agreement was to appoint "a person to assist in making the contract" (for the specifications, possibly), and to direct the building of the engine. The Canal Company, moreover, was to have the privilege of contributing $12,000 towards the building of the engine, while Post and Brooks were to pay the remaining $18,000.

On December 16, 1845, both parties agreed that the Canal Company should build a new reservoir between Plane Nos. 1 and 2 West. Since the reservoir would flow the lands of the Reverend Charles Brooks of Boston, the Canal Company was to buy those lands for their new project. On the same day, the Charles Brooks property was deeded to the MC&BCo. of 1844. Materials necessary for construction were to be taken by the Company free of charge from lands owned by Post and Samuel Brooks at Stanhope.

The agreement between Post-Brooks and the Canal Company was "signed, sealed and delivered" by John J. Bryant, Canal Company cashier, in the presence of William H. Talcott and J.N. Hutchinson. Talcott had been hired as Resident Engineer of the Western Division of the canal earlier in the year.[40]

The water question on the whole line of the canal was serious. How serious, in all of its ramifications, in regard to the use of the Passaic River's tributaries, is indicated by the fact that the Canal Company actually considered — privately, of course — building aqueducts across the Rockaway at Dover *and* at Powerville.[41] *Something* had to be done. Paterson was agitating to have the Canal terminated at that city, and planning to ask the Legislature for the right to make the Passaic River navigable from Paterson to Newark.[42] Perhaps the Canal Company's proposal that Paterson use steam power instead of water was sincerely made, born of the knowledge that as the countryside was turned into towns with factories the water available for everyone and every purpose would diminish.

The Stanhope Reservoir, today's Lake Musconetcong, was scheduled for completion for April 1, 1846. Failure on the part of the Canal Company to live up to that stipulation in the December 16, 1845, agreement with Stanhope Iron Company would result in heavy penalties: payment to Edwin Post, $3.00 for each and every day that his grist mill in Stanhope could not operate for lack of water; $40.00 for each and every day that the upper blast furnace could not operate. Completion of the reservoir, then, for a Canal Company that was already in dire financial straits, must have been high on the list of priorities entitled "work to be done:" preparation of the canal "for a good and efficient navigation" for the 1846 boating season.[43]

On the subject of water supply, Canal Company President Tyler, in his last report to the board of directors on February 11, 1846, commented that the drought of 1845 had[44]

> *possibly tended to strengthen the opinion of those who doubt that it* [*the Morris Canal*] *has an adequate supply of water for an extensive business. If the supply is to be measured by the past season, allowances should certainly be made for the fact, that at the time the canal was filled, in July, all the ponds and springs along the entire line of the canal were dried up, and had to be refilled with water brought from the reservoirs, and that, in some instances, wells situated three or four miles distant, which were dry, were filled by the water draining from the canal. The soakage alone, taken up in this manner, undoubtedly exceeded twofold the amount of water required to fill the canal. Add to this this discharge of all the water from the 17-mile level, by the break at Hackettstown* [*sic*]*, *it will be readily seen that the past season offers an unfavorable criterion in properly estimating the supply of water belonging to the Morris Canal.*

Tyler resigned his position as Canal Company President in February, 1846. He claimed it was because the Company showed "no disposition" to go ahead with the rebuilding of "nine additional planes required." Feeling that his work with the Canal Company was

**Either the 11-1/2-mile level is meant here, or else Hackettstown is wrong.*

finished, he "reluctantly gave up a public improvement which, if completed at that time at a cost not exceeding $386,000, would not only have given large pecuniary benefits to its stockholders, but would have added very substantially" to his "professional reputation." Tyler went on to become president of the Macon and Atlanta Railroad in Georgia, where he remained until 1849, when that road was considered successfully operating. Then he took on the presidency of the Schuylkill and Susquehanna Railroad, 1858-61.[45] Benjamin Williamson apparently filled Tyler's unexpired term, for his name as president appears on a deed on file with Canal Company records.[46]

William H. Talcott, who had been hired earlier as Resident Engineer of the canal's Western Division, was made Chief Engineer and Superintendent. In that capacity, his primary goals were to rebuild the inclined planes and to enlarge the canal's carrying capacity. Talcott was well qualified to handle the job. A native of western New York, he began a career as civil engineer at age 21. His engineering education had been gained through Professor Joseph Henry, later of the Smithsonian Institution. By the time he reached age 30, in 1838, Talcott was employed on the Genessee Valley Canal. Then, in 1841, he was hired as Resident Engineer on a western section of the Erie Canal, during its enlargement. Thus, when he came to New Jersey to work on the Morris Canal's enlargement, he was as well prepared for the task as any man could have been.[47]

A new canal superintendent, V.W. Bridge, was hired to help Talcott police the canal; M.B. Inches, to be an assistant engineer. Bridge, hired in March, 1846, was to receive $1500 yearly, plus 2.5 percent of the profits of the business of the canal after paying all expenses.[48]

Unlike the improvements made during 1840-41 and paid for by post notes, the 1845-60 enlargement was to be financed by means of Canal Company bonds bearing 7 percent annual interest. Of the 23 directors for 1846, five members held absolutely no shares of stock in the MC&BCo., although they held tens of thousands of dollars worth of 7-percent bonds! This listing appeared in the *Paterson Intelligencer:*[49]

Director	Origin	Shares	Bonds (where known)
George Griswold	New York	80	$ 80,900.00
John Clive Green	New York	0	$136,650.00
Jacob LeRoy	New York	50	
Samuel R. Brooks	New York	0	
Theodore Dehon	New York	0	
Joseph B. Varnum	New York	0	
J. Soutter	New York	0	
E.R. Biddle	New York	5-2/3	
Jonathan J. Coddington	New York	33-1/3	
Zebedee Cook, Jr.	New York	100	
J. Strong	Jersey City	1	
Peter McMartin	Jersey City	30	
Henry Southmayd	Jersey City	1	
Frederick T. Frelinghuysen	Newark	10	
Benjamin Williamson	Elizabeth	1	
John J. Bryant	Elizabeth	1	
James J. Scofield	Morristown	1	
Henry McFarlan, Jr.	Dover	29	
Daniel Scott	Powerville	1	
Daniel Tyler	Connecticut	1	
D. Henshaw	Massachusetts	10	
Asa Whitney	Philadelphia	800	$ 5,000.00

Small wonder that the correspondent who placed that list in the newspaper worried about the fate of the "new" Canal Company. Less than "directors" of the Company, these men were its *"creditors"* since the bonds were guaranteed to pay a 7-percent annual yield, while returns on the stock shares were subject to fluctuations in the canal's affairs.[50] Nevertheless, it was these 7-percent bonds that were to finance the rebuilding of the canal.

During this enlargement period, the board of directors requested that the Chief Engineer make monthly statements of work done and costs involved. After the board approved the statements, the cashier was to pay a lump sum to the engineer who would disburse monies to contractors. He was then to submit receipts to the cashier to prove that he had paid the bills.[51] The same engineer was to be responsible for policing the entire canal, being "generally charged with keeping the canal in an effective state."[52]

The Canal Company called for a meeting of its stock and bond holders for March 18, 1846, in order to try to get further "subscriptions of $250,000, making $500,000," which were "needed to carry out the plans of the company and put the Canal in complete working order." The directors hoped the stock and bond holders would subscribe that amount, "and as an inducement, the whole $500,000 of bonds are to be made mortgage, instead of uncovered bonds, and bear 7 percent, instead of 6 percent interest. The above named amount will not suffice to renew the planes and stock the Canal with boats, which would require $350,000 more, but will repair all the planes and make the work in all respects a finished work. The Company will, for the present, depend upon carriers to furnish boats."[53]

J. and C. Collins, contractors for the Hackensack River section of the canal, submitted their account for payment. The Canal Company proposed that the Collins' take $4000 in company bonds, payable in 10 years; $2000 cash; and $6000 in company notes, payable in 1, 2, 3, and 4 months.[54] Collins & Company settled for $3500 in cash; $7100 in two $3550 drafts, payable in 4 and 6 months, respectively; $2727.50 credit on the Canal Company's books, which Collins would "freight out;" and $1000 of the mortgage bonds for the $700,000 loan.[55]

At about this time, Benjamin Williamson resigned as Canal Company President and was replaced by Asa Whitney, whom Wheaton Lane described as a "railroad engineer." A "company of capitalists," unfortunately unidentified, offered to advance the $290,000 "needed to finish and furnish" the canal, *if* the stockholders would agree to sell them "20,000 shares, at $7.00 per share. A considerable number of shares have been pledged under this arrangement, and about a third of the bond-holders, in order to help the matter through, offer to release 10 percent of their claims. The Company have no means with which to pay the next interest, and a foreclosure will probably take the property from the stockholders entirely unless something is done to prevent a sale."[56]

Some stockholders "filed a bill in chancery to show that the debt of the company, which the $700,000 mortgage was executed to secure, was usuriously contracted, and that usury (7 percent) has actually been paid, and therefore show that the bonds and mortgages given to secure that debt, should be given up and cancelled."[57]

As far as the *Paterson Intelligencer* was concerned, this was the *third* time the Canal Company had been brought to ruin.[58] And it had not yet so much as rebuilt *one* of its much-touted new inclined planes.

Plans to begin rebuilding were made by midsummer, 1847, however, when it was "proposed to use wire rope on the new plane in place of the sprocket chain used on the others. The wire rope will cost much less, will be lighter, and, it is believed, be much more durable than the sprocket chain."[59] The "new" plane was to be built at Port Colden on land to be acquired from Abraham Woolston.[60] Actually, the Canal Company "swapped" lands with Woolston, giving him "so much of the company's lands on the north side of Plane No. 6 West as can be spared in view of the present location and construction of the plane." These spare lands were "to be exchanged for lands taken from Woolston for the site of the new plane."[61]

This would be Chief Engineer Talcott's laboratory, where he could start "from scratch," not from "patch" of another man's work. But even with the inclined plane Talcott was to build here at Port Colden, the type modern Morris Canal buffs are familiar with, there could not be an increase of business or tonnage on the canal "beyond what can be transported in boats the company can control." Therefore, President Asa Whitney proposed buying "300 more boats in time for the next season's business."[62] These boats and those already owned by the Company would be pledged or sold or rented to anyone who would guarantee to continue operating them on the Morris Canal.

Since it was not considered desirable to own boats, "it is not believed there will be any necessity for the company to be the owners of boats long after the canal has been put in the condition contemplated. As soon as that is done there is no doubt but that the principal consumers and dealers in coal along the line of the canal will find it to their advantage to own boats enough to transport the coal they consume or sell. Besides, the best boatmen will soon become owners of boats." Until that time, however, when the canal was in good order and boatmen and coal dealers owned their own boats, the board of directors would ask counsel to draw "a second mortgage on the canal and its appendages to secure the payment of the bonds of this company to the amount of $100,000."[63]

Moreover, President Whitney was to open negotiations with the Lehigh Coal and Navigation Company "to ascertain on what terms and conditions that company will furnish boats to run on the Morris Canal as soon as it is improved as contemplated," and "the number of boats the Lehigh Company will engage to so furnish." The Canal Company would, in exchange, "reduce the toll on all coal transported in said boats over the whole line of the canal from Port Delaware to Jersey City to 50 cents per ton, and allow all such coal to be deposited on and shipped from this company's pier and slip at Jersey City free of any charges therefor."[64]

The Canal Company then leased 170 of its scows to Messrs. J.F. and C.F. Randolph. Actually, the boats were pledged as security for a $20,000 loan made to the Company by Randolph.[65] To accommodate all these new larger boats, the Delaware River Ferry channel had to be deepened to a uniform five feet at low water, so that loaded boats would be able to negotiate the crossing.[66]

Some new hinged boats, or section boats, were being built to replace the smaller "flickers" or "stiffs" that had run on the canal before the enlargement. "At the commencement of service in 1846, the Canal Company owned one hundred and eight of the new section boats. These were rented to boatmen, as it was considered that, by keeping down the investment of the individual boatmen, this policy would lead to greater use of the canal."[67] As we saw earlier, boats were also rented or leased out because there had always been a niggling doubt as to the legality of the Company's owning boats. The reader will recall that a separate Morris Canal Transportation Company had been formed for such ownership.

To encourage boatmen to engage in transporting canal freight, the Canal Company "announced that prizes of $200 each would be given to the captains making the quickest trip and the greatest number of trips during the season."[68] The two Cavanaugh brothers, Garret and Mike, of Phillipsburg, took their boats from Port Delaware to Jersey City in four days, and became known as "the fast" boatmen.[69]

All of these "improvements" and "inducements" could not have been urged on rapidly enough to cancel out a similar trend among the Canal Company's former allies, the embryonic railroads. The M&E Railroad had already surveyed and put under contract its Dover extension. Ironically, the New Jersey Iron Company at Boonton, another early ally of the Canal Company, was proposing to furnish the rails for the track extension, made with ore and fuel shipped by canal boat.[70]

Stephens and Condit, early Canal Company allies and beneficiaries, had taken on another partner, Mr. Cox, and were now engaged in transporting freight across the Newark Bay and the Hudson River for the M&E Railroad. Their advertisement is both prophetic and pathetic:[71] They were prepared to transport goods of every description between Newark and New York, "forwarding all goods destined to go by canal or railroad, by such boats and to such points as the owner may designate. As they confine themselves to their old business of *Freighting between Newark and New York,* and do not intend to interfere with the canal boating, they trust goods by their line will be likely to reach their place of destination with less delay than those shipped by lines owning boats on the canal." With these apologies to Morris Canal, Stephens, Condit and Cox moved with the times into the era of the Iron Horse.

President Asa Whitney, like so many other Canal Company directors, was a staunch advocate of the Iron Horse. Originally a New York merchant, Whitney, like so many other Canal Company directors, had become wealthy in the China trade. He is said to have been such a Far West enthusiast that he conceived an idea for a transcontinental railroad.[72] He had set forth that idea in 1831 in a book that advocated a railroad from the Atlantic to the Pacific.[73] In 1844 he had submitted a proposal to Congress for a railroad from Lake Michigan to the Pacific Coast. Just how much Whitney's "transcontinentalism" attracted him to the Morris Canal at this point, or how much of it he brought to bear on the canal, is a subject for much more intensive research. All that can be said here is that President Abraham Lincoln signed the Pacific Railroad Act on July 1, 1862, ten years before Whitney's death. Whitney, who served as

Canal Company President for only one year, 1847, remained a director until 1869.

Whitney, with his 800 shares of Canal Company stock, represented Philadelphia interests, according to the New York *Herald.* That newspaper hoped the stock would stay in Pennsylvania, calling it "the greatest curse in the shape of a fancy Wall Street ever had," and advising investors "to keep clear of this stock," which the *Herald* described as "a shell" that would crush in investors' hands before they were aware of it.[74]

Ephraim Marsh had become a Canal Company director on January 13, 1847, according to the Company's *Minutes.* He was to replace Whitney as president of the Canal Company on April 3, 1848. On that same day, Chief Engineer William H. Talcott became a director.[75]

Among the other directors of the Canal Company for the 1848 boating season were Henry Southmayd, of Jersey City; Nathan Smith, State Senator from Sussex County 1846-7, 1848, and incorporator of the Sussex Mine Railroad, 1848; George W. Edwards, Josiah White's attacker, of the Philadelphia and Reading Railroad; Alfred Van Santvoord, investor in the Catskill and Canajoharie Railroad and owner of the Hudson River Day Line; Caleb H. Valentine, Warren County Judge and part owner of the Hackettstown Blast Furnace; and William Dickey, large landowner of Paterson, and P&HR Railroad backer.

One of the first tasks to be tackled by the 1847 board of directors was to reduce expenses by paring salaries.[76] The president's salary was cut to $500 yearly; cashier's, $1800 yearly; bookkeeper's, $1800 yearly; engineer's, $1000 yearly (plus expenses); superintendent's, abolished.

When the new MC&BCo. of 1844 finally issued an "annual" report, it covered the three years that the new company had been in operation: October, 1844 to March 17, 1848. By then, the canal had cost the new company $542,242 in excess of the $1,000,000 it had paid for the canal. This sum included "its present indebtedness of a little over $500,000, chiefly payable in 1856."[77] The Canal Company's receipts for those three years had been less than $150,000, while its operating expenses and interest had cost just less than $190,000. The canal, as it was then constructed, obviously was not very profitable.

If the canal "could not be made profitable when there was no railroad to Morristown, and none to Paterson," one editor asked, "how insane must he be who supposes that it can be made so, after both of these roads are in full operation," and after a railroad communication had been established "between Easton and New York, and between Stanhope and Dover, and New York?"[78]

The question was, as it still is, legitimate. The Morris and Essex (M&E) Railroad was even then being extended from Morristown to Dover "under the superintendence of Mr. Beach, the well-known engineer."[79] Paterson, too, continued to agitate for more railroad trackage into the state's interior so that the city could become less dependent on the canal for coal deliveries.

Paterson had more to complain of these days than water supply. The Canal Company tolls for 1847 were adjusted so that "through traffic" took precedence over "way trade" on the canal. "The reduction or drawback of toll on coal is 10 cents per ton on all coal passing from Port Delaware to Newark for consumption of Newark and vicinity, and 20 cents per ton on all coal shipped or transported from Newark to Jersey City, New York, or Brooklyn."[80] That meant that coal for Paterson cost about 35 cents a ton more than Newark-vicinity users had to pay. More often than not, coal had to be back-shipped; i.e., shipped first to Newark, then back to Paterson.[81]

The Canal Company was also offering reduced tolls for "iron ore shipped upon the Morris Canal during the ensuing [1847] season and passing the Delaware River."[82] In short, New Jersey ore miners were encouraged to ship their ores west to the iron works in Pennsylvania! "New Jersey iron appears to be just now in good repute. The Crane Iron Company of Allentown, Pa., has recently purchased a rich mine . . . near Newton, Sussex County, for $1000, and the Andover mine in the same county is supplying Cooper's Railroad Works at Trenton."[83] This same Crane Company had also opened the mines in Mount Olive Township in 1846. The Sussex (or Stanhope) Iron Works of Edwin Post leased these Mount Olive mines in 1848,[84] and operated the Gidd Mine.[85]

Until the success of the Lehigh Crane furnace in making iron with anthracite, no one in the Lehigh Valley seems to have thought of much besides transporting anthracite. Now a trend was developing to ship coal straight through to New York and beyond, while New Jersey ores were being shipped to Pennsylvania. This trend was significant.

A second trend was the growing competition between the Morris Canal and its sister anthracite waterways. "The Delaware Division of the Pennsylvania canal, which is a competition for the trade of the Lehigh Valley, extends up the Delaware River from Bristol to Easton, 60 miles. Its dimensions are 40 feet wide at top water, 25 feet at bottom, and 5 feet deep, with locks 96 feet long between quoin posts, and 12 feet wide. These dimensions are adequate to pass boats loaded with 60 or 65 tons." The Morris Canal, as originally built, "could not compete successfully with the Delaware Division of the Pennsylvania canal," unless it were enlarged.[86] Consequently, it *was* enlarged "with a view to enable it to participate in the rapidly growing coal trade from the Lehigh Valley."[87] If the Lehigh Canal could just as well and easily ship coal east on the Morris as south on the Delaware Division, the Morris Canal would get its share of traffic. At the time Whitney made this report, a new outlet lock was being built at Wells Falls, opposite Bordentown. Up until 1848, coal was shipped from the Lehigh down the Delaware Division Canal to Bristol. Coal slated for the Delaware and Raritan (D&R) Canal had to be towed back upriver by Lehigh Coal Company steam tugs in order to enter the D&R.[88] The new outlet lock would thus help divert Lehigh coal away from the Morris Canal and into its rival.

To get that share of coal traffic, the Canal Company sought, and got, a charter supplement entitling it to reduce the number of shares of its capital stock to 10,250 shares and to issue Preferred Stock in the amount of 11,750 shares, "to bear such rates of dividends as the Board of Directors may prescribe, to enable said company to perfect their canal." The act was passed on February 9, 1849. Its purpose, as stated in the preamble, was to connect "the navigation of their canal with that of the State of Pennsylvania." Of course a connection with the Quaker State would have to be approved by that state. The "connection" was to be either a bridge or an aqueduct.

The reasoning behind the bridge-aqueduct was presented to the Pennsylvania Legislature in a pamphlet entitled, "Report of the Committee on Inland Navigation."[89]

> *While persons carrying coal on the Lehigh now have the right to pass out the outlet lock at Easton free of charge, and therefore, when there is sufficient water in the Delaware, which is the case at least nine months in the year, the boats loaded with coal for the Morris Canal, can pass over the river and enter that canal. But it so happens that in times of low stages of water, which often occur in the months of July, August and September, the water in the river, between the outlet lock on the Pennsylvania side and the inlet lock on the New Jersey side, is not of sufficient depth to carry full loads across, and hence boats have to unload parts of their cargoes before crossing the river, and coal has to be carried across in separate boats, and unloaded on the Morris Canal on the other side of the river, in order to make up the full loads on the boats, after they enter that canal.*

The reporting committee suggested that the bridge or aqueduct proposed by the Morris Canal would actually be a "positive benefit to the Delaware division" of the Pennsylvania canal system. Morris Canal-bound boats would not need to use water for lockage, and would not thus deprive the Delaware Division Canal of water to operate.

But was it "low water" or "high hopes" that urged the Canal Company to seek the privilege of constructing that bridge or aqueduct? We recall in this context that Ephraim Morris and William H. Talcott had iron interests in the Lehigh Valley, and that Asa Whitney, with his ideas of a transcontinental, railroad was still a Canal Company director.

We should note here, too, that on the other side of the Delaware River, another Asa, Asa Packer, was building his Delaware, Lehigh, Schuylkill and Susquehanna (DLS&S) Railroad, chartered in 1846. On the board of directors was John N. Hutchinson, witness to the signing of the agreement between the Canal Company and the Stanhope Iron Company, on December 16, 1845.

Not to be ignored, either, is the presence of George W. Edwards among the directors of the Canal Company, for Mr. Edwards represented the P&R Railroad, which must have had some hopes in the Morris Canal.

It appears obvious that it was merely "high hopes" and not "low water" at all that brought about the authorization for a Canal Company bridge or aqueduct across the Delaware River on February 9, 1849. Those hopes were dashed when one month later, on March 11, the State of New Jersey "concurred with the State of Pennsylvania and passed an act" giving the DLS&S Railroad Company "the power to build a bridge for railroad purposes only, at or near Easton, crossing the Delaware River." One month after that, the Pennsylvania lawmakers passed an act granting the DLS&S Railroad Company "immunity from the law requiring railroad companies to have a majority of their directors to be citizens resident of Pennsylvania."[90]

The board of directors of the DLS&S Railroad for 1851 consisted of James M. Porter* (Northampton County, Pa.), president; Christian Pretz of Allentown; Asa Packer of Mauch Chunk; John N. Hutchinson of Philadelphia; Edward R. Biddle of New Jersey and the Morris Canal Company; Dudley S. Gregory of Jersey City, formerly of the Morris Canal Company; and Benjamin Williamson of Elizabethtown, formerly of the Morris Canal Company, now of the Central Railroad of New Jersey.[91]

In 1853, the DLS&S Railroad Company changed its name. It became the Lehigh Valley Railroad. And it was the Lehigh Valley that spanned the Delaware, not the Canal Company.

**It was Mr. Porter who had petitioned the Pennsylvania Legislature to grant the Canal Company the right to build its aqueduct across the Delaware River.*

REFERENCES

1. Wheaton J. Lane, *From Indian Trial to Iron Horse: Travel and Transportation in New Jersey, 1620-1860* (Princeton, N.J.: 1939), p. 239.
2. *Paterson Intelligencer,* December 1, 1844.
3. Lane, *op. cit.,* p. 240.
4. *Paterson Intelligencer,* January 6, 1847.
5. *Ibid.,* December 4, 1844.
6. *Ibid.,* October 30, 1844.
7. George W. Cullom, *Biographical Register of the Officers and Graduates of the U.S. Military Academy . . . .* (Boston: 1892), pp. 222-7 (hereinafter cited as Cullom's *Biographical Register*).
8. *Report of the President of the Morris Canal and Banking Company to the Board of Directors, August 12, 1845,* p. 1 (hereinafter cited as *Report of the President, August 12, 1845*).
9. *Ibid.,* p. 9.
10. *Report of the President of the Morris Canal and Banking Company to the Board of Directors,* February 12, 1846, pp. 1-2 (hereinafter cited as *Report of the President,* February 12, 1846).
11. *Paterson Intelligencer,* February 12, 1845.
12. Morris Canal and Banking Company *Minutes,* February 9, 1845 (hereinafter cited as MC&BCo. *Minutes*).
13. *Ibid.,* April 2, 1845.
14. Cornelius C. Vermeule, Jr., *Final Report of Consulting and Directing Engineer, June 29, 1929* (E. Orange; N.J.: 1929), p. 59 (hereinafter cited as Vermeule, *Final Report*).
15. *Paterson Intelligencer,* April 16, 1845.
16. *Ibid.,* April 2, 1845.
17. *Ibid.,* February 21, 1844.
18. MC&BCo. *Minutes,* June 4, 1845.
19. *Paterson Intelligencer,* July 2, 1845.
20. *Ibid.,* July 16, 1845.
21. *Report of the President and Directors of the Morris Canal and Banking Company of 1844, to the Stockholders* (N.Y.: 1848), p. 5 (hereinafter cited as *Report of the President, 1848*).
22. *Ibid.,* p. 6.
23. *Ibid.,* p. 1.
24. *Ibid.,* p. 7.
25. MC&BCo. *Minutes,* March 10, 1846.
26. *Report of the President,* February 12, 1846, p. 4.
27. *Paterson Intelligencer,* March 19, 1845.
28. *Report of the President,* August 12, 1845, p. 3.
29. *Ibid.*
30. *Ibid.,* p. 1.
31. *Ibid.,* p. 2.

32. *Paterson Intelligencer,* September 3, 1845.
33. *Ibid.,* December 17, 1845.
34. *Ibid.,* December 24, 1845.
35. *Ibid.,* January 12, 1848.
36. *Ibid.,* January 19, 1848.
37. *Ibid.,* March 25, 1846.
38. *Ibid.,* August 27, 1845.
39. *Ibid.,* August 13, 1845.
40. *Report of the Directors of the Morris Canal and Banking Company to the Stockholders* (Jersey City, N.J.: 1869), p. 9.
41. MC&BCo. *Minutes,* September 15, 1845.
42. *Paterson Intelligencer,* October 15, 1845.
43. MC&BCo. *Minutes,* March 10, 1846.
44. *Report of the President, February 12, 1846,* p. 2.
45. Cullom's *Biographical Register,* pp. 227-8.
46. Morris Canal and Banking Company Docket Files (hereinafter cited as MC&BCo. *Docket Files*), Docket #108, No. 27, dated January 13, 1946.
47. Jersey City *Evening Journal,* December 8, 1868.
48. MC&BCo. *Minutes,* March 28, 1846.
49. *Paterson Intelligencer,* January 6, 1847.
50. *Ibid.*
51. MC&BCo. *Minutes,* April 20, 1846.
52. *Ibid.,* April 6, 1846.
53. *Paterson Intelligencer,* March 18, 1846.
54. MC&BCo. *Minutes,* April 6, 1846.
55. *Ibid.,* June 6, 1846.
56. *Paterson Intelligencer,* November 4, 1846.
57. *Ibid.,* December 23, 1846.
58. *Ibid.,* July 15, 1846.
59. MC&BCo. *Minutes,* August 12, 1847.
60. *Ibid.,* August 31, 1847.
61. *Ibid.,* July 11, 1850.
62. *Ibid.,* August 12, 1847.
63. *Ibid.*
64. *Ibid.,* August 31, 1847.
65. *Ibid.,* March 17, 1848.
66. *Report of the President, 1848,* p. 5.
67. Vermeule, *Final Report,* p. 61.
68. Lane, *op. cit.,* p. 241.
69. G.W. Keppler, *Scrapbook,* p. 70. This is an article from the Washington (N.J.) *Star,* July 1943. The exact date is missing, but internal evidence suggests that it was written during the week of July 25, 1943.
70. *Paterson Intelligencer,* December 2, 1846.
71. *Ibid.,* June 17, 1846.

72. Agnes Laut, *The Romance of the Rails* (N.Y.: 1936), p. 321.
73. Edward Hungerford, *Men of the Erie* (N.Y.: 1946), p. 25.
74. Cited in Lane, *op. cit.*, pp. 241-2.
75. MC&BCo. *Minutes*, April 3, 1848.
76. *Ibid.*, February 17, 1847.
77. *Paterson Intelligencer*, April 19, 1847.
78. *Ibid.*, May 4, 1847.
79. *Ibid.*, April 29, 1846.
80. MC&BCo. *Minutes*, June 8, 1847.
81. *Paterson Intelligencer*, September 15, 1847.
82. MC&BCo. *Minutes*, January 13, 1847.
83. *Paterson Intelligencer*, October 20, 1847.
84. Munsell's *History of Morris County, New Jersey, 1739-1882* (N.Y.: 1882), pp. 254-5.
85. *Ibid.*, p. 65.
86. *Report of the President, 1848*, p. 4.
87. Lane, *op. cit.*, p. 265.
88. *Report of the Committee on Inland Navigation, Relative to an Aqueduct over the Delaware River, Together with the Amendment of Mr. Porter* (Harrisburg, Pa.: 1850), p. 6.
89. *Ibid.*, p. 7.

15

A Brand New Look

WHEN William H. Talcott replaced Tyler as Chief Engineer of the Morris Canal, he was faced with the overwhelming task of converting the silted-in prisms, bulging lock walls, and rotting inclined planes into an efficient waterway, competent to handle boats of 60 to 70 tons burden. Under E.A. Douglass, during 1840-41, only the lift locks had been enlarged to 98 by 12 feet. Under Tyler, the prisms were widened but not deepened, and Tyler's remaining attention had been devoted to Planes No. 6 West and No. 8 East. The Stanhope Reservoir was scheduled to be built during the winter of 1845-6, as Tyler was leaving the Canal Company. It must have received a great deal of Talcott's attention, for a new wooden guard lock was installed to allow boat passage in and out of the Stanhope level.[1] Although there is no documentation available, it is almost a certainty that Talcott was also occupied with the erection of the new steam engine being installed at the Stanhope Iron Company's works.

Besides the remaining locks that had not been touched by either Douglass or Tyler, many of the old timber bridges over the canal were in need of substantial repair.[2] But Talcott's most pressing priority was to be the Port Colden Plane.

Tyler had made this plane a summit plane, doing away with Ephraim Morris' lock-plane arrangement altogether, and the double-track required by that mechanism. He kept the old sprocket wheel and chain arrangement, substituting "imperishable," or metal, parts in place of the old wooden ones in the plane machinery. The device, however, was still incapable of passing more than one section of a boat, or more than 50 tons if both sections of the boats were used. The location of the plane house on the elevation had been such that it had been impossible to operate the machinery under a head of more than one-fourth the altitude of the plane. Morris' waterwheel had been at the head of the plane.

Starting "from scratch," on a new piece of land acquired from Abraham Woolston, as noted in Chapter 14, Talcott went to work to correct the deficiencies of Tyler's design. On August 12, 1847, the Canal Company's board of directors approved the plans Talcott had submitted on August 6, and the new construction was begun.[3] One major innovation Talcott made was twisted wire cable, instead of the old sprocket chain. Since sprocket chain required a toothed surface to wind and run on, it follows that the smooth wire rope required a different winding arrangement. Carrying pulleys and a deeply grooved cable drum were the answer. The toothed surface was transferred to the inner side of the rim of the cable drum, where it intermeshed with the cogs or teeth of a smaller driving pinion. This arrangement was somewhat different from that shown in later drawings of the plane mechanism.

The drum, or "cylinder," weighing 6-1/2 tons had created many problems. An accident at the foundry, unfortunately unnamed, had "destroyed the first mould and core for the large cylinder" so that it wasn't finished early enough in the boating season to ship it by water. Transporting the 6-1/2-ton drum had required a 15-horse team "and a wagon of wider tread than ordinary." No teamster was willing to transport the drum for a flat fee, and the Company had to pay $25 per day and provide the wagon, as well. Shipment was delayed "on the way several days, by civil process against the teamster, in consequence of injury done to some public bridges, and in obtaining permission to cross others." A two-day rain had made the roads impassable so that "the last fifteen miles required five days, with the constant aid of eight men during the whole time." If it had been possible to transport the drum by water, the cost would have been about $100; by wagon, it had cost $862.68.[4]

Estimates for the water wheel for Plane No. 6 West had been submitted by the firm of Merrick & Town of Philadelphia. It was called (by Talcott) a turbine wheel, and was to have cost $1,000. However, "it was afterwards decided to use the kind of wheel called the 'Scotch Motor,' which cost $1,760."[5]

The first coil of Roebling's twisted wire rope had been delivered at Philadelphia for $1.25 per 100 pounds. Roebling's works at the time were located near Pittsburgh, and the rope was to have been shipped by way of the Allegheny Portage Railroad. Just as the remaining ropes had arrived at Hollidaysburgh, "about 100 miles of the Juniatta Canal was carried away by a freshet," and the ropes had to be taken back to Pittsburgh, where they had been shipped overland to Chambersburgh, "and thence by Rail Road to Philadelphia." The ropes had arrived at Philadelphia at the end of the 1847 boating season, and to make sure they were delivered to Plane No. 6 West (even if it had to be by land) premium prices had been paid — $379.30, in all.

J.L. Bevins & Co. had been contracted to build the new plane, but on August 7, 1847, the Company recalled the contract and gave it to another, unnamed company on August 23.[6] Mr. Woolston, on whose land the new plane was being built, apparently offered the Canal Company "some difficulty," thus delaying construction of the new wheel pit. By mid-September, "the contract was virtually abandoned, and the work thereafter, prosecuted by days' works, under the supervision of the Company's agents." A large break at the foot of the plane on October 8, 1847, had "carried away a portion of the new work, and nearly suspended its further prosecution, for one week, during the repair of said break."[7]

An accident at the foundry had broken "the large quadrant-curve pipe," which passed under the masonry of the water wheel pit, that accident further delaying completion of the masonry. The workers were offered more pay to encourage them to work harder and faster, and where it was possible and safe to work at night, night crews were employed. The time of year was not the best for work of this sort, since in 1847 the months of November and December had been "inclement," as was January, 1848. The masonry work had had to be "covered with straw, earth, or water, to protect it from frost."[8]

The plane at Port Colden was supposed to have cost $21,000. When built, with all the extra expense of transportation added on, it actually cost $27,168.33.[9] However, the plane was tried on January 27, 1848, and was considered successful, with the machinery operating at only about half its capability.[10]

Once Plane No. 6 West was operating successfully, Talcott was able to concentrate on the minimum work that would have to be done in order to have effective navigation:[11]

Rebuild canal section from Newark to Jersey City	$ 18,500.00
Rebuild 10 Western Division planes @ $21,000	210,000.00
Repair 10 Eastern Division lock planes @ $5,000	50,000.00
Repair 2 Eastern Division summit planes @ $1,000	2,000.00
Rebuild Pompton Aqueduct	7,000.00
Raise banks to hold 5′ water	7,500.00
Deepen Delaware River channel	5,000.00
	$300,00.00

It had been on the strength of that $300,000 estimate that the Canal Company had issued 3,000 shares of its new "Preferred" stock at $100 a share, once the charter amendment went into effect (Chapter 14), on February 9, 1849.[12]

To secure passage of that amendment, the Canal Company agreed to surrender its banking privileges. The original charter had specified a 31-year life for those privileges, so that they would have expired at the end of 1855. The "surrender" was official on June 26, 1849.[13] Three thousand shares of preferred stock were issued, with the Canal Company guaranteeing to investors that "no dividend shall be paid to the holders of the old or consolidated stock, until the holders of the new preferred stock . . . shall at all times have received equal to ten percent per annum in semiannual payments on said stock."[14] At least the work on the canal's improvement could begin.

The first order of business, however, was to increase the number of boats on the canal. In 1845, the Canal Company had bought 100 new scow boats for $37,488.71; in 1846, 7 scows and one flicker for $2,942.82; in 1847, 64 scows for $21,405.44.[15] The 100 boats were the new section-boats, capable of carrying 60 to 65 tons cargo, but it was immediately apparent that those boats could not be passed over the old inclined planes "when loaded with over 50 tons." Even at the planes rebuilt by Tyler during 1845-46, the boats still had to be unhinged and transported one section at a time over the inclines. "The comparatively small cargoes that could be carried; the frequent failures of the old planes, and the detention and hazard incident thereto, rendered it impracticable to induce boatmen owning boats to run on the Morris Canal. They preferred to run on the Delaware Division Canal."

About "100 boats called Flickers, built to run on the Morris Canal before its enlargement," were "competent to carry about 25 tons." But they had been used mainly "in carrying on the local trade along the line of the canal since its enlargement." These flickers, plus the 108 boats the Canal Company owned at the start of 1846, were the only boats available for Morris Canal business that year.[16]

By the end of the 1847 boating season, the Canal Company started to offer bonuses to "all Lehigh boats that should continue in the trade of this canal during the whole season," and one party engaged to put from 20 to 50 more boats into service.[17] During 1848, James F. and Carman F. Randolph contracted to operate a fleet of Morris Canal boats.[18]

In 1849, Richard Lewis, of Washington, New Jersey, loaned the Canal Company $20,000. In exchange he got a "bill of sale" for all Morris Canal boats numbered 4 through 175, with the exception of Boat No. 107, sold elsewhere, and Boat No. 163, sold to W. Mains. The loan was repaid on December 10, 1849, and the "bill of sale" was cancelled. To control and direct those boats bought by Lewis, the Canal Company leased the boats from him, and he acted as the Company's "transportation agent, at a salary of $1,000."[19]

The rate of tolls "was materially reduced" for boats and cargoes travelling the Morris Canal during 1848. The practice resulted in about a 30-percent increase in trade, but showed a "diminished revenue from tolls."[20] In 1849, the Canal Company began to offer "drawbacks" for all "Lehigh Valley property," so that it could be shipped to Newark, Jersey City, and New York "at as low a price as said property can be delivered at those markets via any other route."[21] This practice of preference in toll rates for goods from the Lehigh Valley was to remain in effect for years.[22]

The New Jersey Iron Company had also received reduced toll favors in other years. But in 1849, when that company applied for toll reductions on certain articles made at its Boonton works, the Canal Company refused the request. The Iron Company was notified that the Canal Company board of directors "deemed inexpedient and impracticable" any toll reduction "at the present time" because of "the existing embarrassed circumstances" of the Company.[23] The Iron Company protested, but was told that it "was not and never had been the intention" of the Canal Company "to make any tariff of tolls for more than one year, or to any parties or party at a less rate than that charged to the public."[24]

This bit of Canal Company business is understandable, perhaps, when we realize the almost totally dependency of the Morris Canal on the Lehigh Valley, with its canal and Beaver Meadow Railroad, for its coal cargoes.[25] This dependency explains, too, why the charter amendment of February 9, 1849, was to have permitted the Canal Company to make a more effective communication with the Lehigh Valley by means of a bridge or aqueduct between Phillipsburg and Easton.

Ostensibly, the reason for a bridge or aqueduct was low water in the Delaware River during the dry summer months. If water was low in the Delaware during those months, it is a certainty that the lack of it would be a problem at other points on the canal. Rockaway was such a point. During the driest part of 1849, Colonel Joseph Jackson wrote to the Canal Company asking that they put an aqueduct over *his* river, too. He claimed his works were suffering from the Canal Company's diversion of "the waters of the Rockaway River into the Morris Canal at Dover," which passed those waters by Jackson's works at Rockaway.[26]

The aqueducts at Dover and Powerville that Jackson wanted built were, of course, not built. Neither was the aqueduct over the Delaware River. Instead, the channel in the river bed under the cable ferry between South Easton and Phillipsburg was deepened to five feet at a cost of $942.48.[27]

Low water was a problem in the canal section between Newark and Jersey City, too. Tidal silt kept filling in the channel, so that the steam towboat *Jamaica* was needed to move boats from Newark to the lock at the east side of the Hackensack River. During 1849, that service cost $1,500;[28] in 1850, it skyrocketed to $8,576.07. The largest expenditures, however, were to be made on the improvements to the inclined planes.

Interestingly, when the canal was being built, construction emphasis had been placed on the Eastern Division to accommodate the descending navigation of the iron interests. The "enlargement" construction emphasis was to be placed on the Western Division, for the ascending navigation of the coal interests.

Talcott's estimates for the ten remaining Western Division planes were recorded in a notebook, a copy of which is extant:[30]

Plane	Est. Cost	Plane	Est. Cost
1W	$ 7,640.00	7W	$ 8,487.25
2	8,194.00	8	8,130.00
3	7,303.75	9	8,722.00
4	8,103.25	10	7,970.05
5	8,604.50	11	7,653.00
		Total	$81,008.25

Contracts for work on these planes were given out in July, 1850:

Planes 1, 2, and 8W	Brinknall Mills & Co.*
Planes 3 and 4W	Goolder & Van Dyke*
Planes 5 and 7W	J. & C. Collins
Planes 9, 10, and 11W	Solon Chapin
Water Wheels	James Finlay
Machinery for 10 Planes	Evans Thompson & Co.
Railroad Iron	David W. Wetmore (N.J. Iron Co.)
Wire Rope	J. A. Roebling

Talcott's assistant engineer during this period and until the outbreak of the Civil War was Alexander P. Berthoud. Other assistants

**Spelling could be wrong, since copyist's handwriting is almost illegible.*

were "Messrs. Little and Wykcoff of Hacketstown [sic], L. Dewitt, W. Groff, Jr., Cook Talcott, and W. Herrick of Stanhope, and John Sanford of Dover."[31]

Navigation for the 1851 boating season was delayed until mid-June "owing to the great amount of work in rebuilding the planes West of the summit."[32] New machinery had already been placed at the Stanhope Plane (No. 2 West) by that time, however, for an entry in the *Minutes* of April 7, 1851, indicates that the Canal Company was having "difficulty in procuring a proper title to the land on which the new machinery" had been placed. A survey and map of the property in question, in Morris Canal Company Docket File No. 519-1/2, shows the plane tracks in relation to the lands belonging to the Estate of Charles Lewis (foot of the plane), to James Wills, and to Sussex Iron Company (head of the plane).*

The Stanhope Plane, like the others planned to be built by Talcott, was double-tracked: one set of tracks designated "North Track," the other, "South Track." The survey in Docket File No. 519-1/2 clearly shows the tracks with their labels. Talcott's notebook refers to "North" and "South" tracks at each of the planes in conjunction with computing how much plane cable was to be ordered.

Talcott decided that all the plane houses were to have the same plan.[33]

> *The plan of plane houses is made of 46 feet head and fall. All the planes are the same, therefore the height of house may be varied. The head of each plane above center of w. wheel [water wheel] is as follows with 5' water in upper level:*

No. 1W	*46 feet*	*No. 6W*	*44 feet*
No. 2	*46*	*No. 7*	*46*
No. 3	*41*	*No. 8*	*42*
No. 4	*44*	*No. 9*	*56*
No. 5	*41*	*No. 10*	*40*
		No. 11	*One Story*

**Confusion over the name of this company stems from the fact that it had originally been called Morris & Sussex Mfg. Co. In 1841, the name was changed to Stanhope Iron Co., when the owners incorporated. Still later it was to become the Sussex Iron Company.*

From center of w. wheel to top of pit wall ___ 24′ 9″
From top of pit wall to top of water,
upper level ______________________ 21′ 3″
From surface of water to top of walking
plank in flume ____________________ 2′ 0″
This is a convenient height for upper floor _____ 48′ 0″
It can be varied one or two feet either way. . . . It can be made 3 ft. lower without placing the tender too low — but not more than 3 ft.

The plan of the plane house is made in 3 stories as follows:

1st story	*13′*
2nd story	*12′*
3rd story	*12′*
	37′ to upper eaves

The foundation is to be 5′ below top of pit wall and this makes the floor of 3rd story 20′ above top of s [*surface*] *water, or 3′ 3″ below the walking plank on top of flume. It will answer — at this height, but not lower: For those planes where whole fall is 41 or 42 feet it will be best to take off 1′ 6″ from each story.*

From the plane house down into the ground, where the water wheel rested in its wheel pit, cast-iron pipes were inserted perpendicularly in an excavation through the foundations. The lower pipes, below the water wheel, were made by James Finlay. Above them were pipes made by a Jersey City firm, of which John D. Ward was a representative.[34] Later, Ward would become a director, then vice president, of the Canal Company.

Through the cast-iron pipe would run the drive shaft, connecting the water wheel with the cable drum. These shafts were made of "Greenwood" iron, according to a Canal Company diagram on file with other drawings in Trenton. "Greenwood iron" was made by West Point Foundry, which also made the shafts on which the cable drums revolved. In June 1852, the drum shaft of Plane No. 2 West at Stanhope broke "square off." The iron was described as "sound," but "very coarse grained, so much so that the supervisor says it looked like 'Iron ore' more than like iron!"[35] Apparently such breakage was not uncommon. Two years later, Talcott recommended "as a measure of safety the substitution of wrought iron drum shafts in the place of cast iron on several of the inclined planes."[36]

The cable drums themselves were uniformly 12 feet in diameter, of 37.44 feet in circumference.[37] The face of each drum was grooved to accommodate the wire rope which was wound and unwound on it. The number of grooves varied from drum to drum according to the length of the rope needed to move boats over the inclines. Lengths, of course, varied according to the elevations to be overcome. Plane No. 1 West, for example, with an elevation of 56 feet, required that boats be moved 1,000 feet on a 10-percent grade (one foot up or down, for every 10 feet forward movement). The drum there was given a 7-foot, 7/8-inch face (outer surface) with 30 grooves. Since each groove represented 37.44 feet, 30 grooves would accommodate rope equivalent to 1,123 feet of distance to be travelled on the incline.

Talcott's descriptions of the drum, and one incomplete sketch, indicate that the new planes were powered in a way that differed substantially from the method illustrated in Vermeule's detailed drawing of the Boonton Plane's mechanism. One inner edge of the rim of the drum was toothed with "spur segments" or cogs. Meshing with those cogs was a "driving pinion," which in turn meshed with the bevel gear of the drive shaft of the water wheel. Vermeule's 1929 drawing shows these gears as a system separated from the cable drum. Talcott's sketch clearly shows the driving pinion to have been more an integral part of the drum than of the gear system such as Vermeule depicted.[38] Makers of this "plane house machinery," although said to have been contracted to Evans Thompson & Co., also included J.C. Righter and James Fuller & Company of Rockaway, J. Little of Hackettstown, Templin & Connover of Easton, and Lippincott & Miner of Mauch Chunk.[39]

Pulleys for carrying the wire rope up and down the plane, to and from the drum, and around the huge groove wheels at the head and foot of the plane varied in size from 13 to 36 inches. The three groove wheels at the head of each plane were 11 feet in diameter; the one groove wheel at the foot of each plane, 14 feet in diameter. Pulleys and groove wheels, plus the pulley shafts, were made by the Union

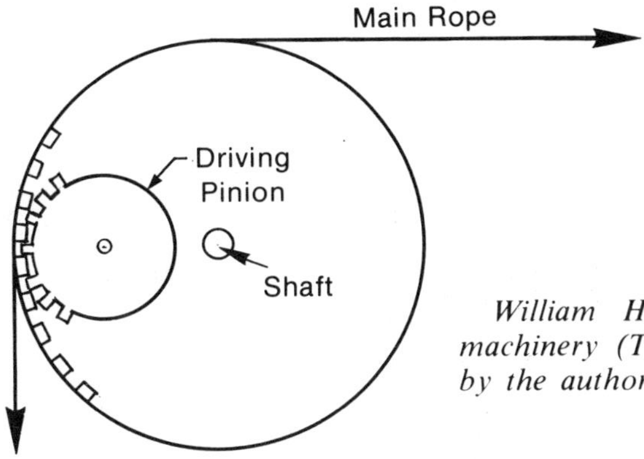

William H. Talcott's inclined plane machinery (Talcott's Estimates, adapted by the author).

Works at Paterson, James Fuller & Company of Rockaway, Templin & Connover of Easton, Lippincott & Miner of Mauch Chunk, and West Point Foundry in New York.[40]

The Company decided in mid-1847, as already noted, to use wire ropes for the inclined planes. Some ropes were ordered from the John A. Roebling works near Pittsburgh, the only company making it in the country at the time.[41] Roebling's rope had been used successfully on the Allegheny Portage Railroad in Pennsylvania, and would now be used on the Morris Canal's inclined planes. There were, however, to be other suppliers of wire rope. Talcott's notes of August 14, 1851, obviously refer to rope made by a firm other than Roebling's, whose product he already knew: he had "never seen any rope like the main rope used, and it is desirable to make as few as possible till they have been tried."[42]

Other suppliers of twisted wire rope were Hazard of Mauch Chunk,[43] and the Newall Co. One reason for ordering from firms other than Roebling's was the fact that many of Roebling's early ropes were defective. The Canal Company's *Minutes* of January 19, 1854, record "that the payments made under the compromise and settlement with John A. Roebling in reference to the defective wire ropes for the 10 new planes, constructed in 1850 and 1851, be taken from account of current expenses and placed to construction account."

Wire rope at this time was not yet "cable," but groups of thin wire

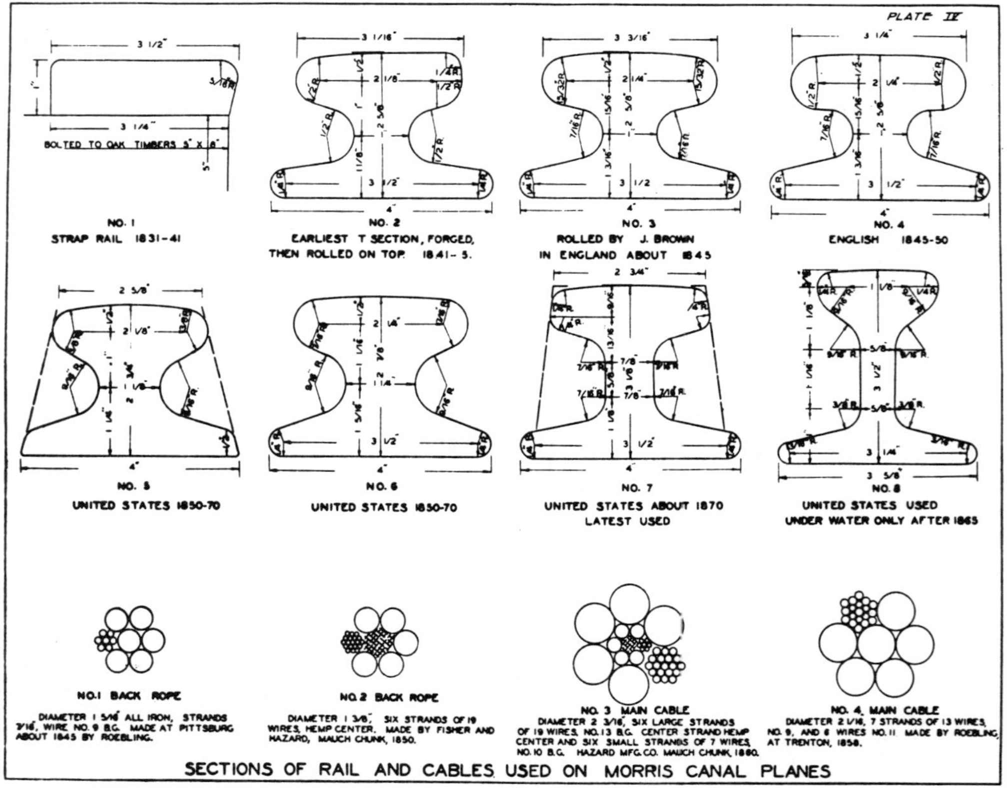

Rail and cable types used on the Morris Canal (C. C. Vermeule, Final Report).

twisted around hemp center strands. The twisted groups were then intertwined to make a stout rope. The numbers of wires involved in each rope varied from 42 to 114. Talcott ordered 650 feet of 49-wire rope with a diameter of 1.375 inches to be used for a ferry rope. This was probably for the Delaware River Ferry, since the condition of the canal from Newark eastward in 1850 (the date of Talcott's report) required that boats be towed by steam to Jersey City. So far as is known, no "ferry," as such, was yet in service over the Passaic and Hackensack Rivers.[44]

The plane rails over which those new wire ropes were to move canal boats were to be placed so that they were 12 feet 2 inches, center to center of the rail. They were laid in both 12- and 15-foot sections, 0.058952 inches apart at the ends to accomodate expansion and contraction. That they were not yet "Tee" rails is suggested by Talcott's estimate that 12,000 spikes were needed "for center of rails, 1″ x 3/4″ x 6″ long, 2 per rail."[45] The estimates he made for the quantity of new rails for the planes totalled 1,876 lengths.[45] The New Jersey Iron Company, as we saw, was the contractor in 1850, and probably during the next few years. But by 1856, Cooper-Hewitt & Company became a major supplier of rails and of the "chairs" that secured the rails in position.[46]

The cradle cars that rode on those rails were built to conform to the new track width and the larger section boats. They were to be 66 feet long, set on eight pairs of wheels made by "McFarlan" and "Whitney." Old wheels were salvaged wherever possible, if they were fit for further use. But only 220 were salvageable, and they were "appropriated to use on planes 3 & ___* E. on Groff's division, and on planes 9 & 10 E. on Hopper's Division."[47]

The cradle cars were equipped with "skids," shod with brake bands and worked with levers. Unfortunately, no description of those brakes exists, merely an order for their components placed with J.D. Ward of Jersey City:[48]

140	skids
296	brake bands
47	long shafts

**The number of the plane is missing in the text.*

47	levers
47	sockets
242	small cranks
54	long links
40-1/2	links (the long end)

The above provided for 27 full sets of brakes, and for 20 half-sets. Brakes were installed in the plane houses as well as on the cars, and Ward probably received that order, too.

At the foot of Plane No. 1 West, Port Morris, a coffer dam had been built "with a suitable gate" to allow the waters of the new Stanhope Reservoir to be drawn off so that the "sheaves, groove wheels, and tracks under water at the foot of the plane" could be "examined, repaired, or replaced, if need be, during the season of navigation."[49] A similar arrangement was made at the foot of the Port Delaware Plane, No. 11 West, where a huge, new oaken mitre gate, 22 feet high and 22 feet wide, was built to keep the Delaware River out when the plane mechanisms needed to be serviced.[50]

On July 11, 1851, Talcott wrote up a "rough estimate of work to be done on Morris Canal before opening of canal in the year 1852":[51]

1. Guard lock at Lake Hopatcong	$ 1,000
2. Widening lock at Lake Hopatcong	6,000
3. Deepening Delaware River channel	10,000
4. Rebuilding Newark plane	25,000
5. Raising banks in 1851	10,000
6. Rock excavation in Bergen Ridge	8,000
	$60,000

The 10 planes on the Western Division were in progress, "on the new improved plan." The aqueduct over the Pompton River had been rebuilt; the work of rebuilding the canal from Newark to Jersey City and the terminal pier was being vigorously prosecuted.[52]

Talcott's 1848 estimates for work to be done had specified $7,500 to raise the canal banks and $5,000 to deepen the Delaware River Ferry. The prices for 1852, $10,000 to raise the banks and $10,000 for the ferry, reflect increasing costs of labor and material. Their presence on the 1851 list also tells us that the work was not yet done.

The board of directors had decided that Planes No. 8 and 12 East, which had been rebuilt on Tyler's plan in 1845 and 1846 with sprocket wheel and chain, were a serious obstruction to navigation. They would have to be rebuilt during 1851-2.[53] Navigation had been seriously and vexatiously delayed at Plane No. 8 East (Montville) due to the frequent breaking of the sprocket chain. Maintenance of those chains was an expensive item — about $3,000 a year. Talcott had estimated that, for $6,000, "the present machinery may be so arranged as to use wire rope, thereby rendering its operation perfectly safe and certain." Nearly all of that $6,000 would be spent "on the plan of the improved plane, and consequently part of such plane when it comes to be entirely rebuilt."

As for Newark, the old machinery would have to be rebuilt, in order to save the season. The board of directors resolved "that the route of the inclined plane No. 12 East on the Morris Canal is hereby fixed upon for rebuilding said plane as follows — commencing at a point in the original route of said canal on the easterly side of Plane Street in the City of Newark and running thence the same courses and distances as the original route of said canal to a point 200 feet westerly from the summit of said plane, and that the machinery and plane house be and hereby is located on the southerly side of and adjacent to the track of said plane — and that all the lands, water, and streams necessary for rebuilding said plane are hereby appropriated for such use — and that the Chief Engineer cause a survey thereof to be made in accordance with the requirements of the charter, and deposited in the office of the Secretary of State."[54]

Of course the work in Newark could not — would not — be completed for years to come. A massive and major revamping of the whole canal through that city was being planned. From the foot of the Newark Plane, No. 12 East, the canal was to be lowered 4 feet 6 inches, depressing it considerably below street level so as to better accommodate street traffic. That meant that the inclined plane, besides having to be rebuilt on the new plan, had to be lengthened to fit the new dimension of the canal prism. Even the "high lock" would have to be lowered 4 feet 6 inches and a new, stone-arched bridge, 200 feet wide, would be built at the Broad Street crossing of the canal. Other stone-arched bridges, slightly less imposing, would have to be built at Mulberry, Washington, Plane, and Halsey Streets.[55]

Beyond the "high lock" toward Newark Bay, there were plans to move the entrance to the canal on the west side of the Passaic River to a point "below the Plank Road."[56] The interesting thing about this entry is the possibility that the Plank Road had not yet been built because of a controversy.

The Newark Plank Road and Ferry Company had been chartered in 1849 to connect Newark with the Hudson River by means of one of the "new" type roads, i.e., "new" to New Jersey. It was to run into difficulty with the New Jersey Railroad and Transportation Company (NJRR&TCo.), however, over who had the right to bridge the Passaic and Hackensack Rivers. The railroad company had been authorized years earlier to build such a bridge, and it was at this time enjoying exclusive monopoly over the route between Newark and Jersey City. The Plank Road Company tried to keep their bridge from coming under NJRR&TCo. control, but failed and, by 1855, the railroad owned half the Plank Road stock. Bridge construction was at that time begun.[57] The Canal Company's plans for crossing the Passaic and Hackensack Rivers were very much tied in with the Plank Road system, but since the controversy was not ended until September, 1857, the enlargement and improvement of the canal from Newark to Jersey City could not get underway until then.[58]

From the "point below the Plank Road" mentioned above, there were three alternate routes proposed for crossing the Passaic River.[59]

I. *By a Through Cut,* requiring two 100-foot-long breakwaters, excavation of the prism, and an 80-foot-long

Plank Road Bridge	$8,400.00
Two sets of ropes to ferry with	200.00
	$8,600.00

II. *With a lock,* requiring the breakwaters, as above, excavation of the prism, the Plank Road Bridge, and a wooden lock. The

cost, as above	$8,600.00
plus the lock	
plus a tender 9 months in a year @ $15 mo.	2,250.00

III. *To pass under Plank Road in the River,* requiring a tight bulkhead from the present outlet to a point 150 feet below the Plank Road and a transverse wing so as to make it all still water the whole distance of 350 feet. Excavation would be done by mud digger, and the bridge would have to be raised.

The plan adopted, finally, was the outlet lock, to be built at a cost of $8,000.[60]

The Passaic and Hackensack Rivers were apparently to be crossed by means of ferry ropes, but just *how*, is not explained in Talcott's notes. Men (drivers) and horses passed free on the bridges, that much is clear;[61] whether or not this early rope-ferry arrangement consisted of rope affixed somehow to the breakwater piling so that boat crews could pull their boats hand-over-hand is *not* clear.

With these major renovations in Newark, the Town Council of that city entered into an agreement with the Canal Company for building a new market in the space over the canal between Broad and Mulberry Streets. Newark had been trying to make such an agreement since May, 1836, after fire had destroyed the old market.[62] That first request had been denied by the Canal Company, but Newark was persistent, asking again in 1846 and in 1850.[63] Finally, in 1852, a formal agreement was made.[64] The market was built, and the second floor of the building was occupied by the committee rooms and council chambers of the Newark Town Council, with the police station and city prison at the east end of the building.[65] The building created the infamous, rat-filled 1100-foot "tunnel" through which canal boats had to travel between the foot of Plane No. 12 East and the "high lock," No. 17 East.

On the Jersey City section of the canal, the lock at the east side of the Hackensack River was moved further down river at a cost of $10,750.[66] A waste gate was installed at Fiddler's Elbow, to be operated with a 5-inch wheel set in a rack, with a 4.5-foot lever to start the gate.[67] At the site of the easternmost terminal, lots under water were reclaimed, "east of Hudson Street, Jersey City and between Essex and Bergen streets extending 320 feet east of Hudson Street." Bulkheads were built surrounding the "reclaimed lots" at a cost of $39,500.[68] Almost half of that money was recovered in 1856 when six lots, 25 feet wide by 100 feet deep, sold for $15,000.[69]

Lock No. 17 East, the "Deep" lock at Newark. The Center Market, at left center, was built in the space over the canal (ETF).

Four lots, 100 feet by 100 feet, were bought by George and Anderson Bloomer for $10,000. Two lots, 50 feet by 100 feet, adjacent to Bloomer's, were bought by former Canal Company engineer Ephraim Morris for $15,000.*

Obviously the Canal Company was anticipating a substantial increase in the volume of its business when its canal was finally enlarged and improved. And the improvement of the Jersey City Basin area was to be the ultimate phase of the enlargement, as the basin area was extended on reclaimed water lots, eastward into the Hudson River and southward into Communipaw Cove, to accom-

**Bloomer's deed is dated June 13, 1856; Morris', July 23, 1856.*

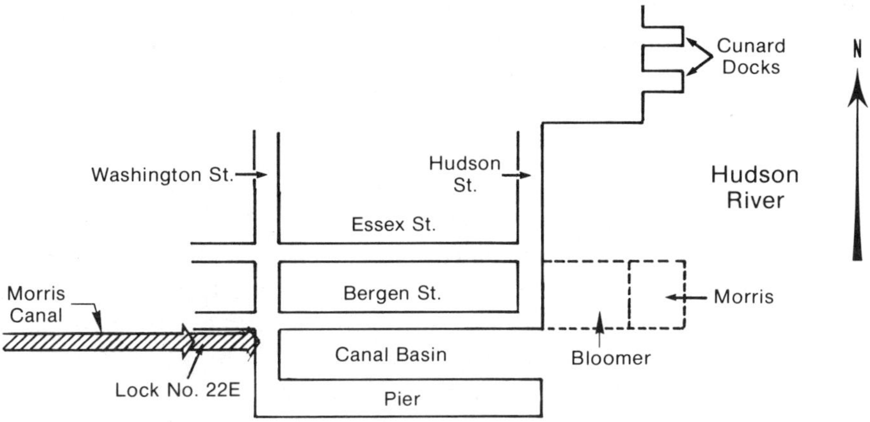

Morris Canal basin property in Jersey City, showing reclaimed underwater lots bought by Bloomer and Morris (prepared by the author).

modate the cargoes brought by ever-increasing numbers of boats.

The Canal Company began to reshape its policy on boats, too. President Marsh was sure that business on the canal by 1852 would already have reached capacity, if the Company had had sufficient boats, and if the Company had been able to control the boats "that were navigating the Lehigh Canal and suitable to be used on the Morris Canal." Accordingly, the board of directors "made arrangements to have built and brought on the Canal over 100 boats, to be under their control, and employed in carrying freight over the entire length of the Canal." They had also "made arrangements with a number of individuals owning boats, to employ them the ensuing year on the Morris Canal."[70]

While the Canal Company was concerned with problems of enlarging its canal to carry even more freight, and of providing facilities at Jersey City to handle that freight, there was a subtle, almost silent continuation of the movement to enlarge and improve the "connecting works" of the canal: the lateral railroads.

In 1850, the Canal Company's board of directors had decided to grant "free of charge, the rights of way through and on any lands belonging to the company for the track" for the Sussex Railroad. The tracks might be placed in "such locations as may not in the opinion of the President of this Company interfere with any of the company's works or any future improvement of the same."[71]

The Sussex Railroad had started out as a mule-powered railroad to move ore from Copper-Hewitt & Company's mine at Andover to the Morris Canal at Waterloo Village, where the ore was placed on canal boats for shipment to Phillipsburg and thence to the Trenton Iron Works. In 1848, the railroad's backers incorporated under David Ryerson, John Wills, Alexander McKain, Nathan Smith, Peter and Edward Cooper, Abraham S. Hewitt, and Andrew A. Smalley. Cooper and Hewitt, the controlling stockholders, later sold their shares to future Canal Company director Thomas N. McCarter and others.

Always seeking more iron ore for their furnaces at the Trenton Iron Company, Cooper and Hewitt became interested in Peter M. Ryerson's Pompton and Ringwood properties. Ryerson, still suffering the effects of the Tariff of 1847, was facing bankruptcy. Hewitt wrote to Ephraim Marsh on July 19, 1853, to find out if the Canal Company would work with Cooper-Hewitt to develop the potential of the Ryerson properties. What Hewitt wanted was to build a railroad from Ringwood to Pompton, so that he could ship his ore to the Pompton Feeder, where canal boats would pick it up for delivery to Phillipsburg. An alternative — very like a threat — was to build furnaces at Ringwood, where the ore was, and receive coal by way of the Erie Railroad, just north of Ringwood.

If Hewitt should decide to build his railroad, he wanted to know if the Canal Company would "make and keep the Pompton Feeder navigable for the same class of boats as the rest of the canal." He also wanted to know what rate of toll would be charged for ore brought from Ringwood to the feeder.[72]

Hewitt also wrote to Thomas Wallace at Ryerson's Pompton Iron Works to ask what that property was worth. Specifically, he wanted to know:

> *1. What is the amount of stock issued?*
> *2. What is the indebtedness of the Co. if any?*
> *3. Of what does the property of the Co. consist?*

4. What do you think this property is fairly worth?

The answer from Wallace indicated that $250,000 worth of stock was issued, but that it was being offered at 25 cents a share.[73] The company had a $31,000 mortgage on its property, which consisted of a rolling mill and a blast furnace, water power, and about 40 acres of land. The total worth, according to Wallace's estimate, was $150,000. Ransom states that Cooper-Hewitt never bought the Pompton property, and that the owner following Peter M. Ryerson was William C. Vreeland. The firm did buy Ryerson's Ringwood property, however, on September 1, 1853.[74]

In 1856, the Canal Company authorized the rebuilding of the Pompton Feeder lock, widening and lengthening it to conform to the rest of the enlarged canal. The river navigation above the lock was to be improved "so as to permit full-sized boats to go up to Pompton Iron Works, the improvements being estimated by the engineer to cost $12,000."[75]

Strong evidence supports the claim that the Pompton Feeder would not have been improved at all if it had not been for an agreement between the Canal Company and Cooper-Hewitt. A letter from Marsh to Hewitt, dated September 2, 1856, reveals that Marsh himself had[76]

> *succeeded in getting the consent of our directors to enlarge the Pompton Feeder Lock, and improve the navigation to the Ryerson Iron Works upon your new proposition to build for the present two miles of railroad and to bring to our canal by that and by teams ten thousand tons of ore per year until the road is built all the way. They require me to obtain the written agreement of your company to that effect when we are authorized to expend the money and make the improvements.*

Only a month earlier, on March 9, 1855, the New Jersey Legislature had also passed a law on behalf of the Canal Company, enabling them to "construct a railroad, with one or more tracks, connecting their canal with the road of the Lehigh Valley Railroad Company, and also with the road of the Central Railroad of New Jersey, at or near Phillipsburgh [sic], and also to construct another railway, with one or more tracks, connecting their canal with the road of the

Warren Railroad Company." None of the "railroads" built by the Canal Company was to be more than one mile in length.

In 1853, the Lehigh Valley (LV) Railroad, the former Delaware, Lehigh, Schuylkill and Susquehanna, had begun to build a bridge "over the Delaware River on the Morris Canal Company's lands without permission." Marsh was authorized by the board of directors "to grant that company permission to put up the bridge on canal lands so as not to interfere with the uses of the canal and its appendages."[77] The condition of that grant was that the railroad company agree that it would construct and maintain a suitable railroad from the Morris Canal basin at the head of Plane No. 11 West, "to connect with the track of their road, at the easterly end of their bridge so that the Morris Canal might participate in the trade of that railroad."[78]

As it turned out, the Canal Company built the connecting works: "Resolved — that the officers of this company be and are hereby authorized to make a Rail Road connexion between the canal and the several railroads which cross the canal at Port Delaware, and also to make a similar connexion with the Warren Railroad at or near Washington."[79]

The LV Railroad was expected to be in operation by July 14, 1856, so that a portion of the Morris Canal boats would be "loaded at Port Delaware." And the Warren Railroad was progressing so well and so fast that the Canal Company was expecting its "speedy completion, and perhaps in time to deliver Lackawanna coal to the Morris Canal during the present year [1855]."[80]

By the time the Warren Railroad was completed, it was leased to the Delaware, Lackawanna and Western (DL&W) Railroad, and in July, 1856, the Canal Company began negotiations with that company about "shipping coal from their railroad at Washington upon the Morris Canal."[81] The DL&W built coal chutes at Port Washington to facilitate the unloading of coal cars and loading of coal boats. During 1856, only 25,375 tons of Scranton coal had entered the Morris Canal at Port Washington; in 1857, the amount had climbed to 44,731 tons; in 1858, to 54,218 tons.[82] In 1860, 127,517 tons of Scranton coal flowed through the Morris Canal.[83]

This was a substantial "enlargement," indeed. The installation of the weigh scale in the tracks of new Plane No. 6 West dates from here. It was ordered installed on July 21, 1856.[84] In fact, the rebuilding of Port Colden Plane as early as 1845-46, before any other plane was touched, would logically have proceeded on the strength of the proposed Belvidere-Port Colden Railroad, chartered in 1836, and expected to be built shortly after that. The fact that William Henry, of Oxford Furnace, of Scranton, and of the Leggett's Gap Railroad (later part of the DL&W), was an incorporator of the Belvidere-Port Colden Railroad tends to confirm this. The Belvidere-Port Colden Railroad was never built, of course, but rather the DL&W-inspired Warren Railroad.

It is perhaps not too far afield to note here that William E. Dodge, one of the principal forces behind the Warren Railroad, had not long since attended the formal opening of the New York & Erie Railroad from the Hudson River and Lake Erie. With him at the May 14, 1851, ceremonies had been Anson G. Phelps and William C. Redfield, a Canal Company director from 1854 through 1857.[85] Hungerford, the Erie Railroad historian, called Redfield a father (or grandfather) of the Erie. In 1829, he had mapped out a route for a "great railway" that was "to connect the canals and navigable waters of New York, Pennsylvania, Ohio, Indiana, Michigan, Missouri, and the adjacent States and Territories, opening thereby a Free Communication between the Atlantic States and the Great Valleys of the Mississippi." In 1854, at Rock Island, Ill., the Mississippi River was bridged for the first time at the spot Redfield had picked out. He was present at the inaugural ceremonies. A Canal Company director at the time, he had been one of the original incorporators of the Erie Railroad.[86] Like Asa Whitney, his vision was far in advance of his times. We wonder what he, or Whitney, for that matter, wanted with the Morris Canal.

Meanwhile, the LV Railroad bridge was completed on September 8, 1855, between South Easton and Phillipsburg.[87] The LV Railroad's annual report for 1855 noted that the Morris Canal Company was building "a branch from the eastern end of the Delaware bridge to their canal and are erecting conveniences for loading coal into their boats at that point; thereby saving them the risk and expense of ferrying their boats across the river and doubling their boating facilities without any increase of capital."[88]

While that double-tiered railroad bridge was being built, an accident occurred in which a canal boat was said to have drifted into a scaffold during a freshet in the river. John W. Murphy, no doubt a contractor, later sued the Canal Company for the cost of his lost scaffolding. He waited to sue until the one witness to the event had died. The witness, the Canal Company's ferryman, would have been able to prove that there was no negligence on the part of his employer or fellow employees. But, his death before the trial gave Murphy an edge, and he won his judgment of $3472.20 from the Canal Company in 1860, when the case was finally settled.[89] This was the first, but not the last, time the very presence of a railroad adversely affected the Canal Company.

Unpleasantness also resulted from the Canal Company's having given a right-of-way to the Sussex Railroad at Waterloo. The railroad bridge across the head of Plane No. 4 West had been built with its underside only 15 feet above the plane tracks. Talcott protested to Cooper-Hewitt & Co., saying that a 17-foot clearance had been decided upon for both the Belvidere-Delaware (Bel Del) and Sussex railroad bridges, and he knew "of no reason for changing that decision at Waterloo.[90] Moreover, horses walking their path up and down the plane were frightened by the trains passing over the railroad bridge. Lest they "fall off down upon the plane, a distance of 10 to 14 feet and be injured or killed," Talcott requested that Hewitt have a "railing along the inner edge of the horse path."[91]

The Bel Del Railroad had been proposed so long ago (1836) that it seems almost to have been forgotten. Not so. Here, in 1854, we see it referred to in Talcott's letter to Hewitt. Hewitt, with his iron works in Trenton, and his iron mines in Sussex and Passaic Counties, had backed the Bel Del so that his ore could be shipped to Phillipsburg by canal boat, and to Trenton by railroad. Moreover, plans to build the Andover (or Cooper's) Furnace at Phillipsburg must have been in the works, for it was completed in 1856.

A right-of-way agreement was also made between the Canal Company and the Bel Del Railroad for lands at Phillipsburg.[92] In exchange, the Bel Del was to build the plane house and basins,

Plane No. 4 West and mule bridge at Waterloo Basin, about 1885-90. The Sussex Mine Railroad bridge crossing is visible above the plane tracks (GWK).

plus certain walls and embankments at or near the plane, and also the towpath from the plane to the Easton-Delaware Bridge. Actually, all these items had been built long before the indenture was given. The railroad's side tracks had not yet been built, however, and a further agreement regarding them was drawn up. The Bel Del was to have the privilege of moving the canal to a point east of its then-present location, and of moving the towpath from the west to the east bank of the canal, from Port Delaware south to Cooper's Furnace.

There were other items of agreement, but most pertinent here is the one allowing the Canal Company to connect with the railroad near the plane house, "between the lower track of the LVRR bridge and any track they may build east of the side tracks of the Bel Del." The Morris Canal cargoes were thus to be able to cross the Delaware River free of tolls. The ferry rope, however, was not to be dispensed with, but was to "remain as now arranged."[93]

When the Canal Company had established its rates of toll for 1855, the following item was entered into the *Minutes* of April 2, 1855: "On property transported on the Morris Canal, the carrying of which is brought into competition with any other route, and the sale of which is made to compete with the same or like articles from other sources, the president may authorize such drawbacks from established tolls as may be necessary to enable such property to be delivered via the Morris Canal."

Thus, before all of the inclined planes were rebuilt, the Canal Company was to feel the pressure of serious railroad competition. Of the Eastern Division planes, only No. 8 East at Montville and No. 12 East at Newark were completed before 1856. Western Division Planes continued under construction during 1855 and 1856. Yet in May, 1856, the Canal Company reported that the canal could "now be navigated by boats of from 60 to 70 tons."[95]

During 1855, Talcott had done a lot of estimating as to the probable cost of rebuilding Plane No. 2 and 3 East at Drakesville. His notebook contains data about two offers, or bids, for supplying water wheels for Plane No. 3 East.[96] One bid was from James Finlay, for his "Scotch motor"; the other, from Emile Geyelin for a "Jonval turbine." Finlay's price was $1480, which, when added to the cost of the rest of the machinery, totalled $2,792.44. Geyelin's bid, $2,210, had been for "the wheel and fixtures complete;" the rest of the machinery was to cost $640.15, for a total of $2,850.15. Talcott carefully noted the $57.71 difference between the two bids, but never indicated which had been accepted by the Canal Company!

Talcott's estimates for Plane No. 3 East were completed by midsummer of 1855, and the board of directors ordered that that plane and its companion, Plane No. 2 East, be rebuilt "on the plan of the planes west of the summit."[97] They were completed by the time the canal opened for navigation in 1856; Plane No. 1 East was completed for the 1857 boating season. Only seven more planes on the Eastern Division needed to be rebuilt "to make the work complete."[98]

Plane Nos. 1 and 4 East were ordered rebuilt on the same day, June 9, 1856,[99] but apparently something held up the work. Early in 1857, Plane No. 4 was again ordered to be rebuilt, along with Plane Nos. 5, 6, 7, and 9 East — "provided the sales of preferred stock and other sources of income will justify."[100] By midyear, construction was underway on all of these planes, so we may assume that the "income" justified the outlay of money. In fact, the Canal Company advertised for "good laborers," offering them "steady employment through the Summer, Fall and Winter at their planes."[101]

"Extensive repairs" were being made from Boonton eastward, where 20 or 25 men were reported "at work in putting down new tracks, enlarging the basins, and facing up the banks anew."[102] At Boonton it was noted that "instead of being able to raise but half a boat at once, and then being obliged to lock it at the head of the plane, they will thereafter raise the entire boat, drawing it over the entire crown of the plane into the water beyond, as is the case in the admirable plane at Newark."[103] From this description, the Boonton and Newark planes had been converted to "summit" planes. The same source claimed that the line of the canal, itself, was being altered, although just how it was altered is not mentioned.

The Beavertown Plane (No. 10 East) and the Bloomfield Plane (No. 11 East) were under construction by mid-1859.[104] At Beavertown, today's Lincoln Park, "a new quarry of excellent sandstone" was opened at the plane.[105] By the end of 1859, only the unfinished parts of Planes 8, 10, and 11 East remained to be done of the "permanent improvements."[106]

There is no doubt that Talcott's original plan or design for the new inclined planes in 1847-8 featured the double-tracked system. Yet in 1863, when it was announced that "the rebuilding of the inclined planes is now complete according to the original plan adopted in 1850," the description of the planes included only three double-tracked systems. "Planes 6 and 9 West, and 12 East constructed with double tracks, and overcoming an elevation of 220 feet, have cost

$101,128.38, being at the rate of $459.67 per foot lift. The other twenty planes, overcoming an elevation of 1,229 feet, and constructed with single tracks, have cost $577,973.85, being at a rate of $470.28 per foot lift."[107]

There are no concrete statements to support the theory that the Jonval turbine installed at Plane No. 3 East made it possible to do without the second track and cradle car, which after all served mainly as counterbalance for ascending boats. It is nevertheless tempting to so theorize.

The permanent improvements to the planes might have been nearly completed, but the rest of the canal was still far from being permanent, still far from being improved, particularly at its eastern extremity. There still remained repairs to several locks. "Originally, the locks had been made of stone masonry, laid with hydraulic cement." When they were lengthened and widened during 1840-41, new masonry had been "laid with lime mortar instead of cement." By 1859, the "lime had been dissolved, and the mortar washed out to such an extent as to endanger the stability of the walls." Repairs to these locks were costly, since they involved "grouting the walls with hydraulic cement, reflooring the bottom of the lock, and making new gates and wickets." Moreover, the repairs had been made during the winter months when the locks "had to be housed and artificially heated night and day from four to six weeks" to protect the new mortar from frost damage. Only five locks had been repaired during the winter of 1859-60; eight more were repaired during the winter of 1860-61.[108]

During 1859, two very serious, very unexpected breaks had occurred under two locks, their locations making it necessary to have 60 to 80 men bailing night and day to keep water from seeping into the lock while repairs were being made. "Imperfect sheet piling in the original construction of the locks" was blamed for the breaks, which had cost $3,100 to repair.[109]

During 1859, the Canal Company had "completed the construction of a lock in the lower part of their line, near the Plank Road Ferry," and were "engaged with a large force of men in deepening the canal between the Passaic and Hackensack Rivers. During the construction of the lock, the boats passed out into the river at Commercial Dock," and were "towed down the river and through the Bay and Kills to Jersey City and New York."[110] The dredging of the canal between the two rivers, and the boat towing service while that work was being done, plus some repairs, had cost "about $17,000." To put more water into the canal level from the Hackensack to the Hudson, the Canal Company built "an extensive and substantial apparatus," a steam-operated pumping station capable of pumping 100,000 gallons of water per minute from the Hackensack River into the canal. The whole apparatus cost $18,817.70.[111]

Some sort of steam-powered machinery had been under consideration for "unloading and reshipping coal at Jersey City" in an economical way.[112] During 1859, $449,478.30 had been spent, a good part of it ($41,503.26) on reclaiming underwater property at Jersey City, on building piers, wharves, and basins for the accommodation of cargoes, and on the pumping station.[113] The steam-dredging of the nine-year accumulation of tidal silt had cost $6,700.[114] Once that was completed, however, the stockholders were told that there was, at last, "a full five-foot depth of water in all the levels" of the canal.[115]

The stockholders had also been told in their 1853 annual report from President Marsh that the Company's waterfront property in Jersey City had "attained a very great value, and that the time has arrived when it should be improved by the Company, or be disposed of to others who would improve it, and the proceeds be made available to the stockholders."[116]

Also during 1853, the Canal Company president and directors were asked to report on:[117]

1. How much of the unimproved property belonging to the company at Jersey City would, in their opinion, be eventually required for canal purposes.

2. What, in their opinion, would be the best method of disposing of so much of said property as will not be required for the purposes of navigation, or other purposes connected with the trade of the canal.

The information reported by Marsh was subsequently used by

some "Wall Street speculators,"* who published a pamphlet entitled: "Statement and map of unimproved property of the Morris Canal and Banking Company at Jersey City." The pamphlet is worth quoting for the light it will shed on the future and fate of the entire canal. Unfortunately, the map referred to was not with the pamphlet available to the author.

> *The accompanying map exhibits the property of the Morris Canal and Banking Company, at Jersey City, prepared by some of the largest stockholders, with a view of submitting the same to the directors of the company for appropriate action in relation to said property.*
>
> *It will be noticed that the map delineates no less than 2505 lots, from which the land to be taken for streets is to be deducted. Many of those it will be seen are wharf lots. The value of the premises cannot at this time be authoritatively declared, but may be approximately settled by comparison with the value of like property on the New York side of the river, directly opposite at a distance of about three-quarters of a mile, or upon the Brooklyn shore, at either of which places the same quantity of territory as is now possessed by the Canal Company, would be of a value in amount greatly exceeding its capital.*
>
> *The property in question is nearly opposite in a straight line, a short distance above the Battery, in the city of New York. The commerce of that city is so rapidly increasing, and the railroad avenues constantly opening for trade to it will very soon render it necessary to seek for additional space for its accommodation; and there is no point of such cheap and easy access as the shores of New Jersey; and hence the property in question is first to be occupied.*
>
> *It was not our intention to present at this time the mode by which the property in question may be made immediately available, but having received the suggestions of a gentleman of eminent practical knowledge on this subject, we lay them before the Company for their approval:*

**The term is Talcott's, from a "Memorial of the Morris Canal and Banking Company to the Honorable, Mayor and Common Council of Jersey City," January 9, 1866.*

1. To fill up Hudson street and make a bulkhead or wharf to the southern boundary of the property, say 2,340 feet.

2. To extend from bulkhead on Hudson street into Hudson River, ten piers 475 feet long, 55 feet wide, and 32-1/2 feet deep, which will give an equal number of slips for occupancy of vessels, subject to wharfage.

The amount of revenue to be derived from these wharves we may estimate by comparison with that obtained from the Cunard Steamship Company for the use of one pier, similar to those described above and immediately adjoining them, which was leased from the Jersey Associates, about ten years since, at an annual rental of $7000, but which at the expiration of the present contract can readily be leased at $15,000 a year, to the same parties. The result to the Canal Company of the building of these piers, can be easily seen, should they all rent at the price named above, of which there is no doubt entertained, it will give the company the sum of $150,000 per annum, equal to 7 percent on two million dollars. Of the valuation of these lots, but little diversity of opinion appears to exist — the various surveyors who have been consulted appraising them at about $3,000,000.

Unfortunately for the Morris Canal, the prospect of owning some of that estimated $3,000,000 worth of real estate opposite New York City also appealed to the town fathers of Jersey City and the Central Railroad of New Jersey.

REFERENCES

1. *Report to the Stockholders of the Morris Canal and Banking Company,* 1862, in the *American Railroad Journal,* no page number (hereinafter cited as *Report to the Stockholders, 1862*).
2. *Report of the President and Directors of the Morris Canal and Banking Company of 1844, to the Stockholders* (N.Y.: 1848), p. 5 (hereinafter cited as *Report of the President, 1848*).
3. *Ibid.,* pp. 7-8.
4. *Ibid.,* p. 27.
5. *Ibid.*
6. *Ibid.,* p. 28.
7. *Ibid.*
8. *Ibid.,* p. 29.
9. *Ibid.*
10. *Ibid.,* p. 9.
11. *Ibid.,* p. 16.
12. *Ibid.*
13. Morris Canal and Banking Company *Minutes,* June 26, 1849 (hereinafter cited as MC&BCo. *Minutes*).
14. *Ibid.,* August 1, 1849.
15. *Report of the President, 1848,* p. 14.
16. *Ibid.,* pp. 6-7.
17. *Report of the Board of Directors to the Stockholders of the Morris Canal and Banking Company* (N.Y.: 1849), p. 4 (hereinafter cited as *Report of the Board, 1849*).
18. Morris Canal and Banking Company *Docket Files,* (hereinafter cited as MC&BCo., *Docket Files*), agreement dated January 1, 1848.
19. *Ibid.,* agreement dated January 11, 1849.
20. *Report of the Board, 1849,* p. 5.
21. MC&BCo. *Minutes,* March 2, 1849.
22. *Ibid.,* April 7, 1851.
23. *Ibid.,* April 2, 1849.
24. *Ibid.,* May 15, 1849.
25. *Report to the Stockholders of the Morris Canal and Banking Company,* 1851, in the *American Railroad Journal,* March 3, 1851, p. 283 (hereinafter cited as *Report to the Stockholders, 1851*).
26. MC&BCo. *Minutes,* August 1, 1849, letter dated July 23, 1849.
27. *Ibid.,* February 1, 1850.
28. *Ibid.*
29. *Report to the Stockholders, 1851,* p. 283.
30. William H. Talcott, *Estimates of Quantities for Planes 1, 2, 3, 4, 5, 6, 7, 8, 9, 10, and 11 West, etc., June, 1850* (hereinafter cited as Talcott's *Estimates*).
31. *Ibid.,* pp. 8, 42.

32. *Report to the Stockholders of the Morris Canal and Banking Company* (N.Y.: 1852) p. 5 (hereinafter cited as *Report to the Stockholders, 1852*).
33. Talcott's *Estimates,* p. 61.
34. *Ibid.,* p. 20.
35. *Ibid.,* p. 17.
36. MC&BCo. *Minutes,* June 26, 1854.
37. Talcott's *Estimates,* p. 14.
38. *Ibid.,* p. 15.
39. *Ibid.,* p. 10.
40. *Ibid.*
41. *Report of the President, 1848,* pp. 8, 56.
42. Talcott's *Estimates,* p. 26.
43. *Ibid.,* p. 29.
44. *Ibid.*
45. *Ibid.,* pp. 11, 38-9, 56.
46. Abram S. Hewitt Papers.
47. Talcott's *Estimates,* p. 38.
48. *Ibid.,* p. 34.
49. *Report to the Stockholders, 1852, p. 5.*
50. Talcott's *Estimates,* p. 100.
51. *Ibid.,* p. 68.
52. *Report to the Stockholders, 1851, p. 283.*
53. *Report to the Stockholders, 1852,* p. 6.
54. MC&BCo. *Minutes,* October 2, 1851.
55. Talcott's *Estimates,* pp. 49-51.
56. *Ibid.,* pp. 91-2.
57. Wheaton J. Lane, *From Indian Trial to Iron Horse: Travel and Transportation in New Jersey, 1620-1860* (Princeton, N.J.: 1939), pp. 164-5.
58. Newark *Daily Advertiser,* October 1, 1857.
59. Talcott's *Estimates,* pp. 91-94.
60. MC&BCo. *Minutes,* July 21, 1856.
61. Talcott's *Estimates,* p. 95.
62. MC&BCo. *Minutes,* May 4, 1836.
63. *Ibid.,* February 21, 1846 and November 7, 1850.
64. *Ibid.,* October 7, 1852.
65. *A History of the City of Newark, New Jersey* (N.Y.: 1913), pp. 621, 622.
66. Talcott's *Estimates,* p. 95.
67. *Ibid.,* p. 21.
68. *Ibid.,* p. 101.
69. MC&BCo. *Docket Files.*
70. *Report to the Stockholders of the Morris Canal and Banking Company* (N.Y.: 1853), p. 6 (hereinafter cited as *Report to the Stockholders, 1853*).
71. MC&BCo. *Minutes,* July 11, 1850.

72. Letter cited in James W. Ransom, *Vanishing Ironworks of the Ramapos* (New Brunswick, N.J.: 1966), pp. 57-8.
73. Abram S. Hewitt Papers, letter dated May 11, 1854.
74. Ransom, *op. cit.,* p. 55.
75. MC&BCo. *Minutes,* July 21, 1856.
76. Abram S. Hewitt Papers, letter dated May 11, 1854.
77. MC&BCo. *Minutes,* May 27, 1853.
78. *Laws of New Jersey,* 1853.
79. MC&BCo. *Minutes,* July 19, 1855.
80. *Report to the Stockholders of the Morris Canal and Banking Company* (N.Y.: 1855), p. 7.
81. MC&BCo. *Minutes,* July 21, 1856.
82. *Report to the Stockholders of the Morris Canal and Banking Company* (N.Y.: 1859), p. 9.
83. Lane, *op. cit.,* p. 244.
84. MC&BCo. *Minutes,* July 21, 1856.
85. Edward Hungerford, *Men of Erie* (N.Y.: 1946), p. 111.
86. *Ibid.,* pp. 6-7, 9, 20.
87. Randolph L. Kulp, ed., *Railroads in the Lehigh River Valley,* (Allentown, Pa.: 1962), p. 41.
88. Charles E. Fisher, "The Lehigh Valley Railroad," *Railway and Locomotive Historical Society, Bulletin No. 42,* February, 1937, p. 11.
89. *Report to the Stockholders of the Morris Canal and Banking Company* (N.Y.: 1861), p. 6 (hereinafter cited as *Report to the Stockholders, 1861*).
90. Abram S. Hewitt Papers, letter dated June 19, 1854.
91. *Ibid.,* letter dated October 13, 1854.
92. MC&BCo. *Minutes,* January 22, 1857.
93. *Ibid.,*
94. *Ibid.,* June 9, 1856.
95. *Report to the Stockholders of the Morris Canal and Banking Company,* in the *American Railroad Journal,* May 17, 1856, p. 313.
96. Talcott's *Estimates,* pp. 103, 105.
97. MC&BCo. *Minutes,* July 19, 1855.
98. *Report to the Stockholders of the Morris Canal and Banking Company* (N.Y.: 1857), p. 6.
99. MC&BCo. *Minutes,* June 9, 1856.
100. *Ibid.,* March 12, 1857.
101. Paterson *Daily Guardian,* July 25, 1857.
102. *Ibid.,* January 4, 1858.
103. *Ibid.,* February 10, 1858.
104. MC&BCo. *Minutes,* May 13, 1859.
105. Paterson *Daily Guardian,* June 25, 1859.
106. *Report to the Stockholders of the Morris Canal and Banking Company* (N.Y.: 1860), p. 17 (hereinafter cited as *Report to the Stockholders, 1860*).

107. *Report to the Stockholders of the Morris Canal and Banking Company* (N.Y.: 1863), pp. 15-16.
108. *Report to the Stockholders, 1861,* p. 14.
109. *Report to the Stockholders, 1860,* p. 16.
110. Paterson *Daily Guardian,* May 31, 1859.
111. *Report to the Stockholders, 1860,* p. 7.
112. MC&BCo. *Minutes,* July 21, 1856.
113. *Report to the Stockholders, 1860,* p. 7.
114. *Ibid.,* pp. 15-16.
115. *Ibid.,* p. 7.
116. *Report to the Stockholders, 1853,* p. 7.
117. MC&BCo. *Minutes,* May 27, 1853.

16

A Brand New Outlook

DURING the period of enlargement, 1845 through 1861, there was a decided increase in activity among the various iron industries along the line of the Morris Canal. By the time Edwin Post and Samuel R. Brooks took over the Morris and Sussex Manufacturing Company and incorporated as the Stanhope Inn Company (February 25, 1841), they had already spent $75,000 on real estate and three furnaces.[1]

"The rise in the price of iron" attracted "the attention of Capitalists to the rich ore beds along the line of the canal, and very large sums" were spent "in the purchase of mines, and in the preparation to mine and send to market large quantities of iron and other ore. The erection of a number of Blast Furnaces along the line of the Canal are likewise in contemplation."[2] One such furnace was contemplated by William Green, who applied to the Canal Company for "some privileges" if he "should build a blast furnace at or near Dover," and Canal Company President Marsh was authorized by the board of directors to make "such an arrangement as is proposed" by Mr. Green, "providing the same can be done without interfering with the canal in any way."[3] It is tempting to conclude that this was the first attempt to establish Port Oram Furnace. William Green and Robert

F. Oram *were* partners at this time [1853] in several Morris County mines, among them the famous Swede's Mine at Dover.[4] Robert F. Oram was to establish his furnace "near Dover" later on, but whether or not Green was still associated with him then is not known.

Then, too, iron works were being built along the Lehigh River Valley: Hockendauqua Iron, Thomas Iron, Glendon Iron, Lehigh Crane Iron — just to name a few of the iron works that were to become customers for New Jersey iron ore shipped by Morris Canal. Many of these same iron companies also owned or leased the iron mines in New Jersey from which their ores were extracted:

Mine	County	Iron Company
Lawrence	Morris	Reading
Dalrymple	Morris	Crane
Bryant	Morris	Bethlehem
Evers	Morris	Saucon
Randall Hill	Morris	Crane
Baker	Morris	Lackawanna
Sterling	Morris	Thomas
Hurd	Morris	Thomas
Baker (Mt. Pleas. area)	Morris	Allentown
Richards	Morris	Thomas
Teabo	Morris	Glendon
Mt. Hope	Morris	Lackawanna
Old Ford	Morris	Glendon
Crane	Morris	Glendon
DeCamp	Morris	Glendon
Willis	Morris	Bethlehem
Upper Wood	Morris	Bethlehem
Waywayanda	Sussex	Thomas
Franklin	Sussex	Lackawanna

Interestingly enough, the Waywayanda mine had once been owned by Oakes Ames, who would later earn a place in the nation's history books for his part in the Credit Mobilier scandal that surrounded the building of the Union Pacific Railroad. The mine was once known as the "Acker Mine," and from it Ames took ore to feed the blast furnace he had built on "Double Pond," today known as Lake Waywayanda.[5] Ames had also owned mines at Franklin Furnace and Sterling Hill.[6]

Despite Ames' involvement in the Credit Mobilier scandal, his death brought forth a eulogy from his friend James I. Blair, the railroad baron of Blairstown, New Jersey. Blair called Ames: "The main pillar that carried the Union Pacific Railroad through."[7]

Most, if not all, of the blast furnaces along the line of the canal, and many in Pennsylvania, received at least some ore by way of the Morris Canal.[8] One of the most important canal customers was the Cooper-Hewitt & Company. The company owned or leased numerous mines in Passaic, Morris, Sussex, and Warren counties, from which ores were sent to "Cooper's Furnace" at Phillipsburg for smelting. The resulting pig iron was then sent to the Trenton Iron Works by way of the Belvidere-Delaware (Bel Del) Railroad. Most of the iron was made into rails "of various shapes and sizes."[9]

Cooper-Hewitt also received ores from Mount Olive Township mines. Dozens of letters among the correspondence of the company confirm that if the Sussex Iron Company was *not* in blast at Stanhope after a major explosion and fire, it certainly was engaged in mining and in transporting ore from the mines of A.A. and Ira Drake and G. Salmon, at Mount Olive.

Cooper-Hewitt & Company also received ore from J.L. Allen at Dover. A letter from Allen, dated November 3, 1856, lists the ore deliveries for the month of October, giving the names of canal boat captains and their boats:

Date	Boat	Captain
October 8	Gen'l Jackson	Whitehead
10	M.C. 193	Paynton
13	Fred Smith	Rice
17	M.C. 335	Edzel
21	John Livet	Ginsinger
22	M.C. 196	Shaffer
23	E. Marsh	Hart
25	M.C. 346	Westerfield
29	M.C. 382	Weller
30	Challenge	Wood
31	M.C. 527	Wolfson

These eleven boats carried a total of 484 tons of ore.

The New Jersey Zinc Company wrote to Cooper-Hewitt & Company on December 3, 1856: "The New Jersey Zinc Co. agree to give you the transportation for twelve months of all the ore the Co. may require from Sterling Hill at their works at Newark and pay you for same Four Dollars and twenty seven cents ($4.27/100) per ton of 2240 pounds weight. The above price to include all charges for weighing tolls, etc., etc. The unloading of the boats at Newark to be done by the New Jersey Zinc Co." The letter is signed by William Redmond.

The Canal Company was also a customer of Cooper-Hewitt. Old inclined plane rail iron was sent to Cooper's Furnace by canal boat during the period of enlargement and sold for scrap or was reformed into rails for the new inclined planes. Of course Cooper-Hewitt also made rails for the Camden & Amboy Railroad,[10] and for the New Jersey Railroad and Transportation Company (NJRR&TCo.),[11] and probably for many others.

Several letters from Canal Company Secretary Lewis Condit during 1856 confirm this. Other letters indicate that while Cooper-Hewitt shipped massive amounts of iron ore over the Morris Canal, they did not receive all their coal from that waterway. Cooper-Hewitt dealt directly with the Lehigh Coal and Navigation (LC&N) Company. A letter from the Lehigh Company president, James Cox, dated March 20, 1856, is revealing on that count: "In consideration . . . of the magnitude of your proposed order, we shall be willing to sell you 10,000 to 15,000 tons of *broken* (coal) . . . deliverable in the boats at your furnaces on the Morris Canal at $3.30 per ton, you paying all charges for ferrying on the river and tolls on the Morris Canal. We shall deliver no coal except by canal."

These terms appear to have been unacceptable, for on April 5, 1856, just a few weeks after Cox's offer, Cooper-Hewitt contracted for delivery at their furnaces, "during the next twelve months, in equal monthly proportions (excepting unavoidable or extraordinary accidents) Eight Thousand tons of James Taggert's West Spring Mountain Lump Coal; at Two dollars and twenty-five cents per ton, for the coal, delivered into the boats or cars at Mauch Chunk, to which the toll and freight, or transportation *whatever it may be,* is to be added. . . . We also extend to you the privilege of increasing the order to thirteen thousand tons . . . ."

Cooper-Hewitt still had big plans for the Ringwood enterprise. Ore raised there was wagoned to Morris Canal boats on the Pompton Feeder. Once the proposed Ringwood Valley Railroad was built, however, it would be shipped by rail to the canal at Pompton and then put on canal boats. Tolls for those shipments were set in 1856, although the railroad was not even begun: "Net tolls on iron ore so brought and transported shall not exceed 45 cents per ton for 5 years next ensuing, and 50 cents per ton for the next 10 years."[12] By 1858, the terms were renegotiated: drawbacks of 15 cents per ton would be deducted from the maximum tolls on the canal during the next 16 years, but in no case would the tolls fall below 45 cents, including the boat tolls.[13]

Such an arrangement about tolls is understandable in the light of the enormous tonnage originating from and/or consigned to Cooper-Hewitt & Company. It is surprising, in fact, that in all of the company's correspondence reviewed by this author there is so little mention of boating problems or shortages or whatever. One letter from Edwin Post to Cooper-Hewitt dated October 26, 1854, mentioned the fact that "Mr. Randolph objects to having his boats loaded with ore because he fears they will be broken by the loading from the shutes [sic]."

Boat No. 457 had been "hired out" by the Canal Company and the Company "should have had nothing to do with her freight," which was destined for Cooper-Hewitt & Company. However, the Canal Company "Boat Agent's assistant paid the boat for the trip by mistake" and the Canal Company tried to recover the money from Cooper-Hewitt.[14]

Short weight was claimed by Cooper-Hewitt on one occasion, causing Canal Company Chief Engineer Talcott to reply:[15]

> *We will do all in our power to prevent stealing of the iron, and in case we can detect any capt. selling or allowing any to be taken off, will keep from him the full price of the iron. I cannot help thinking there is considerable uncertainty about the amt. put on the boats. I am told that some of the cargoes overrun about as much as these fall short. I shall see that each capt. is told before he takes the iron that he will be expected to pay for any that is missing.*

Thievery certainly wasn't uncommon among boatmen. Cornelius Van Winkle, proprietor of the Centreville Hotel in 1856,[16]

> *having missed things lately, kept his top eye open on the Canalers; and one day last week espied a party of boatmen stealing his pork barrel. He soon had the constable of Little Falls, Mr. Jacobus, notified, and the pursuit was made. The boat was overhauled at Mead's Basin, or thereabouts, and Jacobus boarded the vessel and seized the Captain, who screamed for assistance from below. The racket brought up an old man, whose name is Mangle, who proved to be the father of the other two, and who presented a pistol at Jacobus. The constable thought discretion was the better part of valor, and backed off the boat for assistance. He soon returned with a reinforcement from the blacksmith shop, and captured the vessel after a short conflict. Constable Jacobus arrived in town* [*Paterson*] *with the Captain and crew in irons, and the report that they had behaved in the most obstreperous manner on the coast. He also brought down the armament of the vessel, which consisted of a double-barreled pistol — empty, a bullet mould, and a corkscrew. The old man went to jail in default of bail, on a charge of assaulting an officer. The Captain, his son, had five dollars which enabled him to pay his own costs; but the other brother, Peter Mangle, had no money and was locked up for the expenses of the trial.*

About two weeks after this incident, the Paterson *Guardian* reported on December 11, 1856, that burglars broke into the office of Mr. Boyes on the Morris Canal coal dock at Jersey City, too. After generally ransacking the office and scattering papers all over the place, all the compensation the thieves got was "a few pennies."

On a more productive note, in 1856 the Canal Company decided to use steam to propel boats on the Morris Canal. A total of $505.42 was spent on the experiments between March, 1856, and the end of 1858.[17] The Erie Canal was beginning to use steam canal boats during the 1850s, and at one point 53 such boats were being built at Buffalo for canal service in 1859.[18] The Delaware & Hudson Canal, too, was rebuilding a steamboat to test the practicability of using steam tows.[19] It would be 10 years, however, before the steam boat *Gussie* would make its appearance on the Morris Canal, and 20 years before the steam canal boat built by William Baxter at Washington in Warren County would make its debut on May 31, 1878.[20]

The 1857 boating season got underway on March 26 with a promising outlook for business. Receipts for 1855 had been $272,175.93; in 1856 they were $304,805.16, and there was no reason to think that 1857 would not see an increase as well.[21]

A slight "difficulty between the common and preferred stockholders," was "settled by giving to the former $200,000 of preferred shares." The total amount of preferred shares issued was now "$1,100,000, bearing ten percent dividends. The common stock is a little over one million, and the mortgage debt, at six per cent, half a million." Up until 1857, the entire canal had cost "between six and seven million dollars."[22]

By the end of August, 1857, the canal's receipts did, indeed, show the anticipated increase over 1856, "up to the 29th of August, amounting to $190,128.00, an increase of $647 over the corresponding period last year." In fact, the receipts for the last *week* of August had been $10,245.[23]

Early in the boating season, there was an amusing incident on upper Mill Street in Paterson, where a lot on the west side of the street above O'Neill's Coal and Lumber Yard belonged to the city.[24]

> *Mr. D. Quackenbush claims to have certain privileges by permission of the city authorities. Mr. Scott had been landing his brick for the new Waverly Mill on that place, and this morning Mr. Q. warned him to desist from doing so for the future. Mr. S. thereupon took a body of men to the place and proceeded to unload in spite of the other party. Mr. Q. then took a saw and commenced sawing the plank between the boat and tow path. Some pushing followed, and it is said Mr. Quackenbush struck Mr. Scott. While things were thus going on, Mr. Pope came to the place with a reinforcement of quarrymen to the support of Mr. Quackenbush, and Mr. Scott sent for the sheriff to defend his men. To match this manoeuvre Mr. Pope brought forward three barrels of lager, and distributed it*

ad lib. among his partizans. The result was Scott's men withdrew and Quackenbush remained in possession.

There was a "shipwreck" on the canal at Paterson in 1857, too. During April, "a bridge crossing the canal at a point nearly a mile beyond the rocks fell in, taking with it a large quantity of stone and earth that formed one of the buttresses." Then early in May, "the scow *Naomi,* Captain Wolf, laden with about 60 tons of zinc ore, while on the passage from Waterloo to Newark, struck on these sunken rocks, by which a hole was beaten into her bottom and she at once commenced to fill." She was raised, of course, after two days of work by a crew of "ten or a dozen men," but not without causing "quite a number of boats, laden with coal and other articles," to collect in the canal beyond the unfortunate boat.[25]

There was a point on the canal at Paterson where a sharp curve near "the rocks" made it necessary for boats about to enter the passage to be cautious. Captains generally blew their horns there to warn boats traveling in the opposite direction that they were entering the spot. On June 2, 1857, "there was more than the usual amount of swearing among the boatmen" as a boat "laden with nails" entered the passage "while one was entering the other end with a load of dressed stone." One captain accused the other of not signaling, and the ensuing argument continued for "about half an hour before they got their scows disentangled."[26]

In March, 1857, before water was let into the canal for the season, the editor of the Paterson *Daily Guardian* had written that Patersonians lived under the constant threat of canal breaches. He suggested that "at one or more different points on the canal, flood gates might be constructed which would rise and stop the flow of water in case of the bank giving way. Then all the damage which could be done would arise from the water within the quarter or half a mile inside the gates."[27] Within two months of that suggestion, there was a serious breach in the canal bank behind O'Neill's Coal Yard, causing extensive damage. The whole 17-mile level of the canal, from Bloomfield to Lincoln Park, was drained.[28]

Of course, breaches were not reserved exclusively for Paterson. Brookfield's reservoir had given way at Waterloo on May 29, 1857, and "about 100 acres of adjoining property were flooded." The Canal Company dispatched a large number of workmen from Newark and Jersey City to make repairs. Only the "three-mile level" was affected, "from Waterloo to Saxon's Falls."[29] The breach was not repaired until Saturday, June 6, and the newspaper admitted that the damage had been even more serious than had at first been thought, with 300 acres being overflowed. The Canal Company had had a force of 500 men working there 14 hours a day for more than a week. There had been seven breaches in all, "tearing away 300 feet of the bank, and depositing two boats in the meadow."[30]

What had started out to be a successful season suddenly took a turn for the worse by October as the bank panic of 1857 struck. The Canal Company was reported to have virtually closed down and to have discharged "the greater part of their boatmen, and other employees — to the number of about 400 — in consequence of inability to obtain funds to pay the current expenses."[31]

The Canal Company denied that they had let go a single boatman. They had "wisely, as they think, postponed all improvements on their works," however.[32] Employees of the Boonton Iron Works were not so fortunate. The Works shut down on October 24, 1857, and 1200 laborers were discharged.[33] Then, in mid-December, 1857, the canal water level was lowered about 10 inches with no intent to lower it any more than that. In view of the mildness of the winter to that point, there was speculation that boats might run all winter.[34]

Despite the "Panic of 1857," the Canal Company announced a 5-percent semiannual dividend on its preferred stock.[35] The following year, men were at work cleaning out the bottom of the canal prism early in March, with expectations that the water level would be brought up to normal by the end of March.[36] Navigation opened on March 29, with "unusually flattering" prospects for a "heavy business." The Canal Company had made extensive repairs and improvements during the winter of 1857-8, and was expecting a very much increased coal trade.[37] Boats began "passing up and down giving an air of activity" to the canal.[38]

Some of those boats that cleared Stanhope Plane No. 2 West were recorded in the *Clearance Book* for that place:[39]

Boat	Captain	Cargo
Clara	Willis	bark & boats
F. Lindsley	Thompson	flour
Flora	Ammerman	limestone
Henry Howell	Hildebrant	zinc ore
Loantika	Jacobus	corn
North Star	Thompson	flour
Sussex	Miller	wood & posts
May Flower	McElroy	charcoal
I. Burke	Clouse	wood & railroad ties
Success	Mase	iron boilers
Mountaineer	Beider	lumber

Early in 1858, the Commonwealth of Pennsylvania decided to sell — to the Sunbury and Erie (S&E) Railroad — its canals that had not already been sold to the Pennsylvania Railroad. The S&E Railroad, in turn, sold those canals "to companies organized to own and operate them."[40] Not surprisingly, as it turns out, the Canal Company followed this development with great interest. E.W. Clarke & Company of Jersey City were also interested. So much so that Mr. Clarke asked the Canal Company: If his friends should get together "and make up a party and buy the Delaware Division of the Pennsylvania canals, would the Morris Canal and Banking Company lease the said canal for twenty years, and what annual rent they would give therefor." The Canal Company replied that they would check first to determine whether or not such a lease would be legal. If so, they would consider $200,000 yearly, to be paid in two installments a year for twenty years.[41]

The Canal Company's board of directors felt that it was very important to "have a control of the Delaware Division," to prevent the Lehigh Coal and Navigation Company from "obtaining a monopoly of said transportation."[42] Since the outlet lock from the Delaware Division Canal to the Delaware and Raritan Canal had been completed in 1848, and the former canal had just been enlarged to accommodate more and more coal from the Lehigh Valley, both canals had become a serious threat to the Morris Canal's trade. Consequently, when the company that was to buy the Delaware Division Canal was being organized, Morris Canal Company stockholders invested in its stock. Several of those investors were made directors of the newly formed Delaware Division Canal Company of Pennsylvania.

President Marsh told the stockholders in his 1859 annual report that "the effect of this arrangement has been, to secure to this company, an approximation to a proper proportion of tolls upon that branch of their business, and to make the Morris Canal as favorable an avenue as any other, for the transportation of Lehigh coal to supply the New York market, and for re-shipment upon tide water."[43]

What Marsh did not tell the stockholders was that on September 23, 1859, the MC&BCo. entered into its leasing agreement with the partisan Delaware Division Canal Company of Pennsylvania. Once that agreement was made, the Canal Company began to negotiate with the Lehigh Valley Railroad for the delivery of coal to the canal at Phillipsburg, and with the Lehigh Coal and Navigation Company for its 1859 toll arrangements on the Delaware Division Canal.[44]

Since the Morris Canal Company would now benefit from traffic using the Delaware Division and Delaware and Raritan Canals, it follows that new legislation was sought in New Jersey and Pennsylvania to facilitate canal boat navigation from one state to the other. A bill authorized either the Morris Canal or the Delaware Division Canal, or both, "to erect, in lieu of the present wing dams at Wells Falls and Phillipsburg," a device "for the purpose of improving and facilitating canal navigation between Pennsylvania and New Jersey, and for supplying the lower levels of the Delaware Division with water. The bill provides for openings for the navigation, and for supplying the mills of Lambertville with water to compensate for what they will lose. The commissioners to arrange the matters are J.B. Bassinger of Morris; J.W. Allen of Burlington; and W.H. Talcott, of Jersey City."[45]

The effect of the competition between the Morris Canal and its rivals, the Delaware and Hudson and Delaware and Raritan Canals, had been to encourage "through-shipment" of coal. Instead of delivering coal along its route at rates proportionate to the distance from the mines, or the loading point, the Canal Company rushed its coal to tidewater so that it could compete for the re-shipment trade. Consequently, coal was sold more cheaply at its farthest destination than at points nearer to Pennsylvania. "Back-shipping" has already been discussed, but it is reintroduced here in relation to Paterson's complaints about the discriminatory toll system and how it affected that city.[46]

> *The facilities for supplying New York must always increase and improve; but the chances are against the people along the route. They must continue to be consumers, at all events. If there is to be any special favor shown, let it be to those who must be customers, not to those whose patronage is uncertain or likely to be lost altogether. Build up home industry, and cease to be dependent upon the foreign consumer merely is the true policy both for a State and a Canal Company.*
>
> *Certain of our shrewdest railroad managers have begun to recognize the propriety of this course. . . . The railroad can compete with the canal, and a line to connect this city with the Morris & Essex road at Rockaway would be an easily constructed work and one that would pay. By means of that and the Morris and Essex Extension we could connect with the principal coal roads, and thus obtain the indispensable article of coal at New York prices.*
>
> *The Delaware & Lackawanna, the Lehigh Valley, the New-Jersey Central, as well as numerous other railroads, transport coal, and compete with the canals supplying New York. With a railroad open to Easton, passing through Little Falls and Boonton, there is no doubt coal could be carried at rates so as to compete with the canal.*

The Morris and Essex Extension mentioned above reached from Phillipsburg to Hackettstown. It had been surveyed in 1857 and was under construction in 1859. This 23-1/2-mile extension — somewhat omniously — was to follow the Morris Canal.

When navigation was about to open for the 1859 season, the boatmen held a meeting in Newark, at which time they set a "new scale of prices for transportation." The price per ton "through to tide water" was fixed at $1.40.[47] And then the boatmen struck, and at Paterson, they rioted. "The movement," however, was "universal throughout the line" of the canal. The channel was "entirely blocked with boats" so that none could get through.[48]

> *Mr. Samuel Ryerson of Pompton had engaged to deliver a lot of lumber in Newark and went down to that city to receive and deliver it. Upon finding it did not arrive, he followed the line of the canal until near Paterson he saw his lumber boat blocked with the rest. The captain was disposed to proceed, and was not allowed by about one hundred and fifty men who were on guard for the strikers. As Mr. Ryerson was about to extricate his boat, he saw several persons endeavoring to sink her, with others, and upon his resisting, a regular row commenced, whereupon Mr. R. sent for the Mayor, who with a number of officers proceeded to the scene of the riot, and after a parley* [*sic*] *ordered the crowd to disperse. It was found impossible to prevent the blocking by boats, and the canal still remains impassable.*
>
> *One rioter by the name of George Rafferty was arrested, who had, it was said, been engaged in sinking the boats, and he was lodged in jail. This man lives a short distance from Paterson, toward Acquackanonk* [*Passaic*].
>
> *Whilst a constable was taking this chap to jail, Mr. Ryerson, who is over sixty years of age, was struck in the face with a stone, and had five or six teeth knocked out. He was also struck on the head with a piece of rock and somewhat injured.*

At about the same time, the engine house of the Mount Pisgah plane of the Lehigh Coal Company was badly damaged by fire. Coal would not be shipped for about a week.[49] One wonders if the striking boatmen on the Lehigh Canal had anything to do with that fire.

A new freight line was established during 1859, too. A regular line of covered boats was to run from New York through the Morris and Lehigh canals to Easton, White Haven, and Wilkes Barre. This "New Jersey and Pennsylvania Freight Line" was to make intermediate stops, as well, and particularly at Washington, New Jersey, where it would connect with the Delaware, Lackawanna and Western Railroad. At White Haven, it would connect with the Lehigh and Susquehanna Railroad.[50]

Morris Canal receipts for 1859 up to May 20 totaled $40,658 against $26,250 for the same period in 1858, an increase of $14,399.[51] Yet, curiously — and again ominously — the Boonton Iron Works at this time was turning out an average 3,500 casks of nails a day, each weighing 100 pounds, then carting them seven miles for shipment on the Morris and Essex (M&E) Railroad![52]

Paterson was hoping that Cooper-Hewitt's proposed Ringwood Valley Railroad would be built and somehow connected with the M&E Railroad, which was then almost completed to Easton. As far as the Paterson newspaper was concerned, such a railroad should have been undertaken by its own residents "several years ago."[53]

Cooper-Hewitt, who had bought the Ringwood property "in 1854 for little more than $100,000," had taken "about 5000 tons [of ore] annually" from the mine there, "transporting it to the forges and then converting it directly into wrought iron," which was then "sent off, principally, to the company's wire factory at Trenton." The firm had had a survey made for its proposed railroad in 1855, "for the purpose of transporting their ore to the Morris Canal. Its length was 11 miles, the estimated cost being $40,000." The survey was for a railroad from Mead's Basin (Mountain View, Wayne) to the Cooper-Hewitt mines at Ringwood.[54] It was never built.

The New Jersey Central Railroad *was* built, however. Its connection at the Delaware River with the Lehigh Valley Railroad was effected by means of an iron railroad bridge completed in December, 1859.[55] Two of the major railroads that would cross northern New Jersey had thus inched their way across the Delaware River.

The M&E, meanwhile, had decided to extend its road eastward, and had asked its stockholders for $1,000,000-worth of subscriptions for new stock to finance construction of its proposed branch from Newark to Hoboken.[56] The matter had already been decided on August 18, 1859, when the M&E officers met with E.A. Stevens to sign the necessary papers.[57] Whereas the M&E had gone from Newark into Jersey City over the tracks of the New Jersey Railroad, it would now go into Hoboken on its own track. Accordingly, the arrangement with E.A. Stevens was actually an agreement with the Hoboken Land and Improvement Company.

Chartered on February 21, 1838, the Hoboken Land and Improvement Company was the Stevens family's answer to the New Jersey Harbour Company, chartered on March 13, 1837. John C., Robert L., James A., and Edwin A. Stevens, plus Thomas A. Conover and Joshua R. Sands, had formed their company to develop Hoboken as a commercial port. The New Jersey Harbour Company had been organized to develop a spot on the Hudson River between Hoboken and Jersey City by making "a capacious harbour for vessels of all sizes," at Pavonia or Harsimus.[58] The list of its incorporators and directors in the 1837 edition of *Laws of New Jersey* reveals that virtually all were affiliated with the MC&BCo.

Not to be outdone, Samuel Swartwout, former Canal Company director, had incorporated the Bergen Land and Improvement Company on February 24, 1838.[59] The *Laws of New Jersey* for that year reveals that almost all the directors named in the act of incorporation were also Canal Company participants.

An amendment to the Hoboken Land and Improvement Company's charter, passed in February, 1860, authorized a railroad from Newark to Hoboken.[60] This railroad was the outgrowth of a dispute between the New Jersey Railroad and Transportation Company and the Stevens-controlled Camden and Amboy. The dispute is admirably treated by Wheaton Lane in *From Indian Trial to Iron Horse*. For our purposes, it is enough to note the conflict and use that part of it which explains how the M&E Railroad finally got its own berth on tidewater.

Again, however, we must recall that the Canal Company had located its piers and basins at the extreme southern end of Jersey City as it was then formed, effectively blocking any further expansion of that city by claiming ownership of lands under the waters of the Hudson River. The Delaware & Hudson had built enormous coal docks at Weehawken, one-and-a-half miles north of Hoboken.[61] Where else was there to go if the M&E, like all other railroads, wanted to locate opposite New York City?

What waterfront the Canal Company did not occupy in Jersey City was owned by the Jersey Associates. Well aware of the extreme desirability of its location and of the barrier effectively imposed by the Canal Company property, Jersey City began to rebel. Claiming that the streets of the city had been laid out and dedicated as public streets in 1804 by Joseph F. Mangin on his original map, Jersey City should have the right to extend those streets on reclaimed land through or over the choice spot occupied by the Morris Canal Basin. In 1848, Jersey City wanted to extend one of its streets to the Hudson for the purpose of establishing a ferry to New York. The Jersey Associates sought an injunction to stop the city. The injunction was reversed, and the Jersey Associates appealed.

The decision of the Court of Appeals, handed down by Isaac W. Scudder, was that the ferry rights had been vested in the Associates via their deed from Cornelius Van Vorst in 1804. That deed and the Associates' charter had also given them the right to lay out and widen streets. Jersey City, then, could not extend or widen Hudson Street to the Hudson River. Scudder decided for the appellants, revived the injunction, and made it perpetual on July 17, 1850.[62]

In 1859, Jersey City again tried to extend Hudson Street over the Canal Basin, this time to move southward with the city's development, and this time to *bridge* the Basin. The Canal Company, of course, protested. The opinion of Chief Justice Whelply was that since the canal was "bridged at many points along its length without injury or impairment," he could see no reason why a bridge at Hudson Street would do any harm except to deprive the Canal Company "of part of a rent for what is now used as a coal yard." In point of fact, Whelply doubted that the Canal Company's right and title to any of the lands under water conveyed to them by the Associates in 1835 were valid and clear. Furthermore, since the streets had already been dedicated as public streets, the Canal Company had no right to deprive the public of *its* right to follow Hudson Street to tidewater.[63]

What the public would want to follow Hudson Street *for* was precisely what was so upsetting to the Canal Company. It would not be long before the reason would become clear, as seen in Chapter 17.

It would become clear, too, at the end of the 1859 boating season, that further enlargement of the canal would be necessary. If the locks were lengthened, and the depth of the water in the prism were increased, even more freight could be handled on the canal.[64]

It doesn't seem possible that the Canal Company did not, or could not, or cared not to see that even an enlarged canal had become obsolete. As one newspaper remarked:[65]

> *We desire to see a sensible answer given to the question, why the settlers in Kansas should appropriate their means in abundance toward the construction of railroads, while the people of New Jersey trundle to market on the conveyances of half a century ago.*

The enlargement of the Bloomfield Plane, No. 11 East, during the winter of 1859-60, inspired the comment that the plane was now "twice its former capacity, so as to allow an entire boat to be taken up, instead of half of one, as before."[66] At first, this sounds impressive, but we know that any of the still-adolescent railroads could load five times more coal in one car than the Morris Canal could in one boat.

A general depression in the coal market during 1859 and 1860 seemed to indicate that carrying companies, canals and railroads alike, would have to reduce their rates to remain competitive. It would be necessary to increase the capacity of the canal for more tonnage just to keep the revenue equal to previous years.

President Marsh was sure that the canal's water supply would support an increase in tonnage, at least "until the trade upon the canal reaches over a million tons per annum" traveling both ways on the waterway.[67]

Despite the fact that 1860 was the driest year since 1846, when Talcott had begun to measure the amount of precipitation that gathered in Lake Hopatcong, "the canal closed with water enough in this one reservoir to have passed each way over the whole length of the canal more than 800,000 tons in addition to the trade that was carried."[68] The steam "pumping apparatus" that had been installed at the Hackensack River had "removed all complaints of low water on the tide level between said river and Jersey City, at a cost of less than $300 for the whole season."[69] There was no want of water on the Newark-to-Jersey City level, then.

The steam pumping station and basin area at Lock No. 21 East. The apparatus pumped water from the Hackensack River into the canal to keep the water depth constant between tides (BNK).

Seven lift locks, one guard lock, and one tide lock* had already been lengthened to 100 feet clear in the chamber, while sixteen lift locks, four guard locks, and two tide locks remained to be rebuilt on the larger scale. Before any of the five locks at Dover could be rebuilt, however, the Company would have to decide if it would be more efficient to eliminate those locks and substitute one inclined plane. Almost half a mile of distance (40/100) would be saved by a plane, and the straighter, easier-to-navigate route through Dover would save one-third of a day per round trip for each boatman. "With the present trade of the canal [1860], this could amount to a saving of about $11,000 per season, and with a trade equal to the present capacity of the canal, the saving would be at least $25,000 per season."[70]

The Jersey City shipping area had been "enclosed on all sides by a substantial cribwork, thereby securing still water for handling coal on the southside of the company's coal piers; increasing the storage rooms, and otherwise facilitating the trade of the canal."[71]

To further facilitate the trade of the Morris Canal, a new canal had been chartered on February 21, 1860. The citizens of Morristown who obtained the charter applied to the Morris Canal Company for financial help for the necessary surveys "to ascertain the cost and feasibility of such canal, which if made will be practically a branch of the Morris Canal." The Canal Company board of directors decided that they would "pay one half of the expenses of said survey."[72]

The new "branch" canal was to be known as "The Morristown, Hanover and Pequannock Canal." Like the proposed Bottlehill Canal before it, this canal was to have the right "by means of a raceway or canal, to connect the waters of the Rockaway River" with the Morris Canal, "provided, that the waters so taken from the Rockaway river, and all waters taken from the Passaic river, or any of its tributaries, by said company, or its successors or assigns, be not in any way mingled with or emptied into the Morris Canal, but shall be returned into said tributaries or into the Passaic River, above the great falls of the Passaic."[73]

**Lift locks 2, 3, and 4 West; 2, 3, 11, and 12 East, Saxton's Falls Guard Lock, and Jersey City Tide Lock.*

The Morristown-Pequannock Canal was to begin at or near Morristown, run through Hanover* and Pequannock Townships, and intersect the Morris Canal "at the most convenient and accessible point in the Township of Pequannock." So far as is known, construction was never started on this canal, but it is interesting to note the names of the commissioners who were appointed by the act to receive stock subscriptions:

George T. Cobb	Ephraim Marsh
William C. Baker	Augustus W. Cutler
Andrew B. Cobb	William H. Talcott
William G. Lathrop	

By the middle of the 1860 boating season, with the "permanent" improvements made to the canal, the Canal Company directors "with a few invited guests" took a "very agreeable excursion over the entire length of their canal from Easton" to Jersey City. Leaving Easton "in one of their own boats, which was comfortably fitted for the purpose," the group spent its first night at Ephraim Marsh's Heath House on Schooley's Mountain. The second night was spent at Jollie's Hotel in Dover. The third night, the party rested at Congress Hall in Paterson, and Friday, July 20, they arrived at Jersey City "in time for the guests from the south and west to take the trains running in those directions." Many of the directors had never seen the whole canal, and since they were "more favorably impressed than ever with its importance," the trip was a success.[74]

All in all, the whole year, 1860, had been successful. Navigation had ceased by mid-December, "without serious interruption during the whole season."[75] The entire tonnage for the year had been 707,631; the total income, $350,710.63.

The year 1861 got off to a very uncertain start with rumors that the Lehigh Coal Company's mines were exhausted. A moment's reflection made most people realize that "the Boonton Iron Works in New Jersey consume annually more coal than the county town of Morris. The rolling mill at Trenton consumes more than the city, with its 15,000 inhabitants, and the Montour Iron Works more than the city

**Hanover Township at this time included Boonton.*

of Baltimore." Science came to the rescue, however, as geology "determined the great fact that the coal mines of the United States will be certain to hold out."[76]

Even with such an assurance, it was difficult to speculate on the year ahead.[77]

> *In view of the present distributed state of public affairs, and of the general business of the country it will not be expected that an estimate of the trade and income of the current year* [*1861*] *should be attempted, but the preparations made to mine and send to market, anthracite coal from the Lehigh and Wyoming coal fields, the amount of sales already made, and the increased facilities for storing and reshipping coal at Jersey City, would seem to justify the expectation of an active trade on the Morris Canal, producing results satisfactory to the stockholders.*

Whether or not any given stockholder was to be "satisfied" for the next few years would depend entirely upon his bias. The nation was on the brink of "Civil War," and New Jersey was far from unified on the subject of slavery. At the risk of over-simplification, the old East-versus-West Jersey, or New York-versus-Philadelphia division, seems to have held true even for such a ponderous moral issue as physical bondage. For the most part, West Jersey was aligned with Pennsylvania's Quaker ideas of abolition; East Jersey, was heavily sprinkled with "Copperheads," or Southern sympathizers, notably in Bergen and Sussex counties. Few of us realize how very close New Jersey came to seceding from the Union in support of the South's cause. Too many North Jersey industries, particularly textiles, depended upon the South for raw materials and for outlets for their products. And too many North Jersey industries did not consider the moral implications of filling contracts for the South as it armed itself for the "irrepressible conflict" with the North.

The "Civil" War? What was civil about it?

Cooper-Hewitt's Trenton Iron Works was busy turning out "munitions of war for the Southern Traitors" early in 1861. When a "news blast" was leveled against the company, however, Messrs. Cooper and Hewitt "put a stop to all such contracts at the Works, and would not knowingly aid the disunionists."[78] Instead, the firm contracted to supply 70,000 musket barrels of puddled steel to the Field and Horton Company, who would "bore them and fit them with the patent breech and finish them" for the United States Government.[79] It was a far cry from furnishing arms for the "enemy" when a patriotic "speculator" applied for the price of Cooper-Hewitt's Ringwood Works "with a view to offering it to the Government for a national foundry." Messrs. Cooper and Hewitt had, "of course, given no encouragement to any such ideas."[80]

Despite the East-West alignment in New Jersey, volunteers for Lincoln's cause were many and enthusiastic. When Lincoln called for more men for the Union Army on July 7, 1862, groups of volunteers gathered on the banks of the Morris Canal in the Roseville section of Newark. A camp was established at the east side of Roseville Avenue, north of Orange Street, "where a large field, having a gentle slope to the Morris Canal, accommodated an encampment of several thousand men. The canal provided bathing facilities and was used in the morning, after reveille, by all men not on the sick report."[81] Appropriately enough, the encampment was officially named Camp Frelinghuysen, with Colonel Cornelius Van Vorst in command as of July 22, 1862. Both men, Frelinghuysen and Van Vorst, were Canal Company directors.

The New Jersey Volunteers, Thirteenth Regiment, were recruited at Camp Frelinghuysen, where they managed to enjoy the weeks spent on the tented fields in Newark. Starting in the morning with a splash in the canal, the balance of the day was spent in drilling. When they were ready for battle, the troops "marched down Orange to Broad Street, and then to the Passaic River, to the wharf and a steamboat to take them all the way by water to Washington, D.C."[82]

Lincoln's draft policy affected the Morris Canal Company as it did every other walk of life. Talcott's assistant, Alexander P. Berthoud, answered the call of his country when Lincoln's Conscription Act went into effect in 1863. He served as a Colonel at the head of New Jersey's Thirty-First Regiment, commanding such men as William Imla Powers who would later become superintendent of the Morris Canal and Mayor of Boonton, New Jersey.[83] The Canal Company decided "to continue to A.P. Berthoud during the nine months for which he was engaged in the army his regular compensation as an employee" of the Canal Company.[84]

Several Canal Company employees — plane, lock, and level tenders — were drafted, and the Company was hard-pressed due to their absence. Since it was possible — and legal — to "buy" substitutes for draftees, the Canal Company decided to pay "$200 each for plane tenders and foremen and $100 for lock and level tenders, and also advanced $600 upon town bonds towards furnishing substitutes for such drafted men." Such payments were made only in cases where substitutes had already been procured. The "spiritual" life of the Canal Company employees who were *not* drafted was also provided for as the board of directors authorized Marsh "to expend, at his discretion, any sum not exceeding $100 in compensating a missionary employed among the boatmen."[85]

Early in the war, the Confederacy had seized the bituminous coal mines in Virginia "to obtain directly the necessary fuel for purposes of modern warfare."[86] Lest that should happen to Pennsylvania's anthracite mines, the Canal Company joined other companies "in contributing aid to regiments formed to protect the mining regions from the recent invasion by the subscription of $1000" as its quota to help the "Coal Volunteers."[87] The "recent invasion," of course, was the Battle of Gettysburg, July 1-3, 1863.

It is very tempting to conclude that the Morris Canal served a vital function during the Civil War period as a part of the Underground Railroad. Concrete evidence, unfortunately, does not exist, although we might speculate on the basis of what *is* known. One outlet for runaway slaves led overland from Philadelphia "across New Jersey to Jersey City and New York; another outlet from Philadelphia was the Reading Railroad." Again, "on the general map may be noticed a few long stretches of road that had apparently no way-stations. Such lines are usually identical with certain rivers, or canals, or railway systems."[88]

According to the records of the New Jersey State Anti-Slavery Society, some Underground Railroad "agents" in 1843 were:

Newark	Adam Miller & Wm. F. Gardner
Bloomfield	C. Pelonbit

Paterson	Alexander H. Freeman
Parsippany	James B. Grimes
Boonton	James Norris
Succasunna	Maria Allwood
Stanhope	Stephen Grimes

With the exception of Parsippany, all of these stops were on, or very near, the Morris Canal.

Charles F. Hopkins remembered, as a boy of 14, driving slaves in a wagon from the Powerville Inn at Morris Canal Lock No. 11 East to Stockholm. His statement was that the Powerville Inn, operated by his father, was a station on the Underground Railroad.[89] Mr. Ephraim Guard of Succasunna was also said to have been an "agent."[90] At Newark, T.V. Johnson, the proprietor of a freighting service for canal traffic, was a "rabid abolitionist." Reportedly, his home in Newark "became a station on the 'underground railroad,' and he personally conducted many a poor negro on his way to freedom." At one point, Mr. Johnson had had to post guards around his home as protection against the pro-slavery element in Newark.[91] The reader, of course, must draw his own conclusions.

As the "uncivil" war drew to a close, Ephraim Marsh died. Ironically, he had been traveling from Jersey City to Hackettstown on a M&E Railroad train, when there was a collision which detained him "about four hours." His health had not been good, and the trip from the ruined M&E train to Heath House on Schooley's Mountain at midnight "immediately occasioned" his death on Saturday, August 28, 1864.[92]

When it was time for the Canal Company's stockholders to elect a new president, it was "rumored at Newark that Thomas N. McCarter of Newton" would replace Marsh as president of the Morris Canal.[93] In fact, however, it was William H. Talcott who received the nod from the stockholders, and was appointed to fill Marsh's place on September 10, 1864.[94] His salary was to be $1,500 in addition to his salary as the canal's chief engineer and superintendent.[95] At the regular annual elections, Talcott was elected to the position of the presidency.[96] By the end of the year, he would ask for a four-month leave of absence in order to go to Europe for

Powerville Inn in ruins at Lock No. 11 East above Boonton. This was allegedly a station in the Underground Railroad (BNK).

the restoration of his health.[97] The board of directors voted him $2,000 for expenses. He was then 66 years old.

Talcott had told the board of directors that he wanted — needed — to appoint "a competent person who should devote all his time to the supervision of the repairs and improvements and the police of the canal under the title of 'Assistant Engineer and Superintendent.'"[98] Accordingly, when Berthoud finished his service with the U.S. Army, he was hired to fill the spot created for him by Talcott. His salary was $400 a month plus travel expenses.[99]

Talcott would need all the competent help he could muster. The most trying days of the canal's life still lay ahead.

REFERENCES

1. *Laws of New Jersey,* 1841.
2. *Report to the Stockholders of the Morris Canal and Banking Company* (N.Y.: 1853), p. 7.
3. Morris Canal and Banking Company *Minutes,* May 27, 1853 (hereinafter cited as MC&BCo. *Minutes*).
4. Munsell's *History of Morris County, New Jersey: 1734-1822* (N.Y.: 1882), pp. 63, 328 (hereinafter cited as Munsell's *Morris*).
5. James P. Snell, *History of Sussex and Warren Counties, New Jersey* (Philadelphia; 1881), p. 353 (hereinafter cited as Snell's *Sussex*).
6. *Ibid.,* p. 413.
7. Snell's *Sussex,* p. 657.
8. *Report to the Stockholders of the Morris Canal and Banking Company* (N.Y.: 1859), p. 9 (hereinafter cited as *Report to the Stockholders, 1859*).
9. Paterson *Daily Guardian,* March 14, 1859.
10. William H. Talcott, *Estimates of Quantities for Planes 1, 2, 3, 4, 5, 6, 7, 8, 9, and 10 West, etc., June, 1850,* p. 55.
11. Paterson *Daily Guardian,* November 30, 1859.
12. MC&BCo. *Minutes,* July 21, 1856.
13. *Ibid.,* May 21, 1858.
14. Abram S. Hewitt Papers, letter dated October 21, 1856.
15. *Ibid.,* letter dated September 19, 1856.
16. Paterson *Daily Guardian,* November 26, 1856.
17. MC&BCo. *Minutes,* January 4, 1859.
18. Paterson *Daily Guardian,* February 1, 1859.
19. *Ibid.,* February 9, 1859.
20. Newark *Daily Advertiser,* June 1, 1878.
21. Paterson *Daily Guardian,* December 6, 1856.

22. *Ibid.*, May 2, 1857.
23. *Ibid.*, September 5, 1857.
24. *Ibid.*, May 2, 1857.
25. *Ibid.*, May 10, 1857.
26. *Ibid.*, June 3, 1857.
27. *Ibid.*, March 19, 1857.
28. *Ibid.*, May 7, 1857.
29. *Ibid.*, June 2, 1857.
30. *Ibid.*, June 9, 1857.
31. *Ibid.*, October 10, 1857.
32. *Ibid.*, October 12, 1857.
33. *Ibid.*, October 27, 1857.
34. *Ibid.*, December 6, 1857.
35. *Ibid.*, January 21, 1858.
36. *Ibid.*, March 22, 1858.
37. *Ibid.*, March 30, 1858.
38. *Ibid.*, April 14, 1858.
39. *Morris Canal and Banking Company of 1844 Clearance Book*, Stanhope, N.J. The information was obtained from loose pages in the author's possession.
40. George H. Burgess and Miles C. Kennedy, *Centennial History of the Pennsylvania Railroad Company: 1846-1946* (Philadelphia: 1949), no page.
41. MC&BCo. *Minutes*, May 21, 1858.
42. *Ibid.*, September 29, 1858.
43. *Report to the Stockholders, 1859*, pp. 8-9.
44. MC&BCo. *Minutes*, September 29, 1858.
45. Paterson *Daily Guardian*, February 10, 1859.
46. *Ibid.*, June 16, 1857.
47. *Ibid.*, March 30, 1859.
48. *Ibid.*, April 2, 1859.
49. *Ibid.*, April 16, 1859.
50. *Ibid.*, May 7, 1859.
51. *Ibid.*, May 20, 1859.
52. *Ibid.*, May 12, 1859.
53. *Ibid.*, March 14, 1859.
54. *Ibid.*, June 7, 1859.
55. *Ibid.*, December 17, 1859.
56. *Ibid.*, May 7, 1859.
57. *Ibid.*, August 19, 1859.
58. *Laws of New Jersey*, 1837.
59. *Laws of New Jersey*, 1838.
60. *Laws of New Jersey*, 1860.
61. Paterson *Daily Guardian*, March 19, 1859.
62. "Report of the Committee on Streets to the Jersey City Common Council" (Jersey City, N.J.: 1865), pp. 4-8.

63. *Ibid.*
64. *Report to the Stockholders of the Morris Canal and Banking Company* (N.Y.: 1860), p. 17.
65. Paterson *Daily Guardian,* August 22, 1859.
66. *Ibid.,* April 3, 1860.
67. *Report to the Stockholders of the Morris Canal and Banking Company* (N.Y.: 1861), pp. 7-9 (hereinafter cited as the *Report to the Stockholders, 1861*).
68. *Ibid.,* p. 15.
69. *Ibid.,* p. 14.
70. *Ibid.,* pp. 15-16.
71. *Ibid.,* p. 15.
72. MC&BCo. *Minutes,* June 14, 1860.
73. *Laws of New Jersey,* 1860.
74. Jersey City *American Standard,* July 23, 1860.
75. *Report to the Stockholders, 1861,* p. 13.
76. Paterson *Daily Guardian,* March 1, 1861.
77. *Report to the Stockholders, 1861,* p. 8.
78. Paterson *Daily Guardian,* February 6, 1861.
79. *Ibid.,* October 2, 1861.
80. James W. Ransom, *Vanishing Ironworks of the Ramapos* (New Brunswick, N.J.: 1966), p. 82.
81. David Lawrence Pierson, *Narratives of Newark* (Newark, N.J.: 1917), pp. 292-3.
82. *Ibid.,* p. 295.
83. *A History of Morris County, New Jersey* (N.Y.: 1914), Vol. II, pp. 366-368.
84. MC&BCo. *Minutes,* January 22, 1863.
85. *Ibid.,* July 18, 1864.
86. Homer Greene, *Coal and the Coal Mines* (Boston, 1898), p. 39.
87. MC&BCo. *Minutes,* July 21, 1863.
88. Wilber H. Siebert, *The Underground Railroad* (N.Y.: 1968), pp. 120-1, 142.
89. Rockaway (N.J.) *Record,* September 15, 1921.
90. Annie Stelce Hosking and Harriet Meeker, eds., *The History of Roxbury Township* (Netcong, N.J.: n.d.), p. 155.
91. *Biographical and Genealogical History of Morris County* (N.Y.: 1899), p. 65.
92. Paterson *Daily Press,* August 30, 1864.
93. *Ibid.,* January 5, 1865.
94. MC&BCo. *Minutes,* September 10, 1864.
95. *Ibid.,* October 27, 1864.
96. *Ibid.,* April 3, 1865.
97. *Ibid.,* December 18, 1865.
98. *Ibid.,* January 3, 1865.
99. *Ibid.,* February 1, 1865.

17

A Most Uncivil War

EVEN before William H. Talcott assumed the presidency of the Morris Canal Company, he had already become something of a commanding general, whose job it was to conduct a war on several fronts, simultaneously, within a theatre of operations. The Civil War front was a minor skirmish in the Morris Canal's Battle for Survival. Lincoln's "irrepressible conflict" contributed to the canal's casualty list, as its employees and those of the coal mines were drafted to help put down the rebellion in the South. The "railroad front" was becoming even more complicated by the fact that the canal was both dependent upon and in competition with the various anthracite carriers. The battle for Jersey City, shortly to be mounted, would itself be waged on three fronts, one of which, the Hudson Street front, had already begun.

On yet another front, the campaign against capricious Nature was simply impossible to win. It had started at the canal's inception, and continued, with varying degrees of alternating success and failure, against gravity and friction, against weight and pressure, against freeze and thaw, and against countless other physical forces and natural laws. Most costly of all had been the battles against drought and its counterpart, flood. Drought had made necessary the construction of new reservoirs and feeders, and had resulted in bitter contests with waterpower users on the Passaic and Musconetcong Rivers. We saw, in 1841, how floods had "conspired" with insolvency to halt plans to enlarge and improve the canal.

Then, 1852 had brought an "unexpected interruption in the navigation by local rain, at two different periods (most extraordinary in severity), causing a break in the banks at several points in the vicinity of Paterson, and the destruction of an aqueduct over Peckman river." Repairs had cost "about $10,000."[1] Later, in 1866, the same Peckman River Aqueduct, along with the aqueduct over Hurd Brook at Dover, would again need repairs, costing another $19,000 for "rebuilding and widening."[2]

During 1854, "severe storms" in March and April caused two serious canal breaches and other damage due to flooding. Repairs had cost $104,203.30 — in money; in time — a shortened boating season, shortened further that year by "an early onset of winter."[3]

Flooding did not destroy the Little Falls Aqueduct, but alternating freezing and thawing year after year had taken their toll. Seepage from the prism of the aqueduct into the spaces between the brownstone blocks of the structure meant that water could, and did, freeze when winter set in, and thaw when spring came, causing considerable damage and endangering the stability of the whole stone-arched aqueduct. The prism was made watertight during 1861, when grout and cement were used to seal the masonry, and pine planks were formed into a lining which was then thoroughly caulked.[4]

Then, on June 4, 1862, came rains and subsequent floods that brought such devastation to the Lehigh Canal that the Upper Grand Section, from Mauch Chunk to White Haven, was wiped out, never to be rebuilt. The Lower Grand Section, from Mauch Chunk down to Easton, was so badly damaged that no coal could be shipped from Mauch Chunk until October 3, 1862, a period of four months.[5]

The coal trade was interrupted throughout nearly all the anthracite coal fields of Pennsylvania, so that transportation of coal was suspended for 2 days on the Reading Railroad; 12 on the Delaware, Lackawanna & Western (DL&W) Railroad; 17 on the Schuylkill Canal; 34 on the Delaware Division Canal and Lehigh Valley (LV) Railroad; 41 on the Delaware and Hudson Canal; and 122 on the

Lehigh Canal. When coal shipments did resume at Mauch Chunk on October 3, they "continued later than usual" into the winter, "until a very sudden and severely cold turn, on the 6th of December, formed thick ice upon the whole length of the canal. This was subsequently broken up at an expense to the Company of $2,397.05, and all the boats then on the way were taken to their several destinations."[6] The canal then operated until "about the 20th of December."[7]

According to the Canal Company's annual report for the year 1862, an estimated 86 Morris Canal boats that were on the Lehigh Canal at the time of the June flood were lost, many with their crews aboard. The report claimed that the Morris Canal was damaged only at Port Delaware, where the rampaging Delaware River destroyed the protective bulkheading at the foot of Plane No. 11 West and deposited debris there to the depth of 16 feet. The stockholders were assured that, had the Morris Canal been totally dependent upon Plane No. 11 West for its coal freight, navigation would have been suspended "for at least a week." With its connections with anthracite railroads, fortunately, "navigation was not suspended even for a single day, on any part of the Morris Canal" because of the flood.[8]

Despite these official assurances to the stockholders, there is conflicting evidence that the 1862 summer floods *did* damage other parts of the canal. At Cherry Tree Bend, Port Murray, for example, Beatty's Creek, which flowed under the canal through a culvert, was so swollen "late in the spring of 1862, after weeks of heavy rain" that it "became a veritable torrent and the water rushing down the ravine undermined the wall of the culvert." The creek washed away the canal's underpinnings, and "the water gushed forth, flooding the surrounding territory and making a little river of the usually insignificant little creek. For three weeks, night and day, the company worked to repair the break."[9]

Mr. Perry, a storekeeper at Port Murray, recorded such a breach in his "day book" under the date of July 11, 1862, when six boats and their crews were held up and were "laying at break," the break being the canal washout. Perry extended credit to boatmen "Campbell, Dempsey, Godfrey, Atno, Newman, and Quacho" to buy provisions from his store while they were idled at the break.[10]

Note that the Cherry Tree Bend breach of 1862 had taken three weeks to repair. A similar incident occurring on July 12, 1863, at the Bend took 15 days to repair. An extraordinary local rain had swollen a stream (unidentified) "beyond the capacity of the culvert to discharge the water. The culvert was washed entirely away, and the embankments were carried away to the depth of 34 feet below the top of the tow-path."[11] Since Cherry Tree Bend was on the 11-mile level between Saxton's Falls and Port Murray Plane No. 5 West, the level was probably drained, and surely boating was halted at that point. In any case, it is difficult to reconcile the disparity between the repair costs for the two floods. The Canal Company's annual report for the year 1862 gave canal expenditures as:[12]

1. Rebuilding Inclined Planes	$ 520.97
2. Lengthening Locks	1,950.87
3. Piers, Wharves, and Basins at Jersey City	19,796.96
4. Land Damages	1,447.31
5. General Improvement Account	10,518.33
	$34,234.44

No separate cost is given for the repairs at Port Murray. The following year, the annual report put total canal expenditures at $55,724.41, of which $8,533.76 was spent on repairs and "a new cement masonry culvert 150 feet long" at Cherry Tree Bend.

Nature appeared to have won this flood-front skirmish of 1862, but the Canal Company made an extraordinary comeback in that extraordinary year. The canal carried 7,351 tons less total freight than it had in 1861,[14] yet netted $18,896.38 more in revenue.[15]

The 1862 flood that devasted the Upper Lehigh Canal was distinguished from earlier — and later — floods, for when the flood waters receded, the boatmen struck.[16] One newspaper reported a strike in 1862 that was unorganized, because it took place "before the day of the labor union," and that it had merely been the result of "three weeks of enforced idleness" while the Cherry Tree Bend breach was being repaired.[17] Led by a few men — Ben Jackson, Billy Gill, Phantom Quigley — the strikers formulated their demands for higher freight rates, which would mean higher pay. This strike allegedly lasted from July 4 until August 13, 1862, during which time the Canal Company was said to have employed strike breakers in order to get navigation going again. Local, and not so local, authorities worked with the strike breakers, and "every boat that tried to make the trip was accompanied by a sheriff and twenty or thirty deputies to resist the efforts of the strikers to obstruct traffic." Not to be beaten, the strikers organized what was familiarly called the "Black Horse Brigade." At times,

> *as many as two hundred men on horses and mules of every description, a picturesque band, paraded up and down the towpath, driving the muleteers — who were willing to work, off the canal, ducking them sometimes in the water, cutting towlines and performing all sorts of tricks of a humorous and oftimes rather strenuous nature, to hinder the passage of the strike breaker's boats. At one time as a result of this policy the line of idle boats on the eleven-mile level near Port Murray measured more than three miles.*
>
> *John Dean, of Newark, the owner of two or three canal boats, was out of sympathy with the strike and persisted in the attempt to break it. With the usual quota of sheriffs to defend him, he succeeded in reaching Port Colden in safety, but there he was obliged to rest one night, for he dared not venture out in darkness. A few somewhat reckless strikers heard of his determination to remain on guard over his boat during the night and decided to put an end to his efforts against them. Silently they boarded his boat and with augers bored the bottom full of holes. Naturally enough, it sank, but Dean was fortunate and got off without injury.*
>
> *The company found itself unable to operate against the wishes of the boatmen and at length yielded to their demands by granting an increase of ten cents per ton freightage.*

The Canal Company *Minutes* reveal only that there *was* a strike, but during June and July. Since the entry mentioned, in this connection, a report from the Committee on Finance, we can assume that wages were under discussion, but the *Minutes* are silent on details.[18]

The actions of the "Black Horse Brigade" are reminiscent of those of the "Molly Maguires," a group of rather radical rowdies who had been terrorizing the coal regions of Pennsylvania. The Black Horse Brigade, however, is not reported to have operated an assassination ring, as the Mollies were said to have done. Whether or not the author of the article about the Black Horse Brigade knew about unions is, of course, impossible to check. Obviously, he did not know that a "well-defined and relatively successful labor movement" had been underway in the United States since about 1827. The movement dissipated during the depression that followed the Panic of 1837, but soon revived.[19]

In its continuing battle with Nature, the Canal Company was particularly engaged at Paterson, where breaks in the canal bank frequently occurred. Since the canal there was so located as to be above the town's businesses, every breach meant damages to properties on the down-hill side of the waterway. As protection against these breaches, "the owners of the coal, lumber and brick yards" in Paterson, had long been making "extensive excavations, deepening the channel, building breakwaters, piers, wet docks, dry docks, and other fortifications necessary to protect the public against the encroachments of the ocean. We trust," the Paterson *Daily Guardian* said, "these attempts will be successful and that while New Jersey is wasting away on her Southern border she may not diminish on that part of the coast skirting the world renouned [sic] Morris Canal."[20]

The Canal Company decided to take measures to protect those businesses — and to protect itself from costly damage suits — by building "a cement protection wall at Paterson, commencing at Rock Cut, above the city, in order to fully guard against breaks at that point." The wall would be continued "until it connects with similar walls built by private individuals and corporations, at their own expense, all along the high bank immediately above Paterson."[21]

We saw in Chapter 16 that extensive damage had been done to O'Neill's coal yard in Paterson when a breach there on May 7, 1857, completely drained the 17.5-mile level.[22] Later, the Canal Company wanted to widen the canal at O'Neill's so that the channel width there would be 50 feet, allowing room for boats to pull over to the side to unload when boats were passing in both directions. O'Neill agreed to the widening in exchange for the right to use part of the towpath to "lay up a cement wall," to protect his property. The Canal Company not only agreed to that proposition; Talcott sent him the cement for the wall![23]

In a way, Paterson struck back at the canal in October, 1866. A "large reservoir, constructed for the Paterson water works on the high lands above the canal, near Paterson, gave way suddenly, emptying itself into the canal and causing a break which suspended navigation four days."[24] The break occurred at today's New Street Reservoir, located halfway up Garret Mountain.

It was the Flood of 1862, however, that unquestionably brought about significant changes of logistics in the Canal Company's multi-faceted war for survival. Just as the concept of the strike distinguished that flood from all others, so did the concept of the railroad as a more reliable form of transportation emerge from the wreckage above Mauch Chunk on the Lehigh Canal. While other railroads and canals suffered serious delays because of the flood, the LV Railroad had resumed transportation of coal by July 9, 1862, a 34-day delay. The DL&W Railroad, which was supplying the Morris Canal with coal at Port Washington, had suffered only a 12-day delay.[25] Rather than rebuild the upper part of the Lehigh Canal, the Lehigh Coal and Navigation Company extended its Lehigh and Susquehanna (L&S) Railroad from White Haven south to Mauch Chunk, then from Mauch Chunk down to Easton. When the L&S reached Easton on February 3, 1868, it became a principal supplier of the Jersey Central.*[26]

The Morris Canal, like the other anthracite-carrying canals, found itself in the awkward position of being both dependent upon and in competition with the anthracite railroads. Until the Flood of 1862, as we have seen, the emphasis had been on "cooperation" between the canals and the railroads in transporting coal. After 1862, the emphasis shifted to "competition."

President Marsh had forecast that there would be an increased demand for anthracite during 1862, "larger than in any previous

**The Pennsylvania Legislature is said to have vetoed the idea of repairing the Upper Grand Section of the Lehigh Canal, but enabled the L&S Railroad to be extended.*

year," and that "as the price of coal now ranges, the carrying companies should obtain better prices than formerly." In his annual report on the year 1863, he commented that "a general advance in the price of coal and other commodities transported upon the Canal, and in the price of labor and material, justify and require that there should be a corresponding increase in the charges for tolls and transportation which has accordingly been made."[27]

Toward the end of 1863, however, the Paterson *Daily Press* reported that the price of coal had begun to fall with, "speculators feeling, doubtless, that they were pushing matters to a dangerous point. The last monthly [November] sale of coal on the part of the DL&W Railroad Company realized about 75¢ per ton less than the average at which coal was sold in October. So much for the wholesome agitation in favor of throwing off the duty on foreign coal. We notice that papers which have long and earnestly advocated a tariff for protection, join with the *Evening Post* in calling for a repeal of the duty."[28]

Between labor, which increased its outlay of monies, and railroads, which decreased its tonnage — not to mention the coal industry, whose vicissitudes directly affected its very existence — the Canal Company began to lean more and more heavily on the iron industries. Hopefully, "the unusual activity in the iron works along the Morris and Lehigh Canals will necessarily furnish a large amount of transportation of iron and iron ore to the Morris Canal" during 1863, President Marsh reported to the stockholders.[29] As one concession to the iron industry, specifically to Cooper-Hewitt & Company, the "lift lock at the head of the Pompton feeder" was enlarged to accommodate the larger boats that operated on the main canal in 1863, as Mr. Hewitt had requested.[30] This concession was but one strategem to keep the allegiance of the iron industry, a very valuable ally in the fight against the railroads. It will be recalled that Cooper-Hewitt had "mentioned" — threatened, is too strong a term — a deal with the Erie Railroad if the Canal Company did not enlarge and improve the Pompton Feeder with its lock.

Increased tonnage was another strategem employed, and that meant keeping the canal prism filled to its 5-foot capacity. To help, the Company paid $5,317.04 for an Osgood's Patent Dredging

Machine, an under-water excavator, suitable for working in the Canal without interrupting navigation, that was put in operation on July 1, 1863.[31] Paterson's *Daily Press* noted that the dredge was at work in the canal in November, "near the coal yards at the head of Mill Street," and described the device as "an ungraceful hulk" of a boat, about 60 feet long, and steam-driven. The *Daily Press* explained that dredging had previously been "done by hand, before opening the canal in spring," and presumed that the current "difficulty in procurring [sic] labor . . . induced the company to attempt the work by machinery."[32] (There was no indication whether the dredge used anthracite or wood to generate its steam.) By 1865, when the Canal Company bought its second dredge, the cost was $10,532.62, double that of the first one.[33]

Yet another strategem was to increase the canal's capacity by enlarging its locks. That task, begun in 1861, was "suspended at the outbreak of the war. At that time 9 locks had been lengthened, leaving 22 lift and guard locks to be lengthened, in order to allow an increase of ten feet in the length of the boats." Since that time, only one other lock had been enlarged, "leaving now only 21, of which 3 are west of Port Washington, leaving only 18 to be lengthened, in order to give the trade to and from that point the benefit of longer boats and greater tonnage." If the directors adopted Talcott's 1861 recommendation to replace Dover's five locks with one inclined plane, only 13 locks would need to be lengthened. A similar plan, using an inclined plane "in place of the 3 lift locks west of Port Washington, that have not been lengthened," would have left only ten locks to be so enlarged.[34]

It was imperative, as far as Talcott was concerned, to make these improvements quickly. "The Morris and Essex Railroad is being extended to Easton, which, when completed, will open a direct railroad communication between the coal mines and the consumers of coal on the line of the Morris Canal, and also between the Jersey ore mines and Pennsylvania furnaces."[35] The tide of battle was turning. Originally, the canal had been the only *through* transportation route, with a right-of-way clear across the state. The little branch or lateral railroads that earlier had been so anxious to cling to the Morris Canal were now beginning to get through-routes of their own. Only a fool would fail to realize the implications of that prospect, and Talcott was no fool. More than being a very competent engineer, he showed all the qualities requisite to being an excellent administrator.

The substitution of one inclined plane for the three locks west of Washington was never made. Nor was the substitution of one inclined plane for Dover's five locks. Instead, the locks at "Green's," near Phillipsburg, were "thoroughly" repaired during 1865. Locks 4, 5, and 7 East at Dover were "partly rebuilt and lengthened, with new lower and fall gates and improved fixtures for opening and shutting the lower gates."[36] When the lock at Green's, No. 6 West (No. 9 West, new style), was lengthened, William Carty's house on the towpath side of the canal was damaged, for which he was paid $150.[37]

Another of Talcott's recommendations to facilitate boat traffic had been to rebuild the Waterloo Lock "farther from the plane," or remove the lock altogether and extend Plane No. 4 West "to meet the lower level." Boats were frequently delayed at Waterloo because of the "lift-lock being placed so near" the foot of the plane. It took "almost as much time to get a boat into, or out of the car, as is required to make one passage over the plane."[38] Despite Talcott's urging to remedy the "difficulty," no record exists of its having been done.

Wages and labor problems continued to harass the Canal Company throughout 1864. In March, as the canal was being readied for navigation, the Paterson *Daily Press* announced that "the shrill notes of the bugle" would herald the canal's opening before the end of the month, and speculated that only "if there should be no serious trouble in the mining region" would there be "an active season in the coal trade."[39] Besides the threats of strikes, a scarcity of boatmen throughout 1864 made it "impossible to keep all the boats running."[40]

Coal prices were up again, but they did not necessarily reflect high wages for the miners, however. The "usual monthly sale of coal by the DL&W Railroad Company" on July 27, 1864, in New York, brought "on an average $1.75 per ton more than in June, when it fell somewhat."[41] Coal selling for $8.50 or $9.00 a ton in New York was sold for $12.00 a ton in Paterson. "Even the common articles, as potatoes, turnips, cabbage, and the like, can be had cheaper in

New York" than in Paterson, and Patersonians had only themselves to blame. They had "all along insisted upon limiting themselves to one channel of supply, the antiquated Morris Canal, instead of inviting competition by railroad to the coal regions."[42]

The Paterson *Daily Press* reported that newsmen from the Albany *Statesman* had visited the Lehigh and Scranton coal mines late in 1864 to try to "get posted on the coal question," and had found out that[43]

> *the miners charge $2 a ton to get the coal and place it in the cars. In addition to this the company pays twenty-five cents a ton to the owners of the land. The total cost of Lehigh coal when placed in the cars is $2.25 per ton. The mines are only eight miles from Mauch Chunk, and yet coal sells at Mauch Chunk for $8.50 per ton, showing that it is the transportation company, and not the miners, who cause the present high price of coal. To get coal from the mines to Mauch Chunk, eight miles, costs very little indeed, as the whole distance is downhill — gravity doing that work which locomotives are compelled to perform on other roads. Last year coal was mined for $1 a ton. This year coal is mined for $2 a ton. It is now selling at Mauch Chunk for $8 — a rise of $4 in the selling price. The cure for this evil is an opposition railroad from Mauch Chunk to the mines.*

Business was good during 1864, despite prices and miners and what-have-you. The canal opened on March 23 and closed on December 10, for a boating season of 260 days. There were no "Cherry Tree Bend" incidents, or any others to cause any "serious interruption" to navigation. While the total tonnage of the canal exceeded that of 1863 by 5,408 tons, there was an interesting trend in the origins of coal tonnage. Lehigh coal freight fell by 24,571 tons, while Scranton coal tonnage increased 4,988 tons. Even with the increased freight from Scranton, the coal trade was down 19,583 tons from 1863. Newly appointed President Talcott* attributed the falloff

**Marsh died on August 28, 1864; Talcott was chosen to fill out his unexpired term and was elected president in 1865.*

to a "scarcity of boatmen," but he noted that ore tonnage had increased 32,257 tons over 1863.[44]

The Canal Company's policy of renting boats to boatmen meant that there were fluctuations in the amount of income from that source. This would be particularly so if, as Talcott had said, there was a scarcity of boatmen. He recommended that the Company offer to stocholders the opportunity to invest in a "Boat Loan." Such a loan would be "represented by coupon bonds, having twenty years to run, bearing seven per cent interest, payable semi-annually, with a sinking fund reserved out of the boat rents sufficient to meet the principal at maturity." Such a loan would "undoubtedly be disposed of among the Stockholders," and "would be a good safe investment."[45]

The Canal Company had paid "two semi-annual dividends of five percent each," on both their preferred *and* consolidated stocks in 1864.[46] In 1865, preferred stock received its 5-percent dividend at the half-year, as it was guaranteed to do, but consolidated stockholders received only a 4-percent yield. Interestingly enough, there were two places where stockholders could collect their dividends: at the Company's Jersey City offices, and at the office of E.W. Clarke & Co. in Philadelphia.[47] Clarke & Co. were the "party" who had bought up the Delaware Division Canal from the Sunbury & Erie Railroad in 1858.

The 1865 boating season was 258 days, from April 1 to December 14, and was apparently not hampered by a "serious interruption." At least, "navigation" was not interrupted. There was, however, a serious, "almost entire suspension of business" that had to have affected navigation, if in quite a different way. The coal and iron trades had virtually shut down during most of May and June, 1865, because of "the low price of iron, and unsettled condition of the coal market." Miners in Scranton coal regions chose that time to strike, "which nearly stopped the shipment of coal at Port Washington during two or three of the most favorable months of the season." Later in the season, "the coal and iron markets improved" and the miners went back to work, "with a consequent resumption of freight carrying."[48]

The result of this kind of "interruption" to navigation was a drop in total tonnage carried on the canal during 1865. In 1864, the canal had handled 723,927 tons; in 1865, 716,587 tons, or a loss of 7,340 tons. With an increased number of boats to work with, and with "increased facilities for shipping coal at Port Delaware, the 1866 season promised to be better. 1865 saw the introduction of 158 new boats, and 148 were on order for 1866.[49]

The extension of trackage and the building of more coal chutes at Port Delaware had cost $13,199.00 in 1865. Further improvements there were planned for in 1866, but "the high prices of labor and materials, required upon the canal" would probably "prevent, while they continue, any considerable improvement."[50]

The Canal Company did not lower any of its laborers' wages, but "the scarcity of skilled labor throughout the country . . . compelled an advance in the wages of carpenters, masons, blacksmiths, and plane tenders." Experience had taught the Canal Company "that in order to properly control and retain" its plane and lock tenders and brakemen, "they should all be provided with comfortable dwellings owned by the Company and contiguous to their work." In this, his very first engineer's report to the President of the Canal Company, A.P. Berthoud recommended "a limited expenditure each year until the object sought is accomplished." New lock houses had been built during 1865 at "Lock No. 3, West (new style No. 6, Port Colden), and at Powerville."[51] However, in 1852 the chief engineer had been instructed to dispose of the lock house at Newark for no less than $800.[52] Whether or not another house was provided at that time is not known.

Berthoud also recommended that a weigh scale be built at Bloomfield to try stopping what he estimated to be "at least" a 5-percent loss on all shipments east of the Port Colden scale. He didn't know whether or not the losses were due to "the ignorance or dishonesty of shippers," but a weigh scale, in his opinion "would remedy the evil, and no honest shipper would find any fault." He suggested, too, a third Osgood Patent Dredge.[53]

Despite the continuation of the war for survival, and except for "derangements" at Plane Nos. 1 and 6 West, and No. 12 East, each taking about 40 hours to repair,[54] navigation had been almost continual in 1865. There were, however, some "guerilla" incursions at Paterson, described by newspapers as:[55]

Disgraceful: Is it not creditable to that section of our city lying along the Morris Canal that the Canal Company should have been compelled to seek the protection of the city for boats passing through, against the attacks of rowdies and thieves? It appears from a communication read last night at the Board of Aldermen that boats are frequently attacked and the inmates driven off, their property plundered. Sometimes the assailants get up on the bluffs and pelt the boats with stones. The Company are naturally aggrieved at such conduct and ask for a police force to protect them, offering very liberally to be at a reasonable portion of the expense. The matter was referred to the Police Committee for report and we have no doubt they will take proper measures to prevent the repetition of such disgraceful outrages.

By June, the "Police Committee" reported "'progress' and stated they would report fully next meeting."[56] Apparently the Canal Company was not impressed by Paterson's method of dealing with the problem, and sought redress of the State Legislature. As a result, the Governor (Joel Parker) was authorized "to commission special policemen, to be called 'Canal Police'; they are to be appointed and paid by the Company; I intend," said Superintendant Berthoud, "to recommend the appointment of, at least, twenty of our employees, this without additional cost to us to protect ourselves, and those using the Canal, and those living in its vicinity, against the lawlessness of the boatmen."[57] There is no record of such a police force having actually been formed by Governor Parker, but if he *had* taken such action it would have been analogous to having the "home militia" keeping peace among the citizens while war was being carried on on foreign soil!

Other events of this "most uncivil war" period are reflective of more normal Canal Company business. For example, while the State Legislature authorized the formation of a "Canal Police" group, it also authorized several persons to contract with the Canal Company for the use of canal water for a variety of commercial activities. Thus, on March 7, 1865:[58]

James F. Van Doren, of the township of Washington, in the County of Warren be and he is hereby authorized to contract with the Morris Canal and Banking Company for the use of him and his heirs and assigns for a portion of the feed waters of the Morris Canal along plane No. 7 of the Western division of said canal, in said township of Washington, to supply his flour mill near said plane for such time, term, and condition as may be agreed upon by said parties.

By this time, there were many precedents for using canal water in manufacturing. In Chapter 13, we saw how the Stanhope or (Sussex) Iron Company had erected its works at the head of Plane No. 2 West in 1841. In 1844 a forge had been built at Plane No. 1 East, to be run by canal waste water.[59] At Plane No. 12 East in Newark, several industries operated with canal water. One of those, the India Rubber Works, had suffered a terrible fire on May 28, 1857, in which a fireman died.[60] The presence of canal water at the site was apparently not helpful to the fire brigade.

About 1850, Christopher T. Unangst, a Canal Superintendant, leased waste water at Bloomfield's Plane No. 11 East, at a $250 yearly rent. He assigned his right to Jonathan W. Potter, who in turn leased part of his water rights to Robert W. Southmayd and Charles A. McCracken for their Combination Rubber Company at $450 a year.[61] James Searing, lock tender, had had water rights at Newark's Lock 16 East (Lock 15 East, old style) since 1856,[62] although it is not known what the water was used for.

New Jersey Iron Company at Boonton was also operated by water from the canal at the head of Plane No. 7 East. During 1864, those Boonton works were "in full blast during a large portion night and day. The number of men employed in the works, including 125 miners and boatmen, is about 700."[63] According to Munsell, practically all of the residents of Port Oram were English miners "employed by the Boonton Iron Company."[64]

Port Oram's residents might have been employed at the Boonton Iron Company at this point, 1864, but a few years later, they probably entered the employ of a new furnace built at the head of Plane No. 5 East at Port Oram. That was one of the most famous of all users of canal waste water — the iron works that came to be known as Wharton Furnace. Among the Port Oram Iron Company incorporators, on March 31, 1868, was Canal Company President William H. Talcott.[65] Its superintendant was Talcott's son, a prominent citizen of Dover, E.N. Kirk Talcott.[66]

When we consider the renewed emphasis that the Canal Company had come to place on the iron industry — mining and manufacturing — it is not surprising to learn that William Talcott was also one of the founders of the famous Thomas Iron Company, along with the "highly esteemed David Thomas," who, as mentioned earlier, had built the first anthracite blast furnace to make iron successfully. Nor is it surprising to learn that both Ephraim Marsh and Talcott, as well as Marsh's brother John and his son William W., also Canal Company directors, were directors of the Thomas Iron Company.[67]

As for the Port Oram furnace, its incorporators entered into an agreement with the Canal Company on February 19, 1869, which allowed them to:[68]

> *take water from the canal for their boilers, tuyeres or other purposes connected with the works of the Iron Company, to the full capacity of a pipe 8-inches in diameter for and during the term of continuance of said canal at its present location on that level and during the continuance of Stevens' brook into the said canal at the foot of Lock No. 3 E.* [*sic*] *or for such time as said waters may be required for the iron works — $100 yearly rent. Construction of the connection between canal and iron works was to be supervised by William Groff, Supervisor of the Middle Division. All waste water was to be returned to the canal at the foot of Plane No. 5 East by surface carriage.*

All this was still in the future, however. At the end of 1865, Talcott's health was such that he took what amounted to a leave of absence from the affairs of the Canal Company. He left New York on the steamer *Arago* on January 20, 1866, for an extended tour of Europe. He was expected to return in May, and while he was away, John D. Ward of Jersey City, Vice President of the Canal Company, was in charge, with Berthoud superintending the canal.[69]

The 1866 boating season was only 222 days long, somewhat shorter than usual. One reason for that, already mentioned, had been a four-day interruption to navigation caused by a breach at the New Street reservoir in Paterson. Then, too, the water wheel at Montville's Plane No. 9 East had broken in July. When repairs were made, at the end of seven days, navigation was so backed up that it took several days to unclog the accumulated boats above and below Montville.[70]

Following through with the 1865 suggestion to provide housing for valued employees, the Canal Company spent "about $11,000 in 1866" to build houses "for the use of lock-tenders, brakemen, and ferrymen." The Company had come to believe that "the houses were necessary for the proper accommodation of the men near to their work," and that those houses would "tend to insure at those places, the services of such men as are needed for the efficient operation of the Canal." Another item under "expenses" during 1866 was the sum of $9000 for the substitution of steel rails for some of the inclined planes in place of the old iron ones.[71]

Two planes, unidentified, received the new "Bessemer steel rails" as an experiment. The experiment was successful, and the Canal Company decided to buy steel rails for two more planes and part of a third during 1867.[72] Four more planes received steel during the winter of 1869.[73]

"Bessemer Steel": a name to conjure with. It was probably the single most important factor in the development of the nation's great trunk railroads.

But still another process for making steel was being developed during the 1860s by Dr. Pomeroy. According to the September 1, 1866 issue of *Scientific American,* the Pomeroy process made steel directly from the ore without loss of heat so that ingot steel cost little more than common pig iron. Bessemer's process, on the other hand, required that pure pig iron be used in order to get steel of a superior quality.

Pomeroy's process was discussed in two "private and confidential" letters from Samuel Reeves Brooks to the Hon. Robert J. Walker, written in October, 1866, and published as a pamphlet entitled "Letters to the Hon. Robert J. Walker, on the Progress of the Iron

Manufacture and on the Prospects of Steel Manufacture for Steel Rails, Car Wheels, Etc., in Connection With the Franklinite and Zinc Mines of New Jersey."

Samuel R. Brooks we have already met as a Canal Company director and Stanhope Iron Company partner. The Hon. Robert J. Walker was a politician, serving variously as U.S. Senator, as Secretary of the Treasury under President Polk, and briefly as the Governor of the Territory of Kansas. His interests seem to have centered on speculation of all sorts and on schemes for a transcontinental railroad. Judging from the contents of Brooks' letters to Walker, it is possible that both men were backers of the Atlantic and Great Western Railroad of the 1860s, which eventually became part of the Erie system. Moreover, it seems that Brooks was substantially connected with the Sterling Hill and Mine Hill zinc and franklinite mines,* and with Dr. Pomeroy's patented process for making steel. It appears that both Brooks and Walker were interested in establishing a steel works in Newark, for there are maps included in the Brooks pamphlet showing the location of the mines at Franklin, New Jersey, and a proposed site for the New Jersey Zinc Company's works at Newark.

The bulk of the Canal Company's *Annual Report to the Stockholders* for the fiscal year ending February 28, 1867, was devoted to an account of some very complicated goings-on at the Jersey City Basin area. We have already had a foretaste of the bitterness of the battle there between the Canal Company and the Mayor and Common Council of Jersey City over the extension of Hudson Street across the Basin. The bitterness was in no way sweetened by the appearance of the Matthiessen and Weichers Sugar House on the north side of the Little Basin and the omnipresent Central Railroad of New Jersey (CRRNJ) on the south. To understand the intricacies — and the logistics — of this Battle of the Basin, we must understand the historical bases of the several positions. Chapter 2 showed that when the Jersey Associates had bought Paulus Hook from Cornelius Van Vorst in 1804, the property was in effect an "island,"

**At Sterling Hill, Mine Hill, and Franklinite Mines, he would also have been associated with Oakes Ames and thus, perhaps, even with the Union Pacific Railroad.*

which the Associates had thought could be made into a shipping port to rival New York across the Hudson. Because they thought of such development, the Associates had had a map of their property made up by Joseph F. Mangin. The map showed the Paulus Hook area laid out in blocks separated by streets, some of the streets extending — on paper only — out into the waters of the Hudson River on the east, Communipaw Cove on the south.

The first phase of basin construction by the Canal Company at Jersey City had been to build a bulkhead on the south side of Bergen Street, from Washington Street eastward to the end of Hudson Street. The next step was to build a stone-filled cribwork pier in the space that would have been occupied by South Street — had South Street existed anywhere but on paper. Then, the South Street pier was filled in and extended eastward into the Hudson, and the northern wall of the Basin (Bergen Street) was also pushed out Hudsonward. (The reader will recall that Ephraim Morris had bought two of these wharf lots east of Hudson Street in 1856.)

The next phase of construction had sent piers southward into Communipaw Cove at right angles to the South Street Pier, thus creating a pattern of piers not unlike a modern marina. All of this filling and building had been done in the belief that the Canal Company had every right to reclaim submerged lands. The result was that Jersey City's growth was effectively stunted by the presence of the basin and its extensions all along the southeastern edge of the city. The Company had claimed the right to the lands under water, and over the years they had in fact reclaimed many underwater lots by means of cribbing filled with stone and dirt.

Well, if the Canal Company could claim the right so to reclaim lands, the Sugar House proprietors could — and did — make the same claim. Talcott reported to the stockholders, in the 1867 Annual Report, that Matthiessen and Wiechers had "disputed this Company's title to the shipping basin constructed" on their submerged lands.[74] Actually, they did more than dispute the Canal Company's right; they asserted their *own* right. With the help of a powerful steam derrick, they "tore up and destroyed by force . . . steam boats and other boats" in the canal basin, and probably tore into a few piers while they were about it. The Canal Company asked

Messrs. Matthiessen and Wiechers "in a friendly way to desist," but since they refused, the Canal Company had gotten an injunction to force them to stop their guerrilla warfare. That had been in May, 1863.[75]

The Sugar House skirmish was but a prelude to the full-scale war that was in preparation. At the other end of the Morris Canal basin, to the south, the Central Railroad of New Jersey (CRRNJ) was already under construction from Elizabethtown "to some point on Newark Bay."[76] That "point" happened to lie directly across the canal basin. This Battle of the Basin was to be for the MC&BCo. of 1844 what the Battle of Gettysburg was to the Confederate Army: the "high tide" of a very uncivil war. Both the Confederacy and the Canal Company, each representing a way of life that had become obsolete, began fighting for their very existences, ironically enough, during July, 1863.

Bruce Catton describes the 1863 Battle of Gettysburg as[77]

> . . . *an act of fate; a three-day explosion of storm and flame and terror, unplanned and uncontrollable, coming inevitably (as the war itself had come) out of the things that hard-pressed men had done in the light of imperfect knowledge, the end result of actions that moved with an inexorable logic toward a fundamental and astounding goal.*

I submit that the Battle of the Basin, waged between the CRRNJ and Morris Canal, can be described similarly, in language not much less stirring.

What was "inevitable" about it? Why "inexorably logical"?

What King Cotton was to the Confederacy, King Coal was to the Canal Company. Until *after* the Civil War, however, the South had no competitors in the world market for its product. The Morris Canal was not that fortunate. All of the little lateral and branch railroads that the Canal Company had brought into being, had given rights-of-way to, and had bought stock in to help finance construction — all had become important coal carriers in their own right. They no longer needed the Morris Canal. What they *did* need were terminals on the Hudson River opposite New York City, preferably at Jersey City.

The Delaware and Hudson Canal had established a coal terminal at Weehawken in 1859, as we saw in Chapter 16. The M&E Railroad had reached Hoboken by 1860. In 1857, the Erie Railroad had begun its own terminal on Harsimus Cove, between Jersey City and Hoboken, completing it in 1860. The CRRNJ, feeling very left out with its tracks reaching only to Elizabethtown, petitioned for and was given a charter supplement enabling it to extend to New York by bridging Newark Bay. The date was February 23, 1860.

Of all the early railroads nurtured by the Morris Canal, the CRRNJ was the least "friendly." The spirit of competition and the pangs of hostility had been engendered early. The antagonism seems to have grown out of the apparent falling off of friendliness between the Canal Company and the Lehigh Coal and Navigation Company. There had been a corresponding growth of friendliness, or at least of the "mutually beneficial cooperation" we have witnessed, between the DL&W Railroad and the Morris Canal, and between the Canal Company and the LV Railroad.

Let us pause for a very brief overview of the CRRNJ. We saw, very early in this history, how the Elizabethtown and Somerville (E&S) Railroad had been chartered in 1831, with Isaac H. Williamson its first president. Williamson was very friendly to the Morris Canal at that time. However, in 1846, the E&S Railroad went bankrupt and was bought up by "New York capitalists," who reorganized the company. Benjamin Williamson, former Canal Company president and director, was the president of the new E&S Company.[78] In 1847, a newly formed Somerville and Easton (S&E) Railroad was formed, which bought the E&S. Its president was apparently John Taylor Johnston, and one of its principal backers was William E. Dodge. In 1849, the route of the road from Easton to Elizabethtown was renamed "The Central Railroad of New Jersey." By 1852, the CRRNJ had reached Phillipsburg. It had already arranged to intersect the DL&W Railroad by means of the Warren Railroad, chartered on February 12, 1851, to connect the CRRNJ and DL&W roads at New Hampton, New Jersey. The arrangement for that interconnection had been tripartite, with each of the three railroads subscribing for $150,000 worth of stock to finance construction. The Warren and M&E Railroads had also competed fiercely for that connection with the DL&W, but the CRRNJ won the contest. It could hardly have lost. William E. Dodge, director of the Warren Railroad, was also among the directors of the Scrantons' Lackawanna Iron & Coal Company.[79]

There probably would not have been a head-to-head confrontation — no Gettysburg — between the CRRNJ and the Canal Company if the former had not sought a further supplement to its charter authorizing it to place its road in Communipaw Cove, part of New York Harbor, "where safe and convenient navigation by means of steam ferry-boats can be obtained with the City of New-York."[80] Canal Company President Marsh told his board of directors in mid-1863 that he had had "several interviews and some correspondence" with the president of the CRRNJ and that the railroad desired "to effect an arrangement by which they may acquire the right to occupy a portion of the 'South Canal Basin,'" belonging to the Canal Company.[81]

There was, apparently, *no way* that the CRRNJ could acquire any such right on Canal Company property. The Canal Company — so they said — had plans for that part of the basin. On July 21, 1863, the same day that Marsh reported to the directors on his several dealings with CRRNJ President John Taylor Johnston, he presented a letter from William H. Talcott, the Canal Company's Chief Engineer. The letter noted that navigation on the canal suffered frequent interruptions at the Jersey City lock, interruptions which were a "serious inconvenience to the trade and also to the shippers at Jersey City." The volume of trade on the canal both at present and in the future justified the construction of an outlet lock from Mill Creek into Communipaw Cove. The plan, Talcott claimed, had long been contemplated, and he recommended that it be adopted immediately. The Talcott letter was dated June 17, 1863; the battle had been joined.

The CRRNJ's construction continued, and on August 14, 1863, Marsh wrote to that company — not to Johnston, but to the CRRNJ Company:[82]

> *You will take notice that the canal of the Morris Canal and Banking Company is largely fed and supplied by the tide waters which flow up Mill Creek, and that Mill Creek*

is also an important outlet for the canal for purposes of navigation.

That the Morris Canal Company will feel it their duty to resist the placing of any obstructions in New York Bay, or Communipaw Cove, which will tend to impede the flow of water into or from Mill Creek, to interrupt the usual free access to navigation from and into Mill Creek.

Of course the CRRNJ did not cease and desist, but kept right on building. It had every intention of continuing its trackage to a depot, to be built right in the Hudson River, immediately south of and precisely across the canal's "South Basin." And of course the Canal Company sought an injunction to force CRRNJ construction to stop. A Bill was filed in Chancery, which the CRRNJ had to answer. The *Answer* was published and distributed in pamphlet-form.

Since the entire future of the Canal Company stood on the arguments contained in this *Answer,* we will look at it in some detail. A sketch, or series of them, will help clarify the issues and illuminate the text.

The south side of South Street pier was packed with tall-masted coasting vessels, while the inside of the basin formed by the Bergen Street pier and South Street was filled with canal boats. With the CRRNJ tracks and depot cutting off the major part of the South Basin, the Canal Company stood to lose its coastal shipping trade.

The Canal Company's Bill of Complaint was based on — of all things — Mill Creek as a navigable feeder, and on the Company's rights to the lands under water that had been reclaimed.

Sometime during the 1850s, the Canal Company argued, they had intended to make an outlet lock through Mill Creek into Communipaw Bay, but had not done so. Beach had recommended filling in the creek, as we saw, and compensating Mr. Prior for the loss of power to his mill. However, the creek had been left open, being bridged for the horses, mules and men, while boats crossed in "a wooden structure," or "wooden box or fixture."[83] Before the steam pumping station had been built to keep the Jersey City level filled even at low tides, the canal was filled at high tide here, at Mill Creek, and at Fiddler's Elbow, as well as by the Hackensack and Hudson rivers.

The devices or "fixtures" installed were called "sluicegates." They were "forced open when the outside tide came upon a level with the waters inside the canal. The gates would continue open as long as the tide flowed in, and then close when the tide ebbed at high tide and commenced to lower." Thus, water could be retained in the canal "as high as the high tides of the rivers." This method of supplying the canal level was obviated by the steam pumping apparatus, and the CRRNJ did not hesitate to point out that fact in its "Answer," thus disposing of one of the Canal Company's objections.

As for the Mill Creek being navigable, the CRRNJ used one of the Canal Company's own "answers" to a Bill of Complaint filed against them by "Gilbert and Others," who actually *were* using Mill Creek for boating. Since the Canal Company's answer was upheld by the Courts, the CRRNJ pointed out that the Court's decision must hold in the present case. Another of the Canal Company's objections was thus disposed of.

When it came to answering the Canal Company's claims to rights in lands under water, the CRRNJ had a field day. In order to secure its *own* riparian rights, the CRRNJ had, at an expenditure of upward of $180,000, "purchased the entire shore rights," on Communipaw Cove, "along the shore that faces south from the south-east point of Jersey City to Mill Creek, and the entire shore facing east from Mill Creek" to a point several hundred yards south of Mill Creek, "with the exception of a few small pieces."[84] The Canal Company claimed that not only did they use and need to use Mill Creek, but that they also used about 550 feet of shore line between Beach and Henderson Streets for a shipping area. The CRRNJ denied that that section of canal had ever been used "for any such purpose," although the canal had been there for more than 30 years.[85]

The CRRNJ further answered that since the Legislature of New Jersey had given them the authority to build their railroad in Communipaw Bay, the Canal Company, which could claim no such authority for their works, had no cause for complaint. Going all the way back to the original deed from Van Vorst to the Jersey Associates in 1804, the CRRNJ argued that whatever rights had been vested in the Associates by virtue of that deed, had not passed to the Canal Company of 1844 when the Associates had sold lands to them.

All the Canal Company had acquired with those and other deeds to property of the Associates had been "a right of way for the purposes of their canal, and so much land for wharves, locks, etc., as might be necessary for the enjoyment of their canal."[86]

The arguments that clinched the case for the CRRNJ were:[87]

1. The basin area in dispute was not used for canal purposes, but it was leased out as winter quarters for boats and for storage of rafts and timbers at very large rental fees.
2. The Canal Company had never filed in the office of the Secretary of State [N.J.], or any other public place, a survey or location of any pier or basin other than the original Little or North Basin. No record had been found of a 40-acre basin, such as the Canal Company claimed had been constructed in Communipaw Bay, or of any basin, pier, or structure south of South Street.

The CRRNJ intended to occupy 1000 feet of the Canal Company's basin, leaving them "more space than the calls of business do now, or ever can require." The works the CRRNJ was erecting were "essential and absolutely necessary to the enjoyment of their franchise," and if they were not to be allowed to build them, their extension from Elizabethtown to Jersey City was useless. The railroad company had already suffered because of the Court's temporary restraining order. Their "contracts with other railroads, by means of which they have heretofore reached the waters of the Hudson with their passenger business, have terminated." Clearly, the CRRNJ must develop and complete its own route to New York Bay.[88]

The Canal Company had charged that the CRRNJ, "by their officers and agents," were "secretely and clandestinely contriving with the Mayor and Common Council of Jersey City to injure the company." That, the CRRNJ answered, was untrue, and there was "not the slightest foundation for the assertion that the police force of Jersey City was to be used against the company." The CRRNJ denied that anything had been done secretly, "but, on the contrary, aver that everything has been done openly, and in the usual way" in which the Jersey City Council did their business. As for the Canal Company's allegation that CRRNJ President John Taylor Johnston was allowing his company's funds, or property, or its chartered rights to be used for his "private speculation," Johnston issued a firm denial.[89]

Since the CRRNJ's *Answer* had challenged the Canal Company's rights in and titles to the Jersey City properties, it was up to *them* to answer. Probably realizing that they could not prove clear and undisputed title to, or right to build on, submerged lands, the Canal Company chose not to argue the case, and it was withdrawn.[90] Instead, the two companies entered into a written agreement on November 19, 1863, which specified the rights of each party in the submerged lands in New York Bay. The CRRNJ, new riparian owners of virtually the entire shoreline from Bayonne to Powles Hook, conveyed some of its waterfront property to the Canal Company, which in turn conveyed part of its South Basin to the CRRNJ for its depot. In acreage, the Canal Company came out ahead; in value, the CRRNJ was the undisputed winner.[91]

Far from considering the agreement an even exchange, director John D. Ward insisted that his Minority Report, or Protest, be entered in the Canal Company's *Minutes*. So "strongly impressed" was he "with the belief that the property of the stockholders of the Morris Canal and Banking Company will be irreparably injured by confirming this agreement," that he felt "bound as one of the guardians, for the time being, of that property to enter his earnest protest against its confirmation."[92]

1. The agreement "was not directed or authorized" by the Canal Company's board of directors, "but was made and signed without its knowledge and consent."
2. The agreement proposed that the Canal Company "give up without compensation, a large and valuable portion of property," to which there was good reason to believe the Company had good and sufficient title. "The value of the land claimed by the Canal Co. in Communipaw Bay was estimated by the President, a short time since, at half a million dollars, and more than two thirds of the whole worth according to that estimate, $333,333, will be totally lost" to the Canal Company if the agreement were confirmed.
3. The agreement bound the Canal Company "not to resist the extension of Hudson St. and sundry other streets over or through Canal property, and the construction of bridges, over the canal and basins and in such places as the Rail Road Com-

pany may desire without reference to the interest or convenience of the Canal Co. some of which streets and bridges," Ward felt, would "materially interfere with the business of the canal."

4. The agreement would "reduce the Morris Canal and Banking Co. to the condition of a vassal" to the CRRNJ, "by giving to that Co. the right to compel the Canal Co. to expend thousands of dollars in constructing a canal from the mouth of Mill Creek to the remnant of the Canal Basin which the agreement proposes leases to us."
5. The agreement required the Canal Company "to execute deeds, conveying to the Central Rail Road Co. of New Jersey a large amount of land to which they assert the Canal Co. have not a shadow of right." If the CRRNJ really believed its own assertion, Ward reasoned, the proposal was an insult and should be treated as such. If the Canal Company did not have a right to those lands, they should keep them for their own use or at least receive the true value if the lands were to be relinquished.

Ward's protest was duly entered, but to no avail. The agreement with the CRRNJ stood.

Mill Creek Outlet Lock *was* built, during 1863-64. It is not known whether or not the CRRNJ "compelled" that construction, but it is not likely. An outlet from Mill Creek into the Cove would have been convenient for many reasons, not the least of which was the presence of the railroad blocking part of the canal basin to boats trying to enter from the Hudson River.

Like Sherman's march, the CRRNJ's "march to the sea" had been successful. Appomattox Court House lay just ahead.

REFERENCES

1. *Report to the Stockholders of the Morris Canal and Banking Company* (N.Y.: 1853), pp. 5-6.
2. *Report to the President and Directors of the Morris Canal and Banking Company to the Stockholders* (Jersey City, N.J.: 1867), p. 6 (hereinafter cited as *Report of the President, 1867*).
3. *Report to the Stockholders of the Morris Canal and Banking Company* (N.Y.: 1855), pp. 5-6.
4. *Report to the Stockholders of the Morris Canal and Banking Company* (1862), in the *American Railroad Journal*, n.d., n.p.
5. *Report to the Stockholders of the Morris Canal and Banking Company* (N.Y.: 1863), p. 14 (hereinafter cited as *Report to the Stockholders, 1863*).
6. *Ibid.*, pp. 14-5.
7. *Jerseyman*, May 2, 1863.
8. *Report to the Stockholders, 1863*, p. 14.
9. Washington (N.J.) *Star*, September 28, 1934.
10. *Ibid.*, September 5, 1957.
11. *Report to the Stockholders of the Morris Canal and Banking Company* (N.Y.: 1864), p. 14 (hereinafter cited as *Report to the Stockholders, 1864*).
12. *Report to the Stockholders, 1863*, p. 15.
13. *Report to the Stockholders, 1864*, pp. 13-4.
14. *Report to the Stockholders, 1863*, p. 17.
15. *Jerseyman*, May 2, 1863.
16. Morris Canal and Banking Company *Minutes*, July 24, 1862 (hereinafter cited as MC&BCo. *Minutes*).
17. Washington (N.J.) *Star*, September 28, 1934.
18. MC&BCo. *Minutes*, July 24, 1862.
19. John D. Hicks, *The Federal Union* (N.Y.: 1937), pp. 480-1.
20. Paterson *Daily Guardian*, March 19, 1857.
21. *Report to the Stockholders of the Morris Canal and Banking Company* (Jersey City, N.J.: 1866), p. 15 (hereinafter cited as *Report to the Stockholders, 1866*).
22. Paterson *Daily Guardian*, May 8, 1857.
23. Morris Canal and Banking Company *Docket Files*, letters dated July 28, 1860 and July 31, 1860.
24. *Report of the President, 1867*, p. 5.
25. *Report to the Stockholders, 1863*, pp. 14-5.
26. Randolph L. Kulp, ed., *Railroads in the Lehigh River Valley*, (Allentown, Pa.: 1962), pp. 60-1.
27. *Report to the Stockholders, 1863*, p. 7.
28. Paterson *Daily Press*, November 30, 1863.
29. *Report to the Stockholdrs, 1863*, p. 7.
30. *Report to the Stockholders, 1864*, p. 14.
31. *Ibid.*, pp. 14-5.

32. Paterson *Daily Press,* November 30, 1863.
33. *Report to the Stockholders, 1866,* p. 15.
34. *Report to the Stockholders, 1864,* p. 15.
35. *Ibid.,* pp. 15-6.
36. *Report to the Stockholders, 1866,* p. 15.
37. MC&BCo. *Minutes,* November 4, 1866.
38. *Report to the Stockholders, 1863,* p. 15.
39. Paterson *Daily Press,* March 14, 1864.
40. *Report to the Stockholders of the Morris Canal and Banking Company* (N.Y.: 1865), p. 6 (hereinafter cited as *Report to the Stockholders, 1865*).
41. Paterson *Daily Press,* July 27, 1864.
42. *Ibid.,* October 11, 1864.
43. *Ibid.,* November 1, 1864.
44. *Report to the Stockholders, 1865,* pp. 5-6.
45. *Ibid.,* p. 7.
46. *Ibid.,* p. 6.
47. Newark *Daily Advertiser,* July 21, 1865.
48. *Report to the Stockholders, 1866,* pp. 5-6.
49. *Ibid.,* pp. 7, 17.
50. *Ibid.,* pp. 6, 16.
51. *Ibid.,* p. 15.
52. MC&BCo. *Minutes,* October 7, 1852.
53. *Report to the Stockholders, 1866,* p. 17.
54. *Ibid.,* p. 16.
55. Paterson *Daily Press,* May 16, 1865.
56. *Ibid.,* June 21, 1865.
57. *Report to the Stockholders, 1866,* p. 16.
58. *Laws of New Jersey,* 1865, PL 264.
59. Munsell's *History of Morris County, New Jersey: 1739-1882* (N.Y.: 1882), p. 47 (hereinafter cited as Munsell's *Morris*).
60. Paterson *Daily Guardian,* May 28, 1857.
61. MC&BCo. *Docket Files,* Docket #254, File 155.
62. MC&BCo. *Minutes,* June 13, 1856.
63. Paterson *Daily Press,* November 1, 1864.
64. Munsell's *Morris,* p. 310.
65. *Ibid.,* p. 60.
66. Beer's *Atlas of Morris County,* 1868.
67. MC&BCo. *Minutes,* January 19, 1869.
68. MC&BCo. *Docket Files,* Docket #409.
69. *Report to the Stockholders, 1866,* p. 7.
70. *Report of the President, 1867,* p. 5.
71. *Ibid.,* pp. 6-7.
72. *Report of the Directors of the Morris Canal and Banking Company to the Stockholders* (Jersey City, N.J.: 1869), p. 8.

73. *Report of the Directors of the Morris Canal and Banking Company to the Stockholders* (Jersey City, N.J.: 1870), p. 8.
74. *Report of the President, 1867,* p. 9.
75. Bill in Chancery: Morris Canal and Banking Company *vs.* F.O. Matthiessen and others.
76. *Ibid.,* p. 21.
77. Bruce Catton, *This Hallowed Ground* (N.Y.: 1956), p. 250.
78. James P. Snell, *History of Sussex and Warren Counties, New Jersey* (Philadelphia, 1881), p. 111.
79. *Ibid.,* p. 487.
80. "Answer" of the Central Railroad of New Jersey to the Morris Canal's Bill for Injunction, pp. 20-1.
81. MC&BCo. *Minutes,* July 21, 1863.
82. "Answer," p. 30.
83. *Ibid.,* pp. 9, 30.
84. *Ibid.,* pp. 21-2.
85. *Ibid.,* p. 10.
86. *Ibid.,* p. 11.
87. *Ibid.,* pp. 16, 18.
88. *Ibid.,* pp. 23-4.
89. *Ibid.,* p. 27.
90. Jersey City *American Standard,* November 23, 1863.
91. MC&BCo. *Docket Files, Memorial of the Morris Canal and Banking Company,* January 9, 1866.
92. MC&BCo. *Minutes,* January 21, 1864.

18

"Mene, Mene . . ."*

IF Appomattox lay ahead, there were still several battles to be fought, fought knowing, as the Confederacy knew, there was no chance of winning. The Canal Company, too, should have known that it couldn't win its battle with the hostile railroads.

Early in 1864, the American Dock and Improvement Company applied to the State Legislature for a grant of lands under water at Jersey City, and there was lively debate in the Senate on the subject. The Dock Company had offered to pay the State $10,000 annual rent in exchange for the grant, plus 10 percent of their sales and annual rentals.[1] However, the question of submerged lands was very much unsettled, and the Legislature created a Harbor Commission in April, 1864, "to ascertain the rights of the state, and of riparian owners in the lands lying under the waters of the Bay of New York, and elsewhere in the state."[2] The Canal Company appointed a committee of its own to keep an eye on what those Commissioners were up to.

The Harbor Commission prepared a map on which was

*From the Book of Daniel: "Mene, Mene, Tekel Upharsin," meaning "The handwriting was on the wall."

'delineated what they proposed to establish as an exterior line for the wharves, piers, etc., of Jersey City, including the Canal Company's property. When the Canal Company committee examined the map, they found that "the larger portion of the large basin" had been "intended by the Commissioners to be appropriated to public use by separating it entirely from the remaining portion of the Canal property and constructing several piers therein."[3]

Although the Legislature didn't adopt the Harbor Commissioners' suggestions, their report had "created a strong prejudice against the rights of the Company in these lands." Consequently, when the Hudson River Wet Basin and Warehouse Company sought to incorporate in March, 1866, both houses passed the law. The incorporators had asked that the state give them a 60-year lease at an annual rent of $15,000 for all the lands claimed by the Morris Canal Company "in their 'shipping basin,' and all between it and the channel of the Hudson River." Before Governor Marcus Ward could sign the bill into law, however, the Canal Company popped up with an offer to pay the State $20,000 a year for the same property! The Governor didn't sign the bill, and without his signature, the bill failed to become a law.[4]

The Hudson River Wet Basin and Warehouse bill was reintroduced during the 1867 Legislative Session with some modification, but again it failed to pass. Instead, the Legislature passed "a further supplement" to the Canal Company's charter that granted the Company a lease on the submerged Jersey City lands in exchange for $25,000 a year tax-rent, to be paid into the State's Treasury.[5]

The Charter Supplement of March 14, 1867, specified just how much space the Canal Company might occupy in New York Bay. It gave the Company the authority to "construct piers, wharves, docks, basins, warehouses, and other structures," and to "make reasonable rules and regulations for the use thereof, to enable them to carry on a transportation business in and over their canal and elsewhere, which they are hereby empowered to do, and also to afford facilities for commerce." In exchange for that authority, however, the Canal Company's tax-exempt status would not apply or "extend to the property or privileges hereby granted."

The Canal Company was given 90 days to decide whether or not to accept the State's offer to charge them for leasing their own basin. Thomas N. McCarter and Isaac Scudder, both Canal Company directors, and both very able counsels, were asked to review the State's proposal and give their opinions in writing. McCarter wrote:[6]

> *The Legislature throughout the whole charter has seemed to take special pains to limit the property and operations of the Company to the actual and necessary purposes of canal navigation.* [*In fact,*] *it is manifest that the Legislature intended to restrict the operations of the Company to the mere superintendence of a canal and to the collection of tolls — and to guard carefully against the diversion of funds or operations to any other purposes, viz: Under these and other provisions of the charter it has always been a matter of serious doubt whether the company can lawfully own boats, or horses, or engage in the transportation business over their own canal or do anything more than receive tolls from business done by others.*

McCarter reasoned that the supplement entitled the Company "to go into the transportation business, to build piers, wharves . . . and warehouses, and also to afford facilities for commerce." These greatly enlarged powers would legalize "important operations of the company heretofore carried on of at least doubtful legality." Of course, everything not strictly navigation would henceforth be taxable at $25,000 a year.

However, *some* basin is better than *none* — if the title be defective, McCarter decided. Moreover, the state authorities were "beginning to exert in a practical manner the right of the state to these lands under water, and especially as the very tempting location of the property and its great value has already induced large offers from outside speculators for a grant from the State similar to the one now under consideration." In effect, McCarter said: accept and pay; reject and lose all.

Isaac W. Scudder, Canal Company director and counsel, reasoned a little differently. Since the State grants title to lands previously held in questionable right, it would be better to accept the $25,000 yearly burden to be able to fight off the Central Railroad of New Jersey

(CRRNJ) on the south, the Sugar House on the west, and Jersey City all around. Everyone *wanted* the Morris Canal's basin property; Scudder concluded that the proposed "Supplement" decided who *owned* it.

Acting on the advice from these worthy and respected counselors, Talcott and the board of directors decided to approve the Supplement so that the agreement could be made up and placed in the Governor's hands within 90 days.

With a touch of the prophet, Talcott pointed out that[7]

> *the proximity of the terminal points of other large transporting companies, whose connecting lines stretch across the continent, and the admirable situation of the premises in question for commercial purposes, together with the fact that the commerce of New York must seek its appropriate facilities on the shores of New Jersey, have rendered this property eminently desirable in the estimation of capitalists. Without the grant from the State, the Company will be subjected to a constant struggle to maintain their possession of the property, which will be very annoying and expensive, and, it may be, hazardous. It is therefore apparent that the interests of the Stockholders will be better subserved by the possession of large franchises and an unquestionable title.*

Once again, John D. Ward submitted a Minority Report, a protest against accepting the "Supplement:"

1. It would mean complete and irrevocable abandonment of title to lands and franchises acquired fairly, openly, and honestly 22 years ago,* and would be an acknowledgement of the State's ownership of all the work put up by the Morris Canal.
2. For the State to require "rent" of $25,000 while insisting that the Morris Canal keep the navigation and works in good order and that the public be allowed to use it free of charge, yet claiming the reversion of the canal properties at the expiration of the charter, were "all unprecedented, unreasonable, and

**By deeds of January 2, and September 13, 1845, from the Associates of the Jersey Company.*

arbitrary burdens — such as have not heretofore been imposed upon any corporation created by the State of New Jersey."

3. If the State could tax the canal $25,000 a year, Jersey City could tax, too, at $15.80 per $1,100, but not less than $5,642.84 yearly. Such a financial burden would make dividends to stockholders virtually impossible, since business was not likely to increase; nor could toll rates be apt to rise since the canal would be obliged to compete with the railroads. For Ward, the $25,000 tax-rent was a breach of faith on the part of the State.
4. There was no guarantee that this would be a permanent arrangement. The State had shown that the grants of former Legislatures could be "entirely disregarded." What would stop the State from repealing or annulling *this* law?
5. The stockholders should have a say in the decision. It should not be made by the directors alone.

Ward's protest notwithstanding, the Supplement became law, and the Basin of 1867 came into being. So did the burdensome tax-rent of $25,000 yearly, to be paid into the State's Treasury.

By mid-1867, the Sugar House, now known as the New Jersey Sugar Refining Company, asked permission to build a dock or wharf on the Little Basin. Talcott, with resignation born of experience, recommended that it would be in the interest of the Company to grant such permission.[8] The board of directors agreed to let a wharf be built in the "north basin" and, of course, to allow the use of the wharf in that basin.[9]

Once the enemy of the Canal Company, the Sugar Company later joined ranks with them when the Jersey City Board of Works awarded a contract to build a bridge over the basin in 1878. With a bridge, the Sugar House proprietors would lose their dock space at the end of Washington Street, where their barges were loaded and unloaded. Both companies secured injunctions to stop the Board's construction. The Sugar Company wanted to close the end of Washington Street with some kind of float.

As the Canal Company had once been ridiculed in verse, the Sugar Company now was publicly chided:[10]

We are monarch of all we survey
Our rights there are none to dispute.

From the sugar house to the bay
We are lords of the street and the float.

Such sentiments had been directed at the Morris Canal before the Civil War; now the canal's very right to existence had been called into question.

Talcott put his finger on the problem for the future in his 1868 Annual Report to the Stockholders:[11]

> *The question of improving the submerged lands embraced within the bounds of the State Grant has received the earnest attention of the Directors and Officers of this Company. The rapid changes which have taken place in the ownership of shore property in the immediate vicinity, during the past year, indicate that the time is not far distant when this property will be needed for commercial purposes.*

Marsh, in his 1856 Report, had announced that "the Jersey City Improvements, when finished, will cost the company about $35,000. They could be sold at present for $80,000."[12] Expenditures at Jersey City, unfortunately, were not listed in every annual report, nor were they included in the *Minutes.* But even the very incomplete records in this author's possession indicate that a minimum of $400,000 was spent on the "improvements," the years of heaviest spending having been:

1859	$80,000.00
1863	81,923.69
1866	70,306.81

It had been a costly war for survival in Jersey City. By far, the heaviest losses were felt in the Canal Company's prestige, however. Having lost its position of supremacy, the Company was to be encroached upon all sides by many new enemies.

The Jersey City and Fort Lee Railroad, proposed in 1865, would have been built from the CRRNJ's new depot to York Street, through York to Greene, through Greene to Pavonia Ave., through Pavonia to Hudson Street, continuing in Hudson Street to the northern boundary of the city. The Mayor and Council of Jersey City would have to pass an ordinance permitting construction, and Talcott wrote to ask that Mayor Orestes Cleveland withhold his approval of such an ordinance. Since the Fort Lee Railroad was to be built ostensibly for the convenience of users of the CRRNJ's depot and ferry, a route using Grove Street would better serve them and be cheaper. Talcott argued that it had been legally established that Hudson Street did not extend south beyond the canal basin and that the city had no authority to so extend it. Finally, if the Fort Lee Railroad *were* to be built, it would either cut off entirely the access of coasting vessels to Morris Canal facilities, or would require wide, expensive, and somewhat dangerous drawbridges over the basin, at great inconvenience to both canal and railroad traffic.[13]

Cleveland *did* veto the ordinance, and John D. Ward took the time to write him a personal "thank you." Ward was convinced that the ordinance was intended to invest the issue "with the semblance of legality, though destitute of even the shadow of justice, or any pretense that public convenience would be promoted by so gross an inversion of private and corporate rights."[14]

The danger of the Fort Lee Railroad was over — for the moment. But now the town of Bergen's Sewer Commissioners proposed to discharge all sewage collected from the several districts of the municipality into the Morris Canal. Again, Talcott protested: "The daily discharge of several cubic yards of very offensive solid matters into the slack waters of the canal" would very rapidly "form obstructions of such character" as to "either entirely destroy the navigation in the vicinity of the point of discharge, or render its frequent removal necessary." The Canal Company, he avowed, would resist any plan that would "make the Morris Canal the receptacle of the sewage of the Town of Bergen."[15]

And then the Hudson Street Extension surfaced again. This time it was the CRRNJ's undisguised attempt to have that street extended to reach its depot and ferry. The Canal Company sent a memorial to the Mayor and Council of Jersey City, containing an excellent argument against a drawbridge over the basin.[16] Jersey City retaliated by appointing a two-man Committee on Streets (J. N. Davis and A. Drake), whose report cited at length from the pamphlet that had been published by a number of Canal Company speculators in 1853. The Committee on Streets pointed out that the pamphleteers had made it clear that the Morris Canal Company intended to build piers on its $3,000,000 worth of real estate *to rent out*, i.e., for profit. On the other hand, the Committee felt sure that the extension of Hudson Street could be made at no expense to the public, especially since the Canal Company deserved no compensation and could bring no new lawsuit on the issue. The issue had been settled in 1859, they claimed, when the Court decided that Hudson Street could be extended, and in 1863, when the Court decided that the CRRNJ's occupation of submerged lands south of South Street was perfectly legal.[17] Obviously, Hudson Street was the key to winning the Battle of the Basin.

When Jersey City *published* the report of the Committee on Streets, the Canal Company responded in kind, publishing a Memorial on January 9, 1866. The lengthy statement refuted — or tried to refute — the Committee's claims. Finally, the Canal Company appealed to Trenton for help. The appeal to the Session of 1866 was accompanied by a map *showing* how the extension of Hudson Street would virtually destroy the whole canal basin system.[18] It seems that the Morris Canal no longer had influential friends in Trenton. No legislative action was taken; the Canal Company was left to fights its own battle against Jersey City. For the time being, the Company beat a strategic retreat to regroup for future fights on this front.

Meanwhile, the battle against rising costs continued. When Talcott nominated Jacob F. Randolph for the Company's vice presidency, Randolph offered to serve at a reduced salary of $3000 yearly. He was unanimously elected.[19]

The cost of carrying coal now directly reflected the increasing competitive struggle for markets. The 1867 boating season had begun "with such a competition between the avenues leading from the different coal fields, that the cost of carrying coal, to tidewater, was soon reduced to extremely low rates, and so continued to the end of the season." Nor was it for just the "through" trade that the anthracite railroads and canals were competing. The canal lost a part of its "way" trade to the railroads that serviced the same route that the canal did. The cost of labor and materials continued high, increasing the amount of expenditures for repairs. Common labor was in such demand that high wages were being paid elsewhere, and the

Canal Company was losing boatmen at a rate so significant that it had become "impracticable to keep all the Company's boats manned and operating."[20]

More ominous than high freights and low tolls, which had meant a "corresponding reduction in the receipts of the Company," was the falling-off of tonnage. The Canal had just experienced its most successful year, 1866, with receipts totalling $616,350.36, and a total tonnage of 889,220 tons of freight carried. Receipts for 1867 had dropped to $448,199.71, with the tonnage falling off to 822,741 tons, of which only 424,831 tons had been coal. Ore tonnage totaled 289,550 tons during 1867. Of that amount, 143,925 tons were consigned to iron works in the Lehigh Valley; 61,958 left the State by way of the Warren Railroad at Port Washington; and only 83,667 tons were delivered along the way of the canal.[21]

The Morris Canal was increasingly a victim of the capricious coal industry. A crippling strike at the coal mines in mid-1868 almost entirely suspended the coal trade, with corresponding reductions in the Canal Company's "comparative earnings" from week to week.[22] The strike lasted more than seven weeks, and when trade was resumed, the Canal Company had only been able "to obtain rates of transportation which were more remunerative than before, but not sufficient to compensate for the loss which had occurred during the strike."[23]

Talcott was authorized by the board of directors to make an agreement with the Morris and Essex (M&E) Railroad "based upon a division of the trade going into the Lehigh Valley, giving the canal sixty per cent and the railroad forty per cent of the trade."[24] The M&E declined to enter into such an agreement, and the Canal Company appeared to have lost another former ally.[25] Since 1866, the railroad had become a competitor in the coal and iron "through" trade; now it was even a threat to the "way" trade, as well.

And the M&E was expanding, rapidly, in many directions. The railroad applied to the Canal Company "for permission to make a drawbridge across the canal at Drakesville for the Chester Railroad." Then, in 1869, in Paterson, a "branch" railroad was being built from Hoboken to Boonton, and the Canal Company sold the right-of-way to the M&E "over lands formerly used as a canal basin in Paterson

[Gledhill's Basin, near the intersection of Marshall and Barclay streets], and at Brophy's in Little Falls Twp. [today's West Paterson]." The right-of-way, 99 feet wide, was sold for $500. The M&E was to construct and maintain their railroad bridges, sixteen feet above the bottom of the canal, at Paterson, West Paterson, and Boonton, "and make a draw on the Pompton Feeder at least twenty feet wide."[26]

While under construction, the railroad suffered a "terrible blasting accident" at Paterson, near Squirrel Woods.[27]

> *James Oplinger, the foreman of a working gang, and a laborer named Thomas Feeney were engaged in 'squibbing' a hole made for blasting the rock. There was a keg of powder close by, and by some means the fire communicated to it and an explosion took place. Oplinger was severely hurt about the region of the abdomen, loins, and thighs, as well as about the hands and face. His clothes were set on fire, and so great was his pain that he plunged into the canal to extinguish the flames.*

The M&E wasn't the only railroad expanding at the time. The Newark and New York (N&NY) Railroad had tried to buy the Morris Canal in the city of Newark "and use it for track of the road from the gas works to High Street, and connect these with the Newark and Paterson Road. By this way they would go under all the principal streets."[28] Of course the Canal Company wasn't selling, but the N&NY did open for passengers in mid-July, 1869. On July 23, several trains, each carrying between 1000 and 1500 persons, traveled from Newark to the Communipaw Ferry in 17 minutes. Passage was free that day — at least to those who were fortunate enough not to have their pockets picked.[29] The N&NY Railroad was a CRRNJ project.

Since the railroads were expanding, the Canal Company proposed yet another enlargement for the canal. Talcott informed the board of directors that "the extremely low rate of charges for the transportation of coal to tidewater by all our competing routes, shows the absolute necessity of further increasing the tonnage of our boats." There were 15 locks yet to be lengthened to 100 feet so that the size of Morris Canal boats might be enlarged for ten more tons of cargo.

That, Talcott estimated, would reduce freight rates 10 cents per ton. The prism should be dredged so as to be able to hold six feet of water, and the plane machinery should be further strengthened so as to be able to handle 100-ton cargoes. With such improvements, the Canal Company would "be able to retain a full share of the coal trade, at such rate of tolls as will secure the regular payment of dividends on the preferred stock, and in process of time, with the aid of the Jersey City property, a fair return to the consolidated stock."[30]

Another "improvement" that may have been part of Talcott's plan to recapture some of its lost business was a charter supplement granted on April 17, 1868, allowing the Canal Company to build railroads. Of course they were to be "connecting" railroads, linking the canal with railroads within the state, except in Hudson and Essex counties. And, of course, the law restricted them to 2-mile lengths; but they were railroads, and the Canal Company was authorized to operate them.[31] One such railroad would be built, as we will see.

There is no evidence to indicate the Canal Company's thinking about its fleet of canal boats at this time. However, Cornelius Stewart of Lebanon, and Joseph C. Stewart of Washington, New Jersey, formed a freight company in 1868. They proposed to buy from the Canal Company "boats not employed in the New Jersey and Pennsylvania Freight Line," and run them under the name, "Stewart Freight Line."[32] The Canal Company agreed to lease several boats to the Stewarts, the total value of $15,965 to be paid to the Canal Company in ten equal installments. The lease was signed on January 29, 1868, for 32 boats, numbered:*

627	640	1071	1692
631	641	1072	1693
634	642	1073	1694
635	643	1074	1695
636	645	1686	1697
637	693	1687	1698
638	978	1688	1699
639	981	1689	1700

**These 32 boats were valued at almost $16,000, or $500 apiece.*

The Stewart Freight Line agreement allowed drawbacks: half of the tolls on merchandise, lumber, plaster, and bituminous coal shipped westward, and half of the tolls on all freight shipped eastward except charcoal, mineral coal, ice, iron ore, lime in bulk, limestone, posts and rails, railroad ties, zinc ore, and wood. No drawbacks were to be allowed to competitors.

With sincere regret, Vice President Jacob F. Randolph reported to the stockholders that William H. Talcott had died on December 8, 1868:[33]

> *By his energy, skill, and good judgment, a property at one time esteemed almost valueless was rendered useful to the State of New Jersey, and made to contribute largely, as a channel of inland navigation, to the early development of the great coal and iron interests of the country. With a thoroughly disciplined mind, he was capable of concentrating great powers upon the subject to which he gave his attention, and while he did not neglect the most minute detail, he grasped and mastered the whole subject before him. In his death we lament the loss of a distinguished and upright citizen, a man of large scientific attainments, an Engineer of distinguished skill and ability, and one whose administrative talent and power of concentrated energy made others look to him as an unerring leader and guide."*

Talcott had been president of the Company only four years, but had served as Resident Engineer of the Western Division from 1845 to 1846, when he was made Chief Engineer. In all, he had served the Company well and devotedly for 23 years, and Jacob F. Randolph, his vice president, now became president of the MC&BCo. of 1844. He had inherited the "War For Survival" that was all over — but only for Talcott.

The 1869 boating season opened on March 25, but by mid-April, "rumors that a general strike of the miners in the coal regions was probable" adversely affected what otherwise was promising to be a good season. The rumors "not only deterred boatmen from manning" the Company's boats, "but also induced many who had already started to abandon the Canal," and business was checked "very materially." The strike started on May 7 "and continued for about ten weeks, causing for that time an entire suspension of shipments" from the mines. By the time coal was flowing again, about July 1, the Canal Company found it difficult to re-man the boats, "as the season was so far advanced" that "boatmen had found other employment."[34]

As an inducement to boatmen, the Canal Company adopted the system of "purchasing teams and leasing them to boatmen at a given sum per trip," and soon found that the system worked. Tonnage "steadily increased, so that during the month of September and to the 4th of October," the Company's "aggregate receipts were more than during the corresponding time of any previous year, except 1866."[35]

Two days later, on October 6, 1869, Nature again kicked up her heels and sent a "great and unprecedented freshet" that caused serious loss and damage to adjacent canals and to property, particularly in Warren County. The storm caused five breaks in the Western Division of the Morris Canal within a distance of 3 miles of Port Delaware. The largest, about three-quarters of a mile from Cooper's Furnace, was about 500 feet long. For a distance of 300 feet, the towpath embankment and the bottom of the canal, to the depth of 10 feet, were entirely washed away. The work of repairs was completed in two weeks with crib work "in a good and substantial manner at an expenditure of $18,621.30." By October 19, 1869, coal shipments were resumed from Port Delaware.[36]

The "greatest number" of boats to be loaded in one day — 28 of them — were loaded when traffic resumed.[37] However, the Lehigh Canal had also been damaged by the flood and did not send coal to the Morris Canal until November 19, more than six weeks later. Because the Lehigh Canal was not navigable, the furnaces in the Lehigh Valley could not receive ore shipments from the Morris Canal, either, and ore tonnage fell off by some 50,000 tons.[38]

The Canal Company board of directors "met to consider what measures should be adopted to recover from the effects of the break at Port Delaware." It was decided to have Mr. McNair, the engineer, conduct "a preliminary survey for a railway connection at Port Delaware for the purpose of transshipping ore at that point."[39] When Randolph issued his 1870 Annual Report, dated April 4, he announced that a railroad connection was being arranged, "which

Morris Canal boats waiting to be loaded at the Company's coal chutes at Port Delaware, around the turn of the century (BNK).

should be completed at an early day."[40] The charter supplement of April 17, 1868, was to be used. The canal at Port Delaware had been enlarged "to twice its former width for a distance of nearly half a mile," and "to guard against any irregularity in the supply of coal," the Company began to stock-pile 50,000 tons of coal under the "trestle work at Port Delaware."[41]

Another innovation during 1869 was the use of the steam-propelled yacht *Gussie*. She was 50 feet long, 8-1/2 feet broad, and drew 4 feet of water, with a 12-horsepower engine. On June 18, 1869, Canal Company President Randolph, with Secretary John Rogers and a "small but select party," took an excursion on the canal to try out the *Gussie*. They "started from the canal lock at Jersey City and passed through the canal" to Newark, "the bridge being passed by letting down the smoke pipe which worked on hinges. The water was so low that progress was very slow," and the *Gussie* did not reach

Newark "until three hours after the prescribed time. Some invited guests missed the excursion. Passing through the lock and over the inclined plane, the *Gussie* steamed a short distance above the city and returned home. The directors of the Company proposed to make a trip through the entire length of the canal starting on Tuesday and reaching Easton on Friday."[42]

When the cross-state trip took place on June 22, 1869, the touring party spent the first evening in Bloomfield,[43]

> *where they were handsomely entertained at the residence of President Jacob F. Randolph. Wednesday evening was passed at Dover, and last night at Schooley's Mountain. Today (Friday) they will arrive at Easton. The craft on which the party performed their trip excited much curiosity along the line of the canal, it being the first instance in which steam has made the canal go in that section.*

Although this was not the *first* time steam had been used on the canal, earlier experiments having been tried in 1856, the *Gussie* created excitement. A Paterson newspaper called her a "steam tug of a new sort," which was "now going up and down the Morris Canal as an experiment. The mules look with green eyes toward the new-comer."[44] The *Gussie* tooted her steam whistle as she went by the "Singack" brick yards on June 23. "Murray's" brick yard returned the whistle with its own."[45]

The author has been unsuccessful in locating a brick yard belonging to "Murray." However, there were several other brick companies in the Singac area. The Passaic Brick, Tile and Manufacturing Company had been incorporated on April 2, 1868, by Charles Ingles, Henry W. Merrill, Robert J. and John Hopper, and David Roe. Their company was known as the Merrill and Hopper Brick Yard. The act of incorporation gave them the right to make raceways and to take in streams to fill those raceways. Several of these men had earlier formed a company called the Pequannock Peat and Manufacturing Company, whose act of incorporation authorized them to make "a railroad and canal, or either of them, from their said lands to the Morris Canal or the feeder thereof." This Pequannock Peat Company would have been the one located in Lincoln Park's "Bog and Vly." A mule-railroad, built to transport peat to Plane No.

Morris Canal boat ascending Plane No. 10 East at the Lincoln Park — Towaco border, circa 1870-75. The Boonton branch of the DL&W Railroad is in the background. In the foreground are logs that were cut at the sawmill near the front of the plane. The oil painting was done by Samuel Woodruff (BNK).

10 East at the Lincoln Park-Towaco border, can be traced on the map of Morris County in Beers' *Atlas* of 1868.

The Passaic Brick Company's charter did not specify that it might build a navigable canal, but that is apparently how the company interpreted its authority. The "Merrill and Hopper Brick Yards at Singack [sic]" began to build "a branch canal of large size, widened at the different works already in operation, which is connected by a lock with the Morris Canal, enabling vessels to be loaded at the brick sheds thus carrying their cargoes of building material to Paterson, Newark, New York, or wherever desired." The canal was expected to be completed before the end of June, "a large force of men being employed, and the lock is also being constructed of the largest size and 100 feet in length."

The new canal was to be three-quarters of its expected length in 1869, and would be "advanced" through the brick company's "lands to accommodate new yards disposed of along its proposed line." Pat McCormick, the boss of the extension of the canal, named the Singac branch " 'the Early Rising Canal,' about which name there is a joke of Pat's. The canal is the property of the company and will remain such, regularly authorized by charter."[46] By mid-June, 1869, the brickyard canal was nearing completion. It was expected to "make a splendid means of transportation to the various yards."[47]

Perhaps future researchers will successfully locate the Early Rising and its connecting lock, if it has not already been destroyed by highway construction in the vicinity of Routes 23, 46, and 80 in Wayne.

There were other brickyards serviced by the canal in Wayne. The Van Blarcom & Company Brickyard had been on the Pompton Feeder as early as 1859. The canal "provided a great advantage in transporting to market," which enabled the brickmakers "to compete in Bloomfield and Newark as well as Paterson."[48] Another brick company built a "light horse railroad" about a half-mile long to connect its yard "with the bank of the Pompton Feeder of the Morris Canal, to which the bricks are transported on little cars, 1000 (weighing 2-1/2 tons) being a good load for one horse. At the canal the bricks are loaded on boats, and so sent to market — Paterson,

Newark or Jersey City."[49] No mention is made in Canal Company records of the brickyards or their connecting works, however.

A news item appearing in the Washington (N.J.) *Star* announced that "the Morris Canal, under the able management of Jacob F. Randolph, president, has been put in excellent order along the whole line — the bridges are in excellent repair, new coal receiving stations have added to the capacity of storage of large amounts of coal," and "the company is now free of debt, a thing that has never occurred before since its organization."[50]

Historians record the fact that, by the middle of the nineteenth century,[51]

> *the United States had attained second place among the nations of the world in coal production . . . . This enormous increase in the production of coal was coincident with unparalleled industrial expansion, and the latter was, in a large measure, dependent upon the former. Although anthracite stood first as the source of the nation's fuel a century ago, it has long since been displaced by bituminous coal, with which this country has been generously endowed. While in 1850 bituminous coal comprised but one-third of the total production, it had increased to one-half by the year 1870.*

It would increase to four-fifths by 1900, and to about seven-eighths by 1920.

That fact might have spelled prosperity for many years to come for the Morris Canal. Sadly, however, virtually all of the nation's bituminous coal lies beyond the Appalachians, out of the canal's reach. Also beyond the Appalachians lay the Mesabi, Menominee, and Vermillion iron ore ranges, discovered after 1845. It took another canal, the Soo, to bring the ore out of the remote Michigan Upper Peninsula, a canal built by Charles T. Harvey, a Fairbanks Scale Company salesman, and financed by the already wealthy Fairbanks family. The first boat load of ore went through the Soo Locks on August 14, 1855, in the Brig *Columbia,* Judson Wells, master. At almost the same moment, Henry Bessemer arrived in America from England to begin making his patented steel.[52] A new era was already dawning, and the center of the steel industry had begun to move westward.

One of the casualties of the westward shift was Port Oram Furnace. Incorporated in 1868, the furnace did not operate successfully until late 1870 or early 1871. Then, having just gotten going, "after a series of troubles," the furnace was "blown out in consequence of the scarcity of coal."[53] It was apparently started up again only to be idled about 1880. Joseph Wharton revived the Port Oram works, and the furnace was so successful that in 1902 the Town Council voted to honor Wharton by changing the town's name from Port Oram to Wharton. The Wharton Furnace enjoyed a surge of productivity during World War I, but eventually it was taken over by the Replogle Steel Company.

Another of Wharton's enterprises that was relevant to the Morris Canal was the Pennsylvania iron works called Bethlehem Iron Company. According to Clearance Books at Brooklyn Lock, Lake Hopatcong, and Stanhope Plane No. 2 West, a lot of ore left New Jersey's mines to be shipped to Bethlehem Iron. Under Wharton's direction, that company became the Bethlehem Steel Company. Later, Charles M. Schwab, who had headed the Carnegies' Pittsburgh steel works, took over where Wharton left off, and the firm became what it is today: Bethlehem Steel Corporation.[54]

The Carnegies' steel works turned out its first Bessemer steel on August 26, 1875. Superintendent of their "J. Edgar Thomson Furnace" was Bill Jones, a former foreman at the celebrated Cambria Iron Works and a one-time laborer at the Crane Iron Works, in Catasauqua, Pennsylvania.[55] Not only was iron moving westward to become steel; iron *workers* were moving, too, to become steel workers. Even Cooper-Hewitt's Durham Furnace began making pig iron to be turned into steel. One page of their ledger indicates that during November, 1870, Morris Canal boats carried that pig iron:[56]

Date	**Boat**	**Captain**	**Type of Pig Iron**	**Tons**	**Consignee**
Nov. 1	443	Wm. Warner	Grey Forge Pig Iron	95-1/2	N.J. Iron & Steel
Nov. 7	443	Wm. Warner	Grey Forge Pig Iron	94-1/2	N.J. Iron & Steel

Date	Boat	Captain	Type of Pig Iron	Tons	Consignee
Nov. 12	1687	Geo. Spencer	No. 1 Pig Iron	42	Cooper-Hewitt & Co.
Nov. 14	443	Wm. Warner	Grey Forge Pig Iron	97	N.J. Iron & Steel
Nov. 18	1694	R. Thornton	No. 1 Pig Iron	61-1/2	Cooper-Hewitt & Co.
Nov. 22	645	John Smith	No. 2 Pig Iron	60-1/2	Cooper-Hewitt & Co.
Nov. 26	948	W. Ringbold	Grey Forge Pig Iron	92-1/2	Cooper-Hewitt & Co.
Nov. 29	988	J. Williams	Grey Forge Pig Iron	90-1/2	Cooper-Hewitt & Co.

Hewitt's interest in steel later led him into the U.S. Steel Corporation:[57]

633 Burbridge St.
Germantown, Phila.
Apr. 4th 1901

Hon. Abram S. Hewitt,

Dear Sir,

I was very glad to see your name among the directors of the United States Steel Corporation. If the steel interests of the country are to be united it is very satisfactory to have one of such fearless integrity among those in control.

It is especially gratifying to me to have the New Jersey Steel Iron Co. represented. Though I have been with the American Bridge Co. since July I find that I still unconsciously identify myself with the old company.

Yours truly,
Henry S. Prichard

Randolph's annual report to the stockholders for the year ending February 28, 1871, announced that "for the first time in the history of the coal trade, the rate of transportation on coal declined as the season advanced," so that the total tonnage for 1870, although 57,372 tons more than in 1869, had netted $39,696.04 less in receipts. Fortunately, the 1870 boating season had been "without serious interruption," so that expenses were lower than usual, leaving the "net income about the same" as for 1869.

For the first time in the history of the canal, the United States Government had slapped a "surplus profits tax" on the Canal Company's earnings for 1869, so that in addition to having to pay $25,000 tax-rent to the State of New Jersey, $1,263.10 was owing to the Federal Government.[58]

Like taxes, death was also inevitable. During the last five years, the Canal Company "lost by death four of its most important and valued officers." Ephraim Marsh had died on August 28, 1864; Lewis N. Condit, the Company's secretary for nine years, had died on January 25, 1865; William H. Talcott had died on December 8, 1868; John Marsh, Ephraim's brother, had died on January 18, 1870. We can add John Rodgers, who had replaced Condit as secretary and who had died on September 23, 1870.

Rodgers' replacement was John R. Burnett, a long-time Canal Company toll collector at Newark.[59] He would stay with the Morris Company only until 1871, however, when he would enter the retail coal business in Newark, at 198 Washington Street, at the canal.[60]

Repairs during 1870 were made with an eye toward eventually increasing the canal's depth to six feet, so that 100-ton cargoes might be carried. Machinery and other facilities at the Company's boat yards had been increased so that boats could be built and repaired "with greater despatch and less cost than formerly." Four inclined planes had received 100 tons of "Bessemer Steel Rails" during 1870, and a railroad connection "with all the railroad companies at Phillipsburg" had been completed, so that the canal would be able "to receive ore and freight going West and reload in the same cars."[61]

Continuing with its relatively new system of leasing teams of mules to boatmen, the Canal Company bought 125 "western mules" in 1870, bringing the total number of company livestock to 275 "mules and horse for towing boats, most of which are young and able-bodied."[62]

And then, as regularly as the return of the seasons, a coal miners' strike was called on December 1, 1870, and was still in progress on April 3, 1871, when Randolph wrote his annual report. The canal had been ready for navigation since March 25, 1871, but the strike meant that no coal was being shipped.[63] When navigation did get under way, business "boomed." By October 1, 1871, "a total of 2,091 cargoes had passed through the canal," which included "700 sent from Washington," and the Canal Company had in its employ "325 boats, 100 men, and 300 mules."[64] Another source claimed that the Morris Canal had "about 500 boats in operation" in 1871.[65]

A "steam yacht towing boat" was in operation on the canal during 1871, plying back and forth "between City Dock lock and Plane Street." Twenty-five feet long, with an upright boiler, she came "puffing through the dark regions underneath Center Market with the door of her furnace wide open and her stalwart engineer distinctly visible by the glare therefrom." Every day crowds gathered "on the Broad Street bridge of the canal to watch the busy tug's movements." If one day she should "collapse a flue while underneath the bridge, there would be a sudden extinction of humanity" on the bridge, "for which the interested lookers-on would be totally prepared."[66]

Later, the Canal Company put a new tug, the *Wave,* on the canal, to pull or tow boats from the deep lock at River Street to the foot of the inclined plane, "so that its boatmen wouldn't have to 'pole' boats under the Center Market building anymore." It was "dark and dismal under that building, and the boatmen all hate to pass under it. Huge spiders drop down on the boats, and monster rats are heard squealing as they scamper about."[67] But, according to one old boatman* who boated the Morris Canal between 1883 and 1887, there was no steam towboat operating under Center Market. Mules walked on the street from Center Market to the head of Plane No. 12 East, while boatmen and crew leaned against the wall of the canal under the market and pushed the boat through, or used a boat pole to pole themselves through the 1100-foot tunnel.

The canal was "in better condition and repair than ever before" during 1870.[68] However, about June, 1871, a large break or cave-in occurred in the canal near the little town of Broadway, Warren County. The ground was said to have "suddenly collapsed and left a 40-by-60-foot hole." The canal was closed five days while workmen repaired the break. In describing the cave-in, a young Broadway girl had commented "with wicked coolness . . . 'Oh, it's only Broadway going home.'"[69]

**Joseph A. Lum, tape-recorded interview, June 20, 1972, Easton, Pennsylvania.*

In fact, the whole Morris Canal was "going home." The steel noose was continually tightening. The DL&W Railroad had already leased the M&E Railroad on February 9, 1869, and was building its "Boonton Branch" from Hoboken through Paterson in 1870, as we saw. The other iron horses shod with new Bessemer Steel could do no less. The Pennsylvania Railroad acquired the C&A Railroad when it leased the United New Jersey Railroad and Canal Companies in 1871. That same year, the CRRNJ absorbed the Lehigh and Susquehanna, the Lehigh and Lackawanna, the Nesquehoning, and Treskow Railroads in Pennsylvania by a lease from the Lehigh Coal and Navigation Company.

In addition to this Pennsylvania trackage, the CRRNJ would later acquire two small New Jersey roads that were more or less important canal feeders. One, intended originally to join the M&E Railroad, was the Ogden Mine Railroad, chartered on February 19, 1864.[70] The CRRNJ absorbed the line in 1881. The other road, the Mount Hope Mineral Railroad, was originally a horse-powered road that connected the Mount Hope Mine with the canal at Rockaway. It was later superceded by the Dover and Rockaway, which also served Port Oram. This road, too, was absorbed by the CRRNJ.

The Lehigh Valley Railroad had started out, as we saw, as the Delaware, Lehigh, Schuylkill and Susquehanna Railroad. It then acquired the Beaver Meadow, the Hazleton, the Penn Haven-White Haven, and several other Pennsylvania railroads. A second track had been built over the LVRR's bridge over the Delaware River in 1865. That was accomplished by extending the piers and abutments on the New Jersey side of the river about "24 feet up the stream" on lands belonging to the Canal Company.[71]

Perhaps because of the Lehigh Coal and Navigation Company's growing association with the CRRNJ, the Canal Company "appointed a committee to protect the rights of the Company in connection with" a proposed lease of the Delaware Division Canal to the Lehigh Company in 1866.[72]

Up until roughly the Civil War, it was possible to account for the competitiveness among the anthracite canals and railroads by simply referring to a map of the Pennsylvania coal region and noting the market that each carrier sought to service. We can see that by 1870

Port Delaware (Phillipsburg) circa 1860, showing the approach to Plane No. 11 West and the Delaware River. Easton is in the background. The tracks of the Lehigh Valley and The Belvidere-Delaware Railroads and the plane house are in the center of the picture (ETF).

the Southern Field was serviced by the Pennsylvania and Reading Railroads. The Northern Field was serviced by the DL&W and the D&H Railroads. The Middle Field, our primary concern here, was the initial battle ground for the LV Railroad and its arch rival, the CRRNJ.

All of these competitors had grown by the triple process of accretion, acquisition, and aggression. What is more, while they were competitors for the anthracite trade, they had all come to want their tidewater outlet to be the Hudson River at Jersey City, particularly the Morris Canal's basin property there. Knowing this makes it a bit easier to understand the following events.

Despite the fact that in 1870 the Morris Canal was in better condition than it had ever been, and that it was able to transport boats carrying upwards of 70 tons of coal, the Canal Company board of directors met on March 20, 1871, to consider the merits of accepting a supplement to its charter, passed on March 14, 1871, allowing them to lease the canal. The Lehigh Valley Railroad had offered to lease the canal for 999 years, and the offer was accepted, "subject to ratification by a majority interest of the stock of the Company, as required in the supplement named."[73]

President Jacob F. Randolph, realizing "the probabilities of our works passing into other management, and of being enlarged and thereby made a still more important avenue of trade," thought it proper, in his annual report,[74]

> *to make record: That the Morris Canal and Banking Company of 1844 by the purchase of the works and franchises of the old company, has thus far performed the purposes which are laid down for it in the preamble to its charter, viz:* — To secure to the State a great public work, to be a great public benefit and advantage to the people of New Jersey.

REFERENCES

1. Paterson *Daily Press,* March 18, 1864.
2. *Report of the President and Directors of the Morris Canal and Banking Company to the Stockholders* (Jersey City, N.J.: 1867), p. 9 (hereinafter cited as *Report of the President, 1867*).
3. Morris Canal and Banking Company *Minutes,* January 3, 1865 (hereinafter cited as MC&BCo. *Minutes*).
4. *Report of the President, 1867,* pp. 10-11.
5. *Laws of New Jersey,* 1867, P.L. 251.
6. MC&BCo. *Minutes,* April 1, 1867.

7. *Report of the President, 1867,* p. 12.
8. MC&BCo. *Minutes,* July 25, 1867.
9. *Ibid.,* January 22, 1868.
10. Jersey City *Evening Journal,* July 9, 1878.
11. *Report of the President and Directors of the Morris Canal and Banking Company to the Stockholders* (Jersey City, N.J.: 1868), p. 9 (hereinafter cited as *Report of the President, 1868*).
12. *Report to the Stockholders of the Morris Canal and Banking Company* (1856), in the *American Railroad Journal,* May 17, 1856, p. 314.
13. Morris Canal and Banking Company *Docket Files,* letter dated April 28, 1865 (hereinafter cited as MC&BCo. *Docket Files*).
14. *Ibid.,* letter dated June 24, 1865.
15. *Ibid.,* letter dated May 23, 1865.
16. *Ibid.,* letter dated September 18, 1865.
17. "Report of the Committee on Streets to the Jersey City Common Council," 1865.
18. MC&BCo. *Docket Files,* letter dated February 26, 1866.
19. MC&BCo. *Minutes,* April 1, 1868.
20. *Report of the President, 1868,* p. 6.
21. *Ibid.*
22. MC&BCo. *Minutes,* July 22, 1868.
23. *Report of the Directors of the Morris Canal and Banking Company to the Stockholders* (Jersey City, N.J.: 1869), p. 7 (hereinafter cited as *Report of the Directors, 1869*).
24. MC&BCo. *Minutes,* April 8, 1868.
25. *Ibid.,* July 22, 1868.
26. MC&BCo. *Docket Files,* indenture dated December 15, 1869.
27. New York *Times,* July 24, 1869.
28. Newark *Sentinel of Freedom,* September 3, 1867.
29. New York *Times,* July 24, 1869.
30. MC&BCo. *Minutes,* July 22, 1868.
31. *Laws of New Jersey,* 1868, P.L. 1121.
32. MC&BCo. *Minutes,* January 22, 1868.
33. *Report of the Directors, 1869,* p. 9.
34. *Report of the Directors of the Morris Canal and Banking Company to the Stockholders* (Jersey City, N.J.: 1870), p. 5 (hereinafter cited as *Report of the Directors, 1870*).
35. *Ibid.,* pp. 5-6.
36. *Ibid.,* p. 6.
37. Newark *Sentinel of Freedom,* October 26, 1869.
38. *Report of the Directors, 1870,* p. 6.
39. MC&BCo. *Minutes,* November 3, 1869.
40. *Report of the Directors, 1870,* p. 6.
41. *Ibid.,* p. 8.
42. Newark *Daily Advertiser,* June 19, 1869.

43. *Ibid.*, June 25, 1869.
44. Paterson *Daily Guardian*, June 23, 1869.
45. *Ibid.*, June 24, 1869.
46. *Ibid.*, May 29, 1869.
47. *Ibid.*, June 14, 1869.
48. *Ibid.*, August 20, 1859.
49. Paterson *Daily Press*, June 25, 1870.
50. George W.Keppler, *Scrapbook*, p. 89.
51. *The Story of Anthracite* (N.Y.: 1932), pp. 108-9.
52. Stewart H. Holbrook, *Iron Brew: A Century of Ore and Steel* (N.Y.: 1939), pp. 22, 27, 189.
53. Jersey City *Evening Journal*, March 3, 1871.
54. Holbrook, *op. cit.* p. 278.
55. *Ibid.*, pp. 206-9.
56. Abram S. Hewitt Papers. Passaic County Historical Society Museum, Lambert Castle, Paterson, N.J.
57. *Ibid.*
58. *Report of the Directors of the Morris Canal and Banking Company to the Stockholders* (Jersey City, N.J.: 1871), pp. 5-6 (hereinafter cited as *Report of the Directors, 1871*).
59. *Ibid.*, pp. 6-7.
60. *A History of the City of Newark* (N.Y.: 1913), Vol. II, p. 88.
61. *Report of the Directors, 1871*, p. 7.
62. *Ibid.*
63. *Ibid.*
64. George W. Keepler *Scrapbook*, p. 98, an article from the Washington *Star*.
65. *Ibid.*, p. 100.
66. Newark *Sentinel of Freedom*, July 4, 1871.
67. Newark *Sunday Call*, April 16, 1882.
68. *Report of the Directors, 1871*, p. 7.
69. Washington (N.J.) *Star*, June 20, 1946.
70. Paterson *Daily Press*, November 21, 1863.
71. MC&BCo. *Minutes*, July 20, 1865.
72. *Ibid.*, April 2, 1866.
73. *Report of the Directors, 1871*, pp. 7-8.
74. *Ibid.*, p. 8.

19

Hudson River Water Gap

JACOB F. Randolph, Canal Company President, and Theodore F. Randolph, Governor of New Jersey, 1869-72, *and* president of the Morris & Essex Railroad until elected to the governorship: These were the men who decided the fate of the Morris Canal in 1871, and forever after. Related? Perhaps, but a blood relationship is not the point here. What is at issue is the obvious railroad propensity of Governor Randolph, and his obvious interest in the coal trade.

Theodore F. Randolph had been born in New Brunswick, New Jersey on June 24, 1826.[1] His father, James F. Randolph, was the founder of the *Fredonian,* New Brunswick's newspaper, which he then edited for 40 years. Theodore learned both editing and mercantile skills such as accounting, and spent many years in Southern states. He was married in Kentucky in 1852, to the daughter of Congressman N.D. Coleman, of the Maysville district.

In 1853, he moved to Easton, Pennsylvania, then immediately thereafter to Jersey City, "where he engaged in the business of mining coal and transporting iron and ores." By 1859, Randolph had been elected to the New Jersey Assembly and, in 1861, was elected to the State Senate, where he served for three years. During those three years, he advocated "a more equitable system of taxation between corporations and the people."

In 1867, Randolph was elected president of one of the more powerful corporations in the state — the Morris and Essex (M&E) Railroad, a position he resigned when he was elected to the governorship of New Jersey. While he was the chief executive of the M&E, that railroad expanded greatly, adding the Chester Branch, extending the Montclair Branch, and undertaking the Boonton Branch from Hoboken through Paterson. As a result of this growth, the M&E's gross tonnage and receipts nearly doubled in 18 months. Randolph then negotiated the 1868 lease of the M&E to the Delaware, Lackawanna & Western (DL&W) Railroad, whereby stock- and bond-holders were guaranteed "in perpetuity," a 7-percent yield on their investments.

While he was governor, Randolph mounted a successful attack to break the Camden and Amboy's (C&A) monopoly of rail transportation in New Jersey and its control of the State Legislature. Moreover, transit duties, which had been levied on railroads for the numbers of passengers and amounts of freight they carried, were abolished. Instead, a property tax of 0.5 percent was levied on railroads whereby the State's public treasury "was largely benefitted."

The State's public treasury received another boost under Governor Randolph. He established the state's first Riparian Commission, whose labors gave to the state an income of over $3,000,000. The MC&BCo. was one not-so-willing contributor to that fund, as we will see.

As significant as Randolph's affiliation with the M&E Railroad was his close association with Hudson County, Jersey City in particular. His interest in the Riparian Commission must have stemmed from Jersey City's urgent need and desire to expand its physical territory at the expense of the Morris Canal and from the M&E Railroad's just-as-urgent need and desire to keep the Central Railroad of New Jersey (CRRNJ) from becoming *the* dominant railroad on the water front in Hudson County. If other railroads wanted Hudson County waterfront property so that they could compete with the M&E, they were going to have to pay dearly for it.

A "further supplement" to the Canal Company's charter was thus sought in a very special climate. The State Legislature's "anti-monopoly" leanings, as reflected in the kinds of issues being dealt with during Randolph's tenure, were very much influenced by persuasive railroad lobbyists and powerful Hudson County politicking.

The monopoly in the state at the time was the United Canal and Railroad Companies of New Jersey, which embraced the New Jersey Railroad and Transportation Company (NJRR&TCo.), the C&A Railroad, and the Delaware and Raritan Canal. What those companies had "united" *for* was to present a solid front of opposition against any other railroad being built between New York and Philadelphia. And several railroads of Pennsylvania were unquestionably interested in securing such a route across New Jersey to the Hudson River. The CRRNJ, essentially a New Jersey railroad whose Pennsylvania trackage had merely (and recently) been leased, had already established itself in Communipaw Cove, as we saw earlier. The DL&W, essentially a Pennsylvania railroad, had completed its trek to tidewater over its leased New Jersey trackage, thanks to Governor Randolph. And one of its New Jersey leases was on the M&E Railroad, whose president, just resigned, now occupied the governor's chair in Trenton. From that vantage point, he could view with unmatched clarity the struggle between the Philadelphia and Reading and the Pennsylvania Railroads as they tried to get to tidewater terminals on the Hudson by crossing New Jersey through the monopolist's territory.

Governor Randolph had resigned the presidency of the M&E Railroad in 1867, but it is possible that he retained both stock and interest in that road. We can only speculate as to how much he used his influence with the Legislature and his veto power to destroy the "monopoly" and to foster the plans of the Lehigh Valley (LV) Railroad. All that is known is that the Morris Canal was leased to the LV Railroad in 1871, when Theodore F. Randolph was Governor of New Jersey, and Jacob F. Randolph was Canal Company President.

The lease was authorized by an act of the Legislature on March 14, 1871, modifying the Company's charter. The act stated that the lease might be made "either perpetually" or for a shorter time, leaving the details to the lessees and the Canal Company to work out. All stockholders dissatisfied with the new arrangement were given three months to turn in their stock for repayment, provided they gave written notice of their dissatisfaction. Stock thus turned in could be resold by the Canal Company at whatever price the directors might "deem expedient." Stockholder consent to the lease had to be "authenticated" and filed in the State Secretary's office.

By far the most significant provision of the supplement was the State's authorization for the Canal Company to use the surplus water of its canal, "or any of its feeders, not needed for the purposes of navigation, in furnishing and supplying the inhabitants of any city, town or village along the line of said canal, or in the vicinity thereof, with a sufficient quantity of pure and wholesome water for manufacturing or domestic and other uses." Moreover, the Canal Company was empowered "to erect such works, and make such alterations in the said canal as may be necessary on property to enable said company, or its lessee or lessees, to furnish such supply of water from the canal." The rights of the mills and manufactories of the Musconetcong River were protected by the supplement, as were those of the SUM at Paterson.

We have seen the MC&BCo. become a transportation company, almost a manufacturing company, and an investment company. We will shortly see it become a major water-supply company in northern New Jersey.

Shortly after the supplement was passed on May 4, 1871, the lease of the Morris Canal to the LV Railroad was effected for "the sum of one dollar in hand" paid by the latter. *Everything* was leased to the railroad — "all the corporate franchises, rights and privileges, *other than that of being a corporation"* (italics mine) to be held and enjoyed "perpetually hereafter," and including all "personal" property such as boats, mules, dredging machines, — everything "excepting only the furniture of the principal office" of the Canal Company at Jersey City.

The Canal Company was to receive $79,250 "on the first Tuesday of the months of August and February hereafter" to meet its obligation to pay its semiannual dividends of 5 percent on the 11,750 shares of preferred stock, and 2 percent on the 10,250 shares of consolidated stock. The LV Railroad was also to pay the $25,000 annual tax-rent on the Jersey City Basin property as well as all other assessments that came due for the Canal Company. It was also to pay the Canal Company $1500 a year for the latter to maintain its corporate existence. The Canal Company was to issue its stock to the LV Railroad "from time to time at the request" of the railroad company. Finally, the Lehigh Valley was to be allowed to sell "from time to time, either at public or private sale," any real estate held by the Canal Company that was not needed for its "canal and navigation works." The railroad company was, in short, to be able to act in the name of the Canal Company, and to use the Morris Canal's official seal.

It has been said that the canal had been leased in order to eliminate one source of competition in the anthracite trade. Vermeule somewhat cleared the air on that point by explaining that when the railroad leased the canal in 1871, its line ended at Easton, and that "it had no railroad connection with New Jersey or New York, and it was to secure a tidewater outlet that it leased the canal."[2] That is true — as far as it goes. On the other hand, Wheaton Lane claimed that the Jersey City Terminal, with its 1500 feet of frontage, had been the chief advantage accruing to the LV Railroad Company.[3] That, too, is true — as far as it goes. By the time the lease was effected, it doesn't seem likely that anyone seriously considered the Morris Canal to be competition for the Iron Horse. After all, each of the three major anthracite railroads that were competing against each other had a "carrying capacity of not less than ten million tons annually."[4] There was no way for the canal to "compete" in that league.

Before *any* benefits could be derived from the lease of the canal, however, still another "panic," the Panic of 1873, had to be weathered.

The wartime prosperity of the North during the Civil War had been phenomenal. Railroads, which had overbuilt during the 1850s, became profitable investments as they became more and more necessary to transport the materials of war. Railroad mileage expanded rapidly to keep up with Mr. Lincoln's needs. The Morris Canal, as well as other internal waterways, had also experienced a boom as the volume of business increased. We have seen that the decade encompassing the Civil War had been the Morris Canal's peak years.

Most Northern banks, having lost customers in the South because of the war, had more than made up for their "losses" by accommodating the growing industrial and commercial North.[5]

Financing the Civil War at its beginning had been a particular problem for Mr. Lincoln and Secretary of the Treasury Chase. Rather than advocate taxes, Chase dealt extensively with the Philadelphia banking house of Jay Cooke, sometimes paying interest as high as 7.3 percent on loans from that bank. By the end of the war, the U.S. Government was in debt for well over two billion dollars.

Some taxes were imposed during the Civil War, however. Excise taxes were levied not only on a variety of articles, but also on businesses and corporations, professionals such as doctors and lawyers, railroads, savings banks, *ad infinitum.* But the fact remains that the two-billions of dollars owed to Jay Cooke of Philadelphia had virtually financed the war. Cooke received U.S. Bonds as security for his loans, and he resold those bonds to small investors, whose patriotism led them to see the investments as a way of helping the country as well as themselves. Cooke, of course, profited "considerably," since "the government allowed him a commission of one half of one percent on all the sales he made up to ten million dollars, and three-eighths of one per cent on all sales above that amount."[6]

On September 8, 1873, Cooke's banking house announced that it was bankrupt. The firm had been a "heavy investor in the securities of the Northern Pacific Railroad and had resold millions of dollars' worth of bonds to small investors by the same method it had employed during the war for marketing the government issues."[7] By the end of September, 1873, the New York Stock Exchange closed its doors, and the country collapsed into a depression that would last for five years. Manufacturing and trade fell to about two-thirds of their former level.

To some extent, at least, the collapse was natural, the economy having expanded *un*naturally under wartime conditions. Overproduction in agriculture, particularly wheat, had characterized the periods, as had over-expansion of the spider-web network of railroad tracks and overindulgence in speculation. Certainly Jay Cooke alone did not cause the Panic of 1873. Nor was it Treasury Secretary Chase's fault. The key seems to have been the opening of the great

Far West. Asa Whitney's dream was fulfilled on May 10, 1869, at Promontory Point, Utah, when Leland Stanford drove the golden spike that connected the entire nation, east and west, by way of the Union Pacific and Central Pacific Railroads. Opening up the West led to growing more agricultural staples than could be sold anywhere, here or abroad, at large profits. With farmers thus hit in their wallets, their purchasing power declined, which loss was felt by manufacturers and handlers of general merchandise. Industry ground to a halt, which was, of course, a signal for railroad construction to cease.[8] The pattern sounds familiar.

The effects of the "overexpansion" on coal prices prior to the Panic of 1873 were significant. The year 1872 had been unprecedented for coal production, particularly in the Schuylkill region, since there had been no strikes for the first time in five years. With an overproduction of more than 25 percent over the best previous year, prices nosedived to the bankruptcy level. Half the collieries in Schuylkill County shut down by August, 1872, "rather than continue to lose money."[9] Things were so bad that Franklin B. Gowen, President of the Philadelphia and Reading Railroad, called a meeting of anthracite executives: Asa Packer of the LV, Thomas Dickson of the D&H, George Hoyt of the Pennsylvania Coal Company, and Samuel Sloan of the DL&W. There was also, as they say, "a new boy in town" — the CRRNJ, new lessee of the Lehigh Coal and Navigation Company's property. The New York *Tribune* reported on January 17, 1873, that "the coal men agreed to maintain prices for 1873 at an average of five dollars a ton wholesale in New York."[10]

The fact that a group of railroads appeared almost overnight in a position to regulate prices among themselves is in itself significant. They were well aware that they were carrying the bulk of the anthracite from Pennsylvania to tidewater, and that they were in the driver's seat of the coal industry on that account. Still, the LV Railroad's Morris Canal was not yet to be abandoned. As Canal Company President Randolph had said in his 1871 Annual Report to the Stockholders, the canal was in better condition than it had ever been. The terms of the lease of 1871 required the lessees to keep it that way. Besides, the Easton and Amboy Railroad, the LV's New

Jersey "branch," was not completed yet, and both the CRRNJ and DL&W Railroads continued to expand.

At Paterson, where the DL&W's Boonton branch had just been built, parallel with the canal in places, a new bridge was found to be necessary at Barclay Street. Paterson's aldermen had wanted a swing bridge at that place for more than 15 years. Paterson maps for 1835 show only one bridge over the canal, at Mill and Clay Streets. Later, bridges at Van Winkle and Pine Streets were added.

Before the 1874 boating season began, the Canal Company had drawn off the water that had been left in the canal all winter as protection against frost damage to the banks. Paterson, ever suspicious of the Canal Company, ever vigilant for nefarious deeds, suspected that the men who were at work at what looked like cleaning and dredging of the "Morris Ocean" were really engaged in a secret enlargement of the canal. In this way, the Canal Company might succeed in deepening the canal so that the bridges appeared to be raised![11]

What sounds ludicrous on the surface was really a source of concern wherever heavy traffic needed to cross a canal bridge. The higher the bridge over the canal water, the steeper the grade of the road leading up onto the bridge. Such grade crossings were becoming more and more dangerous as traffic increased everywhere along the line of the canal.

Nor was the idea of the Canal Company raising the bridges over the canal as ludicrous as it sounds. On March 14, 1879, the State Legislature passed an act on just that subject. It read, in part:

> *whereas the said company have, from time to time, raised their bridges across said canal, and in one thousand eight hundred and seventy-three adopted an ordinance providing that all bridges thereafter constructed across their canal should be so built that the lowest part of any bridge should be ten feet from the water; and whereas, in pursuance of said ordinance many of their bridges have been since raised, and the company have required the boards of chosen freeholders of the several Counties intersected by said canal also to conform to said ordinance.*

All future bridges to be built by the Canal Company were to conform to their own ordinance. If the Company did raise any bridges for more headway, the approaches to such bridges were to be built up at Canal Company expense, so that the grade did not exceed one foot in ten, and the approaches were to be at least one rod wide.

Of course a "secret" enlargement was not underway, but the idea is not unthinkable. In fact, the Canal Company had announced in its 1876 Annual Report to the Stockholders that the plans formulated in 1866 to further enlarge the canal for 100-ton boats would be followed out under the LV Railroad's management.

More and more Canal Company customers were beginning to "think railroad" to facilitate their shipments. Cooper-Hewitt at Ringwood gave serious thought to having the Erie Railroad take their Ringwood ore from Sloatsburg, New York, to Paterson, where it could then be loaded onto canal boats bound for Easton. It would be easier, after all, for them to transport ore north to Sloatsburg than transport it south all the way down to the Pompton Feeder at Mountain View.[12]

Perhaps it was the threat of the railroads, or perhaps the great uncertainties within the coal industry itself, or both, that caused fluctuations in the canalboat-building business. In 1872, before the "Panic," the C.S. Van Lieuw boat yard at Washington had 60 to 70 employees busy turning out no less than 28 boats that were to be ready by May 1, in time for a good part of the 1872 boating season.[13] Three years later, in 1875, the boat yard was closed; all hands were discharged, except for three men who presumably were kept on to do repair work. Canal Company employees who were kept on their jobs had their wages cut. Mechanics were cut from $2.50 a day to $1.50; laborers, from $1.30 to $1.10.[14]

One source reported that there were 350 boats on the canal in 1880, averaging 70 tons each. By 1890, the number had decreased to 314, but the average tonnage had increased to 77 tons. The total tonnage hauled on the canal in 1880 was 462,636 tons, between one-fourth and one-fifth as much as the Delaware and Raritan Canal. By 1890, it was 503,486 tons, a considerable increase. The Company in that year made net earnings of $33,606.56. The average canal boat was worth about $300.[15]

The following year, in 1891, the Canal Company ordered 40 new boats from the Phillipsburg yard in addition to "the 25 ordered a short time before." The Washington (N.J.) *Star* was moved to comment: "This does not indicate that the canal will be closed very soon."[16] By 1894, the Canal Company owned 317 boats, and was ordering ten new boats from the Phillipsburg yard.[17] Despite the orders for new boats, the simple fact is that the canal, its boats, its appurtenances, its fixtures — all were deteriorating rapidly, partly from just plain old age, and partly from lack of funds.

Bad weather, too, took its toll. The winter of 1874-5 had been "almost unprecedented" during the Canal Company's history "for intense severity, making ice in some places of more than three feet in thickness and freezing the banks so deeply as to prevent us as yet [April, 1875] from resuming navigation, and at least ten days of mild weather will be required before it will be safe to fill our levels preparatory to resuming business."[18]

Part of the Little Falls Aqueduct fell on the night of April 1, 1876, and repairs called for a "temporary conduit" for the canal water to flow through while work proceeded on the arch. The conduit consisted of "a strong wooden trunk, bridge-like in appearance, and passing directly through the waterway of the canal . . . supported on each side of the river, by the abutments of the aqueduct, in such a manner, that the passage-way for boats will remain in perfect safety, even though the whole arch should fall."[19]

Within five months, the aqueduct over the Singac Brook also broke down. This "timber structure about 80 feet in length reaching over a height of from 20 to 30 feet and supported by one pier in the centre," and thus "divided into two spans of 40 feet each," had been the source of "some trouble" by leaking. It had been repaired from time to time, but finally, on Saturday, September 2, 1876, at about 3 o'clock in the afternoon, "a few of the timbers in the bottom gave way and caused a considerable leak." The timbers were replaced temporarily, thus averting for the moment the danger of a large break. By 7 o'clock that night, however, almost the whole bottom of the one span dropped out, draining dry the entire 17.5-mile level between Bloomfield and Lincoln Park. A few boats, being over uneven places in the bottom of the canal when the water drained, settled onto those places and broke apart. One boat had been half-on, half-off, the Little Falls Aqueduct at the time, loaded with coal. Loss of time was serious to boatmen since they carried "contract freightage," and to be even a day behind in a contract was to forfeit so much for delivery.[20]

Navigation was suspended on the Western Division for a week during 1887, when the shaft of the water wheel broke at Plane No. 6 West, Port Colden. Traffic was so backed up that when repairs were finally completed, "one hundred and eleven boats passed over the plane in one day."[21]

When the canal was being readied for the 1878 boating season, the section from Pompton Plains all the way to Jersey City was receiving a thorough bottom-cleaning. Some 300 men were employed in the work, receiving 80 cents a day in wages. At Newark, some men were said to be getting a dollar a day. (At the time, lock tenders received $30 to $40 a month, not much more than the laborers. Of course some lock tenders got their houses free.)[22] It sounds to our modern ears as if it was no longer profitable to work *for* the Canal Company, and was no longer profitable for the lessees to operate it.

Still, the Morris Canal continued to be useful to local interests and industries such as the Mountain View Brick Manufacturing Company during 1879. Steam was used to grind and mix and mold the bricks, as many as 35,000 per day. Fuel, transported by canal boat to make the steam, was *wood,* not coal![23]

The town of Bloomfield used the services of the Canal Company for delivery of 60 canal-boat loads of gravel "for road improvements" in 1881.[24] When we recall that there were, as yet, no dump trucks to cart heavy materials such as gravel, the contribution of the canal boat in such cases is more readily appreciated. Bad weather struck at Bloomfield that same winter, when thick ice enclosed more than 50 coal-laden canal boats there.[25]

In Newark, water power was used to grind flour and feed at a mill on Academy Street. Power was supplied from the top of the inclined plane, No. 12 East, where there was a 30-foot well, into which water was "let through gates opening from the canal." A turbine wheel at the bottom of the well furnished power to the mill. Waste water from the well was "carried down to the lower level through a deep sluice in the same manner as the waste water is carried from the turbine wheel

An excursion boat on the Morris Canal at Kesse's Hotel, Clifton. The spot today is marked by the intersection of Van Houten Avenue and Grove Street. The canal is now part of the Garden State Parkway (BNK).

which works the plane cars. By means of a crank in the superintendent's office [in the plane house], water in the upper level can all be drawn off through the sluice ways any time."[26]

The ever-growing need for iron ore to supply the blast furnaces that, in turn, were trying to keep up with the nation's demand for steel, meant that New Jersey's iron-mining industries were still active for the moment and needing the canal's services.

Various powder works began to move into the canal area, notably near Lake Hopatcong and on the Pompton Feeder in Mountain View. In 1871, the Atlantic Giant Powder Company located at Kenvil. The dynamite produced there was to be used to blast ores from the area's deep shaft mines. This company had originated in California.

From Ulster County, New York, came Lawrence Ferguson, who bought "eight or ten tracts of land near Mead's Basin, in Wayne Township, along the Pompton Feeder of the Morris Canal." Ferguson had had an extensive powder mill in Esopus, New York, and he proposed "to remove his entire establishments to Wayne . . . where he will have convenient water communication by canal with tidewater."[27]

The Laflin & Rand Company, too, were to build their powder works on the Pompton Feeder, in 1875. The works were known as the "Rendrock" Company. Blasting powder was shipped from Rendrock via canal boat until at least World War I, during which at least two shifts of workmen labored daily to turn out the material of war. Powder was shipped at a rate of $2 per ton-mile, in contrast to anthracite, which was still being transported at one cent per ton-mile. Powder that was used to blast the dangerous rocks in the East River at Hell Gate was made at Laflin & Rand's works.[28]

Heaven's gates, in a manner of speaking, were being serviced by the Morris Canal, too, as Sunday School excursions on the waterway became popular during the 1870s. President Randolph inaugurated a program of annual excursions for the Sunday School scholars of Bloomfield's First Presbyterian Church. Boatloads of such "scholars" were conveyed in Canal Company boats to Paterson Falls, where they had a day's outing and picnic.[29] One Sunday School excursion from Bloomfield to the Paterson Falls required three boats to carry an estimated 250 to 300 persons.[30]

Aaron H. Vough, Canal Superintendent on the Western Division, personally organized such excursions from Washington to Stewart's Grove, in Hackettstown. "One of the tricks he used to play was to let the scow loose near the foot of the planes, resulting in considerable of a splash when the boat struck water at the bottom of the incline." The Canal Company provided the excursionists with work scows, the wide-open interior of the scow being ideally suited to carrying large numbers of people on outings. According to one reporter for the Washington (N.J.) *Star,* the free transportation had characterized both the excursions *and* the Canal Company during the 1870s, but that "after a time the canal company got hardboiled and refused to donate their boats for this purpose any longer."[31] With all the financial problems that were plaguing the Canal Company in the 1870s and beyond, it is scarcely surprising to learn that belt-tightening economies had an effect on free excursions.

The big scourge of the 1870s and 1880s was to be the renewed attempt of the CRRNJ to bridge the Morris Canal Basin at Jersey City to connect its depot with the heart of the city. In this attempt, the CRRNJ was joined by Jersey City, itself, which as we saw earlier had been totally frustrated in its attempts to expand southward because of the Morris Canal Basin's presence.

Jersey City had grown, however, in "an act of consolidation" during the late 1870s. It had taken "all the suburban cities" into its embrace. Among those suburbs was "that part of Lafayette which was, and is yet, known as Communipaw," that "narrow neck of land, severed from the lower section of Jersey City by the Morris Canal basin."[32] The Canal Company was required by its charter "to keep open a passageway, from its basin to the river, 150 feet in width, so as to permit of the passage of river craft. This waterway is much used by the Matthiessen & Weicher Sugar Company, whose refineries stand on the basin at the foot of Washington Street, Jersey City, and by the steel-works, which are situated at points further westward."

Communipaw, *The New York Times* observed, was the home of a few fishermen and of many market gardeners. Its chief claim to distinction lay in the fact that it was the eastern terminus of the New Jersey Central Railroad. The paper continued:[33]

A few years ago the railroad company began to feel lonesome, and longed for a connection with the main section of Jersey City. Nothing prevented it but a little water gap owned by the Morris Canal. To fill this in was impracticable for two reasons. First, because it was a work not easy of accomplishment, and in the second place, the clause in the Morris Canal charter commanding that a navigable waterway be kept open prevented it. To bridge over 'the gap,' as the basin is called, was the only practicable scheme that suggested itself. The advantages of a bridge over the Morris Canal were at once pointed out to the residents of Jersey City. It followed that Jersey City began to clamor for a bridge over the gap from the foot of Washington Street. They have made in previous years determined efforts to secure legislation to authorize its construction, but they have been as often defeated. This season the demand that a bridge be constructed at Washington Street was renewed with such vigor that it became an issue not only in local, but in State, politics. When, at the Fall elections, the people of Jersey City nominated their Assemblymen, they required pledges of them that they would support the Washington Street bridge project, and the fact that Mr. Speaker Jackson voted last year for the bill was one of the main features of his canvass for the Speakership of the present Assembly.

But though the clamor for the bridge has been especially loud during the last two or three years, the scheme itself cannot be said to be a new one. It is said that the bridging of the gap was in contemplation when Washington Street was first laid out. When the authorities graded the street they neglected to condemn the Morris Canal's property for a right of way over the basin. They thought they could take that without compensation to anybody. They discovered their error when they undertook to proceed to the erection of the contemplated bridge. Then the Morris Canal Company applied for an injunction, claiming that the City had made no tender of compensation to the company. It was granted, and on agreement the injunction was made perpetual.

Most of the legislation asked for by the agitators of the project has been designed to devise a method of evading the injunction. Against these bills the Lehigh Valley Company, urging that the construction of the bridge might imperil the lease by the State to the Morris Canal, which is worth $25,000 per year to the State, has been the only visible objector. It is said, however, that the railroad company have pooled their issues with the Pennsylvania Railroad, which, naturally enough, opposes anything which the competing corporations regard as of advantage to their business. This year it is said that the Central Railroad is not alone in its struggle. The Baltimore and Ohio people are declared to have made preparations to make their Eastern terminus at the Communipaw depot, and the struggle for the little span of iron over the gap is really a hand-to-hand encounter between two of the great trunk lines of the country.

Soon after the organization of the present Legislature, Mr. Dusenberry, of Hudson, introduced a bill by which, it is hoped, the injunction may be set aside and the work begun.

Mr. Dusenberry's bill, alluded to above, was "designed to place the city [Jersey City] in a position to condemn the Canal Company's franchises and secure a right of way over the waterway." However, it appeared to be not Jersey City that was anxious to have the bill passed, but the CRRNJ and the Baltimore and Ohio (B&O) Railroad. The bill, if passed, would have compelled Jersey City to issue bonds to pay for the Washington Street extension over the Little Basin, an expense "the already over-burdened taxpayers" of that city were unwilling to bear.[34]

The real parties to the controversy, then, appeared to be the CRRNJ and B&O Railroad on the one side, against the Canal

Company, the LV Railroad, and the Matthiessen & Weicher Sugar Refinery on the other. How, or why, it was that ex-Senator Taylor of Newark, ex-Attorney General Gilchrist, Senator Magie of Union County, and Judge Robbins of Monmouth County were interested parties in the conflict is unclear, but they all participated in the lively debates in Trenton.[35]

Out of this controversy between and among the railroads, a "Jersey City Ferry War" erupted. George Waterman and John Mahon, two men who ferried passengers across the Jersey City Water Gap at the canal basin, had testified that a bridge at Washington Street would seriously interfere with navigation. Their affadavits " provoked the ire of the Central Railroad officers," who thereupon started up a rival ferry, with "two fine new boats." The " 'old company' at once put on two new boats," and the war was on. The new line ferried passengers free of charge — a tactical strategy that has won many a war before battle was joined.[36]

The old ferry was described by contemporaries as "primitive," with "three ill-looking flat-bottom and leaky row-boats, manned each with a single oarsman." Primitive as it was, however, it was the only means of transporting passengers to the Central Railroad's depot, and to the trains of the Long Branch Railroad. "Some of the best-known residents of Jersey City" were said to have used the primitive conveyance, frequently having to stand up in the boats, which, although designed to carry only nine passengers plus the oarsman, frequently carried as many as twenty customers.[37] An old seafarer named Christopher Moss has been said to have started ferry service, in 1863, when the Central Railroad was building its "islanded" terminal to the south of the canal basin. Moss ferried local Jersey City laborers who were engaged in the railroad's construction.[38] In 1891, the CRRNJ proposed to run a small steam ferry across the "Gap," but in 1928, the "penny" ferry was *still* operating, after sixty years.[39]

The Central Railroad's presence in Communipaw Bay was a very mixed blessing, indeed, to folks besides the Canal Company. Canal boatmen used the basin wharves for wintering their boats.[40] Later, the CRRNJ was found by the Jersey City *Evening Journal* to be "extorting wharfage fees from canal captains" who were not really under any obligation to the railroad company. When the captains learned from the newspapers that they had been bilked, they sent an appointed delegation of spokesmen to Jersey City's Mayor Orestes Cleveland "to thank him for the manner in which he has delved into the matter and exposed the illegality of the doings." The CRRNJ officials, "frightened by the exposure," tried to get back the incriminating receipts that might have been used as "evidence of the shady transactions."[41] The waters and wharves involved were Jersey City's, not the CRRNJ's or Canal Company's. The incident is mentioned here to illustrate the foothold and influence of the CRRNJ at Jersey City before the turn of the century.

On strictly legal grounds, nothing might be permitted to interfere with navigation on the Morris Canal. So long as the Canal Company's charter remained in effect, the canal must be kept free to operate. Still, the following year, 1880, the "Jersey City and Communipaw Railroad Company" applied to the Supreme Court to appoint commissioners to condemn lands necessary for the construction of a railroad from the steel works at the foot of Warren Street to the Central Railroad Depot in Communipaw. To build that railroad, it would be necessary to bridge "the gap," as it had become known to one and all. Ex-Governor Bedle represented the railroad company; ex-Attorney General Gilchrist and Thomas N. McCarter spoke on behalf of the Canal Company, whose lands it was proposed to condemn. Representing the State was Attorney General Stockton. With all those legal guns pointed his way, Judge Knapp of the Supreme Court "decided to appoint the commissioners and also to certify the case to the Supreme Court for adjudication."[42]

If the CRRNJ could not bridge the "Hudson Water Gap," it would try to tunnel *under* it. A 2150-foot-long tunnel was proposed to begin at Morris Street, Jersey City, slope downward for 720 feet, then go 260 feet under the canal basin, then begin to ascend for 1070 feet to the Central's depot at Johnson Avenue.[43] It was never dug.

The Canal Company had always had difficulty proving title to its lands under water. The LV Railroad tried to remedy that fact for its own purposes. The State Legislature had established a Board of Commissioners to implement the General Riparian Act of 1869, passed at Governor Randolph's urging. The Commission had the power to lease and sell the State's interest in riparian lands. In 1888, the LV Railroad applied to that Commission for a grant of the State's reserved right in the canal's Big Basin, and on July 12, 1889, the grant was made, in fee simple. To New Jersey, the railroad paid $357,142 for that grant, plus "the further sum of $48,000, being $50 per front foot fixed under the statute by the Commissioners for the reversionary interest of the State, the total payment being $405,142." To the Canal Company, the railroad paid $500,000, which was immediately applied in part-payment of the Company's debts, and the Canal Company deeded its interest in the Big Basin to the LV Railroad. On that terminal tract, "and on the mud flat to the west of it, for which the LVRR paid the State $250,000, the LVRR built its tidewater terminal."[44] These lands in the "Big Basin," granted by the State to the Canal Company and conveyed by them to the Lehigh Valley, were *not* supposed to revert to the State of New Jersey upon the expiration of the Canal Company's charter, or upon the abandonment of navigation. We will see that the reverse was true.

Obviously, the Lehigh Valley also had a permanent foothold in the harbor area of Jersey City before the turn of the century. That it had some influence or standing with the courts can be seen in the case of the Lafayette Railroad Company's attempted encroachment on canal property in 1891.

When the Lafayette Railroad Company, a CRRNJ "branch," "built a track on land claimed by the Morris Canal Company south of the Griffing Iron Works" in 1891, the Lehigh Valley "ordered the removal of the track and threatened to have it torn up." The Lafayette Company sought an injunction to stop the Canal Company from interfering with its tracks, and attempted to show that it owned the land on which those tracks had been placed. Canal Company counsel successfully demonstrated that his clients owned the land, having occupied it for forty years. This injunction sought by the railroad company was denied. The newspaper article's lead emphatically and, it appears, somewhat triumphantly announced: "The Central Beaten."[45]

The LV Railroad maintained its foothold in 1893, too, when the CRRNJ attempted to put a new railroad bridge across the canal at the foot of Jersey Avenue. The Jersey City Street and Water Board

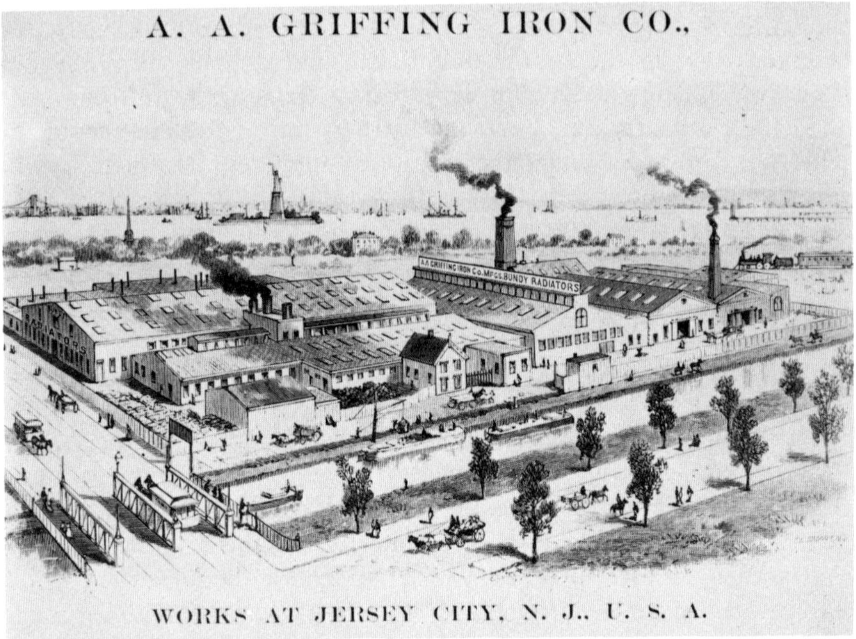

A. A. Griffing's Iron Works on the Morris Canal in Jersey City at the turn of the century. Note the Brooklyn Bridge towers at the extreme left background, the Statue of Liberty, the horse-drawn trolley cars, the passenger train at the extreme right center, and the canal boat unloading at Griffing's (ETF).

had given the CRRNJ permission to remove the old wooden bridge with its approaches too steep for heavy loads moving to and from the railroad yards to negotiate. The new bridge was to be level with Jersey Avenue, to accommodate the CRRNJ's traffic, but with "a means for canal boats to pass."

The new CRRNJ bridge was partially built when the LV Railroad decided not "to grant its rival any advantage," and sent out "a detachment of men to remove the bridge." An LV Railroad locomotive was dispatched, and chains were attached to the bridge pilings as well as to the locomotive. "After all the beams had been scattered, attention was turned to the flatboat which the Central Railroad people brought to the bridge with piling and beams for the bridge. The boat was frozen in the ice, but after a chain was fastened to the boat, the locomotive had little trouble in breaking the icy grip and towing the boat away from the scene."[46]

Of course, canal operations continued while this railroad rivalry was blossoming into full-scale war. But from here on, it is not surprising to see the hostilities break out in far less likely places than the Jersey City portion of New York Harbor.

The canal was still very useful for "excursions" and "outings," and perhaps the more so since steam excursion boats had been put on Lake Hopatcong. Boats picked up passengers from the DL&W Railroad Station at Landing, and from the CRRNJ at Nolan's Point. The "Lake Hopatcong Steamboat Company's" boats landed excursionists at Purroy's Point, Bertrand's Island, Mount Arlington, Altenbrand's Grove, and Ice House Point.[47]

Eventually — inevitably, one is tempted to say — a rival steamboat line appeared at Lake Hopatcong. The Newark *Sunday Call* gave two very interesting and informative accounts of the rivalry between the two companies but, more important, gave an insight into the water problems created in part by the joint Canal Company-Lehigh Valley Railroad Company efforts to become what the Lease of 1871 had said they might: A water-supply company. For the light they throw on future developments, both articles are included, just as they were written.

The first of the Newark *Sunday Call* accounts is a description by a staff reporter; the second is a "letter to the editor," signed with the initials "J.L.D." The reporter headed his account: "Lake Hopatcong; The Steamboat Quarrel and Low Water," dated July 26, 1891.

> *When I visited the Lake last week, I brought my rubber boots with me, expecting that I should have to use them in walking across the damp bottom of what was once beautiful Lake Hopatcong. The exaggerated reports of the rapid decrease in the water gave the idea that by the time one reached Hopatcong, the lake would have entirely disappeared. In this, visitors were happily disappointed. The total fall of the water is about three and a half feet. This is*

particularly noticeable at the southern extremity of the lake at the steamboat landing. The cause of this is the Morris Canal Company. If you ask why the water of the lake has been lowered the agents and sympathizers of the canal company will tell you the past dry season has compelled the company to draw upon the lake to an unusual extent to supply the canal. As it happens, this has not been an unusually dry season, and people who ought to know say the canal company has no use for the extra drain.

The true inwardness of the matter is believed to be the competition between the two steamboat companies which carry passengers to all parts of the lake. One is known as the Lake Hopatcong Steamboat Company; the other is the Hopatcong Steamboat Company. The former has several steam launches which take passengers from the railroad station landing, charging each passenger 50¢ for each trip. The latter company has two large and nicely-equipped steamboats, and passengers are charged only 25¢ each to be taken to their destination, while return tickets are sold for 40¢.

The former company takes passengers through the canal. The boats are small, overcrowded and uncomfortable. The regular visitors, as a rule, patronize the Hopatcong Steamboat Company. To reach this company's landing, it is necessary to walk over the bridge above the railroad tracks. It is but a minute's walk, but as the boats and their dock is not in sight, the strangers often patronize the high-priced boats. The launches which use the canal are a source of revenue to the canal; the other boats are not.

The canal company began drawing water off the lake and left the rival company's boats high and dry. The latter, not to be thwarted in this manner, got a dredger, and began the work of making a deeper channel. This has been successfully accomplished, and as long as the canal company continues drawing water off the lake the dredger will be kept busy widening the channel.

The second article, or "letter to the editor" of the Newark *Sunday Call* of July 26, 1891, was scarcely kinder to the Canal Company, and considerably more accurate in its appraisal of the situation at Lake Hopatcong.

Much has been written about the low water in this lake, but much of it is purely nonsensical, and is undoubtedly the result of jealousy and envy.

The original sale [*sic*] *to the Lehigh Valley Railroad Company was supposed to be part of a large and comprehensive system of supplying water through the canal to Newark, Jersey City, New York, etc. The Lehigh Company has put in a dam and raised the original surface of the water three feet above the ordinary level last year as is shown all around the lake on the rocks, etc. This water was to be held in reserve and used when wanted in dry times. The increased rise covered acres and acres of low land, stumps and rocks, and added very largely to the beauty of the lake, its value for shore frontage, and enlarged its area by a thousand acres, more or less, which to the riparian owners has become a vested right.*

The Hopatcong Steamboat Company built two side-wheel steamboats, the HOPATCONG, with a capacity of 150 passengers and drawing 22 inches of water, and a launch of smaller size, and to avoid the charges of lockage in the canal, the Steamboat Company employed a dredge and excavator to remove the stumps and cut a new channel one and a half miles long from near the Hopatcong railroad depot to deep water in the lake, and this channel has been used by the MUSCONETCONG carrying the passengers and freight between the depot landing transferred to and from the larger HOPATCONG at deep water during the completion of the rest of the channel.

To cripple the new line the water has been drawn down, but it has failed of its purpose, as the new boats draw less water than the propellor launches.

The opposition thus created has been the cause of lowering the water of the lake, exposing thousands of stumps and rocks, endangering navigation, and despoiling the beauty of the lake in many places by exposing stumps and rocks referred to, and in some cases endangering the health of citizens along the bayous.

To prevent all this, proceedings are about to be instituted, and Cortlandt Parker will probably be selected as the champion of the people to restrain the changes of line and surface. Nice questions in law are involved and the whole of New Jersey are interested in maintaining the great beauty of Hopatcong, regardless of a squabble between rival lines of steamboats. J.L.D.

Camden newspapers for June 12 and 26, 1875, reported that the "iron tugboat" *Hopatcong* was launched at the yard of Wood, Dialogue & Co. at Camden. The *Hopatcong* had been built for the Ogden Iron Co. and left Camden on June 26, 1875, in charge of Patrick McCarthy, bound for Bristol and the Delaware Division Canal. At Easton she crossed the Delaware River again and, by way of the Morris Canal's locks and planes, arrived at Lake Hopatcong. She was used to tow "canal barges on the lake from the railroad dock at Nolan's Point to the canal." The whole controversy of the rival steamboat companies can most likely be explained by reference to the fact that the CRRNJ had completed a connection with the Ogden Mine Railroad, which it had absorbed in 1881! Given the fact that the Canal Company immediately lost half its ore tonnage from the Ogden Mine to the CRRNJ by that maneuver, its hostility to the CRRNJ-affiliated HOPATCONG is understandable.

Already enriched from having pirated the Morris Canal's Ogden Mine ore tonnage, the CRRNJ had also tapped the zinc ore traffic from Franklin, New Jersey. In Newark, too, where the New Jersey Zinc Company had its processing plant, the Central built the "Manufacturer's Railroad," a spur from its main track to service the zinc company.[48]

The same railroad was, after all, the rival company whose executive officer had uttered the now famous threat during the Battle of the Basin at Jersey City: "Grass shall grow upon the towpath of the Morris Canal!"[49] For the Canal Company, there would understandably be a certain amount of poetic justice in seeing the Central Railroad, in any of its manifestations, stuck in the mud!

The Canal Company's lessees would indeed be hearing from the Hon. Cortlandt Parker, Esq. "Nice questions of law" were in truth involved. They probably could — and should — be asked again today.

At the time, however, in 1891, the Lehigh Valley was not totally concerned about whether or not steamboats could continue to operate on Lake Hopatcong. It was not very concerned about canal boats continuing to operate on the Morris Canal. In fact, the Lehigh wanted very much to close navigation "for all time" on its canal. In February, 1888, the lessees had "asked the Legislature to sanction the abandonment of the great ditch as a water course. It was claimed in the Senate that the company lost $84,000 last year [1887] by operating the canal and the company now desires to give up navigating it." According to one reporter:[50]

> *The story that the Lehigh Company wants to use the canal for a railroad bed is denied by the men who know and direct the moves of the company, and they say that no matter what will ultimately be made of the bed of the canal, the desire is now merely to stop using the boats. The company's counsel, Thomas N. McCarter, said to a CALL reporter last evening [February 10, 1888] that the Morris Canal is more expensive than other canals on account of the numerous inclined planes which require frequent renewals of expensive machinery and cables and constant care of engineers and repairers. The company can carry coal more cheaply on the railroad than on the canal, and now has the railroad facilities needed. Nothing can be gained by laying rails in the canal bed owing to the circuitous route it takes and various levels abruptly encountered, but were the canal abandoned a great deal of valuable land would be made available for other purposes in every city and town through which the canal passed as well as in the open country all across the State. In Jersey City it is proposed to use a short section of the canal as a roadbed for the Lehigh Valley Railroad in reaching the*

new depot which the company proposes to build adjoining the Central Railroad property . . . .

Despite McCarter's denial that the Lehigh lessees wanted to build a railroad in the canal bed (except for a terminal in Jersey City), the Lehigh Company met with Pennsylvania Railroad representatives at Taylor's Hotel in Jersey City on March 30, 1888. How the secret meeting became known to the press is still a secret, but the Newark *Daily Advertiser* carried the story the next day. The secret talks centered on a proposed combination of the two railroads in building a railroad to Orange, using the canal bed part of the way, presumably in Jersey City, tunneling part of the way, presumably through Bergen Ridge, and elevating the road over the canal through Newark.[51]

This type of rumor was not new. Earlier, a similar report had left Peter Melick, canal superintendent at Newark, to deny rumors that the canal would be abandoned during the winter of 1882. Since the lessees had announced plans to repair locks and planes at the end of the 1882 boating season, how could anyone believe that they intended to abandon?[52] In fact, the "Deep Lock," No. 17 East, at Newark, was so badly deteriorated the previous year that the lessees had had it completely rebuilt.[53] That did not sound as if abandonment were close at hand.

It was becoming increasingly clear, however, despite the denials, that the LV Railroad wanted very much to rid itself of the burden of a canal that was becoming a serious threat to its financial well-being. Particularly burdensome were the taxes levied on the company by New Jersey. Year after year, President Randolph protested those taxes, especially on the Big Basin property that no longer belonged to the Morris Canal Company. Some of that land had been conveyed by the Canal Company to the "Jersey City Terminal Railway Company (now Lehigh Valley Terminal Railway Company), by deed bearing date June 25, 1890." In a formal protest, known as the "Tax Appeal and Protest of 1896," Randolph complained that the State Board of Assessors had evaluated the basin property "greatly in excess of the true taxable value of said property."[54] The taxes were not lifted, however eloquent or just the appeal.

Because the Canal Company's charter forbid abandonment until the charter's expiration, the lessees were forced to support a canal that had become a detriment. Traffic on the waterway hardly justified making even necessary repairs, let alone improvements.

Serious breaches in the canal during the 1880s compounded an already bad situation. In June, 1884, the canal was breached at Seventh Avenue and at Lock No. 18 East in Newark, emptying the canal and idling several manufacturing plants. The Canal Company was liable for damages caused by its failure to supply water according to its contracts with "The Hedenburg Works, the Domestic Sewing Machine factory, T.P. Howell & Co., tannery, the flour mill just above the inclined plane, and the pottery, the licorice establishment, the trunk factory, and Crabb's factory, all on or near Bloomfield Avenue," and all of which drew "water from the canal under contract." With the canal level empty, the factories had had to run on "steam power alone, which adds largely to their expense, for water power is cheaper."[55]

Another serious breach occurred at Mountain View in September, 1888, near the New York and Greenwood Lake Railroad line. The resulting flood of canal water covered "land for nearly two miles, damaging crops, filling the cellars of the farmers, and carrying away the supports of the railroad bridge which crosses the canal within a few feet of the spot where the break occurred." The train had had to stop at Pompton Junction to discharge passengers, who then had to walk "across the bridge to Mountain View Station." Men sent to the site of the breach from Jersey City were both railroad and canal workers. They had arrived at night, too late to get much done in the way of repairs until the morning of September 19, 1888.[56]

But the Morris Canal was losing water from a cause other than breached banks. A hint of that cause appeared in the story of the rival steamboat companies at Lake Hopatcong. A further hint was contained in McCarter's denial that the LV Railroad sought to use the canal bed for a railroad. At the end of his statement, he remarked that "the upper part of the canal" would "probably be used in connection with some water scheme, and it is rumored that arrangements will be made to dispose of the overflow of Hopatcong and the Rockaway River so far as controlled by the company."[57]

And there it is: a "water scheme" to dispose of surplus water, just as the Lease of 1871 had authorized the Canal Company and/or its lessee, the Lehigh Valley, to do, all very legally. A brief flashback to 1873 will be helpful.

A few months before the Panic of 1873 struck, the Canal Company's lessees obtained a charter supplement (April 2, 1873), authorizing them to construct, operate, and maintain a "branch canal with all necessary locks and planes, from a point in the main canal at or near Centreville [Clifton], in Passaic County, to the Passaic River at Dundee Lake, above the dam of the Dundee Water Power and Land Company."

The Dundee Manufacturing Company, as noted earlier, had been a project of Peter M. Ryerson and others, to develop industry along the Passaic River between Clifton and Passaic. The charter of 1832 had been amended on March 20, 1872, long after the bankrupted Ryerson had sold his interests in the company *and* in the Passaic River waters to the SUM.

The Canal Company's — or rather, the LV Railroad's — interest in the Dundee Manufacturing Company is an indication of how far afield the Lehigh Valley's 1871 lease of the Morris Canal had already taken it. And this was but one of many water schemes that would be developed in the coming years.

REFERENCES

1. *Biographical and Genealogical History of Morris County, New Jersey* (N.Y.: 1899), pp. 731-5.
2. Cornelius C. Vermeule, Jr., *Final Report of Consulting and Directing Engineer, June 29, 1929* (E. Orange, N.J.: 1929), p. 63.
3. Wheaton J. Lane, *From Indian Trial to Iron Horse: Travel and Transportation in New Jersey, 1620-1860* (New Brunswick, N.J.: 1939), pp. 245-6.
4. E.B. Thomas, "The Morris Canal, An Economic Waste," Memorial of the Lehigh Valley Railroad Company to the Senate and General Assembly of the State of New Jersey, 1911, p. 8.
5. John D. Hicks, *The Federal Union* (N.Y.: 1937), pp. 664-5.
6. *Ibid.,* p. 667.
7. Louis M. Hacker and Benjamin B. Kendrick, *The United States Since 1865* (N.Y.: 1939), p. 55.
8. *Ibid.*
9. Marvin W. Schlegel, *Ruler of the Reading* (Harrisburg, Pa.: 1947), p. 37.
10. *Ibid.,* p. 44.
11. Paterson *Daily Press,* March 3, 1874.
12. Abram S. Hewitt Papers, Passaic County Historical Society Museum, Lambert Castle, Paterson, N.J., letter from Edward Cooper to Abram S. Hewitt, May 18, 1874.
13. Washington (N.J.) *Star,* January 9, 1947.
14. *Ibid.,* July 8, 1941.
15. *Ibid.,* April 3, 1947.
16. *Ibid.,* January 23, 1941.
17. *Ibid.,* January 27, 1944.
18. Morris Canal and Banking Company *Minutes,* April 5, 1875 (hereinafter cited as MC&BCo. *Minutes*).
19. Paterson *Daily Press,* April 20, 1876.
20. *Ibid.,* September 4, 1876.
21. Washington (N.J.) *Star,* January 1, 1947.
22. Newark *Sunday Call,* March 3, 1878.
23. W. Woodruff Clayton and William Nelson, *History of Bergen and Passaic Counties, New Jersey* (Philadelphia, 1882), p. 558.
24. Newark *Daily Advertiser,* July 7, 1881.
25. Jersey City *Evening Journal,* January 6, 1881.
26. Newark *Sunday Call,* April 16, 1882.
27. Paterson *Daily Press,* February 1, 1873.
28. *Ibid.,* October 9, 1885.
29. Bloomfield *Record,* April 23, 1897.
30. Bloomfield *Saturday Gazette,* July 17, 1873.
31. Washington (N.J.) *Star,* May 22, 1941.
32. *The New York Times,* January 20, 1879.
33. *Ibid.,* February 3, 1879.

34. *Ibid.*
35. *Ibid.,* February 4, 1879.
36. *Ibid.,* March 18, 1879.
37. *Ibid.,* September 7, 1880.
38. Personal communication from Edward T. Francis to the author, undated.
39. New York *Times,* September 11, 1928.
40. Jersey City *Evening Journal,* November 3, 1883.
41. *Ibid.,* May 16, 1891.
42. New York *Times,* July 4, 1880.
43. *Ibid.,* February 25, 1887.
44. Thomas, *op. cit.,* p. 11.
45. Jersey City *Evening Journal,* May 18, 1891.
46. *Ibid.,* January 5, 1893.
47. *Official Railway Guide,* June, 1888.
48. Newark *Sentinel of Freedom,* May 17, 1881.
49. MC&BCo. *Minutes,* April 1, 1867.
50. Newark *Sunday Call,* February 11, 1888.
51. Newark *Daily Advertiser,* March 31, 1888.
52. Orange (N.J.) *Journal,* October 27, 1881.
53. MC&BCo. *Minutes,* April 4, 1881.
54. George T. Werts, *et al., Report of the Morris Canal Investigating Committee on the Abandonment of Navigation on the Morris Canal* (Trenton, N.J.: 1903), pp. 115, 117.
55. Newark *Sunday Call,* June 1, 1884.
56. Newark *Evening News,* September 19, 1888.
57. Newark *Sunday Call,* February 11, 1888.

20

Water, Water, Everywhere

BY a charter of 1791 and a charter of 1824, two private corporations, the SUM and the Morris Canal Company, had been given absolute control over the Passaic River watershed. A look at a map of New Jersey confirms the fact that virtually the whole of the northeastern segment of the state is involved in that watershed. In 1871, with the Lehigh Valley (LV) Railroad Company's lease of the Morris Canal, a Pennsylvania corporation joined the very localized SUM in controlling this vital part of the State's water supply system.

It has been said — and rightfully so — that "without the cooperation of these companies, the success of any sizable scheme to supply water from the Passaic watershed above Little Falls in the early days would have been impossible." The story of their "cooperation" is perhaps the most incredible chapter of the Morris Canal's history. The SUM, the Canal Company, and the LV Railroad "combined in plans for providing a water supply for the various municipalities convenient to the Passaic Valley," in 1881.[1]

In 1882, the New Jersey State Legislature considered and passed a bill to authorize "the formation of private corporations for the purpose of supplying municipalities with water for domestic purposes, upon the mere filing of a certificate setting forth certain general particulars, and providing for the acquirement by condemnation under the State's right of eminent domain of all water courses not of use for any specified purpose." Governor Ludlow vetoed the bill since the right of eminent domain would have been exercised to condemn public rights for the benefit of private individuals. Out of this emerged the first State Water Supply Commission.

In 1883, an attempt was made to have a law passed in the New York State Legislature to allow private corporations to divert water from the Ramapo River, before it entered the Passaic watershed, for the purpose of supplying New York. So much opposition was voiced that the bill did not pass. It was said, however, that the "scheme" had been "fathered by the same interests" who had tried to use the Passaic watershed in 1882.

The State's first Water Supply Commission had compiled an ambitious report "outlining a plan for a joint water supply from the Pequannock River to ten cities and towns, including Jersey City and Newark," and a future plan to extend the supply system to other sections of northeastern New Jersey.[2] The plan was not acted upon, and the growing cities, desperately in need, still did have potable water.

The North Jersey Water Company appeared in 1886, a "private" company which offered to supply Jersey City with water from the Pequannock River. This "private" company was apparently not affiliated with the SUM — LV Railroad faction, for the railroad countered with an offer to supply Jersey City with water at cheaper prices.

In 1887 and 1888, other proposals were put forth to use Passaic River water to supply New York City from a pumping station at Little Falls, the water to be passed under the Hudson through pipes. If that sounds like a novel idea, the reader is referred to Chapter 7, in which appear John L. Sullivan's ideas on the same subject. In 1888, the idea was endorsed "by eminent engineers and New York and New Jersey lawyers and the Medical Society of the County of New York." It was even approved by the SUM, the LV Railroad, the Dundee Water Power and Land Company, Passaic Water Company of Paterson, the Aquackanonk Water Company, the West Milford Water and Storage Company, and the Montclair Water Company.[3]

As far as the Morris Canal's rights in those waters were concerned, the Legislature was told that the State "had authorized the company to discontinue navigation and to sell the water and sources of water supply controlled by it for more important public purposes" such as "supplying cities and towns with pure and wholesome water."[4] Governor Green had indeed signed a bill authorizing the abandonment of the navigation in October, 1888.[5] The scheme *was* started, but never completed, and the cities remained without the water supplies they needed. They were forced to take water out of the Passaic River below Belleville, where water was increasingly polluted. In some cases, as in Newark and Belleville, water was supplied from the Morris Canal, but that water was scarcely less polluted than the Passaic River by the time it got as far as Belleville.[6]

When the scheme to supply New York with water fell through, the syndicated interests turned to Jersey City and Newark with offers to supply those cities with water. Newark insisted on water from the Pequannock River, and in 1889, the syndicate contracted to sell Newark its water from that river, and the North Jersey Water Company sold "its riparian rights to the contracting party." Jersey City "contracted with other private interests for a supply from the Rockaway River" in about 1900. Until its reservoir at Boonton was completed in 1904, Jersey City was supplied with water pumped from Little Falls by the East Jersey Water Company.[7]

What seems like a diversionary exercise is explained by the fact that *all* of these "private" companies *were* the Lehigh Valley Railroad, operating under a variety of names, the most prominent being the East Jersey Water Company.

The East Jersey Water Company had been formed on August 1, 1889, "for the storage, sale and delivery of water and the construction and maintenance of the necessary reservoirs, pipe lines and other works.[8] The Lehigh Valley, as lessees of the Morris Canal, owned the water rights in several of the reservoirs of northeastern New Jersey. These rights it assigned to the East Jersey Water Company, the actual construction part of its water-supply operations. In developing the Pequannock Watershed to supply Newark with water, the East Jersey Company built Oak Ridge, Clinton, Macopin, and Canistear Reservoirs.[9]

Jersey City had been using water from the Morris Canal, from the Passaic River, and from artesian wells. As early as 1844, John D. Ward, "an engineer," had urged the city council to buy or otherwise acquire a water supply.[10] Ward, the reader will recall, became a Canal Company director about 1850, its vice president for three years (1865-1867), and a director again until 1869, when his name no longer appears among the list of members of the board.

In 1873, Newark and Jersey City had set up a Joint Commission on Water Supply, at which time recommendations were made for water to be supplied "from Lake Hopatcong and the Rockaway and Pequannock catchment areas, conveyed by the Morris Canal to a point in Bloomfield, from which an aqueduct would connect with the mains of the Jersey City works." The recommendation was not acted upon, and by 1887 the water conditions in Jersey City had become so bad that "a petition was circulated among the citizens of Jersey City requesting that a chemical expert make a report on the character of the Passaic River water."[11]

The issue hung fire until 1899, when Patrick H. Flynn, a contractor, was hired by Jersey City to supply the water rights, the land, and the plant necessary to provide water to Jersey City from the Rockaway River at Boonton.[12] He had to buy the SUM's rights in that river first, and of course he did. His proposal to Jersey City was for $7.5 million. When he didn't have enough money to fulfill his contract, "the men who supplied Newark with water took the job off his hands."[13] Not surprisingly, "those men" were part of a water trust called the "New Jersey General Security Company," of which the East Jersey Water Company was an affiliate. The Boonton Reservoir was completed, finally, in 1904.[14]

On September 24, 1900, the East Jersey Water Company deeded over to the city of Newark the entire Pequannock water-supply system at a total cost of $6,000,000.[15]

Out of the East Jersey Water Company, in 1908, arose the New York and New Jersey Water Company to supply Bayonne and Kearny, and the Passaic, Aquackanonk, and Montclair Water Companies. The last three, together with the Kearny Water Company, were incorporated on October 23, 1923, as the Passaic Consolidated Water Company.[16]

Now, the East Jersey Water Company was a "water company" in name only, having been given "no statutory or charter authority to traffic in water."[17] However, when Newark had contracted with the East Jersey Water Company for its water supply from the Pequannock River, the contract contained a clause entitling the *construction* company to "retain all of the surplus water from the city's watershed that was not needed by the city." Following this contract, the East Jersey went extensively into the business of furnishing water to many subsidiary companies. It built its big filtration plant at Little Falls, and from that point "it began wholesaling water, although in reality," it was only "selling water to itself under other names." The East Jersey was not only selling Newark's surplus from the Pequannock watershed, but the company was charged with filching "other great quantities of water from the Passaic at Little Falls and from Lake Hopatcong and Greenwood Lake."[18]

Out of the legal furor that arose from this state of affairs came certain basic truths. The SUM and the Canal Company were found to be "solid and united" corporations but it was found that the LV Railroad was a veritable Hydra. In its water supply activities, it was known as the East Jersey Water Company, the Passaic Water Company, the Montclair Water Company, the Aquackanonk Water Company, and so on. In its railroad activities it was also known as the National Docks Railroad Company and the Edgewater Railroad Company.

This "peculiarity" had been brought about in an effort to avoid heavy taxation. In the case of railroads, "mainstems," or main railroads, were taxed less heavily than were "branch" railroads. Obviously, "it was advisable for every railroad to have as many 'mainstems' as possible and several sections of road were organized as 'mainstems,' even though those 'mainstems' might be very short sections of road."[19] The Erie Railroad had done something similar when it formed the Long Dock Company, and it appears that the Lehigh Company had done the same thing with the water companies and for the same reason: The avoidance of taxation.

The Lehigh Valley was particularly hard hit by taxes. By the terms of the 1871 Lease, it was responsible for all assessments made upon the Morris Canal's property that was not used for navigation. Then, too, there was the $25,000 annual rental, or tax-rent, as John D. Ward had called it. As noted in Chapter 19, the Lehigh Company had paid dearly for the right to be rid of that tax-rent. The Canal Company had released all its interests and rights in the Big Basin for $500,000, which money the Lehigh Company applied against the Canal Company's $1,000,000 worth of debts.

Property used for canal purposes had been "perpetually" exempted from taxation by virtue of Section IV of the Canal Company's charter. The exempt status was allowed until 1884, when the State Board of Assessors levied taxes on the Canal Company under the General Tax Act of 1884 for the taxation of railroad and canal property. The lessees protested that tax and began litigation. Finally, in 1890, they paid the tax under protest.

The tax amounted to $150,000 for the 6-year period, 1884-1889. The lessees continued to pay the tax yearly after that, under protest, until finally in 1896, Jacob F. Randolph filed a formal protest and appeal to the State. State taxes for a 19-year period averaged $23,177.50 a year; county and local taxes averaged $8500 a year.

The General Tax Act of 1884, which provided for the taxation of railroads and canals, contained sections (22, 29, and 31) that recognized "special exemptions" from taxation. In recognizing the exempt status of the several railroads and canals, the State offered to surrender its right to take their properties at the times specified in their charters. All the companies had to do was surrender their exempt status and allow themselves to be taxed!

In 1888, the state passed an act authorizing any canal company (in particular, the Morris Canal Company) to abandon its franchise of navigation and cease its operation.* Again, however, the State's generosity was coupled with the provision that if the Canal Company *did* abandon navigation on the Morris Canal, it would have to submit to taxation.

"Doubtless any of these enactments could have been availed of by the Morris Canal and Banking Company for the purpose of extinguishing the State's right to take the canal in 1924 or 1974, but the proprietors did not at that time see their way clear to making use of

**This was the act signed by Governor Green, noted earlier in this chapter.*

these provisions."[20] But the thought occurs to this writer that if the State, and apparently everyone else, considered the canal to be so uterly worthless, why did it continue to hound the Canal Company lessees to surrender the canal's tax exempt status? In any event, the reader can understand why the Lehigh Valley Railroad Company was anxious to avoid paying any more taxes into the State's Treasury.

The decade of 1890-1899 was hard on both the Morris Canal and its lessees. There was a canal boat accident at Plane No. 7 West at Washington to one of the few boats that were still using the canal.[21] A large break occurred "about a quarter mile east of Hackettstown," which emptied the 3-mile level. The breach was "30 feet long and four feet below the bed of the canal," but luckily the water drained off into a creek so that there were no suits for damages to adjoining properties.[22] A freshet in the Lehigh River in 1894 washed out the Lehigh Canal's outlet lock into the Delaware River. Since that lock was an *inlet* lock for the Morris Canal, there was from that point on no further water communication with Pennsylvania.[23] A cloudburst on July 6, 1896, caused extensive damage to the canal from Washington all the way to Port Warren. Claims for property damage were abundant.[24]

On Tuesday, April 20, 1897, Canal Company President Jacob F. Randolph died at his home in Woodfern, in Somerset County, New Jersey. The Bloomfield *Record* noted his passing with more emphasis on his few years of service as Superintendent of the First Presbyterian Church Sunday School, and his inauguration of the annual Sunday School canal boat excursion to the Paterson Falls, than on his 30 years of dedication to the Morris Canal presidency. The Newark newspapers merely noted his passing, with a short death notice. *The New York Times* did not take, or give, any notice at all.

Two years later, in May, 1899, the canal closed at intervals because business was so slack. The demand for coal had become so "light" that it didn't pay the lessees to keep the canal operating on a full-time basis, and in 1898, they had "adopted the plan of closing the canal for half a day at a time for frequent intervals." Some of the canal workmen said they preferred that practice to being laid off for three or four days at a time, although their loss in wages amounted to the same.[25]

During 1899, navigation was twice shut down early in the season. Opening on April 24, the waterway was closed at noon on Thursday, May 11, and did not reopen until Tuesday morning, May 14. By Thursday, May 18, it closed again and reopened at noon Saturday, May 20, to allow boats to try to reach their destinations before the canal closed for the weekend. Two men at each of the inclined planes, plus lock tenders, were laid off temporarily during the shutdowns, at what the Newark newspapers considered to be a considerable saving to the Canal Company lessees.[26]

The railroad of the lessees was also having unusual difficulties. Its own bridge over the canal near Pacific Avenue, Jersey City, had been left open — or was not properly closed — on Monday morning, April 13, 1891. A "bob-tail locomotive used for drilling trains around the yard had been run on the bridge," and bridge, locomotive, and all, "were dumped into the canal together." While a wrecking crew worked all night by flickering lantern-light, hoisting the engine out of the canal and rebuilding the bridge, canal boat traffic stood still, "the teams still attached, waiting for an opportunity to pass."[27]

Railroad bridges over the Morris Canal in Jersey City (ETF).

Early in 1900, Engineer Alfred P. Boller announced that the only way to solve the problem of the DL&W Railroad in Newark, was to eliminate the Morris Canal! The Lackawanna's Montclair Branch presented a problem at the point where it crossed the canal at Newark's Orange Street. As Engineer Boller saw it, there were three alternatives to this problematical grade crossing. The first was to go under the canal. A passage to go under the canal would have to be at least "36 feet below the present grade" of the railroad, an idea he dismissed as "so impracticable that time is wasted to consider it." A second alternative was to go through the canal, a plan that was, of course, "dependent upon the canal abandonment, and as that institution has admittedly outlived its usefulness, and the owners are striving after a legal release, it cannot be but a question of time when it will be no longer a barrier." The third alternative was to go over the canal, a proposal that Engineer Boller dismissed as unjustifiable, since it would adversely affect property rights. He saw it as obvious that there was "only one way possible to reach a practicable solution of the street crossing problems" involved for the DL&W's branch west of High Street, Newark, and that was "by the adoption of a grade line that requires the elimination of the Morris Canal, and as that antiquated water way will in all probability be a thing of the past at no distant day, the Lackawanna Road and the city of Newark should await the time of that relief." He personally recommended waiting.[28]

Still, Newark's grade-crossing problem remained until, finally, a serious accident taking the lives of several school children precipitated some definite action. On February 19, 1902, a trolley car collided with a passenger train at Clifton Avenue with considerable loss of life. The solution to avoid future accidents was to carry the canal boats up over the Lackawanna tracks at Orange Street so that the Clifton Avenue trolley crossing could be depressed. By March, 1904, the LV Railroad agreed that the "fatal crossing" should be abolished. The question was: Who was to pay the $90,000 estimated to do the work? Both the LV Railroad Company and the City of Newark contended that neither of them should have to pay for the change, since it was the Public Service Corporaton and the Lackawanna Railroad Company that would benefit.[29] As it turned

Unnumbered Plane at Orange Street, Newark. Built in 1903 to eliminate a dangerous grade crossing after a trolley accident resulted in several fatalities. This was the only electrically operated plane on the Morris Canal system (BNK).

out, the LV Railroad, M&E Railroad and its lessee, the DL&W Railroad, agreed to bear the burden of expense for construction of the Orange Street Plane. It was to be the only electrically operated plane on the canal system.

An inverted siphon was installed under Orange Street to connect the upstream and downstream parts of the canal. The Orange trolley line rode on rails at Orange Street level, while DL&W trains rode rails in a depressed cut between the up- and downstream segments of canal. The plane, then, simply went from canal level, up over the trolley and railroad, and down to canal level again.

Before the nineteenth century came to a close, the Essex County Park Commission began condemning property along the canal for Branch Brook Park. A 100-foot-square piece of property "on the towpath of the canal, halfway between Fifth and Bloomfield Avenues," owned by "Charles H. Stiles and others," was appraised at $1000. The owners were not satisfied with that evaluation, and appealed to the courts. The Circuit Court, in the persons of Judge Child and a jury, decided in favor of the appraisers.[30] Apparently it was all right to condemn private property for park use, but not for canal use. We will see that idea, rooted in decisions such as this, take hold and grow rapidly in the decades ahead.

Even Newark's Board of Health was donning hob-nailed boots, getting ready to stomp on the canal's backbone. As the canal was being readied for the 1901 boating season, and "prior to the return of water," Health Officer Chandler carefully inspected all of the canal bed in Newark. The local authorities apparently had conducted this inspection annually for some years. Wherever the canal bed was found to be "foul," the Canal Company was compelled to give it a thorough cleaning out. That year, 1901, the section of canal above Lock Street was found to be "in poor condition," requiring "much work" to clean it.[31]

By the end of 1901, the Newark *Sunday Call* was caught up in a series of articles about the "proposed abandonment" of the Morris Canal. One very lengthy article pointed out that[32]

> *in the contemplated abandonment of the canal, and looking over the powers possessed by the company . . ., two prominent factors present themselves for consideration: First, that the canal route was to be forever a public highway; therefore it would seem that in default of enabling legislation, supposing such could be passed, the title of the bed of the canal would, in case of the water-carriage power of the company being given up, revert to the public, and not to the original owners. No precedent has been established on the abandonment of the canals, but in case of turnpikes, even where the title is absolute in the company, the reversion in abandonment is to the public.*

This is an interesting argument. We will be hearing many variations on the same theme before the canal is actually abandoned.

Basing his argument on the disposition of the property that had been used by the old Mount Pleasant, South Orange, Springfield, and Clinton Turnpikes, the author of the article commented that all those roads had been formed by incorporated companies, the titles to the properties having been vested in the incorporators. Yet when the turnpikes were abandoned, they became public highways. Individual ownership had lapsed, and it was only proper, he thought, that it should be so, since "an owner of a building on any turnpike acquires a vested interest in the turnpike for access to his property which a reversion to private ownership would nullify."

Undoubtedly, the abandonment of the canal was desirable, but the author questioned: "What shall be done with the bed?" Answering his own question, he contended that two suggestions offered themselves: "Either a reversion to a common highway or its utilization as a railroad." After all, Newark wanted "internal transportation of freight," and wanted it badly. If a railroad could be built in the canal bed "through the heart of the city," it would be "of untold value to the manufacturing interests." And since there were "no physical objections to the work that money can not overcome," the consideration of such a project would be desirable.

However, the abandonment of the canal would not be totally "consistent with a true appreciation of the progress of events," since canals were not yet "things of the past." The State of New York, for example, was "spending millions on the enlargement of its canals," and the Dominion of Canada was "gradually but surely opening up a line of water transit for large vessels to start from Chicago and run to Europe." Even the National Government was "considering the project of a deep waterway from the lakes to the Hudson," and "the Nicaragua canal" was also receiving national attention. Unfortunately, the divided ownership of the lands on which the Morris Canal was built prevented any such use from being made of it.[33]

The Public Service Corporation, newly reorganized with a new president, former Canal Company director and counsel Thomas N. McCarter, also had an eye on that canal bed. McCarter was invited to address Newark's Board of Trade on November 24, 1903. His topic was the availability of the canal bed in Newark, and the prospect of using that bed for transportation other than canal navigation.

At the time, discussion of those purposes was "if not premature, at least academic," but McCarter discussed them anyway. Newark's Board of Trade was concerned primarily about the 6.8 miles of canal that lay within the city's boundaries. McCarter dutifully pointed out, however, that the rights of countless other persons along the line of the canal were also at stake. In particular, "the rights of the people who have invested millions of dollars in and about the shores of Lake Hopatcong should be considered;" and "the same situation very largely pertains to the owners of property abutting Greenwood Lake." Moreover, there were those persons who had habitually relied on the Morris Canal for their coal shipments.[34]

After examining and explaining the four ways in which the Canal Company had acquired its right of way, McCarter pointed out that lands held in fee or by adverse possession and user for more than 60 years could legally be transferred by the Canal Company to anyone at all upon abandonment. Lands held by condemnation were a different story. The Legislature would have to enact a law authorizing the abandonment of the canal. Then, those condemned lands could be transferred to other parties so long as they were transferred "for another public use, which additional public use, in the language of the lawyers, put no additional burden upon the land." As far as McCarter, president of the Public Service Corporation, was concerned, putting a Public Service trolley in the canal bed "would be no additional burden upon the land." As for the canal lands held under "reversionary" deed, there was no doubt in McCarter's legal mind that "anyone who seeks to use the canal bed for any other public purpose must recondemn it for the purpose for which they seek to use it."[35]

Of the 6.8 miles of canal bed in Newark, McCarter had found that about half of the titles to the land were held in fee simple, and therefore the canal bed could simply be transferred to whomever the Canal Company desired. That left 1.6 miles through the city of Newark that the Canal Company held by deeds specifying that when navigation on the canal ceased, the lands were to revert to the original owners or their heirs. Anyone wanting to use the canal bed through Newark, then, was going to have to recondemn at least one-sixth and probably one-half of it.[36]

The Lehigh Valley Railroad Company, according to McCarter,

was willing to let the Public Service Corporation or the City of Newark use the canal bed in Newark for a trolley line. However, what would the stockholders of the Morris Canal and Banking Company think of that? They would surely want compensation. And they were largely Pennsylvanians.

In Newark itself there would be problems about using the canal bed. Center Market, built in the space over the canal, would have to be maintained, retaining walls would have to be built, and the city dock would have to be released to the city, and so on. Assuming that all of the difficulties could be removed satisfactorily, however, the canal bed could be used "with the very greatest profit." The Public Service Corporation, then "in control of the street railway facilities" of Essex County, had a working capital of $10,000,000 which it was prepared to spend on "the rehabilitation, extension, and improvement of the street railway service," and the improvement of its gas and electric lighting systems.[37]

From Newark, the Public Service Corporation proposed to send out radiating trolley lines to East Orange, Elizabeth, Hoboken, Hackensack, Perth Amboy, and Trenton. When negotiations were completed to use the new tunnel that was then being built under the Hudson River, Public Service lines would connect directly with New York City.[38]

Until the time came when the canal was used as a subway or other means of conveyance, the ever-growing population of Essex County would "never be properly served." The legislative commission which was then considering the problem of abandoning navigation on the Morris Canal would soon be reporting its findings. If that commission of three ex-governors should approve of canal abandonment, Mr. McCarter suggested that it would be to everyone's advantage to work together "for the good of the people of the city of Newark and County of Essex."[39]

The time for a decision on the disposition of the canal would soon be reached. It was, after all, this Newark Board of Trade that had urged the State Legislature to appoint a "board of inquiry" for the purpose of aiding in that decision.[40]

Three former governors of New Jersey: Werts, Griggs, and Voorhees, were appointed according to the act of March 31, 1903.

They compiled what has come to be known as "The Three Governors' Report," in an attempt to answer the various complicated questions involved in abandoning navigation on the Morris Canal. These questions fell into six basic categories:

1. *Lands, waters, water rights, rights of flowing and diversion, riparian rights, rights of way and other easements and privileges held or claimed by the Canal Company or its lessee, the Lehigh Valley Railroad.*
2. *How and in what manner the same were acquired, and upon what terms the same were held, what powers and franchises had been granted to said companies, or either of them, and whether said powers and franchises, or any of them, have become exhausted or have been lost by non-user or lapse of time or otherwise.*
3. *Whether said companies, or either of them, should be permitted to retain the property, rights, powers and franchises pertaining to said canal, or to sell or dispose of the same, to other persons or corporations, so that the same, or any of them, shall be devoted to public uses.*
4. *The fair value of the canal and its appurtenances, and the said property, rights, and franchises, to the said companies, or either of them and also the fair value thereof to this state.*
5. *What supply of potable waters could be gathered and distributed by means of said canal, and from what sources it could be drawn, having due regard to the rights and interests of private persons owning property adjacent to the sources of supply.*
6. *Whether it would be expedient for the State to own or control the water supply of the said canal and to use the same for the purpose of supplying municipalities with water on just and equitable terms.*

The act further specified that the commissioners report their findings to the next legislative session. The Senate concurred in the Assembly's action, and the law went into effect immediately.

The 1903, or "Three Governors' Report," was basically a primer for a generation of legislators whose age alone separated them from the beginnings of the Morris Canal. It had been, after all, 80 years since the lobbyists had kept Trenton abuzz with canal proposals.

By way of educating the legislators, the 1903 Report presented a brief history of the canal and its chartered rights. It inventoried the Canal Company's properties, waters, and rights, and described how they had been come by; discussed whether or not the Canal Company should be allowed to continue owning them or be permitted to sell them; determined how much they were worth to both the Canal Company and the State; suggested how the canal might be used to distribute potable water; and considered whether or not the State should enter the water-supply business.

According to the "Three Governors," the Canal Company, and/or its lessees, held rights of way and other lands by four sorts of titles:

1. *Title by deed in fee simple absolute.* [*This property was by and large not used by the Canal Company for purposes other than navigation and would not pass to the State when it assumed the canal.*]
2. *Title by deed by base fee, or as long as used for a canal or works tantamount thereto.* [*This property, it appeared, would all revert to the original owner upon the vacation of the canal.*]
3. *Title by condemnation for canal purposes under legislative authority.* [*This property would also revert to the original owners unless the State authorized a change of the public use of those lands. Where the new public use would mean additional burdens or damages for the owners, the Senate would have to make additional compensation.*]
4. *Title by gift, occupation, or adverse user.* [*These titles would become fee simple titles (as Class 1) wherever the adverse user or gift was properly established.*]

As for the Canal Company's water rights, the former governors listed only "the important ones:"

1. *The tidewaters of the Hackensack River, pumped into the canal by the company's steam pump apparatus.*
2. *The Pompton Feeder, supplied by the Pompton River.* [*The Pompton River, itself, was fed by the Pequannock,*

the Wanaque, and the Ramapo Rivers. At the time the report was compiled, there was "no interference with the natural flow of any of the three rivers except the Wanaque, at the head of which is Greenwood Lake, which is partly a natural and partly an artificial reservoir." The dam at Greenwood Lake, the lands adjacent to it, and a large part of the lands under the waters of the Lake were owned by the Canal Company in fee simple. This property, the "Three Governors" said, had "some, but no very great value."]

3. *The Rockaway River at Dover, Powerville, and Boonton and the Beach Glen Brook near Rockaway. "Upon the abandonment of the Canal the waters of the Rockaway will simply return unmolested to their ancient channel and flow down to the new water-works of Jersey City at Old Boonton . . . ."*

4. *Lake Hopatcong, 914 feet above see level, with a tributary drainage area of about 24 square miles, which collected a considerable body of water.* [*Hopatcong naturally flowed westward into the Musconetcong River until the canal was built and some of the lake's waters were diverted eastward. "The lands at the outlet, and some other lands, were bought, flowage rights obtained, and the waters raised about eleven feet. The new high water level so established has been maintained for about 75 years." If and when the canal was abandoned, "these waters could be availed of, to some extent, to reinforce the public water supply of Jersey City or Newark, or other municipalities lying to the eastward. It is believed that they would have some value, though no very great value, for this purpose." However, "the value of this supply would be greater for Jersey City than for any other city," for it "could draw the water into the upper Rockaway with little or no expense for piping, whence it would find its way down the Rockaway River into the City's Reservoir at Old Boonton."*]

5. *The Musconetcong River. [The "Three Governors," who were obviously oriented toward the industrialized and heavily populated eastern section of the state, dismissed this source with the brief comment that "upon the abandonment of the canal this water would naturally return to its ancient channel, the Musconetcong River."]*

No legislative action was taken as a result of this report, despite the importance and thoroughness of its conclusions and recommendations.

Newark, of course, had a stake in the disposition of the Canal Company's water rights, since the city was being supplied by the waters of the Pequannock River that had been part of the Canal Company lessee's water inheritance via the Lease of 1871. Newark had earlier begun a policy of buying up lands around its reservoirs to protect them against pollution. By about 1900, this policy made the city the owner of "43 of the 63 square miles of the Pequannock watershed."[41] By 1909, the city had outgrown its water supply and began applying for permission from the North Jersey District Water Commission to build a new reservoir on the Wanaque River, the route by which water was conveyed from the Canal Company's reservoir at Greenwood Lake into the Pompton River and thence into the canal. While the applications were pending, plans were made to enlarge Oak Ridge Reservoir on the Pequannock River.

In order to reach some understanding of why the Lehigh Valley Railroad came to oppose the construction of the new reservoir, it will be helpful to retrace some of the steps leading to the opposition. It is here that the Honorable Cortlandt Parker, champion of the people's rights, who was expected to be heard from during the battle of the rival steamboat companies at Lake Hopatcong, will very definitely be heard.

Since the legislature had failed to act on the report of the "Three Governors" in 1903, several civic organizations got together to try to force the issue of abandonment. The New Jersey State Civic Federation, the New England Society of Orange, and various other groups, asked Parker to draft a bill to be submitted to the legislature. The bill was designed to represent the interests of the people and towns most affected by the canal's existence.

The three former governors had sought "briefs representing various interests involved" in the question of the abandonment of navigation on the Morris Canal. They solicited the opinions of such interests as "Jersey City, Newark, the summer resorts and ice-harvesting fields of Lakes Greenwood and Hopatcong." About 25 such briefs were handed in, representing the interests of the state. Only one brief seems to have been acted upon — "that representing the interests of the corporation which was . . . the lessee of the canal," the LV Railroad.[42] The purpose of the Civic Federation, therefore, was to get a bill passed that would reflect the interests of the *people,* not the railroad company or the state government.

Cortlandt Parker's bill called for the state's attorney general "to institute proceedings in chancery or other proper court of this State, against the Morris Canal and Banking Company, seeking the forfeiture of the charter of said Company by reason of its cessation to operate said Canal and exercise the functions of its charter, and to carry on said proceedings with all possible diligence, and that he shall have power to call to his aid such counsel learned in the law as he may select, in the prosecution of such suit, or of other litigation which may be necessary." Parker's bill would have "prohibited [the Lehigh Company] from exercising any business in this State, either as a corporation for the operation or management of a railroad in this State, or of a canal therein, whether for its own profit or benefit or for that of the Morris Canal and Banking Company, or any assignee of the same, or of the good people of this State."[43]

Parker's bill was *not* intended to force the abandonment of navigation on the Morris Canal. It intended that the canal should be restored to its former good condition, "provided that said Canal shall be, when restored to good condition, managed by the said Morris Canal and Banking Company, and no other person or corporation."[44]

The Civic Federation published pamphlets on this subject, one of which, "A History of the Morris Canal," cited an unnamed but eminent German authority on inland navigation in Europe and North America. The pamphlet explained that, in England, canal building had ceased about 1830, when the railroads arrived. In order to control the canals, the railroads leased them, or significant parts of them. Then, by means of rate-fixing or "other expedients," the canals were "damaged" sufficiently so that they could be bought up cheaply by the railroads. Thus, canal traffic could be stopped altogether, or controlled at will by the railroads. There were no laws in England to prevent this state of affairs until it was too late to save many of the canals. Finally, in 1871, the value of canal property in England had fallen to about one-third of its original worth. When a law *was* passed in 1873, railroads were forbidden to close canals in their possession, or to let them deteriorate.

But Parker's bill also intended to force the LV Railroad to give up its lease, since, according to his reasoning, it had no right to manage a New Jersey concern. The railroad was, after all, a "foreign" company.

Before this sort of thing was allowed to happen in the case of New Jersey's canals, the Civic Federation proposed to do something about it. First of all, there was no reason to close the Morris Canal except, perhaps, in the heavily populated cities such as Jersey City and Newark. In the rural districts, however, the canal had several possibilities. Water power and traffic were but two of those possibilities; potable water was a third and, as we have seen, very significant possibility.

As far as power was concerned, the State would not be bound by the provisions of the Canal Company's charter, but would be free to "make such disposition of the water power involved in the Canal as it deems best." Traffic on the canal need not be confined to the age-old tradition of boat plus mule, but could be expanded to include steam barges or trollies, or even electric motor possibilities. The potable water possibilities were by far the most important.[45]

> *Whether the State will assume its ethical attitude toward the water supply at an early date or a late one, at some date it and its inhabitants will realize that a state's water supply, as over against private corporation ownership of a water supply, stands related as does a duty to a crime. Private ownership of oil may be a curse; the cornering of wheat may be a more or less legitimate move in a commercial game, but private ownership of water supply can never be anything but a crime. And when the State comes to realize that, the*

Morris Canal will be found to be intimately intertwined with the water supply of a large section of country in a variety of ways.

Mr. Parker's bill was brought before the Legislature and passed. The attorney general brought the suit called for by the bill, which quite[46]

naturally brought up the entire water question and included the bringing of further suit against certain private corporations carrying and selling water. This, again, naturally brought out the fact that this business extended beyond the limits of the state and hence introduced into an already complicated situation the problem now acutely before the state in various forms of injunction and certiorari in connection with a determined effort made by various water companies to meet contracts made by them with Staten Island to supply that territory with water from the State of New Jersey.

We have seen, earlier in this chapter, who those various water companies were.

Paterson moved to prevent the "water monopoly" from diverting water from the Passaic River at Little Falls by an injunction late in 1909. The East Jersey Water Company moved to transfer its property to the Passaic Water Company, which supplied Paterson, and to the Aquackanonk Water Company, which supplied Passaic. At the top of the heap, however, was the "water trust," the New Jersey General Security Company, under which all the subsidiary water companies were subsumed.[47] The State of New Jersey began proceedings to determine what rights, if any, the East Jersey had to any waters of the Passaic River in the first place.[48]

Another investigating commission, appointed by the Legislature in 1911, was formed of Governor Woodrow Wilson, Attorney General Edmund Wilson, and Judge Bennett H. Van Syckle. The Commissioners, along with about 20 members of the Legislature, toured the canal by boat from the Jersey City Basin to Easton, leaving on July 1, 1912, and spending the Fourth of July examining the Lehigh Canal at Easton. This committee recommended that navigation be abandoned, and that the Lehigh Valley Railroad pay the State of

New Jersey "the sum of only about seven hundred thousand dollars" and convey to the State its water rights.[49]

Before the Legislature felt itself ready to act on that committee's recommendation, it decided to appoint a new and larger investigating committee and arrange another inspection tour for 1913. This committee included:[50]

Foster F. Birch, Dover, N.J.
Henry M. Doremus, Newark, N.J.
Dr. J.D. Ely, Marlboro, N.J.
Mangold H. Ellenbogen, Paterson, N.J.
Prof. Albert F. Ganz, Hoboken, N.J. (Stevens Tech.)
Charles H. Ingersoll, South Orange, N.J.
Col. William Libbey, Princeton, N.J.
John Q. Blair Reilley, Phillipsburg, N.J.
C. Howard Slater, Jersey City, N.J.
Fred G. Stickel, Jr., Newark, N.J.
John W. Wescott, Camden, N.J.
Carlton Godfrey, Atlantic City, N.J.

On July 1, 1913, the 12-member investigating committee and about 20 of the state's lawmakers left Jersey City Basin for a first-hand examination of the canal. Like the 1903 group of three former governors, this committee was assigned several major areas for research and reporting:

1. The properties which would revert to the State of New Jersey in 1974.
2. The value of the properties and water rights, and the obligations which the State would assume when the canal became state property.
3. The net annual income the State could expect to receive if it assumed ownership of the canal properties.
4. The estimated cost to improve the canal for navigation, if it were found to be practicable for the State to operate any part of the canal, and how much the State could expect to receive in income from its operation.
5. The amount of property the LV Railroad Company would retain and be entitled to, and how much that property would be worth on the market as opposed to what it was worth to the railroad company.

6. The possible forfeit of the charter by the Canal Company or its lessee "by reason of the failure of said Companies, or either of them, to operate or maintain" the canal according to the requirements of the charter.
7. Whatever else the Committee thought would be helpful to the Legislature in making its decision.

The report compiled by the 12-man committee is known as the *Godfrey Report,* after Carlton Godfrey, the chairman. The following text summarizes the committee's findings in each of these seven areas.

Area 1

The Canal Company held its canal properties by four kinds of titles:

1. In fee simple — 32.24 percent
2. By reversionary deed } 60.65 percent
3. By condemnation }
4. By occupancy and adverse possession — 7.01 percent

The titles held under reversionary deeds and by condemnation would revert to the original owners of the property upon abandonment of the canal — unless the State changed the property to some new public use. When the State acquired the canal it would "have the right to use it as a canal or to change it to some other public use without the payment of compensation by reason of such changed use, unless the changed use imposed a greater burden upon the land than would the canal." The State, of course, has the legal power to declare a new public use for the canal by means of an act of the Legislature. There would be little danger of the properties reverting to original owners or heirs so long as the State had that power and exercised it.[51]

Area 2

In trying to estimate the value of the rights-of-way held by the Canal Company, the committee felt that they had "substantially no value" as long as the present use continued. However, a change in public use would *give* value to the canal strip. In presenting their appraisements, the committee pointed out that what the Canal Company had was "only the right of possession" until 1974. The committee's appraisements were based on values that could only be obtained through the State's exercise of its power to change the use of those properties. The Canal Company had no such power, and so the value of the lands depended "almost entirely upon the value which the State alone" could give them. The change of use, however, could not be made until the State acquired possession of the canal property.[52]

The committee's findings in this area, i.e., the value of the property and water rights which the State would acquire when it acquired the canal, were based on the same sort of reasoning about the State being able to change the purpose of the canal property. The water rights had been appraised in 1909-10 by Messrs. Charles Hansel and Morris R. Sherrerd, at a value of $3,000,000. The estimate had been made on "the assumption that Greenwood Lake and other sources of water supply would be permanently preserved for the benefit of Northern New Jersey."[53] The committee questioned whether or not the lake *would* be retained, since Greenwood Lake had been made to flow lands in New York State when the Canal Company had built its dam there. The law passed by the New York Legislature on April 11, 1855, to allow that reservoir to be made had specified that the rights of flowage would cease after the Canal Company's corporate existence had ceased.[54] If the Canal Company's charter was repealed, the right to maintain the dam at Greenwood Lake would cease, "and the State of New Jersey would thereby lose the benefit to be derived from the water supply" of the lake. That water supply had been valued at $1,000,000.

Obviously, the repeal of the charter would serve no useful purpose, for the State would lose substantial rights. The State should acquire the Canal Company's stock and keep the Company itself in existence.

Also included in this second area was the question of the State's right in the Canal Company's Big Basin. The Grant of 1889, made by the State's Riparian Board, of the State's reversionary interest in the basin had been declared invalid by a 1911 court decree. The LV Railroad appealed the decision, and at the time the committee was writing its report, the case was still pending in the courts. Had the 1911 Commission's recommendations been acted upon, the railroad company would have released its right to the return of the $48,000 it had paid for the 1889 grant, which, with interest, totalled about $115,000.

The LV Railroad was still protesting the taxes imposed by the State, having paid, to date, $961,918 — always under protest. The State courts had ruled in favor of the State's right to tax the railroad, so the railroad appealed to the U.S. Supreme Court. That decision, too, was pending at the time the *Godfrey Report* was compiled.

"In the event of a decision favorable to the railroad, a sum considerably in excess of one million dollars would have to be returned by the State and canal property would be exempt from taxation" until 1974. Under the settlement proposed by the 1911 Commission, "the railroad would have discontinued such suit and released any claim, present or future, which it might have to exemption from taxation."[55]

Area 3

The State might not receive any appreciable amount of income from the canal properties, but those properties could "be made to benefit the State in at least three particulars: water rights, parkway," and other rights.

Income from water rights would be small, but the assumption of the canal would make it possible for the State to furnish potable water to the people of New Jersey. The development of a parkway in the canal route west of Paterson would "be of inestimable value in the development of the section of New Jersey" through which the canal passed. Among the "other rights" that would benefit the State would be the "unquestioned right to tax the basin properties and the other lands" the railroad would receive if the 1911 Commissioners' plan were adopted.[56]

Area 4

The *Godfrey Report* was not favorable to the idea of the State improving the canal for operation because it would be too expensive, and the amount of traffic that could be expected to use the canal, even if improved, would not warrant the expenditure.[57]

Area 5

The property that was to be retained by the LV Railroad Company under the arrangement proposed in 1911 would have been the strip of canal from Fiddler's Elbow to and including the Big and Little Basins. However, the 1911 decision of the Court of Chancery declaring the grant of title to the Big Basin had modified that proposal. The right of the railroad company to use the basin until 1974 was unquestioned; only its title to ownership of the lands in that basin had been negated by the court's decision.[58]

This cancellation of title was one of the more unfortunate events that marred the Lehigh Company's history in New Jersey. The Riparian Board had been established by a law of 1869, which act gave that board "the right to convey the lands of the State under water within certain territories at the sum therein specified of fifty dollars for each linear foot of property as measured along the State bulkhead line."[59] On the basis of its "purchase" of 960 feet of frontage for $48,000, the Lehigh Company had invested "about eight hundred thousand dollars" in improvements on the property and had been paying taxes on it.

Then, about 1907, the power of the Board to make this conveyance was questioned, and the State brought suit to have the Lehigh's deed set aside. The Court of Chancery then decided that the Riparian Board did *not* have the right to make any such deed, and that the Lehigh Company should get its money back with interest. The question of refunding the taxes already paid since 1889 was left to another court to decide.

The *Godfrey Report* reflects the interest of the committee in the remarkable expansion of Jersey City's importance as a "great manufacturing and commercial locality." The city was particularly fortunate in having the terminals of several of the nation's great trunk line railroads *and* deep water facilities of New York Harbor. If Jersey City and the LV Railroad could settle their differences so that the latter would be permitted "to lay their tracks along or on the present location of the canal," the committee was sure that "several miles of choice location for factory sites would immediately become available in Jersey City," and that "many thousands of additional employees" would find jobs there, thus turning the southerly part of Jersey City into "a hive of industry."[60]

There was already a plan to build a deep-water canal to accommodate large vessels at the Jersey City waterfront. Trunk railroads connecting property with that canal would be most desirable for Jersey City's future development.[61]

Area 6

As to whether or not the Canal Company had forfeited its charter, the Godfrey Committee had to say "no." Under its charter, no forfeiture on the part of the Canal Company could occur "unless as a result of appropriate legal proceedings." The duty imposed upon the

Company by its charter was "to keep the canal open and in condition for navigation by all who desire to use it." The committee admitted that it had "no knowledge of conditions so evidently and strongly violative" as to warrant such "appropriate legal proceedings."[62]

Area 7

Under the last area of information requested by the Legislature, the *Godfrey Report* volunteered its opinion that the canal in Newark "demanded" some changed public use. Such change and abandonment there was held to be "imperative."[63] Newark should, the committee felt, be given the opportunity to buy the canal properties within its borders. In fact, if the canal were abandoned across the State, all municipalities should be given the opportunity to buy the canal strip within their limits. The necessary legislation should be enacted making it possible for them to do so.

The Godfrey committee considered the proposals of the Morris Canal Parkway Association, a group agitating for the canal to be kept as a continuous water parkway across the state from Belleville westward. The Association's plan had merit because the canal benefited the rural areas where settlement was sparse. The committee felt that "it would be unwise to draw off the water from this canal and leave an open ditch unless such canal bed could be at once changed and devoted to some new use. In addition to destroying the natural beauty of the canal and surrounding country for no particular purpose, the effect of such abandonment without an immediate change of use would be to destroy the continuity of the canal strip, as the individual owners or heirs," whose lands were held by reversionary deeds, "would re-acquire the lands as deeded."[64]

Using the canal as a parkway would mean public access. Since it would be necessary to trespass on private property to gain access in many places, if the canal were left as it was, the idea was not practicable. However, if "a broad boulevard" were built parallel to the canal across the state, the public could access the waterway. Of course, where the canal passed through a town, the town itself would have the choice of whether or not to make its portion of the canal boulevard conform to the rest of the plan.[65]

> *Such a boulevard would open up and develop a part of our State only too little known to our citizens — one of the*

most beautiful, one of the most picturesque stretches of country anywhere in the East.

For miles west of Newark the canal continues to increase in varied scenic beauty — up hill and down dale, ending in a wild crescendo, a veritable riot of beauty west of the Queen of Lakes of New Jersey — Lake Hopatcong, and through the mountainous sections of Sussex and Warren Counties. It is indeed the artificial Rhine of America.

Lake Hopatcong, that "Queen of Lakes," had been amply represented by the most vocal of all Morris Canal critics, Hudson Maxim. The committee expressed its appreciation for his "continuous and invaluable aid."[66] We will return to Mr. Maxim to hear his views on the subject.

In winding up its report, the committee realized that the Canal Company could not be expected to give up its rights in the canal properties without compensation, and without "a satisfactory compromise." Because settlement of the canal question was of vital importance to the whole state, the committee recommended that 24 points be adopted by the Legislature. Not the least of its recommendations was that the properties in the "Basins be deeded and conveyed to the Lehigh Valley Railroad Company," along with "the canal lands from the Little Basin in Jersey City to Fiddler's Elbow."[67]

From Fiddler's Elbow to the Passaic River, the committee suggested, should be controlled by a "Commission hereinafter provided for," with power "to retain said canal in this section or to devote the same to other use or uses and to sell or lease the same or any part thereof." The canal should be abandoned through Newark, and should "be diverted at some point west of the City of Newark and east of Little Falls," with the remainder of the canal to be "retained from such point of diversion to the Delaware River." The proposed canal boulevard should be constructed, "properly beautified and landscaped," whether continuous or not across the state.[68]

The commission which was to be "provided for," was to have the power "to sell, lease or exchange any portions" of the canal lands which were not "in their opinion" necessary "for the purposes of the parkway." They would also have the power "to grant permission for the construction and operation of an electric railway, water mains, gas pipes, sewers, telephone or telegraph wires* or conduits, or any other public service facility through any portion or all, of said Parkway, and to sell or lease a strip of land of suitable width extending from the northwesterly line of the City of Newark to the City of Paterson for electric railway or other public purposes."[69]

The LV Railroad was to drop its suit against the State in the matter of tax exemption and "release to the State its right to receive from the State the consideration, with interest," paid for the Big Basin property. Any of the water rights or rights in water formerly sold or leased by the Canal Company or its lessee were to be bought back by the Canal Company, excepting only the water supply arrangements with Newark and Jersey City.[70]

The Canal Company would also be required to buy back all of its outstanding stock, except that owned by the LV Railroad. The stock was then to be assigned to the State, and the Canal Company's charter was to remain in effect under State, or State-agent, control. The LV Railroad and the Canal Company would then have the privilege, as part of this equitable compromise, of paying to the State "not less than one million dollars in cash," as a settlement price.[71] Half of that sum was to be used to bring the boulevard project into existence. The remaining half-million dollars was to be used "to aid in the construction of a hard pavement road from Camden to the ocean, in the construction and repair of the Ocean Drive, and in the construction and extension of the Delaware River Drive."[72]

A Parkway Boulevard and Canal Commission should be created by the Legislature to implement all of these plans using the monies realized on the compromise settlement with the Lehigh Company. Serving on that commission should be the members of the investigating committee that had compiled the *Godfrey Report,* and members of the State's Highway Commission.[73]

The remainder of the *Godfrey Report* — and it is extensive — set forth the various engineering plans, maps, and illustrations for the proposed changes to the canal. Like the 1903 and 1911 reports, it was not acted on by the Legislature.

**There had been telegraph poles and wires on the towpath since 1864. (Paterson* Daily Press, *November 2, 1864).*

REFERENCES

1. *North Jersey District Water Supply Commission of the State of New Jersey, Report for the Period May 5, 1916 to June 30, 1925* (Newark, N.J.: 1925), p. 218 (hereinafter cited as *North Jersey District Report*).
2. *Ibid.*, p. 219.
3. *Ibid.*; p. 220.
4. *Ibid.*
5. Orange (N.J.) *Journal*, October 27, 1888.
6. *North Jersey District Report*, p. 222.
7. *Ibid.*, pp. 220-1.
8. *Ibid.*, p. 231.
9. *Ibid.*, pp. 224-5.
10. *Ibid.*, p. 227.
11. *Ibid.*, p. 228.
12. *Ibid.*, p. 229.
13. Charles A. Shriner, *Four Chapters of Paterson History* (Paterson, N.J.: 1919), p. 74.
14. *North Jersey District Report*, pp. 229-30.
15. *Ibid.*, p. 225.
16. *Ibid.*, p. 232.
17. Adolph Roeder, "A History of the Morris Canal," Special Report of the New Jersey State Civic Federation, No. 11 (Orange, N.J.: 1910), p. 14 (hereinafter cited as Roeder, *Special Report No. 11*).
18. Newark *Evening News*, January 5, 1910.
19. Roeder, *Special Report No. 11, op. cit.*, p. 13.
20. George T. Werts, *et al.*, *Report of the Morris Canal Investigating Committee on the Abandonment of Navigation on the Morris Canal* (Trenton, N.J., 1903), p. 17.
21. Bloomfield *Citizen*, October 4, 1890.
22. Washington (N.J.) *Star*, May 22, 1941.
23. *Report of the Morris Canal Investigation Committee, Appointed Under Joint Resolution April 12, 1912* (N.Y.: 1913), p. 157 (hereinafter cited as *Godfrey Report).*
24. Morris Canal and Banking Company *Docket Files*, map dated July 6, 1896.
25. Newark *Evening News*, May 19, 1899.
26. *Ibid.*
27. Jersey City *Evening Journal*, April 14, 1891.
28. Newark *Daily Advertiser*, February 28, 1900.
29. *Ibid.*
30. Newark *Evening News*, May 19, 1899.
31. Newark *Sunday Call*, April 7, 1901.
32. *Ibid.*, December 29, 1901.
33. *Ibid.*

34. Thomas N. McCarter, "Address delivered before the Newark Board of Trade, Nov. 24, 1903 . . . ," in *One Phase of a Jerseyman's Activities* (Garden City, N.Y.: 1933), pp. 1-2.
35. *Ibid.*, p. 3.
36. *Ibid.*, p. 4.
37. *Ibid.*, pp. 4-5.
38. *Ibid.*, pp. 6-8.
39. *Ibid.*, p. 9.
40. *Ibid.*, p. 2.
41. *North Jersey District Report*, pp. 225-6.
42. Roeder, *Special Report No. 11*, pp. 4-5.
43. *Ibid.*, p. 7.
44. *Ibid.*, p. 8.
45. *Ibid.*, p. 9.
46. *Ibid.*, p. 12.
47. *Ibid.*, p. 13.
48. *Ibid.*, p. 14.
49. *Godfrey Report*, pp. 3, 8.
50. *Ibid.*, p. 8.
51. *Ibid.*, pp. 9-10.
52. *Ibid.*, p. 11.
53. *Ibid.*, p. 13.
54. *Ibid.*, p. 14.
55. *Ibid.*, p. 15.
56. *Ibid.*, p. 16.
57. *Ibid.*
58. *Ibid.*, p. 18.
59. *Ibid.*, p. 17.
60. *Ibid.*, pp. 18-19.
61. *Ibid.*, p. 19.
62. *Ibid.*, p. 20.
63. *Ibid.*
64. *Ibid.*, p. 21.
65. *Ibid.*, p. 22.
66. *Ibid.*, p. 23.
67. *Ibid.*, pp. 24-5.
68. *Ibid.*, p. 25.
69. *Ibid.*, pp. 25-6.
70. *Ibid.*, p. 26.
71. *Ibid.*
72. *Ibid.*, p. 27.
73. *Ibid.*

21

Where There's Smoke, There's Maxim

WHILE the various investigating committees had been compiling their voluminous reports, the canal continued to operate. It had to. With a puzzling dispassionate calm, the Lehigh Valley (LV) Railroad Company reported to its stockholders in 1905 (and beyond) that, since no authorization had "yet been secured from the Legislature of New Jersey for the abandonment of navigation on the Morris Canal," operation was continuing "under the terms of the lease."[1] Again: "As the State of New Jersey has not yet granted your company's petition for the abandonment of navigation on the Morris Canal, its operation has been continued."[2]

Paterson's Ramblers Club, a group that used the towpath for many of its nature walks, expressed delight in 1905 at seeing the canal readied for navigation one more time:[3]

> *We think with pleasure of the long canal dredging and its evidences that the charming old waterway is likely to last a few years more. We are grateful to our State Legislators, notably John Hunter and Senator McCran, who have assisted in the preservation of this public benefit. While the old turnpikes are becoming unsafe for the pedestrian,*

here, along the fascinating Morris, he can ramble to his heart's content.

The Ramblers supported the Morris Canal Parkway Association of Montclair in its drive to preserve the canal "as a parkway for the people," and declared that "it would be a calamity if an object so worthy were to fail." Where it passed through cities, of course, the canal "could be purified and improved," but no one who had actually seen the Morris Canal "at Brookdale, for instance, or at Mountainview, would ever make such an ill use of language as to call it a ditch."[4]

Also in 1905, the Delaware, Lackawanna and Western (DL&W) Railroad abandoned its coal chutes at Port Washington. That apparatus had for years funneled thousands upon thousands of tons of coal into waiting canal boats.[5] Now it simply was no longer efficient, or economical, or necessary to transship coal at Port Washington, when freight could be carried straight through on the railroad's own right-of-way.

In Newark, the Public Service Corporation continued to press to be allowed to use part of the canal bed for its electric railway, and was growing impatient of waiting for the canal's abandonment through Jersey City. By mid-1905 they had bought "a private right of way between the Passaic and the Hudson Rivers" for their rapid transit electric railway between Newark and Manhattan. What the company wanted now was the canal bed west of Broad Street, "to connect with the private right of way."[6]

"Purely as a private investment," Public Service President Thomas N. McCarter and Uzzal McCarter had at one time bought a piece of Newark property "with a large frontage on Park Place." Apparently the Messrs. McCarter turned that property over to Public Service somewhere along the line, for in 1905 it was announced that that company was planning to build its new electric railway terminal there. The new rail system between Newark and Manhattan was planned to tunnel the Passaic River, but a decision had not been reached in 1905 as to whether to tunnel or bridge the Hackensack. When completed, the railway would transport passengers from downtown Newark to the heart of Manhattan "easily in thirty minutes, and probably in quicker time."[7]

McCarter's public statements about the abandonment of the canal elicited a great deal of public reaction. One such statement about using the canal bed in Newark for a street railroad prompted so "much extravagant comment" in the press that the LV Railroad's President E.B. Thomas felt obligated to memorialize the State Legislature in the company's defense, and to virtually beg for permission to abandon a waterway that had been "useless for half a century." There were, for example, allegations abroad that the Canal Company's lessee controlled "three fourths of the potable waters of North Jersey," and that the company "desired to keep those waters even after navigation was abandoned on the canal." These potable waters were alleged to be worth more than $50 million, while the water rights of Lake Hopatcong alone were said to be worth more than $5 million, and the "old Jersey City Tide-water Basin" was worth $20 million. The fact was, the Lehigh Company protested, "the entire property and rights of the canal from the Delaware to the Hudson" were worth only $3,379,823 according to the State Board of Assessors, and the Lehigh people felt that even that figure was too high.[9]

The Civic Federation alluded to in Chapter 20 had made the claim that the original cost of the canal had been $14 million and that its present value, *without* the Jersey City Basin, was estimated to be $25 million. The Basin alone, the Federation claimed, was worth $30 million, and that figure was a conservative estimate. The total value of the canal properties to the State, they said, was set at $50 million, at least.[10]

That sort of claim the Lehigh Valley Railroad absolutely denied. The railroad company had sustained gross losses of $4,934,409.29 over the last 20 years of operation. In the last 34 years, "not one single dollar" had been earned "upon the outstanding securities, the contract obligations of which" had been "kept inviolate at the expense of the Lessee."[11]

The water rights of the lessees that had been brought to the public's attention were, from the Lehigh Company's point of view, insignificant in value compared to its investment and its subsequent losses. Those rights amounted to "a right to divert a few million gallons of water a day for eight months a year from the Pompton River, a small amount from the Rockaway, and enough from Lake Hopatcong to keep up the summit level and feed both ways therefrom, and some from the Musconetcong." All of these rights in water would "all vanish on abandonment of the canal."[12]

The State Board of Assessors had estimated the value of the canal basin properties to be $400,000, on which amount the Lehigh Company was taxed annually. Of the remaining canal properties, four-fifths would "immediately revert to the former owners on abandonment of navigation." In fact, all the Canal Company would "have left of its original investment of more than $5,000,000" would be the Jersey City Basin, disjointed parts of the right-of-way "which it owned absolutely, dams and lands at Lake Hopatcong and Greenwood Lake, Cranberry Lake and one or two other points, some mules and canal boats and a few parcels of land not used for Canal purposes." With these, the Lehigh Company "must pay off a bonded debt of $500,000, satisfy stockholders holding $2,247,237.50 of stock and scrip, and the claims of the State, to which the Canal, by the terms of its charter, is to go absolutely in 1974." "The old cry is raised," the LV Railroad protested, "that the Railroad Company has ruined the Canal by diverting from it. The fact is the Railroad Company for nearly thirty years after taking the lease of 1871 deliberately diverted to the Canal from 250,000 to 300,000 tons of coal per year to give it something to do."[13]

The simple truth was that "after steel rails were invented and put in use in the [eighteen] sixties, and the Lackawanna Railroad completed through to tidewater, and the Central Railroad of New Jersey built, freights fell to a point which made navigation on the Canal perfectly hopeless."[14] The object of the LV Railroad in making the lease of the Morris Canal "was to afford it protection from and parity with the Central Railroad of New Jersey, which then controlled the route across the State of New Jersey from the Delaware River at Phillipsburg to tidewater. That company, itself largely interested in the transportation of anthracite coal, made it especially difficult for the Lehigh Valley, a Pennsylvania corporation with its railroad terminating in the middle of the Delaware River between Easton and Phillipsburg, to secure fair treatment in the transaction of its business." Fortunately, the Legislature of New Jersey had passed an act in 1855, "allowing the Canal Company to lay rails at Phillipsburg and connect its Canal with the Lehigh Valley Railroad, which was done."[15] When they leased the canal in 1871,[16]

> *it was believed that this lease would act as a regulator and enable the Lehigh Valley to secure equitable treatment from its competitors upon its own business. It soon became apparent, however that the canal was inadequate, either for the volume of business or for the purpose of controlling competition, especially as at this time came the discovery of steel rails resulting in lower rates of transportation and the concurrent rapid development of railroads. The result was that to protect itself the Lehigh Valley was compelled to construct the Easton & Amboy Railroad from Phillipsburg to Perth Amboy in the State of New Jersey. The construction of that line was begun in 1872 and formally opened for business in 1875. Subsequent to that the road was extended from South Plainfield to tidewater on the Hudson River at Jersey City. Instead of the object of leasing the canal having been to do away with competition, the reverse was the condition. It was leased for the purpose of creating competition and protection to the lessee. The changed conditions of the business of the country proved this to be an error.*

On the subject of the water rights of the Canal Company, the Lehigh Valley pointed out that with the lease of 1871 the State itself had given them, the lessees, the right to sell the canal's surplus waters to municipalities. The General Water Act passed by the State Legislature in 1888 enabled municipalities "to contract for a water supply." Simultaneously in 1888, an act was passed which authorized canal companies to abandon navigation. "This was undoubtedly intended to enable the Morris Canal Company to abandon navigation, the uselessness of the Canal having been already demonstrated."[17] Furthermore, the latter act would make it easier for the Canal Company to help solve the problems of getting water for Newark and other towns.

The Newark water supply would not be affected by abandonment of navigation on the Morris Canal. When Newark contracted with the LV Railroad and its construction subsidiary, the East Jersey Water Company, and construction was begun, it had been found that "the Canal Company had little or nothing which was readily available for the work." Another and independent right-of-way was acquired for Newark, and the Pequannock River waters were taken in such a way as not to affect canal navigation.[18] When the canal was abandoned, the Pequannock would continue to supply Newark.

At Lake Hopatcong, the Canal Company owned the dam "and a large tract of land at the outlet of Lake Hopatcong," plus "a good many acres of the land under the waters of the Lake." Property owners had built cottages around the shore line and did not want the dam removed when the canal was abandoned. Instead, they wanted the dam to be kept up forever so that the water would remain at its higher level in the lake. The Canal Company used to draw down the lake from four to six feet in the summer months, "according to the dryness of the season." The Hopatcong property owners preferred that practice to having the dam removed and the lake permanently drained to its old level.[19]

At Greenwood Lake, the Canal Company also owned the dam and "about six hundred acres of the flowed land, and the right to flow again as many, if not all, of the shore owners." The Greenwood Lake property owners also wanted the dam to remain if the canal were abandoned. Hopatcong was less valuable than Greenwood Lake for water supply and water power for mills, only an estimated $250,000. Greenwood Lake, on the other hand, could be made to supply "water supply plants" along the Wanaque or Passaic Rivers. The only such water supply plant so far was the pumping station at Little Falls, which provided "Paterson, Passaic, Bayonne, Harrison, Montclair, etc., a total of some twenty-two million gallons of water per day." Thus, Greenwood Lake was valued at $500,000. The LV Railroad strongly recommended that, upon abandonment, Lakes Greenwood and Hopatcong be "conveyed directly to the State of New Jersey, including the land, dams and all water rights of every nature." Then the State could utilize the lakes "in any way the Legislature may see fit." The already valuable could be made still more valuable.[20]

As for the LV Railroad Company's contract with the New Jersey General Security Company — the "water trust," as it was called by the Civic Federation — it was to be nullified. The LV Railroad had sold to the Security Company the right to sell all of its surplus waters. The latter had already paid the contracted price of $400,000, but the Railroad Company negotiated a release from the contract and refunded the money. The Canal Company and its lessee would be at liberty thereby to convey their water rights to the State "without hindrance upon the abandonment of the Canal."[21]

In conclusion, the LV Railroad described the Morris Canal as "a trivial affair in size, one hundred miles in length, climbing over great hills and reaching nowhere;" in fact, "an utterly hopeless proposition" that could not "be made to pay unless railroads are abandoned."[22]

The Lehigh Valley's losses under the lease of 1871 had been heavy from the start, but with the "profound change in methods of transportation involved in the invention of steam railroads," the losses became "still more grievous." Those changes in transportation, sad to say, "none could have foreseen." Consequently the Lehigh Valley Railroad Company asked[23]

> *that the Legislature pass a bill substantially as recommended by the Commission of Governors, permitting the abandonment of navigation on the Canal and the sale of the remaining fragments of property for the payment of the debts of the Canal Company and the stockholders, as far as the money will extend; and if other conditions are made satisfactory, offers to convey all land and water rights at Lake Hopatcong and Greenwood Lake to the State of New Jersey; and if desired, expressly by deed and convenant to relinquish to the State all its rights to divert waters from any stream or sources whatever in this State.*

The problem, to Lake Hopatcong's Hudson Maxim, was what would happen to those water rights, once they were relinquished. Jersey City wanted the canal properties within its corporate limits for use by industrial railroads. The city of Newark and Public Service wanted the canal bed for a rapid transit system. The Morris Canal Parkway Association wanted a strip park across the state to be used for water sports, boating, outings and the like. The LV Railroad wanted to be rid of the onus of taxation and the constant litigation it had inherited as lessee of the canal. And Hudson Maxim, what did he want? He just wanted to save "Lake Hopatcong the Beautiful," the Queen of all New Jersey's lakes, New Jersey's "Natural Niagara Park" — the site of his extensive real estate holdings.[24]

Hudson Maxim's biographer, Clifton Johnson, portrayed him as the eccentric scientist, sometimes a ridiculous figure, but a creative genius, a brilliant inventor, a profound philosopher, and a prophet of no mean stature. He was also a dreamer. One morning in 1923, he told his biographer that he had not slept well for dreaming that he had been working all night with a truck, filling the Morris Canal. Macculloch's dream had become Maxim's nightmare. He was obsessed with the idea that the canal, even after navigation had already been abandoned in 1923, should not be allowed to "cumber the earth any longer."[25]

Maxim had moved to Lake Hopatcong at the turn of the century, having arranged with the DuPonts (to whom he had sold all his patents and his rights in them, and who owned some powder works on the lake shores) to start a laboratory there for conducting experiments with explosives. As he himself noted, "it wasn't the beauty of the spot" that attracted him to Lake Hopatcong, but the powder works. "If they had been in a desert," he would not have been deterred. He had moved to the lake in order to "have room to grow up and blow up with the country without too much interference with neighbours."[26]

Before he'd been a resident at Lake Hopatcong for very long, Maxim bought "a large tract of lakeside property," and became "the owner of three-fourths of the land in the borough."[27] He had become wealthy and famous through the development of explosives during World War I, and was regarded with awe and respect by the lake community and by the state's legislators, as well. This man was a friend of Thomas A. Edison and Theodore Roosevelt; people listened when he spoke, and he was a natural spokesman in the controversy over what would happen to Lake Hopatcong when navigation ceased on the Morris Canal.

And speak he did. The Legislature was bombarded by Maxim's maxims. To reach a wider audience, he gathered together many of his speeches and published them in booklet form, which he then distributed at his own expense to the Legislature and to the public at large. The preface to the second edition, published in February, 1913, is worth repeating for what it reveals about Maxim, the man.

This book is a plea for the life of New Jersey's most

beautiful lake, and a plea for the life of the communities around that lake, and a plea also for the life of thousands of people in all parts of the State who, if the lake is preserved as a summer and health resort, will continue to find renewal of health and life in visiting it.

This plea is addressed to the Morris Canal Investigation Committee, to the Legislature, and to the people of the State of New Jersey.

Lake Hopatcong is threatened with ruination; the prosperous communities around it are threatened with extinction, and our most precious mountain-lake resort is in imminent danger of being wiped off the map.

Armed with the fore-knowledge of the intentions of the enemy, and having the facts, the truth and justice as my allies, I hope and I trust that I shall be able to convince any citizen of this State that he should join issue with the defenders where also his own interests lie, and help to prevent a great wrong being done to the whole population of the State as well as to the immediate lake communities.

People of the State of New Jersey, let it be known to you that there is a movement to take away from you, without any sort of compensation whatsoever, a very valuable lake and summer resort property which belongs to you.

Citizen of the State of New Jersey, you are a member of a company, an actual shareholder in a corporation, which is the State, in which also every man, woman and child in the State are shareholders. A part of your annual dividends, due and payable to you, are your rights and privileges in your lakes, public parks and summer and health resorts.

There is a scheme afoot to rob you of your best and most valuable mountain-lake and summer resort property, in order to appropriate Lake Hopatcong as a city water supply, which would benefit only about five per cent of the population of the State, at the expense of the rest — in other words it is proposed to rob twenty Peters to pay one Paul, and you and I, reader, are among the Peters.

Now, it is your duty to yourself and to your family, and to every friend and neighbor of yours, that you should lend a helping voice of protest to prevent you and them from being robbed. The reasons why you also should raise your voice in defense of your own and the people's own, are exactly the same reasons which are actuating me to make this plea.

It matters not whether or not my interests may be greater or less than yours — our interests are certainly identical and mutual.

My sense of duty has led me to make this plea. Your sense of duty should, at least, make you read it.

No special activities or efforts are asked of you. If you will read this plea through, your activities and your efforts will take care of themselves.

HUDSON MAXIM

Landing, New Jersey, Feb. 3, 1913

Clearly, Mr. Maxim had axioms to grind.

Most of the short articles contained in the booklet are variations on a theme, a theme completely expressed in the booklet's title: *Lake Hopatcong the Beautiful: A Plea For Its Preservation and Dedication as a Public Park For All the People.* A subsequent booklet published in December, 1913, was entitled, *Morris Canal Abandonment Problems: A Plea for the Abandonment of the Morris Canal and a Plea for the Preservation and Dedication of Lake Hopatcong the Beautiful as a Public Park and Health Resort for all the People.*

Since the articles are repetitious, selected quotations serve to convey the essence of Maxim's thinking on the subject of the Morris Canal. His first booklet, from which the prefatory remarks above are taken, was essentially a protest against the taking of Lake Hopatcong waters for Jersey City's water supply, as the "Three Governor's" Report had suggested. The title reflects Maxim's emphasis on preservation of the lake. The second booklet was published after the State had passed legislation creating the Lake Hopatcong Inter-Municipal Park Commission. Maxim was made president of that commission, and his emphasis shifted to abandonment of the canal *with* the preservation of the lake, as his title suggests. The gist of all Maxim's arguments is that the removal of the dam at Lake Hopatcong on the abandonment of navigation would lower the lake's level by eleven feet. Docks would be left high and dry, and the exposed lake bottom would be a malodorous mess, to say the least.

In the first booklet, Maxim contended that "the accumulation of laws and precedents now upon our statute books is so massive that the study necessary to the highest Chinese scholarship could scarcely master it. No wonder then that laws are often enacted which are unconstitutional or which contradict already existing laws." As he saw the problem at Lake Hopatcong, it was to be a matter of "the relation of the validity of ownership by purchase to the validity of ownership by seizure."[28]

The Canal Company had built its dam at Lake Hopatcong on lands that it bought from the owners. Where the lands had not been purchased outright, the rights of flowage had been obtained, but only over "a very small part of the total area" that the water covered after the dam was built. "By far the larger part of the lands affected by raising the Lake were never purchased by the Canal Company, and no right whatever to flow them was ever purchased by that Company. A large part of the bed of the old lake and shore lands affected belonged at that time to the East Jersey Board of Proprietors." In 1882, however, Nathaniel Niles purchased about one-half of the bed of the old lake and the newly flowed lands, and thus the lands were still owned privately.

The fact remained that the Canal Company's reservoir occupied lands owned by dozens of private landowners. Unfortunately, a "public service institution" such as a canal or railroad could not be dispossessed once it had acquired property even by occupancy or user. Once such an institution ceased to operate, all rights acquired by it reverted to the original owner. "But even the consolation is lately becoming tainted by certain adverse legal opinions to the effect that property which has been put to one public use, may be diverted to another public use without the consent of the actual owner and without compensation to the owner, provided that the new use is no more onerous or disadvantageous to him than the old use."[29]

Such an adverse legal opinion, as Maxim saw it, lay "so close to the margin of constitutionality that the possessors [were] apt to hold it with a tenure somewhat proportionate to the involvement of their personal interests." Such procedure he claimed, was "not justice to the individual property owner. No one pretends or can pretend that it is justice — only that it is law."[30]

The communities around Lake Hopatcong were every bit as much "public service institutions" as the Morris Canal was, by Maxim's reasoning, since they offered to the people of the state "a highly attractive and healthful resort." Any disposition of the waters of the lake, therefore, should be done with respect to and for the rights of those communities, whether the lands and waters and rights had been acquired by purchase or occupancy or user. "When the State shall have acquired the Canal, no new rights will be created or can be created by that acquisition. The rights of the State will be the same as are the present rights of the Canal Company, or the Lehigh Valley Railroad Company as the lessee of the Canal."[31]

The second booklet printed and distributed by Mr. Maxim did not contain this opening argument.

In 1911, Ford Kurtz, Civil and Hydraulic Engineer, had submitted a report on the proposed electrification of the Morris Canal. His task, as it had been assigned to him, was to determine "the engineering possibility and feasibility of operating the boats, locks and planes of the Morris Canal by electricity generated from its own flow of water."[32] In Europe, such mechanical haulage of canal boats had already been successfully used for a number of years, and similar experiments had been tried on both the Erie and Lehigh Canals. Kurtz had studied the results of those experiments and had also examined "the section of canal between Montville and Boonton, including Planes 7, 8 and 9 East and Locks 9, 10, 11 and 12 East." Whatever water power developments were proposed for that section of the Morris Canal, he considered perfectly feasible.[33] He concluded[34]

> *that if the Morris Canal were put in a state of good repair with full depth of water at all points and an additional net reservoired area of 10.4 square miles similar to present reservoired areas be secured, it would be possible even in the driest series of years to be expected to maintain naviga-*

tion for a traffic of 1,000,000 long tons per annum with an average haul of 62.75 miles; and that the entire canal could be electrified from the flow required for navigation, without the diversion of additional water.

The proposed improvements Mr. Kurtz was hired to consider consisted "mainly of the application of electrical power to the hauling of the canal boats, the operation of all parts of the planes, the operation of lock gates and valves and the lighting of the locks and planes to provide for night navigation."[35] He found that "the extraordinary rise and fall of 1674 feet in the water level of the Morris Canal," which had "hitherto been a disability" in operating the canal, could now, "through the modern harnessing of electricity," be "transformed into an advantage so great," that to continue talk about abandonment without considering the possibility of electrification "would obviously be a blunder that might deprive the people of this State of a most valuable water transportation."[36]

In answer to these findings and recommendations, Hudson Maxim insisted that "In order to render Mr. Kurtz' plan for the electrification of the Canal feasible, it would require even additional reservoirs for impounding the feed water, and all of this water, in addition to the water drawn from all the other feedwater sources of the Canal, it is proposed to employ for the transportation of a million tons of freight per annum, on an average haul of 62.75 miles — a little more than half the length of the Canal." Maxim set the value of that quantity of water at at least $5 million, while the value of the Lake Hopatcong "summer and health resort properties which would have to be sacrificed" would surely be worth at least another $5 million. At such an expense, "at least fifty cents a ton for every ton of freight that passed over the Canal, even if a million tons per annum should be carried," would have to be charged by the Canal Company, or by the State of New Jersey, if it should decide to maintain the canal as a canal.[37] All in all, Maxim felt that the State would gain from hiring the railroads to bring the coal straight through to Jersey City, and not transfer it to the Morris Canal at Phillipsburg.

It has been said that Maxim's "large tract of land," as he modestly put it, consisted of "600 acres with three miles of lakefront."[38] He had, and demonstrated that he had, every intention of seeing to it

The Morris Canal in Jersey City, labeled by the Paterson Morning Call "An odorous eyesore of stagnant water" (The Godfrey Report).

that his three miles of waterfront continued to front on the water. As far as Maxim was concerned, the canal had been made obsolete by the overwhelmingly superior railroads. There was no point to reconstructing it on a larger scale to attempt to operate it profitably. "Snail-paced Canal transportation with handling of freight by mule-power and hand-power" simply couldn't "compete with the steam-drawn freight train and with steam-operated machinery in the handling of cargoes."[39]

Maxim hosted the Morris Canal Investigating Committee when that group toured the canal in July, 1913. On July 1, the party left the canal basin at Jersey City on the first leg of its journey to Phillipsburg. They traveled by mule-drawn scow to the Hackensack River, trollied to Newark because the canal between the two rivers was not fit to use, then scowed again to the head of Lock No. 17 East, where they were met by a steam-powered, propeller-driven excursion boat, the *Mystic Shrine.*

Among the committee members was Fred G. Stickel, Jr., who kept a travelog of the tour, noting such things as the condition of the canal, the scenery, and the traffic on the waterway.[40] At this time, the canal through Jersey City was in abominable condition, being filled with refuse and sewage, not to mention partly decayed animal corpses. By the time Bloomfield was reached, however, Stickel noted that the water was clear and that it did not become foul again until Phillipsburg was reached. However, the thick growth of weeds in the prism was enough to grab at the propeller near Little Falls. Even supposing 500,000 tons of coal per year could be transported over it, the canal's closest rail competitor, the New York, Ontario and Western Railroad, carried 2,000,000 tons of coal to New York in 1910. The Central Railroad had carried 9,000,000 tons that year, while the undisputed leader, the Pennsylvania Railroad, had moved 66,000,000 tons![41]

More important, even if the canal *could* be made to carry more tonnage, a larger canal with larger boats would require still more water, and there simply wouldn't *be* enough water to keep it going for a full season of navigation.[42]

> *New Jersey's most beautiful lakes would be depleted during the months when the water is most needed in them,*

and instead of their continuance as health-giving, prosperity-making summer resorts, they would be reduced to miasmatic lagoons and swamps for the spreading of mosquitoes and contagion.

Cranberry Lake, Greenwood Lake, and beautiful, splendid Lake Hopatcong, that now blossom with health and teem with life and joy and laughter through all the good old summer time, would with depopulated and desolated shores, lie reeking in the sun, stinking to heaven for vengeance.

Northern New Jersey's most cherished rivulets and lakes would be ruined by the diversion of their water from its God-intended uses, in order to feed a navigable, seeping cesspool, five feet deep, forty feet wide, and a hundred miles long. [*The seepage and leakage from the canal alone required more water*] *than five such lakes as Lake Hopatcong to supply the water lost.*

As for the Morris Canal's inadequacy as a coal-carrier, Maxim pointed out that there were many coal dealers in New York City who retailed as much as 1000 tons of coal a day, an amount which would require 15 canal-boat loads a day. To further illustrate his point, he noted that the Brooklyn Rapid Transit Company used 1000 tons of coal per day; the New York Steam Power Company, 600 tons; the sugar and oil refineries, from 500 to 1000 tons daily; while the Edison Company, which was now using bituminous coal, formerly used 2000 tons of anthracite daily. If 10 thousand canal boats, each carrying 65 tons of coal, passed through the Morris Canal each year, the total tonnage would be 650 thousand — 1780 tons a day — "not quite enough coal to supply two ordinary dealers in New York, and not quite enough to supply one concern like the Edison Company."[43]

It was obvious to Hudson Maxim that the Morris Canal Parkway Association had failed to consider the fact that in keeping the Morris Canal as a water parkway "throughout its entire length, almost as much water would be required to keep it full as would be needed to feed and operate the Canal if it were to be reconstructed. It would require more than double the quantity of water which could be supplied by Greenwood Lake, Cranberry Lake, Bear Pond, Stanhope

Reservoir, Green Pond and Lake Hopatcong combined, in order to supply the feed-water to keep full such a parkway canal. Consequently, practically the same water-shed area would be required for such a dead canal as would be required for a live and operating canal."[44]

Maxim was most vocal, however, on the matter of using Lake Hopatcong water for potable purposes. If the lake were to be declared a reservoir for drinking water, the shores would have to be depopulated because of the danger of sewage seeping into the water. Boating, swimming, and fishing would have to be prohibited. It had been proposed "that a compromise be effected, whereby the Lake may be drawn down from three to four feet for potable purposes, and that no onerous restrictions be placed upon the use of the Lake as a summer and pleasure resort." It had also been said that it would be twenty years before the State would want the lake for such purposes. However, Maxim asked the Legislature, why should the residents at Lake Hopatcong compromise? "All of the rights to the waters of Lake Hopatcong which we granted to the Canal Company either do or do not revert to us upon the abandonment of the Canal, and these rights either are, or they are not, wholly ours absolutely; and if they are, why should we compromise? And if they are not ours, then why should we be compromised with?"[45]

Even Cornelius C. Vermeule, Civil Engineer, and member of the Morris Canal Parkway Association, agreed with Maxim on the subject of using Lake Hopatcong as a source of potable water. In a letter to Maxim, dated February 18, 1913, Vermeule noted that the potable water supply of the State was "abundant and amply sufficient, without any necessity of sacrificing the beautiful development of Lake Hopatcong." As Maxim had clearly shown, "an equally efficient reservoir can be obtained elsewhere at a cost which would represent only a small fraction of the loss which the property owners at Lake Hopatcong would suffer from having the lake applied to this purpose."[46]

Like the commissions that had preceded it, the Godfrey Commission filed its report; like those that had preceded it, it was not acted upon. By the beginning of 1915, Jersey City's Board of Commissioners had prepared a booklet entitled *Jersey City's Interest in the*

Morris Canal, containing the city's proposal for the abandonment of the canal. Copies of the report were sent to the Governor and the members of both Houses of the Legislature, and to the LV Railroad Company and the Morris Canal Company.

It was clear, to Jersey City at least, that the presence of the Morris Canal and the various railroads had checked the development of that city. It was equally clear that unless the city had some of its own waterfront property, it would not be able to encourage manufacturers to settle there. The canal and its two basins in Jersey City were estimated to be worth $4 million, and the street rights of the city in the basins were worth $500 thousand. The city therefore offered to pay to the State $3.5 million for a grant of those lands in fee simple when the canal was abandoned. The city would issue bonds to that amount at 4 percent interest and deposit them in the State's Treasury in exchange for half of the water rights and all of the lands belonging to the Canal Company through the city. The Commissioners proposed that the Essex Street Pier be retained by the city as a municipal dock for the development and expansion of factories. The remainder of the canal strip would be used for a trunk sewer and a belt-line railroad. Like the Godfrey Commission's report, this one, too, was received, digested, but not acted upon by the Legislature.

By November, 1915, the LV Railroad Company had lost its tax appeal case for the basin property that Jersey City wanted so much to own, and a new appeal was begun. Almost $1 million was at stake, the suit having already been pending for eight years.[47]

Everyone, every agency, every town, every city along the canal's path was impatient to settle the question of abandonment. While everyone agreed that abandonment was desirable and even necessary, there was no agreement as to the *terms* of abandonment.

Public Service, meanwhile, weary of waiting for its right-of-way, began constructing its new railway terminal in downtown Newark. To alleviate traffic congestion during construction, the company rerouted trolleys away from Broad and Market Streets, crossing Plane No. 12 East at High Street on a specially designed set of cross tracks that accommodated both trolleys and inclined plane cars. The new tracks were installed in 1915, so that traffic could flow smoothly to the new terminal under construction.[48]

World War I intervened at about this point, and except for a temporary and small-scaled revival of transportation activity on the canal, the waterway continued to decline. The powder works along the canal and the Pompton Feeder were busy turning out powder caps for the war effort. Soldiers from Paterson's Fifth Regiment guarded the towpath and lock at Lake Hopatcong.[49]

By the time the war was over, Hudson Maxim was a very wealthy man. He gave up making explosives, but he did *not* give up making war on the Morris Canal, and continued to work toward its abandonment. His booklets had contained a very valuable and informative map of northern New Jersey indicating the watershed areas involved in supplying the Morris Canal and showing, as he was quick to point out, that Lake Hopatcong furnished only about 4 percent of the total. By far the largest watershed area was that of the Passaic River, particularly the Pequannock, Ramapo, and Wanaque feeder rivers. Maxim must have been pleased to learn that, instead of using Lake Hopatcong for potable water supplies, the North Jersey District Water Supply Commission had decided to build its newest reservoir on the Wanaque River.

The Commission's decision was a response to Newark's urgent request for a larger water supply for its rapidly growing population and industrial community. In Chapter 20 we saw that Newark had begun negotiating for more reservoir acreage in 1909. By October 31, 1918, the city and Commission signed a contract for construction to begin. By late 1920, the Wanaque Reservoir got under way as land was bought and bids were solicited.

Earlier, in 1917, the SUM had tried to stop construction on the reservoir, but failed. Although one source puts the date at February 1918, it seems that on June 17, 1922, the Lehigh Valley and the Canal Company jointly began a suit in Chancery, praying for an injunction against the Water Supply Commission,[51] and claiming that diverting 50 million gallons of water daily from the Wanaque River would render the Morris Canal unnavigable, since the Wanaque emptied into the canal by way of the Pompton Feeder. By its charter amendment of 1836, the Canal Company, and therefore the lessee, was entitled to the flow of that river, which originated in its reservoir at Greenwood Lake.

The State was forced to defend itself against this position of the Canal Company and its lessees. Obviously, if the Wanaque Reservoir project, a great public work, "could be blocked by the corporations, all future projects in the vicinity would likewise be blocked, or the costs tremendously increased." The North Jersey District Water Supply Commission put its legal and engineering staffs to work examining the physical condition of the canal, and the legal position of the Canal Company and its lessee.[52]

> *It was found that no traffic whatever was being carried on the canal, that its waterway was filled with sand and weeds, that its structures were rotting away, that the few canal boats existing for the purpose of navigation were assembled at the two ends of the canal and were falling to pieces; that in fact the canal was in every respect abandoned except for the infrequent passage of the company's boats in order that the company could claim it was maintaining its charter rights.*

It was quite obvious to the State's Commission[53]

> *that the principal value left in the canal, in the opinion of the company, was the water rights which they claimed to have obtained from the people by the ancient charter, and that the whole proposition was being exploited by the companies with a view to obtaining a payment in connection with any of the proposed public water supply developments by the Commission in the District.*

After carefully checking the figures submitted by the Canal Company indicating how much water was necessary to operate the canal, the Commission's engineers reported to Vice-Chancellor Foster that "no such quantity would be required." Furthermore, "such quantities of water as the company might rightfully claim from storage in Greenwood Lake could be passed through the Wanaque Reservoir and on to the canal by facilities which the Commission proposed to install at Wanaque Dam."[54]

Notwithstanding the Commission's careful check and thorough report, on September 12, 1922, Vice-Chancellor Foster admitted the injunction. The North Jersey District Water Supply Commission was effectively restrained from "impounding, confining, . . . or changing the natural flow of the waters of the Wanaque River." The injunction did *not,* however, stop construction of the dam, since no water was actually being impounded behind it. The Canal Company and its lessees were victorious, yet the reservoir continued to be built.

What did it *mean* to win that victory? The LV Railroad obviously was not concerned about using the water of the Wanaque River to operate the very canal they wanted so much to abandon. And abandonment is precisely the victory that was won with that injunction. An act of March 11, 1922, created and empowered a three-man commission composed of William H. Speer of Jersey City, Frank H. Sommer of Newark, and Louis A. Focht of Trenton, to negotiate with the Lehigh Company for an abandonment settlement. Judge Speer, chairman, served a short while, then resigned and was replaced by Jersey City's Edward L. Young.

On April 5, 1922, the three-man Morris Canal Commission asked for and received a "Statement of Facts" from LV Railroad Company Vice President E.H. Boles. One of the questions the Commission asked was what the Canal Company lessees were "willing to give up, and what, if anything, they desire to retain."[55] In his April 27, 1922, answer, Vice President Boles said that the canal and railroad companies were "willing to transfer to the State all water rights owned and controlled." They were also "prepared to surrender to the State all the canal property west of the Hackensack River, embracing the valuable strip of land (a substantial part of which is owned in fee by virtue of original acquisition) through the heart of the city of Newark." The railroad company wanted "to retain and receive the conveyance of the State's interest to the property in Hudson County east of the Hackensack River; also a strip of land in Phillipsburg running from Green's Bridge to the Delaware River, about two miles in length."[56]

There were thus actually four parcels of land the LV Railroad wanted to retain: The Big Basin, the Little Basin, the canal bed from the Little Basin to the east bank of the Hackensack River, and the Phillipsburg property, totalling $298,112 in value.[57] The State could have all the rest. But it was in surrendering the water rights of the two companies that the State was truly to benefit. These rights were "the largest individual item of value in connection with the Morris Canal," and were most valuable in the Pequannock, Ramapo, and Wanaque watersheds, involving rights of flowage and diversion which the Companies had maintained "at great annual expense."[58]

The Lehigh's Vice President, Boles, wanted to point out "that one great thing can be said in favor of the existence up to this time of the Morris Canal as a canal — that is to say, if, before the municipalities of Northern New Jersey had fully realized the importance of their needs as to potable water, the Morris Canal had been abandoned and its rights of flowage and diversion had consequently been lost, only to be acquired again at great expense, these municipalities would have capital changes imposed upon them in the near future running into many millions of dollars. Moreover the Canal Company and its lessee have caused to be gathered and kept together and are now able to transfer through the State primarily to its various municipalities for immediate tangible use the last remaining supplies of pure potable water in northeastern New Jersey." The rights in all of that pure potable water at an estimated value of $12,500,000 were to be surrendered to the State of New Jersey. Ninety-three miles of the canal bed was also to be surrendered to the State. All stocks, bonds, and scrip of the Canal Company would also be turned over to the State.[59] "A mere statement of this proposition demonstrates its fairness," Mr. Boles thought. The inclusion of the stocks, bonds, and scrip in the proposition had been done on the assumption that the State would "desire to preserve the Canal Company charter." While the Lehigh Valley did not wish to be thought presumptuous, they certainly suggested "that the State should consider carefully the advisability of preserving the charter in view of the rights carried with it respecting waters and reservoirs."[60]

Perhaps the Legislature showed signs of reluctance to act on this latest commission's abandonment recommendations. Whatever was happening, or not happening, in Trenton was almost certainly related to the LV Railroad's injunction to stop Wanaque Reservoir construction. The Canal Company's charter was, after all, a very big stick. The lease of 1871 was also an effective weapon, since it had given the lessees the absolute right to engage in the water-supply business. The injunction was admitted on September 12, 1922. By November 29, 1922, the Morris Canal Abandonment Commission had reached an agreement with the Lehigh Valley Railroad Company on the terms and conditions under which the Morris Canal would be turned over to the State of New Jersey. The injunction against the reservoir was immediately dissolved.[61]

The act of March 11, 1922, before the agreement was reached, had not only called into being the negotiating commission, but had also specified that the property the State was to receive was to be controlled and managed by the Board of Conservation and Development, subject to further action by the Legislature."[62] That "Board" is today's Department of Environmental Protection.

Among the lands to be conveyed to the State by the November 29, 1922, agreement was the Little Basin at Jersey City. In exchange, the LV Railroad was to receive the State's right and title in part of the Big Basin, the canal strip from the center of Washington Street in Jersey City to the east bank of the Hackensack River, and the canal strip in Phillipsburg from Green's Bridge to the Delaware River.[63]

The LV Railroad agreed to pay to the State $875,000 in five equal annual installments, with 5-percent interest due on the unpaid balance each December 15. The Little Basin was to be deeded to the State, and the railroad company agreed to extend its branch that ran from Warren Street in Jersey City to the Little Basin when requested to do so by the Board of Conservation and Development. Part of the Big Basin, too, was to become State property, as was the canal strip from the Hackensack River to Green's Bridge, Phillipsburg.

With the lands went the Canal Company's and the Lehigh Company's rights to and in waters. The only exceptions to that part of the agreement were the water supplies of Jersey City and Newark. Those cities were to retain their water-supply arrangements with Patrick H. Flynn and the East Jersey Water Company, respectively.

All canal-related property was to be conveyed to the State. At its own expense, the Lehigh Company was to maintain the Morris Canal *as* a canal until March 1, 1923. All of the Lehigh's shares of canal stocks, bonds, and preferred scrip were to be surrendered to the State by December 30, 1922. All stocks, bonds, and scrip owned by others were to be bought in by the railroad company and surrendered on or before March 1, 1923. If the company proved unable to buy in the stock, the State would take legal steps necessary to condemn it.

Meanwhile the Lehigh had to continue to pay the dividends on the stocks and bonds as they had guaranteed in the lease of 1871. All leases, licenses, documents, maps, records, deeds — everything was to be delivered to the State. The Lehigh Company was to withdraw its Supreme Court suit protesting the taxes it had had to pay the State. The agreement was signed by E.E. Loomis, President of the Canal Company, and E.H. Boles, Vice President of the LV Railroad Company, as well as by the members of the three-man Commission.

On February 28, 1923, the Legislature vested in the MC&BCo. all of the property and property rights (other than stocks, bonds, and scrip) that had just been acquired from the Lehigh Company. The Canal Company was to serve as trustee for the State of New Jersey. The stock of the Canal Company was vested in the Board of Conservation and Development on March 19, 1923. On the same day, the Legislature provided for a board of directors, whose eight members were to hold annual stockholders' meetings. The directors were not authorized to sell canal property or rights, but they could lease them. All leases so made, however, were to be subject to 30-day terminations upon written notice, should the State want to terminate the leases.

Obviously the railroad company president's advice about keeping the Canal Company charter alive had been taken very seriously.

On March 13, 1924, navigation was legally abandoned on the Morris Canal and on its feeders. The rights of the Canal Company "to impound and divert waters of lakes, ponds, and streams, and the property and rights vested in the said company in Lake Hopatcong, Lake Musconetcong (otherwise known as Stanhope Reservoir), Cranberry Lake, Bear Pond, the pond on the Musconetcong River at Saxton falls, and Greenwood Lake" were to be "retained by said company in trust for the State of New Jersey, for the public use of conserving the public waters of the State and shall be and are hereby dedicated to such public use."

Chapter 229, Laws of 1924, passed on March 13, 1924, authorizing the abandonment of navigation on the Morris Canal, contains some very interesting sections. Section 4, for example, authorized the Canal Company to sell "any and all" of the property held in trust for the State, other than the rights and property to be set aside and retained for public use at Lake Hopatcong, and so on. Municipalities were to be given first opportunity to buy the canal strips or fragments within their boundaries. "When portions of the canal are wanted by a municipality," one newspaper explained, if a request is made previous to January 1 next [1925], the department [of Conservation and Development] can grant the request on reasonable terms in case there is set forth the purposes for which the property is wanted and that is approved by the Governor, subject to the review of the Public Utilities Commission."[64]

However, if a municipality did not ask for the property, then it could be sold to another municipality that had acquired property within its own boundaries, provided the property formed a single, contiguous tract capable of development for the purposes set forth. "This is the plan that would make it possible for the purchase of the canal right-of-way for a high-speed electric line from Newark to Paterson, with a boulevard for a portion of the distance. It is flexible enough to permit of co-operation by the municipalities through which this portion of the canal runs, or for Newark or any other municipality to acquire the whole strip." In the event that municipalities did not exercise their option to buy canal property within a reasonable time, the property could be offered "at either public or private sale to the highest bidder" so long as the municipality through which the canal passed might "alone or jointly with other municipalities bid for the property. This is a further safeguard against the capturing of the property by private corporations unless the municipalities concerned are willing to consent." That, the Newark *Evening News* felt, was "fair enough," for it was "a guaranty that the useful property" would not "be left idle in the hands of the state."[65] The irony here will not be lost on today's Morris Canal buffs. The canal basin property in Jersey City is a case in point.

Another interesting part of Chapter 229, Laws of 1924, is Section 5, which specified that if lands were sold to the highest bidder, as explained above, "upon any such sale, any previous dedication of the property so sold to any public use shall cease and be of no further force and effect, and the purchaser shall be entitled to take and hold such property free from any previous dedication to any public use." This, despite the fact that 100 years earlier, Section XXV of the Act of December 31, 1824, which incorporated the Morris Canal, specified that the canal right-of-way was declared a public highway, to remain so "perpetually." But then we have seen how the definition and the spirit of the word "perpetual" changed over the years.

The $875,000 to be paid by the LV Railroad Company to the State was to be used, first of all, to defray the cost of dismantling the canal. And any surplus monies after dismantlement was to be appropriated to the use of the Board of Commerce and Navigation of the State of New Jersey. For what purpose? For a brand new canal! The New Jersey Ship Canal, projected to connect Raritan Bay with the Delaware River at Bordentown.[66]

REFERENCES

1. *Annual Report of the Lehigh Valley Railroad Company, to the Stockholders, 1905.*
2. *Annual Report of the Lehigh Valley Railroad Company, to the Stockholders, 1906.*
3. Joseph Rydings, *Country Walks in Many Fields* (Paterson, N.J.: 1934), p. 66.
4. *Ibid.*, pp. 229-30.
5. Washington (N.J.) *Star,* May 25, 1944.
6. *New York Times,* June 21, 1905.
7. *Ibid.*
8. E.B. Thomas, "The Morris Canal, An Economic Waste: Memorial of the Lehigh Valley Railroad Company to the Senate and General Assembly of the State of New Jersey. Statement by The Lehigh Valley Railroad Company. February, 1911." p. 1.
9. *Ibid.*, pp. 1-2.
10. Adolph Roeder, "A History of the Morris Canal," Special Report of the New Jersey State Civic Federation, No. 11 (Orange, N.J.: 1910), p. 3.
11. Thomas, *op. cit.*, p. 2.
12. *Ibid.*
13. *Ibid.*, pp. 2-3.
14. *Ibid.*, p. 7.
15. *Ibid.*, p. 13.
16. *Ibid.*, pp. 13-14.
17. *Ibid.*, pp. 14-16.
18. *Ibid.*, p. 16.
19. *Ibid.*, p. 17.
20. *Ibid.*, pp. 17-18.
21. *Ibid.*, p. 19.

22. *Ibid.*, p. 22.
23. *Ibid.*, p. 24.
24. Hudson Maxim, *Lake Hopatcong the Beautiful: A Plan for its Preservation and Dedication as a Public Park for all the People* (N.Y.: 1913), p. 42 (hereinafter cited as Maxim, *Lake Hopatcong The Beautiful*).
25. Clifton Johnson, *Hudson Maxim, Reminiscences and Comments* (N.Y.: 1924), p. 18.
26. *Ibid.*, p. 215.
27. *Ibid.*, p. 221.
28. Maxim, *op. cit.*, pp. 5, 6.
29. *Ibid.*, pp. 6-7.
30. *Ibid.*, p. 7.
31. *Ibid.*, pp. 7-8.
32. Ford Kurtz, *The Electrification of the Morris Canal* (N.Y.: 1911), p. 2.
33. *Ibid.*, p. 22.
34. *Ibid.*, p. 2.
35. *Ibid.*, pp. 18-9.
36. *Ibid.*, p. 1.
37. Maxim, *Lake Hopatcong the Beautiful*, p. 19.
38. *Lake Hopatcong Historama* (Landing, N.J.: 1955), p. 19.
39. Hudson Maxim, *Morris Canal Abandonment Problems: A Plea for the Abandonment of the Morris Canal and a Plea for the Preservation and Dedication of Lake Hopatcong The Beautiful as a Public Park and Health Resort for all the People* (N.Y.: 1913), p. 9 (hereinafter cited as Maxim, *Abandonment*).
40. Fred S. Stickel, "Along the Morris Canal," *New Jersey History*, Summer, 1971, pp. 93-114.
41. Maxim, *Abandonment*, p. 12.
42. *Ibid.*, pp. 16-17.
43. *Ibid.*, p. 19.
44. *Ibid.*, p. 25.
45. *Ibid.*, p. 31.
46. *Ibid.*, p. 51.
47. Newark *Evening News*, November 15 and 17, 1915.
48. E.T. Francis and T.F. Gleichmann, Jr., "Unusual Trackwork," Pine Creek Railroad Division, N.J. Museum of Transportation, Inc., Allaire, N.J., April, 1970, 4 pages.
49. George W. Keppler Collection, "Notes," August 9, 1917.
50. Cornelius C. Vermeule, Jr., *Final Report of Consulting and Directing Engineer*, June 29, 1929 (East Orange, N.J.: 1929), p. 65.
51. *North Jersey District Water Supply Commission of the State of New Jersey, Report for the Period May 5, 1916 to June 30, 1925* (Newark, N.J.: 1925), p. 240 (hereinafter cited as *North Jersey District Report*).
52. *Ibid.*
53. *Ibid.*, pp. 240-1.

54. *Ibid.*, p. 241.

55. Morris Canal and Banking Company *Docket Files,* Trenton, N.J., letter from the Lehigh Valley Railroad Company to the Commission for the Abandonment of the Morris Canal, April 27, 1922, p. 4.

56. *Ibid.*, pp. 5-6.

57. *Ibid.*, pp. 6-7.

58. *Ibid.*, p. 4.

59. *Ibid.*, pp. 4, 7.

60. *Ibid.*, p. 8.

61. *North Jersey District Report,* p. 241.

62. *Laws Providing for the Acquisition of the Morris Canal by the State of New Jersey and Its Final Disposition* (Trenton, N.J.: 1924), pp. 6-7.

63. *Ibid.*, pp. 8-9.

64. Newark *Evening News,* March 12, 1924.

65. *Ibid.*

66. *Laws of New Jersey,* 1923, Chapter 151, and *Laws of New Jersey,* 1924, Chapter 79.

22

The Very Mortal Remains

CHAMPAGNE flowed in Trenton the night the canal abandonment bill was passed, and there was "great rejoicing." The same had been true on New Year's Eve, 1824, when the canal's act of incorporation had passed. The Passaic County members of the State Legislature were among those celebrants of 1922. One of them was said to have introduced the abandonment bill, but one Paterson newsman wondered: "Why the jubilation?" All it meant was "the loss of business for Paterson and the wiping out of some beauty spots that nature lovers had enjoyed for many moons."[1]

Why, indeed, the jubilation? Well, Newark was eventually to get its right-of-way for a street railway; Jersey City, at least part of its own waterfront to develop; the State, its coveted reservoir system, while yet allowing Lakes Hopatcong and Greenwood to keep their dammed water. The Lehigh Valley Railroad Company, at long last, was going to be rid of its albatross. The company's report to the stockholders for the year 1922 noted that the agreement made with the State of New Jersey not only secured the property in the "Big Basin" at Jersey City, and the right-of-way it wanted at Phillipsburg, but also "relieved" the company "of taxes and maintenance of canal property for which it had no use," with a saving of $300,000 a year.

Somewhat embarrassingly, after all the abandonment hullabaloo of a quarter century, *no* one seems to have had much use for the canal property between Newark and Phillipsburg, once the State had carved out its water supply parcels. Disposing of the canal properties was not going to be an easy task or one quickly done. There were those problematical properties held by deeds containing reverter clauses, constituting more than half of the canal lands. Understandably, the State was concerned about them.

In 1921, the Legislature had asked State Attorney General McCran for an opinion as to whether or not the original owners, or their heirs, would have any rights of reversion in canal lands if the State should acquire the canal in 1924, per the charter's provision, then abandon navigation and utilize the canal "for some other purpose, such as the operation of an electric line." If so, the lawmakers wanted to know if they could provide for the cutting off of such rights by condemnation.

McCran's answer is interesting, since it indicates the climate of legal thought, and because it illustrates one step in the long process of refining the rationale of acquiring the canal property *before* the charter provision became effective. As he put it:[2]

> *If the deeds do not indicate the intention to limit the Canal Company by a condition, but use words importing another intention and the State should abandon the canal for canal purposes, it would not, in my opinion, operate to cause the property to revert to the original owners or their heirs; nor would this occur if, as suggested in the question, the property is to be utilized as a public highway, such as an electric line, even if such a condition has been expressed by the original grantors.* [*Moreover*], *notwithstanding, a condition subsequent may be found to exist in some of the deeds . . . , though under the rule the land may revert to the reversioner, nevertheless, if the land is continued for the use of a public highway for the transportation of goods, etc., then the rights of the reversioner consist only in receiving compensation for the additional burden placed upon the land by the change of its use and such damage as might also result therefrom, if any.*

This is the kind of legal thinking that formed the basis for the Act of March 11, 1922, which authorized the State to acquire the Morris Canal "in whole or in part, and all the stock of the Morris Canal and Banking Company and the rights of all stockholders in said company and in said canal property and water rights, and all or any part of the right, title and interest of the Lehigh Valley Railroad Company in said canal property and water rights by virtue of its lease of said canal from the Morris Canal and Banking Company, or otherwise . . . ."[3]

At the end of the act's long preamble are two statements, the first one claiming that it was "desirable, among other things, that the State of New Jersey should acquire the said canal property and rights, in whole or in part;" the second stating that the acquisition of those properties was "deemed necessary for the public uses of the state." It was this act which specified that all canal property thus acquired should be controlled and managed by the Board of Conservation and Development. In a very real sense, this act created

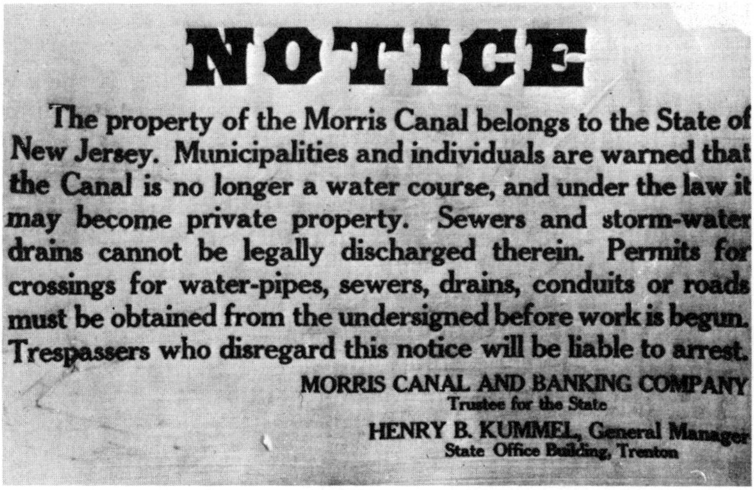
NOTICE

The property of the Morris Canal belongs to the State of New Jersey. Municipalities and individuals are warned that the Canal is no longer a water course, and under the law it may become private property. Sewers and storm-water drains cannot be legally discharged therein. Permits for crossings for water-pipes, sewers, drains, conduits or roads must be obtained from the undersigned before work is begun. Trespassers who disregard this notice will be liable to arrest.

MORRIS CANAL AND BANKING COMPANY
Trustee for the State
HENRY B. KUMMEL, General Manager
State Office Building, Trenton

Notice of the State's ownership of canal properties issued by the Morris Canal and Banking company (PCHS).

the "Morris Canal and Banking Company of 1923." It was this act, too, that had created the three-man committee to negotiate a settlement with the Lehigh Valley (LV) Railroad.

However, we have seen in Chapter 21 that only certain *parts* of the canal properties and rights were to remain dedicated to public use. The State's "official" line was expressed in McCran's opinion, and that opinion embodied the State's policy on disposing of canal properties. What it wanted — the water rights — the State kept. What others wanted, i.e., special interest groups such as Public Service, the State "rededicated" to public use. Even the Delaware, Lackawanna and Western (DL&W) Railroad was a beneficiary. On March 13, 1924, the Legislature passed an act authorizing the sale of any canal property adjacent to a railroad to the railroad company owning that adjoining trackage. Thus, the Boonton Branch of the DL&W could and did build its new railroad station, in the canal bed on Marshall Street at Paterson.

To decide how to dispose of the canal properties, and to whom they should be disbursed, Governor Silzer appointed a committee in April, 1923, to make appropriate recommendations. He chose five men to represent the interests of the State:

Albert W. Drake, Plainfield (Chairman)
Edward D. Duffield, Newark
Samuel P. Leeds, Atlantic City
William Osgood Morgan, Montclair
Cornelius C. Vermeule, East Orange.

To represent the Board of Conservation and Development, new managers and "stockholders" of the Morris Canal, he chose:

Percival Chrystie, High Bridge
Owen Winston, Mendham Township (Gladstone).

The municipalities in each of the counties along the line of the canal chose their own representatives:

Hudson Maxim, Sussex County
H.M. Riddle, Warren County
Frank Sommer, Essex County
James Wilson, Passaic County
David Young, Morris County
Edward L. Young, Hudson County.

Mr. Edward Young was so seriously ill that he was "unable to participate in any way in the work of the Committee."[4]

The committee decided to hold public hearings in various towns along the canal, to get "the benefit of the views and suggestions of all desiring to be heard." From May 29 through August 2, 1923, hearings were held first at Jersey City, then at Newark, then Paterson, then Dover, and finally at Washington. All of the opinions heard by the committee at those hearings were duly reported to the Governor and the Legislature. "In Jersey City, the governing body expressed a wish to take over the waterfront property known as the Little Basin and develop it as a municipal property, while the tenants of the property, who hold leases at the pleasure of the Canal Company or the State, thought the fact that they have invested a considerable sum of money in improvements thereon should be recognized, and that they should consequently be allowed to continue in possession for a reasonable length of time."[5]

Newark and Paterson meetings were well-attended by citizens of Belleville, Bloomfield, Aquackanonk and Clifton, who wanted to see an electric railway put in the canal bed between Paterson and Newark. In Newark, the section of canal lying from the Deep Lock to the Passaic River at Lock No. 18 East was eyed by the Chamber of Commerce as a possible connecting belt-line freight railroad. No one had offered a practical plan for the construction of such a railroad, however. Public Service Railway Company had offered to cooperate with the new owners of the canal bed, *whoever* they might be, so that the railway company could operate a trolley line in it.

At Paterson, the consensus was that the city should acquire the canal bed through that town for *some* public use, possibly a highway, possibly an aqueduct, between Paterson and Pompton. But, west of Paterson there was no demand for a railroad, trolley line, or highway in the bed of the canal. The state already had roads paralleling the canal most of the way. Trolleys were already running virtually the whole length of and close to the canal. The DL&W and Central Railroad of New Jersey (CRRNJ) also, for the most part, duplicated the canal's path, and a third railroad would hardly be needed.

Actually, the public hearings and the report of the committee appointed by Governor Silzer recommending what to do with the canal property were an anticlimax. Frankly, no one knew quite *what* to do with the waterway now that they had been successful in destroying it. There was a lot of hemming and hawing and scratching of heads during the difficult decision period. No one really *wanted* the canal bed throughout much of the line.

Slowly, a set of procedures emerged, once the public hearings were concluded. The right-of-way and adjacent lands belonging to the Canal Company from the Hackensack River to Green's Bridge, Phillipsburg, including the Pompton and Hopatcong feeders, were to be dedicated to highway use, "not for the purpose of constructing a State highway thereon, but in order to enable the several municipalities to apply it to local highway uses if they wish to do so. If any part of the right of way is not so used within a reasonable period," the committee recommended that "it should be abandoned as a highway and disposed of" in one of several other ways.[6]

The section of canal lying in Kearny, between the Hackensack and Passaic Rivers, was a case in point. South of the canal, an entire industrial community had come into being since the canal's extension to Jersey City during the mid-1830s. On the north side of the canal, the Lincoln Highway had been built. In other words, the canal's right-of-way lay between the industries and the highway that would now have to service them. Kearny should be given the opportunity to acquire and improve the property. Failing that, individual adjacent property owners should be given the right to do the improving.[7]

The committee found that the canal's continuity across the State would be completely valueless unless it could be put to a profitable use. With the coming of the automobile and its increasingly popular use, there was no likelihood that either trolley lines or railroad trackage would expand. The canal's route, moreover, was "entirely too crooked " and the grades at the inclined planes were too steep for such use. The cost of straightening and leveling for a highway would be prohibitive, and unjustifiable, since other roads already serviced almost the entire canal line. The committee felt, however, that it would not be "desirable for the State to withhold this right of way from improvement by local municipalities and property owners over a long period of time because of a remote possibility that some future use may be found for it."[8]

To keep the various towns through which the canal passed from being able to "improve" their canal properties while the State waited for a time in the undetermined future when somebody might think of a suitable use for it would be unfair. Unless it were filled in or otherwise taken care of, the canal bed would soon become an obnoxious nuisance. Where needed for public use, it should be so used; "otherwise it should pass into the hands of private owners, preferably adjacent land owners, either by lease, sale or reversion, as early as possible."[9]

In considering the canal's retention as an aquatic parkway, the committee had found that it would take entirely too much water to keep the prism filled — an estimated 31 million gallons a day. "In view of the large and increasing demand from our cities for these potable water supplies," the committee felt that this water necessary for a water parkway would not be "permissible, however desirable it may be from an aesthetic standpoint."[10] This statement seems rather strange when we recall how the Canal Company and its lessee tried to prove that it needed the waters that the Wanaque Reservoir would impound, and the State quickly pointed out that no such need existed. And even stranger: When Hudson Maxim wanted to prove that the canal ought to be abandoned, he cited the enormous amount of water that it would have to take from Lake Hopatcong to stay in existence — as if the lake alone supplied the canal. But on the other hand, when he wanted to convince the Legislature of the absurdity of using Lake Hopatcong for potable water, he reversed his argument to prove that the lake and its watershed provided only 4 percent of the total water used by the canal.

Strange, too, that in Newark, where agitation to get rid of the malodorous Morris had been very vocal, suddenly the canal was seen as a local boon by the Real Estate Board of that city. President of that board, Dr. Charles F. Kraemer, told reporters that the abandonment of the canal would be "a most happy solution to the troublesome problem," despite the fact that "the terms desired by the State officials" in acquiring the canal were "not quite up to their expectations." However, it had been "due to the unflinching firmness" of the members of the abandonment commission that "better terms were obtained for the State than had ever before been offered." One thing was certain: the canal could not "be permitted to remain an open ditch." But just how it might be used so as to serve both "the interest of the State and the communities" through which it passed would certainly "tax the wisdom of legislators."[11]

The effect of the abandonment in Newark was expected to be "beyond comprehension," and would only "be appreciated in years to come when the prospective transportation plans have been put into operation." The canal route west from Mulberry Street, Dr. Kraemer thought, would be "of highest service for a high speed electric line," while the canal from the "Passaic River to the Pennsylvania Railroad crossing at the foot of Market Street" was suited for a "Municipal Port railway" that had originally been developed at Port Newark. It was proposed to extend that railway to the canal bed at the Lincoln Highway, and then continue it westward in the canal bed to the Pennsylvania crossing. This would "bring to all the lands fronting on the Passaic River freight rail service, enhancing land values and immediately make them sought after by a class of industries that are clamoring for rail and water facilities."[12]

Early in April, 1923, Henry Kummel, General Manager of the Canal Company and a member of the Board of Conservation, prepared a memorandum on canal abandonment for the use of the committee charged with recommending what disposition should be made of canal property. The memo was partly a condensation of the findings of the Godfrey Commission of 1912, and was intended to keep certain important facts before the "disposal" committee's eyes at all times.

Under its so-called "excess diversion charge," the State was collecting "$1.00 per million gallons for water diverted for potable purposes from surface sources," namely, reservoirs. On this basis, and assuming that 125,000 gallons per square foot of watershed are discharged from these reservoirs to maintain the flow of the streams, Focht estimated for the Godfrey Commission that if the canal were abandoned and the waters of the reservoirs above this amount were made available, and could be used for potable purposes, the State might derive an annual income of $15,227 from this source. The estimate, however, took "no account of the value of the right of diversion for canal purposes," and was thus an underestimation of the potential income for the State.[13]

Noting that the Act of March 11, 1922, had vested in the State the dam and waters of Lake Hopatcong, and specified in Section 6 that "the lake level shall be maintained at high watermark," Kummel wondered how that provision could "be reconciled with the rights of the mill owners of the Musconetcong," and with "the present charter obligation of the Morris Canal and Banking Company to maintain the canal, which must be lived up to in order to preserve the water rights in all these reservoirs." This difficulty Kummel saw as "an interesting problem." In any event, he was sure that "any plan for abandonment of the canal must provide that these rights do not revert because to recover them from riparian owners after reversion would involve the expenditure of very large sums of money."[14] As it turned out, it was even more important than Kummel realized for the State to provide for its own ownership of those riparian rights.

Perhaps the strangest incident relative to riparian rights took place in 1923, while the State of New Jersey was trying to make up its collective mind as to what it should do with the Morris Canal bed. The United States Government was threatening to take away from New Jersey the very riparian rights that had been in dispute!

On June 10, 1920, the Congress of the United States had passed the Federal Water Power Act. A commission of three was appointed to implement that act: Secretary of War John W. Weeks, Secretary of the Interior Hubert Work, and Secretary of Agriculture Henry C. Wallace. The act itself was one of many similar pieces of legislation aimed at undermining and/or controlling the power of the monopolies that had become a fact of life at the end of the nineteenth and beginning of the twentieth centuries. These monopolies, composed of private corporations, owned most of the nation's natural resources. Nowhere was that more amply demonstrated than in the coupling of railroads and anthracite operators in an effort to control the industry, and in the coupling of railroads and petroleum producers for the same purpose.

This popular demand for social control of these monopolies resulted in several laws to attempt that control. Before a private monopoly was permitted to develop in the field of hydroelectric power, the Federal Water Power Act was enacted. Its avowed purposes were to create a Federal Power Commission; to develop the nation's water power; to develop and improve the nation's navigation; and to provide the use of the public lands for those purposes. It contained three items of vital interest to the State of New Jersey. The first item stated that licenses issued under the Federal Water Power Act were to be in effect "for a period not exceeding fifty years," and could be revoked under certain specific conditions. They could also be surrendered or altered, but "only upon mutual agreement between the licensee and the commission after ninety days' public notice." What persons would build a hydroelectric installation under a license that was to be valid for only fifty years? Probably the same kind of persons who would invest millions of dollars in developing a canal that was "licensed" for only 150 years and was predestined to revert to the licensing state, entirely free of charge!

The second item was that the licensee would be required to "pay to the United States reasonable annual charges . . . for the purpose of reimbursing the United States for the costs of the administration of this Act; for recompensing it for the use, occupancy, and enjoyment of its lands and other property," and to help prevent the licensees from expropriating "excessive profits" to themselves! As a part of that second item, all "combinations, agreements, arrangements, or understandings, express or implied, to limit the output of electrical energy, to restrain trade, or to fix, maintain, or increase prices for electrical energy or service are hereby prohibited."

The third item was that the United States, upon two years' written notice to the licensee, might "have the right upon or after the expiration of any license to take over" the licensed project. If the Government did not take over the project upon expiration of the license, the Federal Water Power Commission might reissue a license to the original licensee under whatever conditions the then existing laws dictated! In times of national emergency, of course, the United States would have the right to enter upon and take possession of any such project "for the purpose of manufacturing nitrates, explosives, or munitions of war . . . ."

Licensees who failed to comply were to be deemed guilty of a misdemeanor and fined not more than $1000. Every month that the licensee failed so to comply would be considered a new and separate offense, punishable anew. The U.S. Attorney General, upon request of the Secretary of War (one of the three commissoners), could institute legal proceedings to revoke the license of the noncomplying licensee. Once the license was revoked, the Government, or other person, agency, or corporation might buy the project, which the court was empowered to sell.

The State of New Jersey made it quite clear that it had no intention of applying to the Federal Water Power Commission for a permit or license or *anything,* on the ground that "the acceptance by the State of any license which such Commission could grant . . . would necessarily involve the abandonment by the State of substantially all of its rights to develop and utilize the potential water power at each of the points and places above mentioned, and likewise the right to grant to a licensee the right to develop and utilize the potential water power at any of the places herein enumerated."[15]

What were those "points and places" where the State planned to develop water power?

"There is opportunity for the development of considerable water power at a number of planes west from Lake Hopatcong and above Saxton Falls on the Musconetcong River, and it is the intention of the State of New Jersey to utilize the canal property acquired for the purpose of water power development, as well as the development of its comprehensive plan hereinafter mentioned, for the conservation of potable waters.

"In addition to the development contemplated by the State of New Jersey of the water power of the canal and its feeders, it is contemplated by the State of New Jersey to develop the water power possibilities of the Delaware River."

Now, the State Legislature, by an act of 1915 (Chapter 241, Pamphlet Laws), had given to the Board of Conservation and Development, new virtual owners of the canal and its water rights, the rights and powers which had earlier been vested in the State Water Supply Commission, "except in so far as they are restricted by Section 22 of Chapter 71, P.L. 1916," which later act created two water-supply districts in the State of New Jersey, known respectively as North Jersey Water-Supply District and South Jersey Water-Supply District.

"Under the authority of the said several acts" that had created and empowered these water-supply agencies, *the State had "been developing the potable water resources of the State, reservoirs being constructed, water works being built at an expenditure of many hundreds of millions of dollars* (italics mine)."[16] The spuriousness of the State's argument will be apparent to the reader who recalls that it was the LV Railroad Company which developed these resources.

On the other hand, the United States Government had "not made any expenditure, not incurred any liability, nor done any work for or on account of such canal construction and maintenance." Nor had it "at any time made or incurred liability for any expenditure or done any work for the improvement of the navigability of any of the waters constituting the canal system."[17]

Yet, here was this United States Government, in New Jersey, trying to exercise its authority and powers through its agents, Weeks, Work, and Wallace. The actions of the Federal Water Power Commission were jeopardizing "the conservation policy of the State of New Jersey over its potable waters, and the operation of the same within the State of New Jersey for the purpose set forth" in the Federal Water Power Act would "work irreparable injury to the reservoirs and water works constructed, and in the future to be constructed, as well as deprive New Jersey of the revenue to be derived from the development and conservation of the water resources of the State."[18]

The State owned lands under water from New York Bay to the Delaware, and the monies received by the State from long-term leases on those lands amounted to about $4,000,000 a year. The Riparian Board, created by the State in 1864, had given way to the Board of Commerce and Navigation, created by an act of 1915 (Chapter 242, P.L. 1915). The Federal Commission's activities in New Jersey consisted of exercising its rights "over the riparian lands of the State of New Jersey for power purposes," rights which the Commission claimed had been conferred upon them by the U.S. Government. New Jersey denied the constitutionality of the Commission's rights.

The Federal Commission's exercise of authority deprived "the State of New Jersey of its sovereign right, not conceded to the Federal Government, except for the purpose of navigation, over its riparian lands and waters," and would, moreover, "deprive the trust fund created under the Constitution of the State of New Jersey for the free schools of the said State, of the revenue derived from that source [leases of lands under water], dedicated by the Legislature of the State of New Jersey perpetually for free education."[19]

The word "perpetually" apparently had taken on a new meaning. One thinks, here, of the LV Railroad's lease "in perpetuity" of a canal that had been chartered for a maximum of 150 years.

The State's argument was based on the contention that the Federal Water Power Act had no application in New Jersey. If the terms and provisions of that act were *made* to apply to the State, then New Jersey wanted that act declared unconstitutional. If the United States Attorney General Harry M. Dougherty intended to enforce the act, that would "seriously and substantially impair the value" of the State's property and property rights, "to its irreparable damage."[20]

The State's Attorney General, Thomas McCran, had very carefully prepared his brief for the case in the U.S. Supreme Court. Article XVI stated that the defendants, Dougherty, Weeks, Work, and Wallace, had been "charged with the duty in their respective spheres of action of enforcing acts of Congress under the Constitution, and it is for the purpose of enjoining them from acting by an abuse of power without lawful and constitutional authority, or by wrongful and erroneous interpretation thereof, to enforce within the State of New Jersey the so-called Federal Water Power Act of June 20, 1920." The State, therefore, had brought suit against the defendants "under the executive direction of the Governor of the State of New Jersey," because "the rights and interests" of the people *and* the State were "involved in the premises."[21]

The very provisions of the Federal Water Power Act were interpreted by State Attorney General McCran as being based on a dual purpose: "Federal control of the water powers of said State, their development and use, securing an income therefrom payable to the United States, and transferring the ownership thereof to the United States, which is an unlawful and unconstitutional purpose."[22] Those provisions, "as so threatened to be administered, would deprive the State and the people of the said State of their lawful ownership of such valuable real properties and valuable interests therein and transfer, or provide for the transfer of the same, to the United States, without due process of law."[23]

The irony here is delicious. The writer will be forgiven, perhaps, for enjoying *this* argument from a State that had itself only recently posed an identical threat to the LV Railroad Company lessees, as was now posed to it by the Federal Government! *This* argument about properties and water rights recently wrested from the LV without payment and under a questionable procedure! And the properties themselves acquired in large part by condemnatory action which *at best* was questionable in the light of Articles IV and V of the U.S. Constitution!

We have run the gambit with the Morris Canal. We have witnessed a most significant transition in American history, whereby real power shifted from "big business" and so-called "rugged individualism" to big government and "ruthless legislation." The "public good" is indeed a stern task master over the rights of the private sector, whether it be individuals or corporations. The right of "eminent domain" may well be the headiest power of all.

Pressures from Jersey City, desiring to develop its own waterfront; from rival railroads, seeking waterfront terminals; from Newark, wanting to expand its street railway facilities; from various municipalities, desperately needing potable water — all these pressures had been brought to bear for more than a quarter of a century on the State's lawmakers, to no avail. After all, if the State decided not to exercise its option to buy the canal at a fair price by the end of 1924, it simply had to sit tight and wait until the charter expired at the end of 1974. And this State, over the years, had exhibited a strong predilection for waiting — or perhaps procrastination is a more accurate term.

Once again, a look at a calendar of events will be helpful:

June 10, 1920	Federal Water Power Act
November 23, 1920	Wanaque Reservoir construction begun
March 11, 1922	State negotiating commission appointed
June 22, 1922	LV-MC&BCo. seek injunction at Wanaque
September 12, 1922	Injunction issued
November 29, 1922	LV-N.J. agreement signed; injunction dissolved

February 28, 1923	Canal property to be transferred to MC&BCo. trustees
October 16, 1923	N.J. files suit against Federal Water Power Commission
March 12, 1924	N.J. Ship Canal Act (ship canal to be built from proceeds from sale of canal)
March 13, 1924	Navigation abandoned, all canal reservoirs held public

Without the external pressure of the Federal Water Power Act of 1920, and the attempt of the Water Power Commissioners to enforce that law in New Jersey, it seems doubtful that the State would have acted so quickly to abandon the Morris Canal so as to assume its water rights. Perhaps we are justified in asking with tongue in cheek, from our 1983 vantage point: "Do two riparian wrongs make a riparian right?"

As soon as the water was drawn off the canal, the Board of Conservation and Development was to be "empowered to consent to the diversion of water formerly used for navigation purposes for the use of municipalities, when approved by the Governor and the North Jersey District Water Supply Commission." Drawing off the water was the first desirable thing to be done, "except in portions where that would create more of a nuisance" than already existed. As soon as the properties of the old Canal Company became available, if the municipalities and counties were alert to the opportunity, they might "make an asset of the property" that had "long been a liability."[24]

Cornelius C. Vermeule, Civil Engineer, was appointed on April 7, 1924, to direct the work of dismantling the canal.[25] His immediate task was to examine the canal to determine how much it was going to cost the State to dismantle the canal-related structures which had been built during its hundred years — the bridges and aqueducts, dams and culverts, plane houses, waste gates and planes, and locks. The 87 miles of canal to be dismantled were divided into sections varying from ten to twenty miles in length. Since no funds were to become available for the work until July 1, 1924, Vermeule's staff busied themselves making surveys, plans and estimates, "careful studies for drainage," a "priority list of dangerous bridges" that must be removed or rebuilt, and establishing a field office at Dover. John B. McBride of Sussex became the Resident Engineer in May, 1924.[26]

There were 140 highway and road bridges maintained by the Canal Company, as well as three foot bridges and three change bridges. Individuals, municipalities, or counties maintained 71 other highway and road bridges over the canal, and 10 foot bridges. Railroad and trolley bridges numbered 32. The State Highway Commission began the dismantlement by removing a trolley culvert that went under the canal on Route 46 at Ledgewood. Dubbed "The Hole-In-the-Wall," this blind culvert had always been dangerous. Several railroad companies then asked permission to remove their bridges across the canal, particularly those which were drawn-span bridges. Vermeule's engineering staff supervised the removal of such bridges, so that the drainage of the canal remained unobstructed.[27]

Work on Lake Hopatcong dam began in September, 1924, after earlier test borings and inspections had shown that the dam had

"Hole-in-the-Wall": The Morris County Traction Company's trolley tracks in a culvert on Route 46, Ledgewood. The culvert goes under the elevated Morris Canal (ETF).

serious structural weakness. Work on the new dam was completed by July 1, 1825, and the Hopatcong Gate House, which regulated the flow of water from the lake into the Musconetcong River, was given a bronze plaque containing a historical legend:

MORRIS CANAL

Chartered by the State, December 31st, 1824.
Built for boats of twenty-five tons, 1827-1831.
Enlarged for barges of seventy tons, 1840-1845.
Leased to Lehigh Valley Railroad Company, 1871.
Acquired by the State, November 29, 1922.
Abandonment of navigation authorized, 1924.

This Canal was of great service to Northern New Jersey for half a century. In its best year, 1866, it transported 889,229 tons of freight. Lake Hopatcong, its principal reservoir, was raised by a dam erected at this site in 1827.

Throughout 1924 and into 1925, surveys, plans, maps, and detailed engineering drawings were prepared. At each canal bridge that had to be removed, for example, a "survey of existing conditions and photographs" were made. Notations were made as to property lines, placement of utility poles, water mains, and gas pipelines. Reconstruction plans were drawn up, and estimates were made for each of the sites.[28]

Utility companies, of course, had to be notified where it was found to be necessary to relocate pipes and poles. Moreover, the "Canal Company of 1923" had to notify the utility companies that the leases they had enjoyed under the LV Railroad Company's management were terminated. In the city of Paterson alone, many leases had been granted to Public Service, The Edison Electric Illuminating Company, the New York Telephone Company, and the Passaic Water Company. The new Canal Company, i.e., the Board of Conservation and Development, assigned the leases to the city of Paterson, and then notified the lessees of the change.[29]

In 1896, the Lehigh Valley had leased to the New York Telephone Company the right to place telephone poles at either end of the Pompton River Aqueduct, and one on the towpath about half way between the aqueduct and Chapel Hill Road, Lincoln Park. Then in 1914, another agreement allowed telephone poles on canal property at Mead's Basin, Mountain View, opposite Gabriel's Hotel. Similar agreements with Public Service Electric Company in 1914 allowed power cables to be strung across the canal near the recently abandoned DL&W Station in Mountain View.[30]

Before 1924 was over, the section of canal containing the magnificent brownstone aqueduct at Little Falls was put under contract. By April, 1925, the dismantling crews were ready to commit what Alvin Harlow was to dub "an act of vandalism:" They had prepared the aqueduct for demolition by dynamite.

During 1885, the top and sides of the prism of the aqueduct had been taken off "and a wooden flume was put in, resting on timbers of yellow pine a foot square and twenty feet long." All during April, 1925, workmen had been removing those timbers, "and they were found to be in perfect condition, due to the constant dripping of water over them."[31]

The arch itself originally had been "designed to carry a weight of eighty tons to ten feet." In spite of the fact that everyone feared that the arch seemed ready to collapse, a crane operator drove a 28-ton crane to the center of the aqueduct during the dismantling operation. The crane was stationed at that point for several days while the arch was partially dismantled in preparation for dynamiting.

Single stone blocks used in the aqueduct's construction in 1829 had been found to weigh "as much as fifteen tons, estimated according to ratio of steam power necessary to lift them." No one bothered to speculate on or marvel at what sort of power had been used in the original construction. A towpath had been built along the downstream side of the aqueduct, fenced in with "an iron railing for the safety of horses and men." Part of that walkway was "still clinging to the stones."[32]

In dismantling the interior of the 125-foot span of the arch, the workmen found "ten enormous tie-rods, joined by huge hand-forged chains," which were said "to be in practically as good condition as when they were turned out of the blacksmith shop link by link." The chains varied in length, but their average weight was given as 200 pounds, with average lengths of fifteen feet. From near the keystone of the arch, workmen removed "a stone tablet, bearing the names of the engineer and workmen on the original job, which was tedious

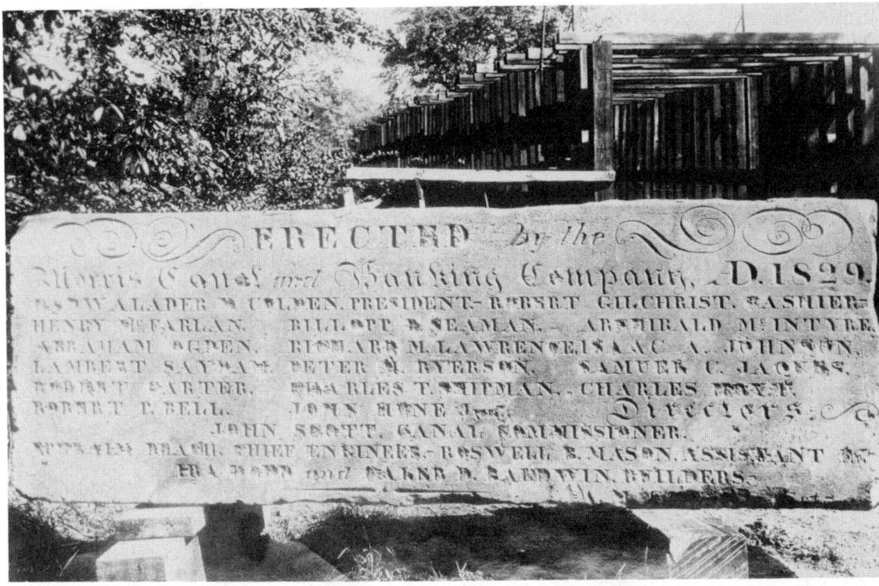

Brownstone block at the Little Falls Aqueduct, commemorating the persons responsible for erecting this Morris Canal structure in 1829 (ETF, Bloomfield Public Library).

because practically everything was done by hand or by horses." The stone was "removed to Trenton, as the property of the commission and the state."

Dismantlement of the arch had been largely carried on from the top of the arch, and most of the dirt and rock was removed from that vantage point until, finally, only the arch itself remained. Dynamite had already been used "several times" to remove the earth and rock, and holes were drilled in the remainder of the arch in preparation for the powder that would blow the Little Falls Aqueduct to kindgom come.[33]

In early May, 1925, a dynamite charge was set and exploded to destroy the arch. Nothing happened! The decrepit, dangerous arch withstood the blast! Another charge, perhaps more strategically placed, finally accomplished the destruction.

Dynamiting the Little Falls Aqueduct in May, 1925 (RUL, photograph by ETF).

Before the inclined planes were completely dismantled, a Canadian engineer studied their design and operation "with a view to their use on a canal in Sao Paulo, Brazil."[34] Much earlier, in the late 1800s, Japanese engineers had also visited the Morris Canal's planes. They took back with them some basic ideas for the Biwa, or Kioto-Tu Canal, connecting the harbor of Osaka with Lake Biwa, Japan's largest lake. In addition to navigation, this canal provided water for irrigation, for human consumption, and for factory power. Begun in 1885 and completed in 1890, the Biwa Canal boasted one inclined plane of 118-foot lift, 1815 feet long.[35]

By April, 1926, Trenton announced that Dover, Boonton, Rockaway, and Paterson had bought the canal rights-of-way through their towns, and the State expected to receive "approximately $33,000" for the sale. All the Dover canal land west of Sussex Street was sold to that town for $16,000; Boonton bought its canal properties for $10,000; Rockaway acquired about 12 acres for $2,000. About 7 acres of canal lands in Paterson went to the Morris and Essex Railroad Company for $3,000, while the city of Paterson assumed ownership of a section of canal bed for $1.00. In the borough of Washington, the New Jersey Power and Light Company bought a tract of 9.34 acres for $3,500.[36]

Even at this early date, April, 1926, the recently drained canal prism was a trash receptacle, as people began to throw refuse and tin cans into it. Of course, in Newark proper, this practice had begun even earlier. As workmen were raking down the sides of the canal prism to fill in the ditch, they were covering "the waste and off-scourings of ninety years." A Newark newspaper columnist noted that "items like beer bottles and tin cans" predominated among the miscellaneous debris, and he thought it would be a good idea to catalogue all the items being covered up by the workmen "for the benefit of scientists and archaeologists" who might dig around in the canal bed a thousand years from now. "People who dig up around Indian camps and Egyptian tombs call the stuff 'culture layers,'" he said, "and somebody may yet discover that the old canal was a brewery site."[37]

The city of Newark was not concerned about archaeologists, however. Throughout 1926, a controversy raged about the use of the canal bed for a rapid transit, or subway line. The canal bed was *suited* for a subway, but it would not serve a large enough segment of the population to make it an ideal rapid transit system. That was the professional opinion of Colonel Thomas W. Sullivan, who had been chief engineer of subway construction in Boston.[38] That position was attacked by the "Roseville Canal Tube Extension Committee" through its counsel, Louis Auerbacher, Jr., who claimed that the canal bed was made to order for high speed tube trains from Newark to Roseville. As far as Auerbacher was concerned, the canal bed was ideally suited for a "high speed line," and contrary to what Mr. Sullivan thought, it would serve "a vast territory with a large, increasing population." He was confident that if the "tube trains" were extended from Newark along the canal bed, there would be a "remarkable transformation" of the city, making of it a "modern metropolis of great size and importance."[39]

By the beginning of 1927, most of the canal properties in Newark had not yet been acquired by the city and, consequently, the proposed railways were not built. The lower end of the canal had been filled in and paved over to become "part of a river road, the city's new Passaic Street."[40]

Plans had been made during 1926 for the counties of Morris and Sussex to rebuild the dam at Waterloo, making it a concrete highway bridge between the two counties. Adjacent property owners were not in favor of the project, however, and the plans were abandoned. The Canal Company sold the properties at Waterloo, and maintenance of the dam was to be assumed by the new owners.[41]

Dismantlement was carried out by and large with an eye toward preserving water rights where they had been enjoyed by various manufacturers along the canal. This was true at Boonton, above the inclined plane, where water had been taken from the canal to generate electric power. At Stanhope, the flow of water from Lake Musconetcong was to be maintained between Lock No. 1 West and Plane No. 2 West to service the Singer Manufacturing Company. At Phillipsburg, too, where the Lopatcong Creek and the canal had run together from the foot of Plane No. 10 West to supply the mill of the Messrs. Green, arrangements were made to continue to supply the mill with water power.[42]

At Greenwood Lake, the Canal Company had to purchase still more land to build a new dam, since the old stone masonry dam would have been too expensive to repair. The Canal Company's deed from the Ryersons (Chapter 11) had entitled Jacob M. Ryerson to continue to use the waters of Greenwood Lake to power the Long Pond Iron Works. The agreement on file in Trenton among the Canal Company records contains a reversionary clause, but apparently that, too, was "perpetual" according to the State's interpretation of that word.[43]

It is interesting to note that the canal's passing was marked by Newark's Central High School in 1926 when students decided to place a "historical tablet" on the Kresge department store. The canal was to be commemorated forever by the words:[44]

> *Along this route for nearly a century ran the Morris Canal, New Jersey's largest and most important artificial waterway. It carried the products of New Jersey's fields, mines and factories from the Delaware to the Hudson. Erected by the Central High School, December 15, 1926.*

The State also placed historical tablets at various sites along the canal, notably at Musconetcong Dam, Saxton Falls, the Colfax Bridge over Pompton River, and on an enormous boulder at the Pompton Dam. The Pompton plaque was stolen sometime after 1941; the author has a photograph of the tablet on its "perpetual" stone backdrop, taken by George W. Keppler on September 12, 1941.

Commemorative plaque at the Pompton Feeder Dam. The plaque has been stolen since this picture was taken in 1941 (GWK).

Commemorative plaque at Saxton Falls (BNK, photograph by Gus Hengle).

The Pompton Feeder Dam and Pequannock Spillway were among the last canal-related structures to be put under contract for reconstruction. By May, 1929, the work of rebuilding was completed far enough for the contractor to submit a final estimate for work done.[45]

On Saturday, May 11, 1929, George W. Keppler noted in his diary that workmen had begun to tear down "the last Plane House on the Morris Canal, Plane No. 12 East at Newark."

When Cornelius C. Vermeule, Jr., issued his *Final Report of Consulting and Directing Engineer,* June 29, 1929, he mentioned the then "present uses" of the canal bed.[46] At Stanhope and Boonton, short stretches of canal remained full "as head races" for water power. At Rockport, between Hackettstown and Port Murray, the State retained a section of canal for use by the Fish and Game Commission as a breeding round for aquatic fowl. Wharton had bought the canal properties within its corporate limits, and below Plane No. 2 West at Stanhope about 4500 feet of canal had been bought privately for the purpose of "supplying water to a gravel pit."

At Saxton Falls the State maintained not only the dam across the Musconetcong River, but also part of the guard lock, without its gates, and a short stretch of prism, to be used as a swimming hole. At Dover, however, the entire canal was filled in, and what had been Dover Basin, the largest canal basin on the system, because a park and playground.

The canal bed from Pompton to Little Falls was to be used for the water conduit which would transport water from the new Wanaque Reservoir. The towpath, guard bank, and guard lock of the Pompton Feeder were to become part of the Passaic County Park system.

The DL&W Railroad bought part of the canal bed at Paterson for a new railroad station, as we have seen, while Paterson planned to use another part of the canal for a storm sewer.

Bloomfield had also anticipated using its part of the canal for a storm sewer, but changed its plans in favor of road development to alleviate increasing traffic problems in the Broad Street area. Since the canal paralleled Broad Street, it was hoped that a road in the right-of-way would eliminate a "bottleneck" between Maple and James Streets caused by a traffic light and rush hour traffic. By the end of 1937, however, the normal traffic jams of commuters and of

Bloomfield High School sports fans pouring out of Foley Field after a game were still a major problem.[47] The Garden State Parkway occupies the spot today.

As of June 29, 1929, Vermeule reported that the net cost of dismantlement had been $1,730,465.14, exclusive of "work executed by the regular maintenance organization." A "Summary of Structures Treated" included:[48]

Bridges removed by Company		143
Bridges removed by Others		49
Bridges maintenance assumed by Others		40
Bridges remaining under other Control		27
	Total	259
New Bridges built (where waterways could not be obstructed)		8
Locks dismantled		26
Locks taken over by Others		6
	Total	32
Planes dismantled		19
Planes taken over by Others		4
	Total	23
Drainage Cuts		156
Large Masonry Culverts Removed		5
Aqueducts Removed		8
Dams and Spillways Built		15

Of the planes, locks, and bridges that remained under the control of "others," some were still being dismantled or removed, "particularly within the limits of the City of Newark," as Vermeule closed his report.

And so it was gone — "officially," at least. The very newspapers that had been most vocal in criticizing the canal now did a complete turnabout and began to carry articles bemoaning its loss. Before the first month of dismantlement was over, the Newark *Evening News* noted that "throughout the 102 miles of its meandering from Phillipsburg to Jersey City," the canal, "in the year that marks its centenary [1924]," was an "empty ditch," in which "here and there the water in low spots awaits the process of absorption and evaporation to disappear entirely." No more would the youngsters of northern New Jersey be able to ask their mothers: "May I go out to swim?" The *Evening News* speculated, probably "ninety percent of the boys who could swim" learned to swim in the Morris Canal. But that was no doubt true everywhere along the canal, for every town seems to have had its "swimming hole" in the old waterway.[49]

Walt McDougall, writing in the Newark *Sunday Call* confessed to being startled at returning to Newark in May, 1924, to see the old Centre Market over the canal already gone, and the canal itself drained, showing itself as a "hideous, repulsive ditch." It suddenly occurred to him that he would never again have occasion to criticize the Morris Canal. Somewhat chastened by that prospect, he turned his talents toward praising the old canal as he remembered it.[50]

> [*His*] *memory pictured what once seemed that elongated stretch of alluring waterway in my boyhood days, its walls of squared, mossy stones overhung by tall trees, bordered by fair gardens faced by comely dwellings, its slow-moving, bluff-browed boats poled languidly along by tanned supermen using strange and provoking profanity in a dialect known only to beings fitted by nature to cope with canal mules.*
>
> [*His imagination let him go beyond the inclined plane at High Street,*] *and along the yellow towpath toward the west through long-vanished green fields flecked with thistles and yellowbirds, under high, weather-worn bridges that linked uplifted roads together, past dark pine and hemlock swamps and silvery mill ponds mirroring silent mills asleep under elms and willows, ever winding, ever clearer and more alluring, toward the unexplored regions were the dim blue mountains melted into the sunset skies, unknown lands, the very names of which were mystical.*

Thus, Mr. McDougall remembered the Morris Canal, and that "medieval moat" that had cut Newark in two for all the years of his life. He missed it, but only after it had gone.

Others, such as Joseph Fulford Folsom, expressed themselves in poetic form. Folsom, who had compiled many a history for the Lewis Historical Publishing Company, left what can only be described as a eulogy in his "Morris Canal Memories:"[51]

The old canal is empty now,
The locks and planes are still,
The boatman's horn no more shall blow
Nor echo from the hill.
How oft afar at candle light
I heard his signal sound:
"Open the lock" before the night,
The Nancy homeward bound.
Up glides the gaunt old specter craft
The shadowed banks along;
While o'er the fields the breezes waft
The concerntina's song.
"Suwanee River" was the lay,
O! plaintive twilight theme;
Far up the level dies away
The cadence like a dream.
Old waterway! in boyhood tried
And every season true,
For summer's swim and winter's glide
There was no friend like you.
Sweet August days, and bermbank green!
Yet could I hope to lie
Without a care in hours serene,
And see the boats go by.
The towpath watch across the pool,
Where heat waves dance and blend,
And laugh to hear the big jack mule
Come stubbing round the bend.
Thus let me have one livelong day,
As in the years of old,
Ere up the level I away
Toward the sunset gold.

Joseph Fulford Folsom

Elderly William Jewel, black tender of Newark's Lock No. 16 East, had worked for the Canal Company since 1883. When the canal was abandoned, he told a reporter that he was sorry to see it go because he felt as if he belonged to it. No longer young enough to get any other kind of job, he simply sat on the porch of the lock tender's house gazing "out upon the Morris Canal of which he is one of the fixtures doomed in the abandonment."[52] William Jewell's story is just one of many such stories of Canal Company employees who had literally devoted their lifetimes to working on the canal, and who were, in a sense, abandoned when the canal was.

The Newark *Sunday Call's* Walt McDougall had said that someday the "romantic traditions" of the old canal would be "gathered into a book, when its course becomes a thing of myth." There would be plenty to write about, for the Morris Canal had[53]

> *its strange, forbidding houses haunted by dark tales, its queer hotels, patronized by rough coarse men, its own saloons harboring hard-drinking fighters, its tough, loud-voiced women and its own silent, wary children aloof from the world on that slow stream, its obstinate mules and patient, harassed horses that for some seven months in each year moved slowly up and down . . . then vanished to hibernate until another spring.*

Another reporter had speculated that "sometime two hundred years hence, an industrious historian, after perhaps years of research, may set out to run again the course followed by the Morris Canal. He will tell of how it has become here a railway; there converted into part of a building development; in some other place, something else. And (oh, great find), here and there he may discover traces of the old bed itself, brush-grown but still discernible."[54]

There is no reason to wait so long. There is much yet "discernible" of the old canal, and there is a host of "romantic traditions," still alive, to tell about its hundred miles and its hundred years. But those are the stories and the memories of which another book may be written.

REFERENCES

1. Paterson *Morning Call,* in George W. Keppler Collection, *Scrapbook,* p. 99.
2. Morris Canal and Banking Company *Docket Files* (hereinafter cited as MC&BCo. *Docket Files).*
3. *Laws of New Jersey,* 1922, Chapter 212.
4. *Laws Providing for the Acquisition of the Morris Canal by the State of New Jersey and Its Final Disposition* (Trenton, N.J.: 1924), p. 3.
5. *Ibid.,* p. 9.
6. *Ibid.,* pp. 13-14.
7. *Ibid.,* p. 14.
8. *Ibid.,* p. 15.
9. *Ibid.,* p. 16.
10. *Ibid.,* p. 17.
11. Newark *Sunday Call,* December 3, 1922.
12. *Ibid.*
13. Henry B. Kummel, *Memorandum Regarding the Morris Canal* (Trenton, N.J.: 1923), p. 5.
14. *Ibid.,* pp. 6-8.
15. Bill of Complaint, U.S. Supreme Court: New Jersey *vs.* the Federal Water Power Commission, October Term, 1923, pp. 10-11.
16. *Ibid.,* p. 12.
17. *Ibid.,* p. 9.
18. *Ibid.,* pp. 12-13.
19. *Ibid.,* p. 17.
20. *Ibid.,* p. 23.
21. *Ibid.,* pp. 23-24.
22. *Ibid.,* p. 25.
23. *Ibid.,* p. 28.
24. Newark *Evening News,* March 12, 1924.
25. Cornelius C. Vermeule, Jr., *Final Report of Consulting and Directing Engineer, June 29, 1929* (E. Orange, N.J.: 1929), p. 8 (hereinafter cited as Vermeule's *Final Report*).
26. *Ibid.,* pp. 8-9.
27. *Ibid.*
28. *Ibid.,* p. 11.
29. MC&BCo. *Docket Files* assignment dated December 23, 1926.
30. All agreements are in the MC&BCo. *Docket Files,* Trenton, N.J.
31. Newark *Evening News,* April 25, 1925.
32. *Ibid.*
33. *Ibid.*
34. Vermeule's *Final Report,* p. 71.
35. *Engineering News,* April 13, 1893, p. 34.
36. Newark *Evening News,* April 15, 1926.

37. Newark *Sunday Call,* June 8, 1924.
38. Newark *Evening News,* February 1, 1926.
39. *Ibid.,* February 4, 1926.
40. Newark *Sunday Call,* February 6, 1927.
41. Vermeule's *Final Report,* pp. 14-15.
42. *Ibid.,* p. 15.
43. *Ibid.,* pp. 16-17.
44. Newark *Sunday Call,* November 28, 1926.
45. Vermeule's *Final Report,* p. 13.
46. *Ibid.,* pp. 19-20.
47. Bloomfield *Independent Press,* October 8, 1937.
48. Vermeule's *Final Report,* pp. 22-3.
49. Newark *Evening News,* May 3, 1924.
50. Newark *Sunday Call,* May 18, 1924.
51. *Ibid.,* January 18, 1925.
52. Newark *Evening News,* December 14, 1922.
53. Newark *Sunday Call,* May 18, 1924.
54. *Ibid.,* May 3, 1924.

Appendix A

TABLE OF TRADE, TONNAGE, AND INCOME
MORRIS CANAL: 1845-1902

(compiled from charts in the Three Governors' Report of 1903, pp. 124-129)

Year	Tonnage				Income
	Total	Coal	Ore	Misc.*	
1845	58,259	12,567	5,802	39,890	$ 18,997.45
1846	109,505	41,142	17,073	51,290	51,212.39
1847	155,559	61,951	28,314	65,294	67,687.09
1848	204,682	82,159	46,922	75,601	65,531.75
1849	234,305	103,482	53,722	77,101	90,220.48
1850	239,682	98,100	60,055	81,527	98,224.25
1851	281,707	137,237	62,690	81,780	110,730.56
1852	358,797	180,189	77,132	101,476	140,202.28
1853	467,288	222,582	104,181	140,555	187,449.71
1854	543,269	267,864	117,191	158,214	245,431.54
1855	533,204	290,730	113,294	149,180	272,125.23
1856	563,413	303,400	133,845	126,168	313,026.15
1857	536,362	284,298	125,805	126,259	286,668.61
1858	554,034	342,725	108,684	289,883	275,019.44
1859	638,019	350,331	155,467	132,221	305,537.93
1860	707,631	404,464	186,064	117,103	350,710.63
1861	619,369	414,939	103,163	101,267	291,846.37
1862	612,018	324,297	169,108	118,613	303,154.39
1863	718,519	402,446	191,667	124,406	374,601.76
1864	723,927	382,863	223,924	117,140	590,393.26
1865	716,587	416,189	174,838	125,560	600,584.30
1866†	889,220	459,175	290,165	139,880	616,350.36
1867	822,741	424,831	289,550	108,360	449,652.46
1868	744,412	350,192	269,672	144,548	423,767.30
1869	650,200	281,626	227,696	140,878	431,245.80
1870	707,572	360,140	220,079	127,353	391,549.76

	Tonnage				
Year	Total	Coal	Ore	Misc.*	Income
1871	629,044	329,584	185,482	113,987	283,725.34
1872	685,191	356,157	191,597	137,435	397,181.72
1873	634,710	299,214	213,447	112,049	323,631.51
1874	491,816	413,918	14,802	63,095	271,436.17
1875	451,045	250,047	99,607	101,390	270,216.62
1876	399,613	290,588	67,447	41,577	217,982.59
1877	417,819	305,897	58,736	53,184	170,855.11
1878	363,348	233,560	85,705	44,083	156,156.46
1879	408,046	323,924	56,123	27,999	137,553.11
1880	503,486	427,606	44,897	30,893	215,667.21
1881	534,743	323,366	134,871	75,506	221,016.59
1882	459,846	304,458	82,478	72,909	193,789.25
1883	423,495	323,477	35,885	64,132	158,167.49
1884	414,097	307,847	29,372	34,829	148,602.01
1885	364,554	301,654	21,803	41,095	$101,487.41
1886	389,462	315,187	36,576	37,698	87,166.61
1887	378,567	266,276	65,584	46,706	109,364.98
1888	372,049	307,847	29,372	34,829	101,476.20
1889	413,073	300,340	67,988	44,744	134,608.66
1890	394,432	297,417	54,889	42,125	120,283.00
1891	370,109	312,968	21,978	35,161	315,200.49
1892	288,011	252,568	5,213	30,229	247,724.31
1893	259,282	236,510	—	22,771	196,052.53
1894	300,092	287,309	1,987	10,796	228,564.76
1895	270,931	259,778	—	11,153	208,557.71
1896	235,168	211,954	—	23,213	167,254.74
1897	231,870	208,480	—	23,390	167,411.70
1898	191,287	166,673	—	24,613	136,924.01
1899	173,555	159,456	—	14,099	137,626.21
1900	125,829	119,479	—	6,350	111,072.10
1901	122,786	115,391	—	7,395	105,502.66
1902	27,392	20,411	—	6,981	52,076.31

Notes:

*Miscellaneous tonnage included: bar and pig iron; blooms and billets; brick; castings and machinery; charcoal; clay, earth, sand, stone, etc.; dry goods, groceries, etc.; flour, feed, meal, etc.; hoop poles and rails; iron rails; lime; lime stone; lumber and cement; nails, spikes, etc.; plaster; timber and logs; wood, bark, etc.; and sundries.

†This was the canal's peak year.

Appendix B

MORRIS CANAL AND BANKING COMPANY
KNOWN CONTRACTORS AND SUPPLIERS

Name	Where Worked; What Supplied
Alyea, Jacob	Newark; malleable iron parts
Andrew, Capt. M.	Jersey City Extension
Bailey, Pitcher Co.	Jersey City Extension
Baldwin, Aaron	Planes No. 9 and 10 East
Baldwin, Caleb & Ira Dodd	Planes No. 11 East; Lock No. 16 East; Little Falls Aqueduct
Bartlett, Gamaliel & A. A. Drake	Planes No. 1 and 2 West
Barker, Levi	Boatbuilder
Beebee, F. W.	Newark
Berback [Co.]	Chain for planes
Birnie & Sharp	—
Bishopp, Alfred	—
Blackwell & McFarlan	Dover; cast iron parts
Brinknall Mills & Co.	Planes No. 1, 2 and 8 West
Britten, Mr., & William Shields	Planes No. 5 and 6 West
Broadwell, Asa	Lime
Chapin, Solin	Planes No. 9, 10, 11 West
Collins, J. & C.	Jersey City Extension; Planes No. 5 and 7 West
Condit, Nathaniel & Timothy	Montville; waterwheels (wooden), Planes No. 8, 9 and possibly 10 East
Cooper, Mr.	Weigh scales, Lock No. 16 East
Cooper-Hewitt & Co.	Phillipsburg, Trenton; iron rails
Davis & Brooks Co.	Jersey City Extension; iron parts
Dupres May & Co.	Jersey City
Evans Thompson & Co.	Plane machinery
Fairbanks Scale Co.	Newark; weigh scale, Plane No. 12 East
Finlay, James	Waterwheels (metal)
Ford, Charles	Iron plate, Plane No. 10 East
Freeman, ?	Iron plates, Little Falls Aqueduct
Fuller, James	Rockaway; iron castings
Garrett, Patrick	Drakesville; Jersey City Extension

Name	Where Worked; What Supplied
Gaynor, John	—
Geyelin, Emile	Turbines
Goolden, Jacob	Phillipsburg; bridges (1830)
Goolden & Van Dyke	Planes No. 3 and 4 West
Green & Hall	—
Gridley, Timothy	Newark
Harned, ?	Boat builder
Hazard, Erskine	Mauch Chunk; wire cable
Hill & Vreeland	Jersey City Extension
Hoagland Iron Works (Union Foundry)	Rockaway; iron castings
Jackson, W.	Philadelphia; rope
Jefferson, Pierce	Dover; repairs
Kinney, A. W.	Newark; malleable iron parts and piping
Lacy	—
Lippincott & Miner	Mauch Chunk
Little, John	Hackettstown; plane house machinery
Logan, C. & Co.	Liverpool; rails
McFarlan & Ayres	Dover; iron castings
Mason, Arnold, & Co.	Jersey City Extension
Mason & Pratt	Greenwich section; Jersey City Extension
Merrick & Town	Philadelphia; turbines
Newall Co.	Wire cable
O'Brien Pursell & Co.	Jersey City Extension
Orick, Wm. R.	Boat builder
Peru Iron Co.	New York; chain
Pope, Samuel	Paterson
Pregnall, William	Dover; boat builder
Prior, John	Jersey City; boat builder
Reilley	Rochester; boat builder
Righter, J. C.	Rockaway; iron parts
Roberts, L.	Jersey City; basin and docks
Roberts, Samuel & Co.	Jersey City Extension
Roebling, John A.	Wire rope and cable

Name	Where Worked; What Supplied
Shields, Thomas & Isaac	Planes No. 6 West
Taylor, George	Jersey City Extension
Templin & Conover	Easton; plane house machinery
Union Works	Paterson (later Rockaway)
Van Emburgh, C.	Newark; dock
West Point Foundry	Plane house machinery parts
Wetmore, David (New Jersey Iron Co.)	Boonton; rails
Williams, James	Greenwich

Appendix C

MORRIS CANAL AND BANKING COMPANY KNOWN OFFICERS AND DIRECTORS, 1825–1871

Name	Residence	Position*	Years of Service
Alsop, Joseph N.	NY	D	1844, 1846
Armstrong, Thomas	NJ	D	1870, 1871
Astor, Henry	NY	D	1825, 1826
Astor, William	NY	D	1827, 1828
Bacott, Robert C.	NJ	D	1839
Barretto, Francis	—	D	1826, 1827
Bayard, Charles P.	PA	D	1867 through 1871
Bayard, William, Sr.	NY	P	1825, 1826
Bayard, William, Jr.	NY	D	1825, 1826
Bell, Robert P.	NY	D	1826 through 1832
Biddle, Edward R.	PA	D	1837
		VP	1838
		P	1839 through 1841
		D	1846 through 1848
Bishop, Japhet	—	D	1844
Blandy, John F.	PA	D	1869 through 1871
Borrow, Wm.	—	D	1840
Boyd, J. C.	—	D	1841
Britten, Stephen P.	NJ	D	1827
Brooks, Alfred	—	D	1841
Brooks, Samuel R.	NY	D	1832 through 1853
Bryant, John J.	NJ	C, D	1844, 1848
Cadwallader, Thomas	PA	D	1837, 1838, 1840
Campbell, Duncan	—	D	1826, 1827
Carter, Robert	NY	D	1826 through 1832
Chambers, Thomas	—	D	1841
Clark, Chester	—	D	1844
Coddington, Jonathan J.	NY	D	1844 through 1853
Colden, C. D.	NY	P	1826 through 1834
Colket, Coffin	PA	D	1857, 1861, 1863 through 1866
Cook, Zebedee, Jr.	NY	D	1844
			1846 through 1848

Name	Residence	Position*	Years of Service
Cooper, Lewis	PA	D	1870, 1871
Crane, Israel	NJ	D	1826 through 1833
Crary, J. S.	NY	D	1836
Cromelien, George	PA	D	1863 through 1871
Darcy, John S.	NJ	D	1836
Davis, Thomas B.	NY	D	1834, 1835
Dehon, Theodore	NY	D	1846
Devine, Mark	PA	D	1861, 1863, 1869 through 1871
Dickenson, D.	NJ	D	1847, 1848
Dickerson, Philomon	NJ	D	1836
Dickey, William	NJ	D	1847, 1848 1852 through 1855, 1857
Dickson, Levi	PA	D	1854, 1855, 1857, 1861, 1863 through 1868
Draper, Simeon, Jr.	NY	D	1837, 1838
Earp, Robert	PA	D	1831, 1832
Eckford, Henry	NY	D	1825, 1826
Edwards, G. W.	PA	D	1847, 1848
Ellis	—	D	1838
Farnum John	PA	D	1852 through 1855
Farnum, Paul	PA	D	1852 through 1855 1857
Fisher, Charles H.	PA	D	1861
Fiske, George B.	—	D	1840
Frelinghuysen, Theodore	NJ	D	1827
Frelinghuysen, F. T.	NJ	D	1841, 1844 1846 through 1848 1852 through 1855 1857, 1861 1863 through 1871
Gilchrist, Robert	NJ	C	1826
Goble, Luther	—	D	1841
Goodhue, Jonathan	NY	D	1836, 1837
Gouverneur, Samuel	NY	D	1826
Gratz, Simon	PA	D	1869 through 1871
Green, Enoch	—	D	1832 through 1834

Name	Residence	Position*	Years of Service
Green, John C.	NY	D	1844 1846 through 1848
Gregory, Dudley, S.	NJ	D	1838
Griswold, George	NY	D	1836 through 1838 1840 1844 through 1848
Haggerty, John	NY	D	1836
Harris, Joseph C.	PA	D	1852, 1853
Hart, Samuel	PA	D	1861
Haven, Woodward	—	D	1844
Hayne, J. W.	—	D	1844
Henderson, David	—	D	1838
Henriques, David	—	D	1834
Henshaw, D.	MA	D	1836
Hicks, Henry W.	NY	D	1835 through 1838
Hogdgon, Alex L.	PA	D	1869 through 1871
Hollis, Peter C.	PA	D	1863 through 1871
Hone, Isaac	NY	D	1830, 1831
Hone, John, Jr.	NY	D	1829
Hoyt, Charles	—	D	1829 through 1832
Hoyt, J.	—	—	—
Hurd, Jacob	NJ	D	1839
Irick, John S.	PA	D	1867 through 1871
Irving, Washington	NY	D	1836 through 1840
Jackson, Daniel	NY	D	1836 through 1838
Jackson, Washington	PA	D	1837, 1838
Jacques, Samuel	NY	D	1829 through 1834
Johnson, Isaac A.	NJ	D	1829 through 1835
Joseph, Joseph L.	NY	D	1836, 1837
Judd, Samuel	NY	D	1832
Keene, James	PA	D	1861 1863 through 1866
King, C. W.	NY	D	1840, 1841
Kinney, Abraham	NJ	D	1840
Koons, Charles	PA	D	1864 through 1868
Lardner, Alex	—	D	1839
Lawrence, Augustus H.	NY	D	1827
Lawrence, Richard M.	NY	D	1827 through 1831

Name	Residence	Position*	Years of Service
LeRoy, Jacob	NY	D	1846
Little, Jacob	NY	D	1844
Lord, D. N.	NY	D	1841
Lord, Edwin	NY	D	1835 through 1838
Low, Daniel	NY	D	1835
MacCulloch, George P.	NJ	D	1826
Mane, Samuel	—	D	1840
March, Thomas	NY	D	1831
Marsh, Ephraim	NJ	D	1847, 1848, 1852 1853 through 1864
Marsh, John	NJ	D	1852 through 1869
Marsh, Rolph	NJ	D	1847 through 1855 1857
Marsh, William W.	NJ	D	1857, 1861 1863 through 1871
McCarter, Thomas, N.	NJ	D	1854 through 1871
McFarlan, Henry, Sr.	NY	D	1826 through 1830
McFarlan, Henry, Jr.	NJ	D	1830 through 1832 1844, 1846 1850 through 1855 1857
McIntyre, Archibald	NY	D	1829 through 1835
McLane, Louis	NY	P	1836, 1837
McMartin, Peter	NJ	D	1844 1846 through 1848 1852
McMartin, W.	—	D	1841
Megear, Thomas J.	PA	D	1869 through 1871
Miller, William S.	—	D	1835
Mirandole, E.	—		1841
Moss, John	PA	D	1837
Murray, James B.	NY	P D	1834, 1835 1836 through 1838
Northrup, Joseph, Jr.	—	D	1839, 1841
Norton, Frederick A.	—	D	1835
Oakley, Thomas J.	NY	D	1833
Ogden, Aaron	NJ	D	1838
Ogden, Abraham	NJ	D	1827 through 1832
Ogden, D. B.	NJ	D	1825 through 1827

Name	Residence	Position*	Years of Service
Orne, James H.	PA	D	1870, 1871
Parker, James	NJ	D	1835
Parsons, Andrew	NJ	D	1838 through 1841
Patten, William	PA	D	1870, 1871
Peterson, Mathew	—	D	1834
Post, Edwin	NY	D	1837, 1840
Post, Jotham	—	D	1835, 1840
Post, Waldron B.	—	D	1838, 1840
Randolph, Jacob F.	NJ	D	1865 through 1867
		VP	1868
		P	1869 through 1871
Rankin, John	—	D	1844
Rathbone, William P.	—	D	1825, 1826
Redfield, John H.	NY	D	1857, 1861 1863 through 1868
Redfield, William C.	NY	D	1854, 1855, 1857
Redmond, Samuel	NY	D	1832, 1833
Rhinelander, William	NY	D	1834
Robb, Thomas	NJ	D	1870, 1871
Ryerson, Peter M.	NJ	D	1827 through 1832 1836 through 1839
Salisbury, H.	NJ	D	1863 through 1866
Scofield, Joseph H.	NJ	D	1844 1846 through 1848
Scott, Daniel	NJ	D	1846
Scott, Elijah	NJ	D	1844
Scott, John	NJ	D	1825
Scudder, Isaac	NJ	D	1852 through 1855 1857 1861 through 1871
Seaman, Billopp B.	NY	D	1826 through 1835
Shipman, Charles T.	NJ	D	1826 through 1832
Shotwell, Henry R.	NJ	D	1852 through 1855 1857 1861 through 1871
Smith, Nathan	NJ	D	1847, 1848, 1850
Smith, Sheldon	NY	D	1832
Smith, Thomas H.	NY	D	1826
South, George W.	NJ	D	1861 1863 through 1869

Name	Residence	Position*	Years of Service
Southard, Samuel	NJ	P	1838
Southmayd, H.	NJ	D	1846, 1848, 1850, 1852, 1853
Soutter, J. T.	NJ	D	1846, 1848, 1850, 1852, 1854
St. Felix, J. R.	NY	D	1834, 1835
Stewart, Cornelius	NJ	D	1854, 1855, 1857 1861 through 1871
Strader, John, Jr.	—	D	1844
Stone, Asaph	—	D	1841
Strong, J.	NJ	D	1846
Suydam, Lambert	NY	D	1827 through 1830
Swartwout, Samuel	NY	D	1834, 1835
Swift, Joseph G.	NY	D	1825, 1826
Talcott, William H.	NJ	D	1847 through 1850 1853 through 1855 1857 through 1864
		P	1865 through 1868
Talman, J.	—	C	1826
Tileston, Thomas	—	D	1844
Travers, John	NJ	D	1836 through 1838
Tyler, Daniel	CT	P	1844 through 1846
Valentine, Caleb H.	NJ	D	1847, 1848
Van Santvoord, A.	NJ	D	1847, 1848, 1850 1852 through 1855 1857
Van Vorst, Cornelius	NJ	D	1838, 1841
Varick, J. V. B.	NY	D	1832 through 1835
Varnum, Joseph B.	NY	D	1844 1846 through 1848 1850 1852 through 1855 1857, 1861
Vermilye, Thomas	NY	D	1825, 1826
Wall, Garret D.	NJ	D	1836, 1837, 1840
Walter, Alfred	—	P	1897
Ward, John D.	NJ	D	1850, 1855, 1857, 1861, 1863, 1864, 1868, 1869
		VP	1865 through 1867

Name	Residence	Position*	Years of Service
Waterman, Isaac	PA	D	1850
			1852 through 1855
Welling, Richard	—	D	1840
Whitney, Asa	PA	D	1846, 1848, 1850, 1852
			1853 through 1869
		P	1847
Whitney, Stephen	NY	D	1836 through 1838
Williams, Richard R.	NY	D	1831
Williamson, Benjamin	NJ	P	1844 (resigned immediately)
		D	1844, 1846
Williamson, Isaac H.	NJ	D	1836, 1837
Williamson, Isaiah V.	NJ	D	1850
			1852 through 1869
Willink, John A.	NY	D	1844
Wilson, Blakely	NJ	D	1867, 1868
		VP	1869 through 1871
Wistar, Isaac J.	PA	D	1867, 1868
Wood, W. R.	—	D	1841
Woodruff, E. B.	NJ	D	1853 through 1871
Wright, William	NJ	D	1852 through 1866
Yates, Charles	—	D	1840
Yates, Henry	NY	D	1837, 1838, 1840
Zabriskie, Christian B.	NJ	D	1836, 1837

*Key for abbreviations: C, cashier; D, director; P, president; VP, vice president.

Appendix D

MORRIS CANAL AND BANKING COMPANY KNOWN STOCKHOLDERS, 1831

(compiled from a list of those who owed money on shares)

Name	Shares	Name	Shares
Adams, E.	2	Guillerau, J. & Son	10
Arrowsmith, N.	1	Halsey, J. D.	1
Bayard, William	10	Harris, William H.	1
Beach, Ephraim	70	Hart, Samuel	3
Berry, J. W.	1	Hurd, Jacob	1
Best, Andrew	1	Jackson & Smith	3
Betting, Hiram	2	Jackson, William	10
Blauvelt, C. S.	1	Jacobs, William	1
Bowlsby, Thomas	1	Kellogg, E.	5
Boyd, James	1	King, H. and R.	1
Britten, Stephen P.	5	Lawrund, J. M. D.	50
Burke, F.	1	Limeback, Charles F.	2
Burtow & Laurie	4	Martin, N.	1
Carr, John	1	Matthews & Cook	5
Churtey, John	4	McCabe, F.	1
Coe, Jared	1	McMillin, George	4
Condit, Sidney	1	McIntyre, A.	2
Coursen & Co.	5	Meeker, D.	5
Crane, O.	1	Muir, Thomas	5
Cruger, Alfred	5	Murphy, John	10
Darcy, John S.	3	Noyes, J. W.	5
Davis, William A.	40	Nutman, James	1
Dickerson, A.	1	Ogden, D. B.	200
Dickerson, J.	5	Parsons, Andrew	5
Dodd, Stephen	2	Rankin, William	4
Doran, Edward	5	Randolph, Charles	16
Douglass, E. A.	5	Remson, John H.	5
Fairchild, D. H.	1	Roach, Philip	5
Fanton, John	1	Ruckle, John	1
Ferguson, John	10	Sandford, D.	1
Funneman, A.	1		

Name	Shares	Name	Shares
Scofield, J. W.	2	Tappan & Cook	5
Shaw, Benjamin	1	Thompson, Mark	1
Shemway, Jacob	5	Van Blarcom, ?	1
Sherman, John	1	Vannatta, A.	1
Sherrerd, John M.	5	Van Winkle, E.	2
Slover, John W.*	5	Vermilye, Thomas	700
Smith, E. T.	1	Vreeland, H. C.	1
Smith, Joshua	2	Ward, E.	1
Smith & Baldwin	3	Ward, J. S.	3
Soper, John R.	3	White, James	2
Stansberry, A. H.	1	White, L.	1
Stayhl, Robert	1	White, William	2
Stone, William L.	20	Wilson, N. and J.	2
Sutton, Herman	2		

*This could be Glover.

Bibliography

MORRIS CANAL AND BANKING COMPANY MATERIAL

ANNUAL REPORTS

1826 — *Annual Report Upon the Morris Canal for the Year Ending July 1, 1826.* New York: William Davis, Jr., 1826. A second version of this report was printed by Jacob Mann, Morristown, NJ, 1826.

1827 — *Report to the Directors of the Morris Canal and Banking Company; Made by the President, Cadwallader D. Colden, May 1st, 1827: With a Statement of the Work Then Done Upon the Canal, A Report from the Chief Engineer, and a Communication from Governor Clinton.* New York: William Davis, Jr., 1827.

1828 — *Report of the Joint Committee of Council and Assembly Appointed to View the Morris Canal and Inclined Planes.* n.p., 1828.

1832 — "Report of a Joint Committee of the Directors and Stockholders of the Morris Canal and Banking Company, with a Communication from the Directors who were of the Joint Committee, to the Stockholders. August 23, 1832." Reprinted in the *American Railroad Journal,* September 29, 1832, pages 626-27.

1835 — *Report of the President and Directors of the Morris Canal and Banking Company, to the Stockholders.* New York: James Van Norden, January 1835.

1836 — *Report of the President and Directors of the Morris Canal and Banking Company, to the Stockholders.* New York: James Van Norden, January, 1836.

1837 — *Report of the President and Directors of the Morris Canal and Banking Company, to the Stockholders.* New York: James Van Norden, March, 1837.

1840 — *Report to the Stockholders of the Morris Canal and Banking Company.* n.p., December, 1840.

1845 — *Report of the President of the Morris Canal and Banking Company to the Board of Directors.* Thomas Snowden, Printer. August 12, 1845.

1846 — *Report of the President of the Morris Canal and Banking Company to the Board of Directors.* Thomas Snowden, Printer. February 12, 1846.

1848 — *Report of the President and Directors of the Morris Canal and Banking Company of 1844, to the Stockholders.* New York: Sibell and Mott, March 17, 1848.

1849 — *Report of the Board of Directors to the Stockholders of the Morris Canal and Banking Company.* New York: Sibell and Mott, April 2, 1849.

1850 — [*Report*] *to the Stockholders of the Morris Canal and Banking Company.* New York: Sibell and Mott, April 1, 1850.

1851 — "Report to the Stockholders of the Morris Canal and Banking Company." Reprinted in the *American Railroad Journal,* March 3, 1851.

1852 — [*Report*] *to the Stockholders of the Morris Canal and Banking Company.* New York: W. E. and J. Sibell, April 5, 1852.

1853 — [*Report*] *to the Stockholders of the Morris Canal and Banking Company.* New York: W. E. and J. Sibell, April 4, 1853.

1854 — [*Report*] *to the Stockholders of the Morris Canal and Banking Company.* New York: W. E. and J. Sibell, April 3, 1854.

1855 — [*Report*] *to the Stockholders of the Morris Canal and Banking Company.* New York: W. E. and J. Sibell, April 2, 1855.

1856 — "Report to the Stockholders of the Morris Canal and Banking Company." Reprinted in the *American Railroad Journal,* May 17, 1856.

1857 — [*Report*] *to the Stockholders of the Morris Canal and Banking Company.* New York: Latimer Bros. & Seymour, April 7, 1857.

1859 — [*Report*] *to the Stockholders of the Morris Canal and*

Banking Company. New York: Latimer Bros. & Seymour, April 4, 1859.

1860 — [*Report*] *to the Stockholders of the Morris Canal and Banking Company*. New York: Latimer Bros. & Seymour, April 2, 1860.

1861 — [*Report*] *to the Stockholders of the Morris Canal and Banking Company*. New York: Latimer Bros. & Seymour, April 1, 1861.

1862 — "[Report] to the Stockholders of the Morris Canal and Banking Company," [1862] Reprinted in the *American Railroad Journal,* n.d.

1863 — [*Report*] *to the Stockholders of the Morris Canal and Banking Company*. New York: Latimer Bros. & Seymour, April 6, 1863.

1864 — [*Report*] *to the Stockholders of the Morris Canal and Banking Company*. New York: Latimer Bros. & Seymour, April 4, 1864.

1865 — [*Report*] *to the Stockholders of the Morris Canal and Banking Company*. New York: J. O. Seymour & Co., April 3, 1865.

1866 — [*Report*] *to the Stockholders of the Morris Canal and Banking Company*. Jersey City, NJ: Times Printing House, April 2, 1866.

1867 — *Report to the President and Directors of the Morris Canal and Banking Company to the Stockholders*. Jersey City, NJ: Edward Jardine, April 1, 1867.

1868 — *Report of the President and Directors of the Morris Canal and Banking Company to the Stockholders*. Jersey City, NJ: Edward Jardine, April 6, 1868.

1869 — *Report of the Directors of the Morris Canal and Banking Company to the Stockholders*. Jersey City, NJ: Times Printing House, April 5, 1869.

1870 — *Report of the Directors of the Morris Canal and Banking Company*. Jersey City, NJ: John H. Lyon, April 4, 1870.

1871 — *Report of the Directors of the Morris Canal and Banking Company*. Jersey City, NJ: John H. Lyon, April 3, 1871.

OTHER CANAL COMPANY MATERIAL

Clearance Book, Brooklyn Lock, Lake Hopatcong, NJ, 1869-1873. A *Morris Canal and Banking Company of 1844* log. Entries include the names or numbers of the boats and masters; their points of origin and destination; their cargoes, weights and rates of toll; and the names of the parties who paid the tolls. Part of the George W. Keppler Collection, in the author's possession.

Clearance Book, Stanhope, New Jersey. A *Morris Canal and Banking Company of 1844* log. Miscellaneous loose pages with various dates in the 1850s. Part of the George W. Keppler Collection, in the author's possession.

Docket Files. An extensive file of maps, deeds, and some correspondence regarding Canal Company lands, acquisition, litigation, water rights, etc. Morris Canal and Banking Company Files.

"Letter from the Lehigh Valley Railroad Company to the Commission for the Abandonment of the Morris Canal. April 27, 1922." Morris Canal and Banking Company Files.

Minutes. Four volumes, 1829–1841, 1844–1880. Morris Canal and Banking Company Files.

Rules and Regulations, March 1, 1904. Photocopy of some excerpts in the author's collection.

Thomas, E.B. "The Morris Canal, An Economic Waste: Memorial to the Senate and General Assembly of the State of New Jersey. Statement by the Lehigh Valley Railroad Company. February, 1911." Morris Canal and Banking Company Files.

PAMPHLETS, MANUSCRIPTS, AND MISCELLANEOUS DOCUMENTS

Barker, Jacob. *Letters Developing the Conspiracy Formed in 1826 for His Ruin.* New York: n.p., 1827.

Beach, Ephraim. Pay, pension, and service records, War of 1812. General Services Administration, Washington, DC.

Brooks, Samuel Reeves. *Letter to the Hon. Robert J. Walker, on the Progress of the Iron Manufacture and on the Prospects of Steel Manufacture for Steel Rails, Car Wheels, Etc., in Connec-*

tion with the Franklinite and Zinc Mines of New Jersey. New York: Stockholder and Mining Record Print, 1866. New York City Public Library.

Hewitt, Abram S., Collected Papers. Voluminous correspondence, much of which reflects the relationship between the Cooper-Hewitt enterprise and the Canal Company. Passaic County Historical Society Museum, Lambert Castle, Paterson, NJ.

Keppler, George W., Canal Collection. Extensive collection of all manner of canal memorabilia, in the author's possession.

Kummel, Henry B. *Memorandum Regarding the Morris Canal.* Trenton, NJ: MacCrellish and Quigley, 1923. Prepared for the use of New Jersey legislators in making up laws for the abandonment of navigation on the canal, and for the State's acquisition of canal properties. In the Keppler Collection.

Laws of New Jersey. Bound volumes of laws enacted by the State Legislature; bound by years and by legislative sessions. Passaic County Law Library, Court House, Paterson, NJ.

Laws Providing for the Acquisition of the Morris Canal by the State of New Jersey and its Final Disposition. Trenton, NJ: State Department of Conservation and Development, 1924.

Litigation:

"Answer" of the Central Railroad of New Jersey to a Bill in Chancery of New Jersey between the Morris Canal and Banking Company (Complainants) and the Central Railroad of New Jersey and others (Defendants). September, 1863. Engineering Societies Library, New York City.

In Chancery of New Jersey: *Bill of Injunction, Between the Morris Canal and Banking Company, complainants, and Francis O. Matthiessen, and others, defendants.* Jersey City, NJ: John H. Lyon, 1863. Engineering Societies Library, New York City.

In the Court of Appeals of the State of New York: *In the Matter of the appraisal of the damage of Peter Townsend and William H. Townsend (Appellants), the owners of lands required by the Morris Canal and Banking Company (Respondents), for the purpose of a Reservoir for thier Canal.* Engineering Societies Library, New York City.

Law Opinion [of Chancellor Isaac H. Williamson] *in the Case of the Paterson Manufacturing Society Against the Morris*

Canal and Banking Company. Paterson, NJ: T. Warren, 1845.
Supreme Court of the United States, October Term, 1923: *State of New Jersey, Complainant, vs. Harry M. Daugherty, Personally and Individually and as Attorney-General of the United States; John W. Weeks, Personally and Individually and as Secretary of War; Hubert Work, Personally and Individually and as Secretary of the Interior; and Henry C. Wallace, Personally and Individually and as Secretary of Agriculture; and John W. Weeks, Hubert Work, and Henry C. Wallace, as Members of and Constituting the Federal Power Commission, Defendants*. n.p., n.d., Rutgers University Library, Special Collections.

Loomis, E.E. (President of the Lehigh Valley Railroad). *The Morris Canal: A Statement to the People of New Jersey outlining the position of the Lehigh Valley Railroad Company in regard to the proposed abandonment of a century-old waterway which has long since outlived its usefulness*. n.p., October 1, 1921.

Reasons for Granting to the Morris Canal and Banking Company An Increase of Banking Capital and the Loan of the State Bonds. No author, publisher, or date. Copy at Rutgers University Library, Special Collections, has words cut out of the title; New York's Engineering Societies Library copy is intact, but not dated, etc.

Records of Christ's (Presbyterian) Church, Catskill, New York.

Records of Newark (includes records of Newark's Town Council and the Third Presbyterian Church). Morris County Free Library.

Register, Minutes and History of the First Presbyterian Church, Morristown: 1742-1885. Morristown, NJ: Banner Steam Print, 1885.

Rutgers Family Bible and Prayer Book. Morristown Public Library.

Schmerber, Louis J. "Faith of Our Fathers." Unpublished manuscript novel, written in 1942, about the Morris Canal during the Civil War; in the present author's possession.

Shaffer, Franklin C. *Diary*. Unpublished, miscellaneous impressions of Newark vacationer at Lake Hopatcong in 1885. Lake Hopatcong Historical Society.

Steele, J. Dutton (CE). *Letter to the President of the Philadelphia & Reading Railroad Co. on the Canal and Railroads for Trans-*

porting Anthracite Coal. Philadelphia: Moss, Brother & Co., 1855. New York City Public Library.

Talcott, William H. "A List of Prices of Work From Sec. No. 60 to Little Falls, Erie Canal Enlargement, 1842." Handwritten notebook in New York City Public Library.

________________."Estimates of Quantities for Planes 1, 2, 3, 4, 5, 6, 7, 8, 9, 10, and 11 West, etc. June, 1850." This 116-page handwritten copy of the original in the Jersey City Public Library is dated July, 1876. The copyist was apparently an employee of the Lehigh Valley Railroad Company, lessees of the canal at that time.

REPORTS AND SURVEYS

Beach, Ephraim. *Report on the Survey of a Route for the Proposed Susquehanna & Delaware Railroad.* New York: Elam Bliss, 1832.

Lehigh Coal and Navigation Company. *Report of the Board of Managers.* n.p.: January 9, 1832.

Lehigh Valley Railroad Company Annual Reports to the Stockholders, various years. New York Public Library.

Morris Canal, With Synopsis of Report on Proposed Industrial Development Railroad. Jersey City, NJ: Port Development Committee, December 31, 1914.

North Jersey District Water Supply Commission of the State of New Jersey: Report for the Period May 5, 1916 to June 30, 1925. Newark, NJ: n.p., July 1, 1925.

Pennington, A. C. M. *Report of the Special Committee of the House of Assembly, on the Embarrassments of the Country. Read Nov. 7, 1837.* Trenton, NJ: James T. Sherman, 1837.

Report of the Commissioners Appointed by the Legislature of the State of New Jersey, for the Purpose of Exploring the Route of a Canal to Unite the River Delaware, Near Easton, With the Passaic, Near Newark, With Accompanying Documents. Morristown, NJ: Jacob Mann, 1823.

Report of the Committee Appointed by Hon. George S. Silzer, Governor, to Recommend What Disposition Shall be Made of the Property of the Morris Canal and Banking Company. Trenton, NJ: MacCrellish & Quigley Co., State Printers, 1923.

Report of the Committee on Inland Navigation, Relative to an Aqueduct Over the Delaware River, together with the Amendment of Mr. Porter. Harrisburg, PA: J.M.G. Lescure, 1850.

Report of the Committee on Streets to the Jersey City Common Council. Jersey City, NJ: n.p., 1865.

Report of the Morris Canal Investigation Committee Appointed Under Joint Resolution of April 12, 1912. New York: Isaac H. Blanchard Co., n.d. (Internal evidence indicates 1913 as the publication date.)

Sullivan, John Langdon. *Report on the Origin and Increase of the Paterson Manufactories and the Intended Diversion of Their Waters by the Morris Canal Company, also on Post Rails as the Means of Cheap Conveyance Throughout New-Jersey, of Bringing Susquehanna Coal to the Iron Mines and Forges, and to Supply Paterson and New-York: Also on a Method of Supplying the City of New-York with Water from the Great Falls of the Passaic.* Paterson, NJ: Day & Burnett, July 1, 1828.

Vermeule, Cornelius Clarkson. *Geological Survey of New Jersey,* Vol. III, "Report on Water Supply." Trenton, NJ: John L. Murphy Publishing Co., 1894.

Vermeule, Cornelius C., Jr. *Final Report of Consulting and Directing Engineer, June 29, 1929.* East Orange, NJ: Abbey Printing Co., 1929.

Werts, George T., John W. Griggs, and Foster M. Voorhees. *Report of the Morris Canal Investigating Committee on the Abandonment of Navigation on the Morris Canal.* Trenton, NJ: n.p., 1903.

BOOKS AND PERIODICALS

A History of the City of Newark, New Jersey, Embracing Practically Two and a Half Centuries: 1666-1913, 3 volumes. New York: Lewis Historical Publishing Co., Inc., 1913.

A History of Morris County, New Jersey. New York: Lewis Historical Publishing Co., Inc., 1914.

Albion, Robert G. *The Rise of New York Port.* New York: Scribner's, 1939.

An Estimate of the Revenue Which May be Derived from the Morris

Canal; Extracted from a Report made in 1823, to the War Department of the United States, by General Bernard, and Colonel Totten, of the Engineer Corps. With Notes. New York: William Davis, Jr., 1827.

Atwood, Albert W. "Today on the Delaware, Penn's Glorious River." *National Geographic Magazine,* Vol. CII, No. 10. July, 1952, pages 1-40.

Barrett, Walter. *The Old Merchants of New York City.* New York: Carleton Publishers, 1862.

Beach Family Magazine. Alfred Holmes Beach and Cora Beach, editors. Vol. I, Nos. 1, 2, 3, and 4, 1926; Vol. II, Nos. 2, 4, 1927. Casper, Wyoming: n.p.

Beck, Henry Charlton. *Tales and Towns of Northern New Jersey.* New Brunswick, NJ: Rutgers University Press, 1956.

———————. *The Roads of Home.* New Brunswick, NJ: Rutgers University Press, 1956.

Beckwith, Robert R. *A History of Little Falls.* Centennial Edition. Wayne, NJ: Louis J. Vorgetts, 1968.

Berce, William E. *Under the Sign of the Eagle.* Wayne, NJ: Louis J. Vorgetts, 1964.

Bergen County, New Jersey: 1630-1923, A History of. Frances A. Westervelt, ed. 3 volumes. New York: Lewis Historical Publishing Co., Inc. 1923.

Biographical and Genealogical History of Morris County. New York: Lewis Historical Publishing Co., Inc., 1899.

Biographical Encyclopedia of New Jersey of the Nineteenth Century. Philadelphia: Galaxy Publishing Co., 1877.

Bloomfield, New Jersey. Free Public Library of Bloomfield, eds. Bloomfield, NJ: The Independent Press, 1932.

Boardman, Fon W., Jr. *Canals.* New York: Walck Publishers, 1959.

Boonton Years: 1867-1967. Maudie Fisher, ed. Compiled for the Boonton Centennial Committee by *The Citizen* of Morris County, 1967.

Borough of West Paterson, 1914-1964. West Paterson, NJ: Golden Anniversary of West Paterson, 1964.

Bowen, Ele. "The Prose, Poetry and Scenery of the Canal Regions of Pennsylvania." *Graham's Magazine,* Vol. XLV, No. 1, July, 1854.

Boyer, Charles. *Early Forges and Furnaces of New Jersey.* Philadelphia: University of Pennsylvania Press, 1931.

"Brindley, James." *Encyclopedia Britannica* (1960 edition), Vol. 4, page 150.

Burg, Amos. "Britain Just Before the Storm." *National Geographic Magazine,* Vol. LXXVIII, No. 2, August, 1940, pages 185-212.

Burgess, George H. and Miles C. Kennedy. *Centennial History of the Pennsylvania Railroad Company: 1846-1946.* Philadelphia: Pennsylvania Railroad Company, 1949.

Byram Township: Township of Lakes: 1798-1973. No author; no publisher; no date; no pagination.

Cadman, John Wm., Jr. *The Corporation in New Jersey: Business and Politics: 1791-1875.* Cambridge, MA: Harvard University Press, 1949.

"Catawissa Railroad." *Harper's New Monthly Magazine.* Vol. XXV, June to November, 1862, pages 20-28.

Catton, Bruce. *This Hallowed Ground.* New York: Doubleday and Co., Inc., 1956.

Cawley, James and Margaret. *Exploring the Little Rivers of New Jersey.* Princeton, NJ: Princeton University Press, 1942.

__________________. *Along the Delaware and Raritan Canal.* Rutherford, NJ: Fairleigh Dickinson University Press, 1970.

Chapin, John R. "Among the Nail Makers." *Harper's New Monthly Magazine.* Vol. XXI, No. 122. July, 1860, pages 145-164.

__________. "Artist's Life in the Highlands." *Harper's New Monthly Magazine.* Vol. XX, No. 119, April, 1860, pages 577-598.

Chevalier, Michel. *Histoire et Description de Voies de Communication aux Etats Unis.* Paris: n.p., 1840.

Clark, W.H. *Railroads and Rivers.* Boston: L. C. Page & Co., 1939.

Clayton, W. Woodford, ed. *History of Union and Middlesex Counties, N.J., with Biographical Sketches of Many of Their Prominent Men.* Philadelphia: Everts & Peck, 1882.

________________, and William Nelson. *History of Bergen and Passaic Counties, N.J.* Philadelphia: Everts & Peck, 1882.

"Coal and the Coal Mines of Pennsylvania." *Harper's New Monthly Magazine.* Vol. XV, No. 88, September, 1857, pages 451-469.

Colden, Cadwallader D. *An Answer to Mr. John Sullivan's Report to the Manufacturing Society in New Jersey.* New York: William A. Davis, 1828.

____________________. *The Life of Robert Fulton.* New York: Kirk & Mercein, 1817.

"Company and Corporation Law." *Encyclopedia Britannica* (1960 edition). Vol. 6, pages 144-154.

Cowley, Malcolm. "Low Bridge and Lock Ahead." *Charm* Magazine, July, 1926.

Cranmer, H. J. "Internal Improvements in New Jersey: Planning the Morris Canal." *Proceeding of the New Jersey Historical Society.* Vol. 69, No. 4, October, 1951.

Crater, Warren B. *New Jersey Central Album.* n.p.: 1963.

Cullum, George W. *The Biographical Register of the Officers and Graduates of the United States Military Academy at West Point, New York, from Its Establishment, in 1802 to 1890, with the Early History of the United States Military Academy,* 3 volumes, 3rd edition. Boston: Houghton Mifflin Co., 1892.

Cunningham, John. *Century of Progress.* Paterson, NJ: First National Bank of Passaic County, 1965.

__________________. *Newark, New Jersey.* Newark, NJ: New Jersey Historical Society, 1966.

__________________. *This Is New Jersey.* Revised edition. New Brunswick, NJ: Rutgers University Press, 1968.

Daniels, Dorothy. *Cut and Engraved Glass, 1771-1905.* New York: M. Barrows & Co., 1950.

Diamond, Rachel F. *One Hundred Years Around the Green: A Brief History of Bloomfield, New Jersey, 1812-1962.* Bloomfield, NJ: Bloomfield's 150th Anniversary, 1962.

Dickens, Charles. *American Notes.* New York: Hurst & Co., n.d. (A postscript in the edition owned by this author is dated May, 1868.)

Domett, Henry W. *History of the Bank of New York: 1784-1884.* 4th edition. Cambridge, MA: Riverside Press, 1884.

Drago, Harry Sinclair. *Canal Days in America: The History and Romance of Old Towpaths and Waterways.* New York: Clarkson N. Potter, 1972.

Dunbar, Seymour. *A History of Travel in America,* 4 volumes. New York: Greenwood Press, 1968 (first reprint of the Bobbs-Merrill 1918 edition).

Dupuy, Ernest R. *Men of West Point: The First 150 Years of the United States Military Academy.* New York: William Sloane Associates, 1951.

______________. *Where They Have Trod: The West Point Tradition in American Life.* New York: Frederick Stokes, 1940.

Edgerton, Major General Glen E. "An Engineer's View of the Suez Canal." *National Geographic Magazine,* Vol. CXI, No. 1, January, 1957, pages 123-140.

Edmunds, Walter D. *Drums Along the Mohawk.* Boston: Little, Brown & Co., 1939.

________________. *Erie Water.* Boston: Little, Brown & Co., 1933.

Elmer, Lucius Q. C. *Reminiscences of the [N.J.] Bench and Bar During More Than Half a Century.* Newark, NJ: Martin R. Dennis, 1872.

Fisher, Charles E. "The Lehigh Valley Railroad." *Railway and Locomotive Historical Society, Bulletin No. 42,* February, 1937.

Folsom, Joseph Fulford, ed. *Bloomfield Old and New.* Bloomfield, NJ: Centennial History Committee, 1912.

Francis, E. T. and T. F. Gleichmann, Jr. *Unusual Trackwork.* Allaire, NJ: Pine Creek Railroad Division, New Jersey Museum of Transportation, April, 1970.

Freeman, Leslie E., Jr. "The New Jersey Railroad and Transportation Company." *Railway and Locomotive Historical Society, Bulletin No. 88,* May, 1950, pages 100-159.

Fulton, Robert. *Treatise on the Improvement of Canal Navigation; Exhibiting the Numerous Advantages to be Derived from Small Canals, and Boats of Two to Five Feet Wide, Containing from Two to Five Tons Burthen.* London: I. and J. Taylor, 1796.

Getty, R. R. "Motorboating Through the Morris Canal." *Motorboat Magazine,* September 25, 1910.

Gill, Mildred Lawrence. *Denville Days: 1600-1955.* Denville, NJ: Weaver Printing Co., 1955.

Goodrich, Carter. *Government Promotion of American Canals and Railroads: 1800-1890.* New York: Columbia University Press, 1965.

_______________, Julius Rubin, H. Jerome Cranmer, Harvey H. Segal. *Canals and American Economic Development.* New York: Columbia University Press, 1961.
Gordon, T. F. *Gazetteer of the State of New Jersey.* Trenton, NJ: Daniel Fenton, 1834.
Gould, Frederick H. "The Decadence of an Old Canal." *Booklover's Magazine,* June, 1905, pages 826-835.
Graham, F. Stewart. "The Locomotives of the Delaware, Lackawanna and Western Railroad." *Railway and Locomotive Historical Society, Bulletin No. 72,* July, 1948.
Greene, Homer. *Coal and the Coal Mines.* Boston: Houghton Mifflin & Co., 1898.
Hacker, Louis M., and Benjamin B. Kendrick. *The United States Since 1865.* New York: F. S. Crofts & Co., 1939.
Hanson, Kenneth R. "An Aerial Survey of the Remains of the Morris Canal." *Proceedings of the New Jersey Historical Society, No. 312,* January, 1963, pages 10ff.
_________________. "New Jersey's First Anthracite Blast Furnace." *New Jersey History,* April, 1961.
Hare, Jay V. "History of the Reading." *The Pilot,* Vol. XIV, No. 3, March, 1913.
Harlow, Alvin F. *Old Towpaths: The Story of the American Canal Era.* New York: D. Appleton & Co., 1926.
Harlow, Lewis A. "The Bunker Hill Monument Traveled the First Railroad in America." *Yankee* Magazine, June, 1968.
Havighurst, Walter. *The Long Ships Passing.* New York: Macmillan Co., 1942.
Helmer, William F. *A Rip Van Winkle Railroad.* Berkeley, CA: Howell North Books, 1970.
Herrmann, Frederick M. "Anti-Masonry in New Jersey." *New Jersey History,* Summer, 1973, pages 149-165.
Heydinger, Earl J. "Railroads of the Lehigh Coal and Navigation Company: Group VI, the Little Schuylkill, the E.M.; L.S.&S.; Catawissa; T.H.&N.; D.S.&S." *Railway and Locomotive Historical Society, Bulletin No. 108,* April, 1963.
_______________. "Little Schuylkill & Susquehanna Addenda." *Railway and Locomotive Historical Society, Bulletin No. 109,* October, 1963.

_______________. "Railroads of the Lehigh Coal and Navigation Company: Group IX, The Mauch Chunk Railroad." *Railway and Locomotive Historical Society, Bulletin No. 110,* April 1, 1964.

Hicks, John D. *The Federal Union: A History of the United States to 1865.* New York: Houghton Mifflin Co., 1937.

Higgins, Ralph Berringer. "An Old Waterway Rediscovered." *Saturday Evening Post Magazine,* February 13, 1915.

Hill, Forest G. *Roads, Rails and Waterways: The Army Engineers and Early Transportation.* Norman, OK: University of Oklahoma Press, 1957.

History and Directory of Warren County, New Jersey. Washington, NJ: Shampanore & Sons, 1929.

History of Jersey City, New Jersey: A Record of Its Early Settlement and Corporate Progress. Compiled for the *Evening Journal* by Alexander McLean. Jersey City, NJ: Press of the Jersey City Printing Co., 1895.

History of Morris County, New Jersey: 1739-1882. New York: W. W. Munsell & Co., 1882. (Cited as Munsell's *History of Morris County,* and as Munsell's *Morris.*)

History of Roxbury Township. Annie Stelce Hosking and Harriet Meeker, eds. Netcong, NJ: Pyramid Press, Inc., n.d.

Holbrook, Stewart H. *Iron Brew: A Century of Ore and Steel.* New York: Macmillan Co., 1939.

_______________. *The Story of American Railroads.* New York: Crown Publishers, 1947.

Hollister, H., M.D. [sic]. *History of the Lackawanna Valley,* 2nd edition. New York: C. A. Alvord, 1869.

Hopatcong Historama. Landing, NJ: Lake Hopatcong Yacht Club, 1955.

Hosack, David. *Memoir of DeWitt Clinton.* New York: J. Seymour, 1829.

Hulbert, Archer Butler. *The Great American Canals.* Cleveland, OH: Arthur H. Clark Co., 1904.

_______________. *The Paths of Inland Commerce: A Chronicle of Trail, Road, and Waterway.* New Haven, CT: Yale University Press, 1920.

Hungerford, Edward. *Men of Erie.* New York: Random House, 1946.

Jailer, Mildred. "The Museum in St. Johnsbury." The *Royle Forum,* No. 135. Paterson, NJ: John Royle & Sons, Inc. June 15, 1971, pages 11-14.

James, Marquis. *The Life of Andrew Jackson.* New York: Bobbs-Merrill, 1938.

Johnson, Allen, and Dumas Malone. *Dictionary of American Biography.* 20 volumes and index, 1928-1936. n.p., n.d.

Johnston, E.A.J., and Herman E. Krooss, *The Origins and Development of the American Economy.* New York: Prentice-Hall, Inc., 1953.

Johnston, Jay. "Waterways to Washington: The C&O Canal." *National Geographic Magazine,* Vol. 117, No. 3, March, 1960, pages 419-438.

Jones, Chester L. *The Economic History of the Anthracite Tidewater Canals.* Publications of the University of Pennsylvania in Political Economy and Public Law, No. 22. Philadelphia: University of Pennsylvania Press, 1908.

Kimbark, L. S., ed. *A History of Beavertown and Lincoln Park, N.J.* Lincoln Park, NJ: L. S. Kimbark, Beavertown Historical Society of Lincoln Park, Inc., Mayor and Council, Borough of Lincoln Park, NJ, 1971.

Kulp, Randolph L., ed. *Railroads in the Lehigh River Valley.* Revised edition. Allentown, PA: Lehigh Valley Chapter, National Railway Historical Society, December, 1962.

Kurtz, Ford. *The Electrification of the Morris Canal.* New York: n.p., November, 1911.

Lane, Wheaton J. *From Indian Trail to Iron Horse: Travel and Transportation in New Jersey, 1620-1860.* Princeton, NJ: Princeton University Press, 1939

Lang, Barbara Carper. "New Jersey's Morris Canal." *Royle Forum,* No. 104. Paterson, NJ: John Royle and Sons, Inc., September 15, 1963, pages 7-10.

Langstaff, John Brett. *New Jersey Generations: Macculloch Hall, Morristown.* New York: Vantage Press, 1964.

Laut, Agnes C. *The Romance of the Rails.* New York: Tudor Publishing Co., 1936.

Leary, Lewis. *Washington Irving.* University of Minnesota Pamphlets on American Writers No. 25. Minneapolis: University of Minnesota Press, 1963.

Lee, Francis Bazley, ed. *Genealogical and Memorial History of the State of New Jersey,* 4 volumes. New York: Lewis Historical Publishing Co., Inc. 1910.

Le Gallienne, Richard. "An Old American Towpath." *Vanishing Roads.* New York: G.P. Putnam's Sons, 1915. Also, *Harper's New Monthly Magazine,* Vol. 127, June to November, 1913, page 706.

Letters from John Pintard to His Daughter Eliza Noel Pintard Davidson, 1816-1833, 4 volumes. New York: New York Historical Society, 1940.

Longworth's American Almanac, New York Register, and City Directory for the 47th Year of American Independence. New York: Longworth, 1822.

Lucas, Walter Arndt. *From the Hills to the Hudson.* New York: Railroadians of America, 1944.

Lyon, Isaac S. *Historical Discourse on Boonton.* Newark, NJ: *Daily Journal* Office, 1873.

McCarter, Thomas Nesbitt. "Address Delivered Before the Board of Trade on November 24, 1903, on the Availability of the Bed of the Morris Canal Within the Limits of the City of Newark, for Transportation Purposes." *One Phase of a Jerseyman's Activities.* Garden City, NY: Country Life Press, 1933.

McCormick, Richard P. *New Jersey From Colony to State, 1609-1789.* New Brunswick, NJ: Rutgers University Press, 1964.

McDade, Matt C. "New York State's New Main Street." *National Geographic Magazine,* Vol. CX, No. 5, November, 1956, pages 567-618.

MacGill, Caroline E. and others. *History of Transportation in the United States before 1860.* Prepared under the direction of Balthasar Henry Meyer. Washington, DC: Carnegie Institute of Washington, 1917.

Maxim, Hudson. *Lake Hopatcong the Beautiful: A Plea for its Preservation and Dedication as a Public Park for all the People.* Second edition. New York: McConnell Printing Co., February, 1913.

_______________. *Morris Canal Abandonment Problems: A Plea for the Abandonment of the Morris Canal and a Plea for the Preser-*

vation and Dedication of Lake Hopatcong the Beautiful as a Public Park and Health Resort for all the People. New York: McConnell Printing Co., December, 1913.

Morton, Eleanor. *Josiah White: Prince of Pioneers.* New York: Stephen Day Press, 1946.

Municipalities of Essex County, N.J.: 1666-1924. Joseph F. Folsom, Benedict Fitzpatrick, and Edwin P. Conklin, eds., 4 volumes. New Jersey: Lewis Historical Publishing Co., Inc., 1925.

Myrose, Elizabeth R. and Claire B. Kitchell. *Along the Whippanong.* Denville, NJ: Haase Publications, Inc., 1966.

Newark, N.J., Illustrated. Newark, NJ: W. A. Baker, 1893.

New Jersey Biographical Encyclopedia. New York: n.p., 1885.

Ogden, Mary Depue, ed. *Memorial Cyclopedia of New Jersey,* 3 volumes, Newark, NJ: Memorial History Co., 1917.

Payne, Robert. *The Canal Builders.* New York: Macmillan Co., 1959.

Pierson, David Lawrence. *Narratives of Newark.* Newark, NJ: Pierson Publishing Co., 1917.

Platt, Charles D. *Dover History.* Dover, NJ: n.p., 1914.

Prominent Families of New York. New York: The Historical Company, 1897.

Rankin, Edward S. *The Running Brooks and Other Sketches of Early Newark.* Somerville, NJ: Unionist-Gazette, 1930.

Ransom, James M. *Vanishing Ironworks of the Ramapos.* New Brunswick, NJ: Rutgers University Press, 1966.

"Report of the Committee of the Franklin Institute, on the Inclined Plane of Professor James Renwick, etc." *Franklin Journal and American Mechanics' Magazine,* Vol. II, No. 5. November, 1826, pages 259-265; and December, 1826, pages 321-332.

Riggs, Albert R. *That's The Way It Was.* Bellefontaine, OH: Carter Printing Co., n.d.

Riggs, Herbertha S. "Waterloo and the Morris Canal." *Daughters of the American Revolution Magazine,* No. 102, January, 1964, pages 42-43.

Rivinus, Willis M. *A Wayfarer's Guide to the Delaware Canal Between Bristol and Easton, Pennsylvania.* Buck's County Park Foundation, 1964.

Roeder, Adolph. *A History of the Morris Canal,* Special Report of the New Jersey State Civic Federation, No. 11. Orange, NJ: n.p., 1910.

Rupp, I. Daniel. *History of Northampton, Lehigh, Monroe, Carbon, and Schuylkill Counties.* Harrisburg, PA: Hickok & Cantine, 1845.

Rydings, Joseph. *Country Walks in Many Fields.* Paterson, NJ: Paterson *Morning Call,* 1934.

Savory, Theodore. "The Mule." *Scientific American,* December, 1970, pages 102-109.

Schlegel, Marvin W. *Ruler of the Reading: The Life of Franklin B. Gowen, 1836-1889.* Harrisburg, PA: Archives Publishing Co., 1947.

Scott, William W. *History of Passaic and Its Environs.* New York: Lewis Historical Publishing Co., Inc., 1922.

Shaw, Ronald E. *Erie Water West: A History of the Erie Canal, 1792-1854.* Lexington, KY: University of Kentucky Press, 1966.

Shaw, William H. *History of Essex and Hudson Counties, New Jersey.* Philadelphia: Everts and Peck, 1884.

Sheldon, G. W. "The Old Shipbuilders of New York." *Harper's New Monthly Magazine,* Vol. LXV, 1882, pages 223-241.

____________. "The Old Packet and Clipper Service." *Harper's New Monthly Magazine,* Vol. LXVIII, 1884, pages 213-231.

____________. "The Old Shipping Merchants of New York." *Harper's New Monthly Magazine,* Vol. LXXXIV, 1892, pages 457-471.

Shriner, Charles A. *Four Chapters of Paterson History.* Paterson, NJ: Lont and Overkamp Publishing Co., 1919.

________________. *Paterson, New Jersey, Illustrated.* Paterson, NJ: n.p., 1890.

Seibert, Wilbur H. *The Underground Railroad.* New York: Arno Press and *The New York Times,* 1968. (Reprint of Macmillan's 1898 copyright.)

Sinclair, Gladys Mellor. *Bayonne Old and New.* New York: Marantha Publishing Co., May, 1940.

Smalley, E. V. "Striking Oil." The *Century Magazine,* Vol. XXVI, No. 3, July, 1883, pages 323-339.

Smith, Walter Buckingham. *Economic Aspects of the Second Bank of the United States.* Cambridge, MA: Harvard University Press, 1853.

Snell, James P. *History of Hunterdon and Somerset Counties, New Jersey,* 2 volumes. Philadelphia: Everts & Peck, 1881.

______________. *History of Sussex and Warren Counties, New Jersey.* Philadelphia: Everts & Peck, 1881. (Reprinted in 1971 by Howard E. Case, Sussex, NJ)

Stevenson, David. *Sketch of the Civil Engineering of North America.* London: n.p., 1838.

Stickle, Fred G., Jr. "Through the Morris Canal." *New Jersey History.* Newark, NJ: New Jersey Historical Society, Summer, 1971, pages 93-114.

Stokes, I. N. Phelps. *Iconography of Manhattan Island, 1498-1909, Compiled From Original Sources.* New York: Robert Dodd, 1918.

Stories of New Jersey. Federal Writers' Project of the Works Project Administration for the State of New Jersey. New York: M. Barrows & Co., 1938.

Stuart, Charles B. *Lives and Works of Civil and Military Engineers of America.* New York: Van Nostrand, 1871.

Sullivan, A. M. *New Jersey Hills,* Prairie City, IL: James A. Decker, 1940.

"The Lake Biwa, or Kioto-Tu Canal." *Engineering News,* April 13, 1893, pages 34ff.

"The Morris Canal Crossing of the Lackawanna Railroad. *Engineer-Record,* Vol. 51, June 15, 1905, page 686.

The Story of Anthracite. New York: Hudson Coal Co., 1932.

Thrift Magazine. Bloomfield, NJ: Bloomfield Community Trust Co., March, 1936. (Material relative to Ephraim Morris, the Morris Mill, etc.)

Tolles, Frederick B. *George Logan of Philadelphia.* New York: Oxford University Press, 1953.

Trollope, Frances. *Domestic Manners of the Americans.* London: George Routledge & Sons, Ltd. 1927 edition.

Trumbull, L. R. *A History of Industrial Paterson.* Paterson, NJ: Carleton M. Herrick, 1882.

Veit, Richard. *Old Canals of New Jersey: A Historical Geography.* Little Falls, NJ: New Jersey Geographical Press, 1963.

Vermeule, Cornelius C., Jr. "Early Transportation In and About New Jersey." *Proceedings of the New Jersey Historical Society,* Vol. IX, No. 2, April, 1924, pages 106-124.

Vernon, Arthur. *History and Romance of the Horse.* Boston: Waverly House, 1939.

Wacker, Peter O. *The Musconetcong Valley of New Jersey: A Historical Geography.* New Brunswick, NJ: Rutgers University Press, 1968.

Waggoner, Madeline Sadler. *The Long Haul West: The Great Canal Era, 1817-1850.* New York: G. P. Putnam's Sons, 1958.

Wakefield, Manville B. *Coal Boats to Tidewater,* Revised Edition. Grahamsville, NY: Wakefair Press, 1971.

Weiss, Harry G., and Grace M. Weiss. *Some Early Industries of New Jersey.* Trenton, NJ: New Jersey Agricultural Society, 1965.

__________________. *The Early Lotteries of New Jersey.* Trenton, NJ: Past Times Press, 1966.

Wharton, J. G. *The Commercial Position of New York and a Way to Improve It.* New York: n.p., 1860.

Wildes, Harry E. *Twin Rivers: The Raritan and the Passaic.* New York: Rinehart & Co., Inc., 1943.

Williams, Maynard Owen. "The Suez Canal: Short Cut to Empires." *National Geographic Magazine,* Vol. LXVIII, No. 5, November, 1935, pages 611-632.

Winfield, Charles H. *History of the County of Hudson, New Jersey, From Its Earliest Settlement to the Present Time.* New York: Kennard & Hay, 1874.

Wright, Carroll D. *The Industrial Evolution of the United States.* New York: Scribner's, 1913.

Yoder, C. P. *Delaware Canal Journal: A Definitive History.* Bethlehem, PA: Canal Press, Inc., 1972.

Name Index

Subject Index

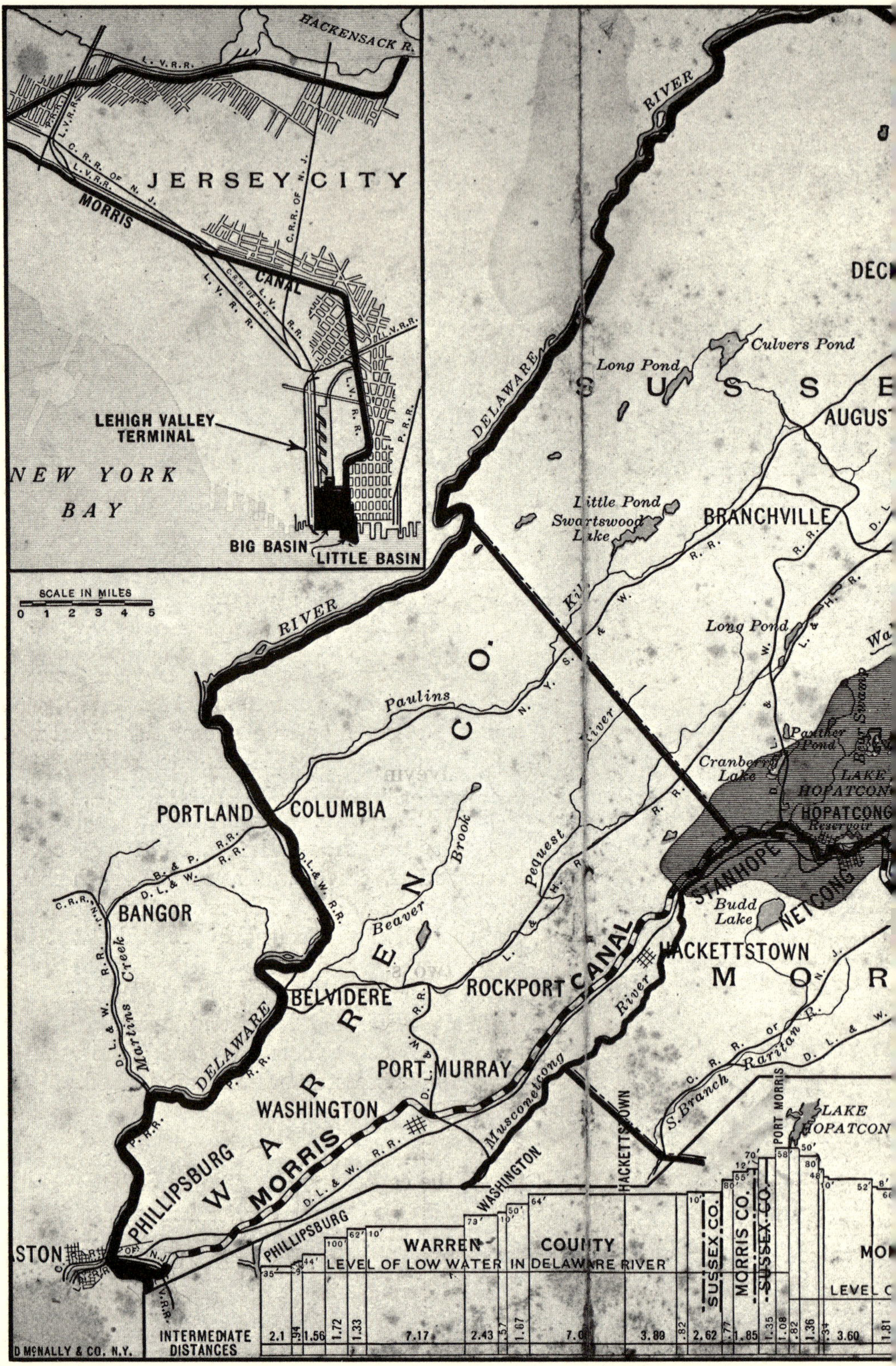

HACKENSACK R.
L.V.R.R.
JERSEY CITY
C.R.R. OF N.J.
MORRIS
CANAL
LEHIGH VALLEY TERMINAL
NEW YORK BAY
BIG BASIN
LITTLE BASIN
SCALE IN MILES
0 1 2 3 4 5
DELAWARE RIVER
SUSSEX
Long Pond
Culvers Pond
AUGUST
DECK
Little Pond
Swartswood Lake
BRANCHVILLE
Long Pond
Paulins Kill
Pequest River
Cranberry Lake
LAKE HOPATCONG
HOPATCONG
STANHOPE
NETCONG
Budd Lake
HACKETTSTOWN
PORTLAND
COLUMBIA
BANGOR
Martins Creek
Beaver Brook
BELVIDERE
ROCKPORT
WARREN CO.
PORT MURRAY
WASHINGTON
PHILLIPSBURG
MORRIS CANAL
Musconetcong River
S. Branch Raritan R.
PORT MORRIS
LAKE HOPATCON
WARREN COUNTY
LEVEL OF LOW WATER IN DELAWARE RIVER
SUSSEX CO.
MORRIS CO.
SUSSEX CO.
INTERMEDIATE DISTANCES
2.1 .94 1.56 1.72 1.33 7.17 2.43 .57 1.67 7.0 3.80 .82 2.62 .77 1.85 1.35 1.08 .82 1.36 .34 3.60 1.81
D McNALLY & CO. N.Y.